G. Geymonat (Ed.)

Constructive Aspects of Functional Analysis

Lectures given at the
Centro Internazionale Matematico Estivo (C.I.M.E.),
held in Erice (Trapani), Italy,
June 27-July 7, 1971

Springer

FONDAZIONE
CIME
ROBERTO CONTI

C.I.M.E. Foundation
c/o Dipartimento di Matematica "U. Dini"
Viale Morgagni n. 67/a
50134 Firenze
Italy
cime@math.unifi.it

ISBN 978-3-642-10982-9 e-ISBN: 978-3-642-10984-3
DOI:10.1007/978-3-642-10984-3
Springer Heidelberg Dordrecht London New York

Reprint of the 1st ed. C.I.M.E., Ed. Cremonese, Roma, 1971
With kind permission of C.I.M.E.

Printed on acid-free paper

Springer.com

CENTRO INTERNAZIONALE MATEMATICO ESTIVO

(C. I. M. E.)

II Ciclo - Erice - Dal 27 giugno al 7 luglio 1971

CONSTRUCTIVE ASPECTS OF FUNCTIONAL ANALYSIS

Coordinatore : Prof. G. Geymonat

CENTRO INTERNAZIONALE MATEMATICO ESTIVO

(C.I.M.E.)

A. V. BALAKRISHNAN : I

A CONSTRUCTIVE APPROACH TO OPTIMAL CONTROL

Corso tenuto ad Erice dal 27 giugno al 7 luglio 1971

A CONSTRUCTIVE APPROACH TO OPTIMAL CONTROL

0. Introduction.

Except for linear problems, it is difficult, if not impossible, to obtain explicit solutions for optimal control problems. The closest we get to a general 'solution' is the Maximum Principle of Pontrjagin. But important as this result is, it only provides us with necessary conditions for a (any) postulated solution. Unfortunately, many control problems do not have an optimal solution. Consider for instance this trivial example:

$$\dot{x} = u \; ; \qquad x(0) = 0$$

Minimize:

$$\int_0^1 x^2(t)dt$$

subject to the constraint that the control $u(t)$ must be equal to $+1$ or -1 a.e. The minimal value is zero, but this is attained for $u(t) \equiv 0$ and of course this is impossible. On the other hand

$$u_n(t) = \frac{\text{Sin}\,\pi nt}{|\text{Sin}\,\pi nt|}$$

provides us with a sequence of admissible controls which approximate the infimum arbitrarily closely. The sequence $\{u_n(t)\}$ of course converge in the weak sense in $L_2[0,1]$ to zero, but unfortunately $u_n(t)^2$ converges to one, and of Course there is no optimal control.

In his recent book [1], L. C. Young has pointed out the fallacy in proving necessary conditions for a possibly non-existent solution. He cites a paradox of Perron that this leads to: consider the problem of finding the largest positive integer. If we assume there exists a solution, say N, then clearly $N \geq 1$; on the other hand, we must have that

$$N^2 \leq N$$

or,

$$N(N-1) \leq 0$$

or,

$$N-1 \leq 0$$

which combined with

$$N-1 \geq 0$$

shows that $N = 1$!

To resolve this difficulty, Young introduces the notion of a 'relaxed' or 'generalized control' and proves the existence of an optimal control in this class, and derives the maximum principle valid for such 'functions'. In the present work we shall go one step further and show how to actually construct - 'compute' - a sequence of approximating controls which converge to an optimal 'generalized control' and which then satisfies the maximum principle. The computational technique is of more than theoretical value; and in fact has proved to be practically useful as well.

Relaxed controls play an essential role in this approach. We begin with a simple exposition of the theory of relaxed controls [Young [1]], because it is of some independent interest as well.

1. Relaxed Controls

Let U be a compact set in Euclidean space E_m. Let H denote the L_2-space of functions $u(t)$, $0 < t < T < \infty$. Let $u_n(t)$ be any sequence of measurable functions such that

$$u_n(t) \in U \quad \text{a.e.}$$

Then we can find a subsequence (renumber it $u_n(\cdot)$) such that $u_n(\cdot)$ converges weakly to $u_0(\cdot)$ say in H. Let $p(\cdot)$ be any polynomial over E_m. Then

$$p[u_n(t)]$$

also contains a weakly convergent subsequence. What is the limit? Unfortunately it is not

$$p[u_0(t)]$$

as the example

$$u_n(t) = \frac{\text{Sin } nt}{|\text{Sin } nt|}$$

shows, taking $p(u) = u^2$. At the simplest level the 'generalized curves' [we shall continue to use the term generalized 'controls' because we shall need this notion only with the controls] may be regarded as providing a means to straighten out this situation.

Consider now the product space $\Omega = I \times U$ where I denotes the interval $[0, T]$. Then Ω is compact metric and let $C(\Omega)$ denote the Banach space of continuous functions over Ω with range in E_m. Let $f(t, u)$ denote such a function. Then observe that for any Lebesgue measurable function $u(t)$ such that

$$u(t) \in U \quad \text{a.e.}$$

we have that

$$\int_I f(t, u(t))dt$$

defines a continuous linear functional on $C(\Omega)$. We know that there must be a countably additive set function μ (of finite variation) defined on the Lebesgue subsets of Ω such that

$$\int_I f(t, u(t))dt = \int_\Omega f(t, u)\, d\mu(t, u)$$

and it is clear that μ is an atomic measure with a unit jump at $u(t)$ for each t. That is to say, on any product set of the form $\Delta \times B$

$$\mu(\Delta \times B) = \text{Lebesgue measure of the set } [t \mid u(t) \in B]$$

For any polynomial $p(\cdot)$, we note that

$$\int_U p(u)\, d\mu(t;u)$$

is Lebesgue measurable in t. A generalized control is simply a measure on (the Lebesgue subsets of) $I \times U$ such that

$$\int_U d\mu(t;u) = 1 \quad \text{a.e.}$$

and

$$\int_U p(u)\, d\mu(t;u)$$

is Lebesgue measurable in t. Alternately, for our purposes it is more natural to define it as a 'family' of probability measures ('control measures') $d\mu(t;u)$ over U such that

$$\int_U p(u) d\mu(t;u)$$

is Lebesgue measurable in t. Thus defined it is not difficult to show that

$$\int_\Omega f(t;u) \, d\mu(t;u)$$

defines a continuous linear functional on $C(\Omega)$. Moreover

$$\int_U f(t;u) \, d\mu(t;u)$$

is Lebesgue measurable in t.

Let now $u_n(t)$ be the sequence we began with, $u_n(t)$ converging weakly in H to $u_0(t)$. Let $f(t)$ be any m x m matrix function, continuous on I and $p(\cdot)$ be any polynomial with domain and range in E_m. Then we can write

$$\int_I f(t) \, p(u_n(t)) dt$$

as

$$= \int_\Omega f(t) \, p(u) \, d\mu_n(t;u)$$

where $d\mu_n(t;u)$ is the corresponding sequence of measures. Now by the weak compactness of measures we know that (independent of $f(\cdot)$ and

$p(\cdot)$) we can find a subsequence (renumber it $d\mu_n(\cdot)$ again) which converges to a measure $d\,\mu_0(t;u)$:

$$\operatorname*{limit}_{n} \int_I f(t)\, p[u_n(t)]\, dt = \int_\Omega f(t)\, p(u)\, d\,\mu_0(t;u)$$

Working with a further subsequence, we know that

$$\operatorname*{limit}_{n} \int_I f(t)\, p[u_n(t)]dt = \int_I f(t)\, v(t)\, dt$$

where $v(t)$ is Lebesgue measurable and since $f(t)$ is arbitrary, it follows that

$$v(t) = \int_U p(u)\, d\mu_0(t;u)$$

Thus if we agree to define

$$p(u_0(t)) \doteq \bar{p}[u_0(t)] = \int_U p(u)\, d\,\mu_0(t;u)$$

where the bar indicates use of 'generalized control', then we do indeed have that if

$$u_n(t) \rightharpoonup u_0(t)$$

then

$$p(u_n(t)) \rightarrow p[u_0(t)]$$

Example

Let us illustrate this with a simple example for m = 1.

Let

$$u_n(t) = \frac{\operatorname{Sin}\pi nt}{|\operatorname{Sin}\pi nt|} \qquad 0 < t < 1$$

what is the limiting generalized function? Note that $d\,\mu_n(t;u)$ for each t has a jump at + 1 or - 1. Hence

$$\int p(u)d\mu_n(t;u) = a_n(t)\ p(1) + (1-a_n(t))p(-1)$$

where

$$0 \leq a_n(t) \leq 1$$

Hence

$$\int_\Omega f(t;u)d\mu_n = \int_0^1 a_n(t)f(t;1)dt$$

$$+ \int_0^1 (1-a_n(t))f(t;-1)dt$$

$$\rightarrow \int_0^1 a(t)f(t;1)dt$$

$$+ \int_0^1 (1-a(t))f(t;1)dt$$

Hence the limiting measure μ is such that

$d\mu(t;u)$ has a jump at $+1$ of $a(t)$

and a jump at -1 of $(1-a(t))$

Now

$$\int_0^1 \int_U u \, d\mu_n(t,u) \longrightarrow \int_0^1 a(t)dt - \int_0^1 (1-a(t))dt$$

Hence

$$\int_0^1 a(t)dt = \frac{1}{2}$$

Also

$$\int_\Delta \int_U u \, d\mu_n(t,u)$$

$$= (+1) \int_\Delta a_n(t)dt + (-1) \int_\Delta (1-a_n(t))dt$$

$$= \int_\Delta u_n(t)dt$$

$$\longrightarrow 0$$

Hence we must have:

$$a(t) = \frac{1}{2}$$

Hence

$$p[u_n(t)] \longrightarrow \int_U p(u)\, d\mu(t;u)$$

$$= \frac{1}{2}\, p(1) + (1 - \frac{1}{2}) p(-1)$$

Thus the limiting measure is a "chattering" between the values 1 and -1 with equal probability. Note that

$$u_n(t)^2 \longrightarrow \frac{1}{2} + (1 - \frac{1}{2}) = 1$$

which is correct.

Generalization is fairly transparent at this stage. For example, for the extension to the immediate case

$$u_n(t) = \text{one of } m \text{ values}, \quad u_1, \ldots u_m$$

and

$$u_n(\cdot) \text{ converges weakly to zero}$$

we have:

$$\int_U p(u)\, d\mu_u(t;u) \longrightarrow \int_U p(u) d\mu(t;u) = \sum_1^m a_k(t) p(u_k)$$

To determine the functions $a_k(t)$, we may note that

$$\sum_1^m a_k(t) = 1, \quad a_k(t) \geq 0$$

$$\sum_1^m a_k(t)\, u_k = 0$$

$$\sum_1^m a_k(t) u_k^2 = \text{limit } u_k(t)^2$$

$$\sum_1^m a_k(t) u_k^{m-1} = \text{limit } u_k(t)^{m-1}$$

giving us m equations to determine the m unknowns. The length of the time interval, so long as it is finite, obviously plays no role.

The weak limit of "ordinary controls" thus leads to a generalized control. Conversely, we have the following important result due to Young: Any generalized control can be approximated in the weak star topology of linear functionals on $C(\Omega)$ by ordinary controls. [Ordinary controls are weak-star dense in the class of generalized controls.]

[Of course the weak-star limits of generalized controls are quite obviously generalized controls.]

2. The Basic Technique

Let us illustrate our technique with reference to a simple control problem:

$$\text{Minimize:} \int_0^T g(t;x(t);u(t))dt$$

where

$$\dot{x}(t) = f(t;x(t);u(t)); \quad x(0) = x_o$$

and the control u(t) is constrained to be in a restricted class of functions (called 'admissible' controls). We replace this problem by the non-dynamic epsilon problem:

$$\epsilon > 0$$

Minimize:

$$\frac{1}{2\epsilon} \int_0^T \|\dot{x}(t) - f(t, x(t), u(t)\|^2 dt + \int_0^T g(t, x(t), u(t))dt\cdot$$

over the class of state functions x(t), absolutely continuous with $x(0) = x_o$ and the class of admissible controls. We present a constructive technique for solving this problem which as ε goes to zero approximates the original problem as closely as desired. The construction exploits the maximum principle indirectly; in fact the Hamiltonian arises in a natural way in the process.

See [2] for the bibliography and related work.

A. V. Balakrishnan

A Basic Estimate

We begin with the immediate question: how well does the epsilon problem approximate the original control problem? This question is of course of primary importance for computation, and it is interesting that we can answer it without the need for any of the usual assumptions of control thoery, even including the conditions that assure unique solution to the differential equation. We can also consider as general a class of control problems as necessary. However, in order not to confuse the main ideas with too much generality, we shall confine ourselves to the following class of problems (the extension to more general problems involving other types of phase plane constraints being readily made):

$$\text{Minimize} \quad \int_0^T g(t;x(t);u(t))\, dt \tag{2.1}$$

subject to:

$$\dot{x}(t) = f(t;x(t);u(t)) \quad \text{a.e.} \tag{2.2}$$

$$\phi(t;x(t);u(t)) = 0 \quad \text{a.e.} \tag{2.3}$$

where x(t) is absolutely continuous and satisfying additional conditions at the end points $t = 0$, and $t = T$. The end-point T is finite but of course not necessarily fixed. The control u(t) is Lebesgue measurable and subject to additional constraints, if any. We shall refer to such controls as "admissible" controls. It should be noted that not every admissible control necessarily yields a trajectory x(t) satisfying all the conditions, (2.2), (2.3) and the end conditions. However it would be natural to assume that there do exist admissible controls that lead to such trajectories. (Even this condition can be dispensed with for our purposes in this section.) Nor shall we need to impose any smoothness conditions on the functions $f(.)$, $g(\cdot)$ and $\phi(.)$. We shall only assume that they are Lebesgue measurable and such that the integral in (2.1) is well-defined for each (finite) T.

The epsilon problem is now formulated as follows:

Let

$$h(\epsilon;x(\cdot);u(\cdot);T) = \frac{1}{2\epsilon}\int_0^T \|\dot{x}(t) - f(t;x(t);u(t))\|^2 \, dt + \frac{1}{2\epsilon}\int_0^T \|\phi(t;x(t);u(t))\|^2 \, dt + \int_0^T g(t;x(t);u(t)) \, dt \quad \ldots. \quad (2.4)$$

Minimize $h(\epsilon;x(\cdot);u(\cdot);T)$ over the class of (absolutely continuous) trajectories x(t) subject to the given end conditions (any other "phase plane" constraints can clearly be added); and admissible controls u(t). We add the condition [F]:

$$\|\dot{x}(t) - f(t;x(t);u(t))\|^2 + \|\phi(t;x(t);u(t))\|^2 \leq m < \infty$$

where m is a fixed positive constant independent of epsilon. This condition is not necessary if for example:

$$\text{Inf} \quad g(t;x;u) > -\infty \tag{2.5}$$

(as in time-optimal problems, see section 4)

The condition (F) is certainly a natural one in that we are, after all, trying to approximate the case $m = 0$. The need for such a condition may be seen by considering the simple example:

$$\dot{x}(t) = u(t) \; ; \; x(0) = 0 \quad ; \quad |u(t)| \leq 1$$

Minimize:

$$\int_0^1 [u(t)^2 - x(t)^4] \, dt$$

Here the epsilon problem without the finiteness condition will have minus infinity for the infimum while the control problem has zero for the infimum. Unless otherwise stated, this condition will be part of the epsilon problem in what follows.

Again in order not to complicate the exposition too much, we shall assume that the infimum of the epsilon problem is attained by a finite final time T_ε, in the sense that

$$h(\varepsilon) = \text{Inf} \; h(\varepsilon;x(\cdot);u(\cdot);T) = \lim_n \; h(\varepsilon;x_n(\cdot);u_n(\cdot);T_\varepsilon)$$

where T_ϵ is finite, and $x_n(\cdot)$, $u_n(\cdot)$, is a "minimizing" sequence for the epsilon problem. For such a minimizing sequence, let

$$d(\epsilon) = \text{Lim inf } \frac{1}{2}\int_0^{T_\epsilon} (||\dot{x} - f(t;x_n(t);u_n(t))||^2 + ||\phi(t;x_n(t);u_n(t)||^2)dt$$

$$G(\epsilon) = \text{lim sup } \int_0^{T_\epsilon} g(t;x(t);u(t))\, dt.$$

Then of course

$$h(\epsilon) = d(\epsilon)/\epsilon + G(\epsilon)$$

Let us now define

$$\delta(\epsilon) = \text{Sup } d(\epsilon)$$

$$g(\epsilon) = \text{Inf } G(\epsilon)$$

where the infimum (and supremum) is taken over the class of all minimizing sequences. While $\delta(\epsilon)$ is finite because of condition (F), $g(\epsilon)$ may well be minus infinity in general. Under the usual conditions on the dynamics, we shall see however that $g(\epsilon)$ will be finite. We have of course:

$$h(\epsilon) = \delta(\epsilon)/\epsilon + g(\epsilon) \tag{2.6}$$

It is natural now to define g(0) to be the infimum for the control problem, assuming it is definable. Then

$$h(\varepsilon) \leq g(0) \tag{2.7}$$

With these definitions we can state the following theorem concerning the approximation:

Theorem 2.1

Suppose $g(\varepsilon)$ is finite for some ε_0. Then $g(\varepsilon)$ is finite for every ε less than ε_0, and moreover as $\varepsilon \to 0$, $\delta(\varepsilon)$ is monotone non-increasing and $g(\varepsilon)$ is monotone non-decreasing.

Further:

$$\lim_{\varepsilon \to 0} \delta(\varepsilon)/\varepsilon = 0 \tag{2.8}$$

if g(0) is definable (not equal to plus infinity).

Proof

Let ε be less than ε_0. We have then, as an elementary analysis on sums of limits shows:

$$\delta(\varepsilon)/\varepsilon + g(\varepsilon) = h(\varepsilon) \leq \delta(\varepsilon_0)/\varepsilon + g(\varepsilon_0)$$

and similarly:

$$\delta(\varepsilon_o)/\varepsilon_o + g(\varepsilon_o) = h(\varepsilon_o) \leq \delta(\varepsilon)/\varepsilon_o + g(\varepsilon)$$

Since every quantity is finite on the left we can freely transpose to obtain:

$$\frac{\delta(\varepsilon) - \delta(\varepsilon_o)}{\varepsilon} \leq g(\varepsilon_o) - g(\varepsilon) \leq \frac{\delta(\varepsilon) - \delta(\varepsilon_o)}{\varepsilon_o} \tag{2.9}$$

and since ε is less than ε_o, these relations are consistent only if

$$\delta(\varepsilon) \leq \delta(\varepsilon_o)$$

$$g(\varepsilon) \geq g(\varepsilon_o)$$

Hence $g(\varepsilon)$ is finite. Moreover since the argument can now be repeated with

$$\varepsilon = \varepsilon_1, \quad \varepsilon_o = \varepsilon_2, \quad \varepsilon_1 < \varepsilon_2 \leq \varepsilon_o$$

the required monotonicity follows. Let $g(0+)$ denote the limit of $g(\varepsilon)$ as ε goes to zero. From (2.7), since $g(0)$ is not plus infinity, $\delta(\varepsilon)$ must converge to zero. Again with $\varepsilon < \varepsilon_1 < \varepsilon_o$, we have

$$g(\varepsilon) - g(\varepsilon_1) \geq \frac{\delta(\varepsilon_1) - \delta(\varepsilon)}{\varepsilon_1}$$

and letting ϵ go to zero in this we obtain that for $\epsilon < \epsilon_o$,

$$\delta(\epsilon)/\epsilon \leq g(0+) - g(\epsilon) \leq g(0) - g(\epsilon) < \infty \qquad (2.10$$

and in particular then $\delta(\epsilon)/\epsilon$ goes to zero.

Remark 1

It should be noted that infimum of the epsilon problem has been sought in the class of admissible controls. This is natural since, freed of having to satisfy the differential equation constraint, any admissible control can be used. On the other hand this means that in general the optimal control will be a relaxed control. In particular $g(0+)$ may well be less than $g(0)$, the latter being usually sought in the class of ordinary controls, as we assume herein also. An example is given in [3] where $g(0+) = -1$, while $g(0) = 0$. However we shall see that the infimum for the control problem allowing relaxed controls will be $g(0+)$, at least under the usual conditions. But the main point is that in the epsilon problem relaxed controls appear of necessity.

Remark 2

As shown in [3], (2.9) and (2.10) actually hold for $d(\epsilon)$ and $G(\epsilon)$ (even though the latter may depend on the particular minimizing sequence chosen!)

Corollary Assume $g(0) < +\infty$ and that $g(\varepsilon_0)$ is finite for some $\varepsilon_0 > 0$. Then, $h(\varepsilon)$ is monotone non-decreasing and omitting at most a countable number of points in $0<\varepsilon<\varepsilon_0$, we have

$$h'(\varepsilon) = (-1)\delta(\varepsilon)/\varepsilon^2 \tag{2.11}$$

and

$$g'(\varepsilon) + \delta'(\varepsilon)/\varepsilon = 0 \quad \text{a.e. in } 0<\varepsilon<\varepsilon_0 . \tag{2.12}$$

Proof For $\varepsilon<\varepsilon_0$, both $g(\cdot)$ and $\delta(\cdot)$ are monotone, and hence continuous except for a countable number of points and differentiable a.e. Now

$$h(\varepsilon + \Delta) - h(\varepsilon) \leq (\delta(\varepsilon)/(\varepsilon + \Delta) + g(\varepsilon)) - (\delta(\varepsilon)/\varepsilon + g(\varepsilon))$$

$$= \delta(\varepsilon)\,(1/(\varepsilon + \Delta) - 1/\varepsilon)$$

(showing monotonicety)

while

$$h(\varepsilon + \Delta) - h(\varepsilon) \geq \frac{(\delta(\varepsilon + \Delta)}{\varepsilon + \Delta} + g(\varepsilon + \Delta)) - (\delta(\varepsilon + \Delta)/\varepsilon + g(\varepsilon + \Delta))$$

$$= \delta(\varepsilon + \Delta)\left(\frac{1}{(\varepsilon + \Delta)} - \frac{1}{\varepsilon}\right)$$

or, (2.11) follows. But omitting a set of measure zero:

$$h'(\varepsilon) = \delta'(\varepsilon)/\varepsilon + g'(\varepsilon) - \delta(\varepsilon)/\varepsilon^2$$

from which (2.12) follows.

3. Fixed End-Point Problems

In order to introduce the basic ideas in the epsilon technique, it is convenient to begin with what is perhaps the simplest class of control problems: Fixed end-point problems with fixed initial condition, and bounded controls.

Problem I:

$$\text{Minimize } \int_0^T g(t;x(t);u(t))\,dt \quad + \varphi(x(T)) \tag{3.1}$$

where T is fixed and finite and

$$\dot{x}(t) = f(t;x(t);u(t)) \quad \text{a.e.} \; ; \quad x(0) = x_1 \text{ fixed} \tag{3.2}$$

$$u(t) \text{ Lebesgue measurable}$$

$$u(t) \in U \quad \text{a.e., } U \text{ being compact} \tag{3.3}$$

It will be assumed in addition that

$f(t;x;u)$, $g(t;x;u)$, $\varphi(x)$ are C^1 in x, continuous in all variables, and further condition G holds:[+]

$$\text{(G):} \qquad [x, f(t;x;u)] \leq c(1+\|x\|^2) \quad \text{for } u \text{ in } U, \; 0 \leq t \leq T \cdots \tag{3.4}$$

We note immediately that the infimum, denoted g(0), is finite. The epsilon problem is formulated as follows:

Let

$$h(\epsilon;x(\cdot);u(\cdot)) = \frac{1}{2\epsilon}\int_0^T \| \dot{x} - f(t;x(t);u(t)) \|^2\,dt + \varphi(x(T))$$

$$+ \int_0^T g(t;x(t);u(t))\,dt \tag{3.5}$$

[+]This condition as well as [3.3] can be relaxed as in [15] for example - we forego this generalization in the interest of simplicity of exposition, especially since it is not an intrinsic limitation on the approach.

Minimize $\dot{h}(\epsilon;x(\cdot);u(\cdot))$ over the class of controls $u(t)$ Lebesgue measurable, $u(t) \in U$, and also over the class of absolutely continuous ('state') functions $x(t)$ with $x(0) = x_1$. (It is clear that additional phase plane constraints can be added here if necessary.) In addition the condition F' is imposed:

$$(F'): \qquad \int_0^T \|\dot{x}(t)\|^2 \, dt \leq m < \infty \tag{3.6}$$

The condition F' is a slight weakening of condition F, which is possible because of the smoothness properties of the functions assumed. Thus let $x_n(\cdot)$, $u_n(\cdot)$ be a minimizing sequence for the epsilon problem. Condition F' implies that $x_n(t)$ is uniformly bounded in $0 \leq t \leq T$ and hence both $\delta(\epsilon)$ and $g(\epsilon)$ (which now includes the $\varphi(\cdot)$ term) are finite. Again, it is readily seen that condition F implies F'. For let

$$\dot{x}_n - f(t;x_n(t);u_n(t)) = z_n(t) \tag{3.7}$$

Then, using (3.4):

$$\begin{aligned} |[\dot{x}_n, x_n]| &\leq |[x_n, f(t;x_n(t);u_n(t)]| + |[x_n, z_n]| \\ &\leq c(1 + \|x_n\|^2) + m\|x_n\| \\ &\leq D(1 + \|x_n\|^2) \end{aligned} \tag{3.8}$$

and by the usual analysis (Gronwall lemma) this implies that $x_n(t)$ is uniformly bounded. (If the initial condition $x_n(0) = x_1$ is generalized to $\varphi_o(x(0) = 0$, we must then require that the set

$$[x \mid \varphi_o(x) = 0]$$

is bounded, for this result to hold as well as for g(0) to be finite.)

To solve the epsilon problem we take the following elementary route. Let an admissible state function x(t) (that is, absolutely continuous and satisfying $x(0) = x_1$ and (3.6)) be chosen. To minimize (3.5), we simply minimize the integrand. Let

$$m(\epsilon;t;y;x) = \underset{u \in U}{\text{Min}} \left(\frac{1}{2\epsilon} \| y - f(t;x;u) \|^2 + g(t;x;u) \right) \tag{3.9}$$

The minimum is clearly attained since U is compact and the functional is continuous. It is readily seen further that $m(\epsilon;t;y;x)$ is continuous in all the variables. Now

$$h(\epsilon;x(\cdot);u(\cdot)) \geq \int_0^T m(\epsilon;t;\dot{x}(t);x(t))\, dt + \varphi(x(T))$$

so that

$$h(\epsilon) \geq \text{Inf} \int_0^T m(\epsilon;t;\dot{x}(t);x(t))\, dt + \varphi(x(T))$$

where the infimum is taken over the class of admissible state functions x(t). To reverse the inequality in (3.10) we have only to note that we can find an admissible control u(t) such that (a.e.):

$$m(\varepsilon;t;\dot{x}(t);x(t)) = \frac{1}{2\varepsilon} \| \dot{x}(t) - f(t;x(t);u(t)) \|^2 + g(t;x(t);u(t))$$

This is obvious if the minimum of (3.9) is attained at a unique point in U. Otherwise we invoke the "half-way principle of McShane and Warfield", as in Young [1].

Let $x_n(\cdot)$ be a minimizing sequence for

$$h(\varepsilon) = \text{Inf} \int_0^T m(\varepsilon;t;\dot{x}(t);x(t))dt + \varphi(x(T))$$

Let $u_n(\cdot)$ be a corresponding admissible control sequence. Now it is readily seen that $x_n(\cdot)$ is equicontinuous. Hence we may, by renumbering if necessary, assume that $x_n(t)$ converges uniformly to $x_0(t)$ say. Further we can see that $x_0(t)$ is absolutely continuous and we may assume that the sequence (again by renumbering as necessary) $\dot{x}_n(t)$ converges weakly to $\dot{x}_0(t)$. Also $x_0(t)$ is an admissible state function. But the sequence of controls converge, in general, only in the sense of relaxed controls. [Indeed to establish the existence of a relaxed optimal control for the epsilon problem, as in the original control problem, takes "no more than a routine exercise in using the Ascoli theorem and the diagonal process" (McShane [4]) only more so in the present case!]

The main point is, however. that the optimal control for the epsilon problem must be sought in the class of relaxed controls. Because the ordinary admissible controls are weak-star dense in the class of relaxed controls, the infimum of the epsilon problem, over relaxed controls, is the same as that over ordinary controls. Moreover for Problem I, the infimum over relaxed controls is also the same as that over ordinary controls. We can now obtain a constructive approach to the maximum principle by allowing for relaxed controls.

Approach to the Maximum Principle

Let x denote the element in appropriate product space:

$$x = f(t;x;u), \quad g(t;x;u)$$

Let μ denote a probability (regular) measure on the Lebesgue subsets of U. As ranges over the class of all such "control" measures, the points

$$x = \int_U f(t;x;u) \, d\mu(u) \, , \quad \int_U g(t;x;u) \, d\mu(u)$$

describe the closed convex hull of the set

$$\left\{ f(t;x;u) \, , \quad g(t;x;u) \quad ; \quad u \in U \right\}$$

Let us agree to use the notation:

$$\tilde{f}(t;x;u) = \int_U g(t;x;u) \, d\mu(u)$$

$$\tilde{g}(t;x;u) = \int_U g(t;x;u) \, d\mu(u)$$

if a "control measure" (the terminology borrowed from Young [11]) is intended. It is convenient to denote the closed convex hull by $C(t;x)$. Let

$$r(\epsilon;t;y;x;u) = \frac{1}{2\epsilon}\|y - f(t;x;u)\|^2 + g(t;x;u) = r(\epsilon;t;\chi)$$

$$\tilde{r}(\epsilon;t;y;x;u) = \frac{1}{2\epsilon}\|y - \tilde{f}(t;x;u)\|^2 + \tilde{g}(t;x;u) = r(\epsilon;t;y;\chi)$$

Then

$$m(\epsilon;t;y;x) = \operatorname*{Inf}_{u \in U} r(\epsilon;t;y;x;u)$$

Let

$$\tilde{m}(\epsilon;t;y;x) = \operatorname*{Inf}_{\mu} \tilde{r}(\epsilon;t;y;x;u) = \operatorname*{Inf}_{\chi \in C(t;x)} r(\epsilon;t;y;\chi)$$

where now the infimum is taken over the class of all control measures, μ. We note that $\tilde{m}(\epsilon;t;y;x)$ is continuous in t, y and x. Let $\tilde{x}_n(t)$ be a minimizing sequence of admissible state functions for

$$\operatorname{Inf} \int_0^T \tilde{m}(\epsilon;t;\dot{x}(t);x(t))\, dt + \varphi(x(T))$$

where the infimum is taken over the class of all admissible state functions. If for each t the optimal control measure that attains the minimum of $r(\epsilon;t;\dot{\tilde{x}}(t);\tilde{x}(t);u)$ is unique, or by invoking the McShane-Warfield half-way principle otherwise, we note that there is a relaxed control such that

$$m(\epsilon;t;\dot{\tilde{x}}_n(t);\tilde{x}_n(t)) = r(\epsilon;t;\dot{\tilde{x}}_n(t);\tilde{x}_n(t);\tilde{u}_n(t))$$

$$= \frac{1}{2\epsilon}\int_U \dot{\tilde{x}}_n(t) - f(t;\tilde{x}_n(t);u)\|^2 \, d\mu_n(u;t)$$

$$+ \int_U g(t;\tilde{x}_n(t);u) \, d\mu_n(u;t)$$

$$\int_U p(u) \, d\mu(u;t)$$

Because the ordinary controls are weak-star dense in the class of relaxed controls, it readily follows that

$$h(\epsilon) = \text{Inf} \int_0^T \tilde{m}(\epsilon;t;\dot{\tilde{x}}(t);x(t))dt + \varphi(x(T))$$

Now it is readily seen that $\tilde{x}_n(t)$ can be taken to be uniformly convergent to $\tilde{x}_0(t)$ say, $\tilde{x}_0(t)$ absolutely continuous, and

$$h(\epsilon) = \int_0^T \tilde{m}(\epsilon;t;\dot{\tilde{x}}_0(t);\tilde{x}_0(t)) \, dt + \varphi[\tilde{x}_0(T)]$$

Next to obtain the maximum principle, we begin by noting that

$$\tilde{r}(\epsilon;t;y;\chi)$$

is a differentiable convex functional on the compact convex set $C(t;x)$. Hence the infimum is attained, and denoting such a point by χ_0 we know that for any point χ in $C(t;x)$ we must have: [omitting a set of measure zero in t]

$$\frac{d}{d\theta}\ \tilde{r}(\epsilon;t;y;\ \chi_0+\theta(\chi-\chi_0))\Big|_{\theta=0} \geq 0 \tag{3.11}$$

writing

$$\chi_0 = \tilde{f}(t;x;u_0)\ ,\quad \tilde{g}(t;x;u_0)$$

$$\chi = \tilde{f}(t;x;u)\ ,\quad \tilde{g}(t;x;u)$$

the left side of (3.11) is readily calculated to be:

$$\left[\frac{y - f(t;x;u_0)}{\epsilon}\ ,\ (\tilde{f}(t;x;u_0) - \tilde{f}(t;x;u))\right] + \tilde{g}(t;x;u) - \tilde{g}(;x;u_0) \geq 0 \tag{3.12}$$

Or,

$$[\Psi, \tilde{f}(t;x;u_0)] - \tilde{g}(t;x;u_0) \geq [\Psi, \tilde{f}(t;x;u)] - \tilde{g}(t;x;u) \tag{3.13}$$

where

$$\Psi = (y - \tilde{f}(t;x;u_0))/\epsilon$$

We note that (3.13) is also a sufficient condition for χ_o to be the infimum, and is already recognizable as the Maximum Principle.

$$\text{Lim} \int_o^T \int_U g(t;\tilde{x}_n(t);u)\, d\mu_n(u;t)\, dt$$

$$= \int_o^T \int_U g(t;\tilde{x}_o(t);u)\, d\mu_o(u;t)\, dt \tag{3.14}$$

and since we may take $\dot{\tilde{x}}_n(t)$ converging weakly (in the $L_2(0, T)$ sense) to $\dot{\tilde{x}}_o(t)$, we have also:

$$\lim \int_o^T \int_U \|\dot{\tilde{x}}_n(t) - f(t;\tilde{x}_n(t);u)\|^2\, d\mu_n(u;t)\, dt$$

$$\geq \int_o^T \int_U \|\dot{\tilde{x}}_o(t) - f(t;\tilde{x}_o(t);u)\|^2\, d\mu_o(u;t)\, dt \tag{3.15}$$

Hence

$$h(\varepsilon) = \int_o^T \tilde{m}(\varepsilon;t;\dot{\tilde{x}}_o(t);\tilde{x}_o(t))\, dt + \varphi(\tilde{x}_o(T)) \tag{3.16}$$

Letting

$$\Psi(\varepsilon;t) = (\dot{\tilde{x}}_o(t) - \tilde{f}(t;\tilde{x}_o(t);\tilde{u}o(t)))/\varepsilon \tag{3.17}$$

we see that the optimal control $\tilde{u}_o(\cdot)$ (which we repeat is now, in general, relaxed) is characterized by the Maximum Principle:

$$[\Psi(\varepsilon;t),\ \tilde{f}(t;\tilde{x}_o(t);\tilde{u}_o(t))] - \tilde{g}(t;\tilde{x}_o(t);\tilde{u}_o(t))$$

$$= \text{Max}\, [\Psi(\varepsilon;t),\ \tilde{f}(t;\tilde{x}_o(t);u)] - \tilde{g}(t;\tilde{x}_o(t);u) \tag{3.18}$$

where the maximum is now taken over the class of control measures. Finally a routine first variation analysis shows that $\Psi(\epsilon;t)$ must satisfy. Also note that for ϵ sufficiently small, strict inequality will hold in (F').]

$$\dot{\Psi}(\epsilon;t) + \tilde{f}_1(t;\tilde{x}_o(t);\tilde{u}_o(t))^* \quad \Psi(\epsilon;t) = \tilde{g}_1(t;\tilde{x}_o(t);\tilde{u}_o(t));$$

$$\Psi(\epsilon;T) + \varphi_1(\tilde{x}_o(T)) = 0 \qquad \ldots. \qquad (3.19)$$

where

$$\tilde{f}_1(t;\tilde{x}_o(t);\tilde{u}_o(t)) = \int_U f_1(t;\tilde{x}_o(t);u) \; d\mu_o(u;t)$$

$$\tilde{g}_1(t;\tilde{x}_o(t);\tilde{u}_o(t)) = \int_U g_1(t;\tilde{x}_o(t);u) \; d\,\mu_o(u;t)$$

where

$$f_1(t;x;u) = \nabla_x f(t;x;u)$$

$$g_1(t;x;u) = \nabla_x g(t;x;u)$$

$$\varphi_1(x) = \nabla_x \varphi(x)$$

∇_x denoting gradient with respect to x.

We finally note that by Caratheodory's Theorem [5], the control measure $d\mu_0(u;t)$ can be characterized by a finite number of atomic parts for <u>all</u> t. This result was also noted by Gamkrelidze.

It is quite straightforward now to get the Maximum Principle by letting epsilon go to zero. Let us use $x(\epsilon;t)$, $u(\epsilon;t)$ to denote $\tilde{x}_o(t)$, $\tilde{u}_o(t)$, for the solution to the epsilon problem (the optimal control $u(\epsilon;t)$ being relaxed). Then it is readily seen from the estimate (2.8) that $\delta(\epsilon)$ goes to zero so that $x(\epsilon;t)$ is equicontinuous and we can consider a convergent subsequence converging to $x(0;t)$ say, uniformly in t. Since $\dot{x}(\varepsilon;T)$ is bounded, so also then is $\Psi(\varepsilon;T)$, and hence also $\Psi(\varepsilon;t)$, the functions $f_1(\cdots)$, $g_1(\cdots)$ being continuous. Moreover, there exists an optimal (relaxed) control, and we may take weak limits in (3.18) and (3.19) to obtain the maximum principle for the control problem, in the form given by McShane [4] . Also, note that

$$g(0+) = g(0)$$

Remark 1

Note that we have incidentally shown that

$$\delta(\varepsilon)/\varepsilon^2 < \infty$$

since

$$\int_0^T \| \Psi(\varepsilon;t) \|^2 \, dt < \infty$$

Remark 2

$$h(\varepsilon) = \operatorname*{Lim}_n \int_0^T m(\varepsilon;t;\dot{x}_n(t);x_n(t)) \, dt + \varphi(x_n(T))$$

$$\geq \operatorname*{Lim}_n \int_0^T \tilde{m}(\varepsilon;t;\dot{x}_n(t);x_n(t)) \, dt + \varphi(x_n(T))$$

and hence with $x_n(t)$ converging uniformly to $x_o(t)$, and $\dot{x}_n(t)$ converging weakly to $\dot{x}_o(t)$, we have:

$$h(\varepsilon) = \int_0^T \tilde{m}(\varepsilon;t;\dot{x}_o(t);x_o(t)) \, dt + \varphi(x_o(T))$$

$$\geq \int_0^T m(\varepsilon;t;\dot{x}_o(t);x_o(t)) \, dt + \varphi(x_o(T))$$

where the strict inequality may hold.

Remark 3

It is convenient to use the notation:

$$r(\epsilon;t;y;x;u) = \frac{1}{2\epsilon} \|y - f(t;x;u)\|^2 + g(t;x;u)$$

Let us fix t and let:

$$m(\epsilon;t;y;x) = r(\epsilon;t;y;x;u_o)$$

and let us assume now that the minimal point u_o is unique.

Then letting

$$m(\epsilon;t;y + \theta h_y;x + \theta h_x) = r(\epsilon;t;y + \theta h_y;x + \theta h_x;u_\theta)$$

we note the the usual inequalities:

$$r(\epsilon;t;y + \theta h_y; x + \theta h_x;u_\theta) - r(\epsilon;t;y;x;u_\theta)$$

$$\leq m(\epsilon;t;y+\theta h_y;x + \theta h_x) - m(\epsilon;t;y;x)$$

$$\leq r(\epsilon;t;y; + \theta h_y;x + \theta h_x;u_o) - r(\epsilon;t;u;x;u_o)$$

Since u is compact, every sequence u_θ contains a convergent subsequence and the limit must be a minimal point and hence equal to u_o. Hence:

$$\frac{d}{d\theta} m(\epsilon;t;y + \theta h_y;x + \theta h_x)\Big|_{\theta=0} = \frac{d}{d\theta} r(\epsilon;t;y+\theta h_y;\theta+h_x u_o)\Big|_{\theta=0} \tag{3.20}$$

and the left side is zero as soon as

$$\frac{d}{d\theta}\, r(\epsilon;t;y + \theta h_y;x + \theta h_x;u_o)\Big|_{\theta = 0} = 0$$

Remark 4

Let $x(\epsilon;t)$, $u(\epsilon;t)$ denote an optimal solution of the epsilon problem, and suppose we see how well we can do in the original control problem if we took $u(\epsilon;t)$ for the control. Let $x_o(\epsilon;t)$ denote the solution of

$$\dot{x} = f(t;x(t);u(\epsilon;t)) \quad x(0) = x_1$$

(note that $u(\epsilon;t)$ may well be a relaxed control but because of condition (3.4), the differential equation still has a unique solution, see Young [11]). Let

$$y(t) = x_o(\epsilon;t) - x(\epsilon;t)$$

Then

$$y(0) = 0; \ \dot{y}(t) = f(t;x_o(\epsilon;t);u(\epsilon;t)) - f(t;x(\epsilon;t);u(\epsilon;t)) - \epsilon\,\Psi(\epsilon;t)$$

from which it follows that

$$\frac{d}{dt}\,[y(t),y(t)] \leq M([y(t),y(t)] + \epsilon)$$

or $y(t)$ goes to zero uniformly in t with epsilon.

Also

$$\left|\left(\int_0^T g(t;x_0(\epsilon;t);u(\epsilon;t))\,dt + \varphi(x_0(\epsilon;T)\right) - \left(\int_0^T g(t;x(\epsilon;t);u(\epsilon;t))\,dt + \varphi(x(\epsilon;T))\right)\right| \leq k \sup |y(t)| \to 0$$

Problem II

Minimize $\int_0^T g(t;x(t);u(t))\,dt + \varphi(x(T))$.

Where T is fixed and finite, subject to:

(3.2), (3.3), (3.4) and in addition:

$$\phi(t;x(t);u(t)) = 0 \text{ a.e.} \tag{3.21}$$

where $\phi(t;x;u)$ is continuous in all the variables, and continuously differentiable in x.

Let g(0) denote the infimum, assumed to be less than plus infinity; in other words there is at least one solution to (3.2) satisfying (3.3) and (3.21). In extending the class of controls to include relaxed controls we require that the latter satisfy:

$$\tilde{\phi}(t;x(t);u(t)) = \int_U \phi(t;x(t);u)\, d\mu(u;t) = 0 \text{ a.e.} \tag{3.22}$$

A relaxed control satisfying (3.22) is not necessarily approximated by ordinary controls satisfying (3.21). Thus the infimum over relaxed controls may well be strictly less than g(0), as the following simple example shows:

minimize $x_1(1)$ subject to:

$$\dot{x}_1 = x_2^2 - u^2$$

$$\dot{x}_2 = u$$

$$\dot{x}_3 = x_2^4$$

$$x_1(0) = x_2(0) = x_3(0) = 0$$

$$|u| \leq 1$$

$$\phi(t;x;u) = x_2 = 0$$

It is apparent that g(0) is zero, while the infimum over relaxed controls is (−1).

The epsilon problem is formulated as that of minimizing (2.4) over the class of admissible state functions (that is, absolutely continuous, satisfying F', and $x(0) = x_1$, and over the class of controls $u(\cdot)$ satisfying (3.2) and (3.3). We note that the estimate (2.8) applies. In particular we shall see that $g(0+)$ is the infimum for problem II over the class of relaxed controls. Let us again use the notation:

$$m(\varepsilon;t;y;x) = \underset{u \in U}{\text{Min}} \frac{1}{2\varepsilon}(\|y-f(t;x;u)\|^2 + \|\phi(t;x;u)\|^2) + g(t;x;u)$$

Then clearly

$$\text{Inf} \int_0^T m(\varepsilon;t;\dot{x}(t);x(t))\, dt + \varphi(x(T)) \leq h(\varepsilon)$$

where the infimum is taken over the class of admissible state functions and $h(\varepsilon)$ denotes the infimum for the epsilon problem. That actually equality holds is readily seen again by invoking the McShane-Warfield halfway principle. Let $x_n(t)$ be a minimizing sequence of admissible state functions, and let us denote the limit (of an appropriate sub-sequence) by $x_o(\varepsilon;t)$.

Approach to the Maximum Principle

Let μ denote a regular probability measure (or control measure) on the Lebesgue sets of U. Let $C(t;x)$ denote the set of points χ in the appropriate dimension product space:

$$\chi = \tilde{f}(t;x;u), \quad \tilde{g}(t;x;u), \quad \phi(t;x;u)$$

where as before:

$$\tilde{f}(t;x;u) = \int_U f(t;x;u)\ d\mu(u), \quad \tilde{\phi}(t;x;u) = \int_U \phi(t;x;u)\ d\mu(u)$$

as μ varies over the class of all such measures. Then $C(t;x)$ is the closed convex extension of the set

$$\left\{ f(t;x;u), \quad g(t;x;u), \quad \phi(t;x;u), \quad u \in U \right\}$$

Let us use the notation:

$$r(\epsilon;t;y;x;u) = r(\epsilon;t;y;\chi) = \frac{1}{2\epsilon}(\|y-\tilde{f}(t;x;u)\|^2 + \|\tilde{\phi}(t;x;u)\|^2) + \tilde{g}(t;x;u)$$

and let

$$\tilde{m}(\epsilon;t;y;x) = \inf_{\chi \in C(t;x)} r(\epsilon;t;y;\chi)$$

Then as before, it is readily seen that:

$$h(\epsilon) = \operatorname{Inf} \int_0^T \tilde{m}(\epsilon;t;\dot{x}(t);x(t))\ dt + \varphi(x(T))$$

$$= \int_0^T \tilde{m}(\epsilon;t;\dot{x}_0(\epsilon;t);x_0(\epsilon;t))\ dt + \varphi(x_0(\epsilon;T))$$

Again letting:

$$\tilde{m}(\epsilon;t;y;x) = r(\epsilon;t;y;\chi_0)$$

we have for any χ in $C(t;x)$ that [omitting a set of measure zero in t]

$$\frac{d}{d\theta} r(\epsilon;t;y;\ \chi_o + \theta(\chi - \chi_o))\Big|_{\theta = 0} \geq 0 .$$

leading to the Maximum Principle:

$$[\Psi, \tilde{f}(t;x;u_o)] + [\phi_o, \tilde{\phi}(t;x;u_o)] - \tilde{g}(t;x;u_o)$$

$$\geq [\Psi, \tilde{f}(t;x;u)] + [\phi_o, \tilde{\phi}(t;x;u)] - \tilde{g}(t;x;u)$$

where

$$\Psi = (y - \tilde{f}(t;x;u_o))/\epsilon \quad ; \quad \phi_o = \tilde{\phi}(t;x;u_o)/\epsilon$$

Corresponding to $x_o(\epsilon;t)$ there is a relaxed optimal control $d\mu_o(\epsilon;u;t)[\sim u_o(\epsilon;t)]$ and by a routine variational analysis we have letting

$$\Psi(\epsilon;t) = (\dot{x}_o(\epsilon;t) - \tilde{f}(t;x_o(\epsilon;t);u_o(\epsilon;t)))/\epsilon$$

$$\phi_o(\epsilon;t) = \tilde{\phi}(t;x_o(\epsilon;t);u_o(\epsilon;t))/\epsilon$$

that $\Psi(\epsilon;t)$ must satisfy:

$$\dot{\Psi} + \tilde{f}_1(t;x_o(\epsilon;t);u_o(\epsilon;t))^* \Psi - \tilde{\phi}_1(t;x_o(\epsilon;t);u_o(\epsilon;t))^* \phi_o(\epsilon;t)$$

$$- \tilde{g}_1(t;x_o(\epsilon;t);u_o(\epsilon;t)) = 0 \qquad (3.23)$$

$$\Psi(\epsilon;T) + \varphi_1(x_o(\epsilon;T)) = 0 \quad \ldots\ldots\ldots \qquad (3.23a)$$

where

$$\tilde{\phi}(t;x_o(\epsilon;t);u_o(\epsilon;t)) = \int_U \phi(t;x_o(\epsilon;t);u) \; d\mu_o(\epsilon;u;t)$$

$$\phi_1(t;x;u) = \nabla_x \phi(t;x;u)$$

and other notation is the same as before for Problem I. We have the epsilon maximum principle:

$$[\Psi(\epsilon;t), \; \tilde{f}(t;x_o(\epsilon;t);u_o(\epsilon;t))] + [\phi_o(\epsilon;t), \; \tilde{\phi}(t;x_o(\epsilon;t);u_o(\epsilon;t))]$$
$$- \tilde{g}(t;x_o(\epsilon;t);u_o(\epsilon;t))$$

$$= \text{Max} \; [\Psi(\epsilon;t), \; \tilde{f}(t;x_o(\epsilon;t);u)] + [\phi_o(\epsilon;t), \; \tilde{\phi}(t;x_o(\epsilon;t);u)]$$
$$- \tilde{g}(t;x_o(\epsilon;t);u) \tag{3.24}$$

Let us now consider the situation as epsilon goes to zero. The main difference from the earlier treatment for Problem I is that we must now show that

$$\phi_o(\epsilon;t)$$

converges as epsilon goes to zero. First of all we may take $x_o(\epsilon;t)$[for suitable subsequence] converging uniformly in t to $x_o(t)$ say, and the corresponding controls to converge (in the weak-star topology) to

$d\mu_o(u;t)$ say. Then we note that (since $\delta(\epsilon)$ converges to zero) that

$$\dot{x}_o(t) = \tilde{f}(t;x_o(t);u_o(t)) \quad ; \quad x_o(0) = x_1 \tag{3.25}$$

and that

$$\tilde{\phi}(t;x_o(t);u_o(t)) = \int_U \phi(t;x_o(t);u)d\mu_o(u;t) = 0 \tag{3.26}$$

the latter following from the fact that for each continuous function $f(t)$, we must have:

$$\lim \int_o^T \int_U [f(t), \phi(t;x_o(\epsilon;t);u)] \, d\mu(\epsilon;u;t)$$

$$= \int_o^T \int_U [f(t), \phi(t;x_o(t);u) \, d\mu_o(u;t)] = 0$$

where we have denoted by $d\mu(\epsilon;u;t)$ $[\sim u_o(\epsilon;t)]$ the control corresponding to $x_o(\epsilon;t)$. Also

$$g(0+) = \int_o^T \tilde{g}(t;x_o(t);u_o(t)) \, dt + \varphi(x_o(T))$$

and $g(0+)$ is also then the infimum for the control problem in the class of relaxed controls satisfying (3.22).

Suppose now that $\int_o^T \| \phi_o(\epsilon_n;t) \|^2 \, dt < M < \infty$

for some sequence ϵ_n going to zero. Let us denote the weak limit of a subsequence by $\phi_o(t)$. Then since $\Psi(\epsilon_n;t)$ satisfies the linear equation (3.23) and $\Psi(\epsilon_n;T)$ is necessarily bounded, it follows that $\Psi(\epsilon_n;t)$ is bounded. Hence

$$\sup \int_0^T \| \dot{\Psi}(\epsilon_n;t) \|^2 \, dt < \infty$$

and hence $\Psi(\epsilon_n;t)$ is equicontinuous, and we may renumber, as necessary, to have $\Psi(\epsilon_n;t)$ to converge uniformly in t to $\Psi(t)$ say. Clearly the first term on the left of (3.24):

$$[\Psi(\epsilon_n;t), \tilde{f}(t;x_o(\epsilon_n;t);u_o(\epsilon_n;t))]$$

converges† weakly to

$$[\Psi(t), \tilde{f}(t;x_o{*}t);u_o(t))]$$

The next term in (3.24):

$$[\phi_o(\epsilon_n;t), \tilde{\phi}(t;x_o(\epsilon_n;t);u_o(\epsilon_n;t))] = \| \tilde{\phi}(t;x_o(\epsilon_n;t);u_o(\epsilon_n;t)) \|^2/\epsilon$$

converges to zero a.e., since the integral over [0, T] goes to zero. On the right side of (3.24), both $\tilde{f}(t;x_o(\epsilon;t);u)$ and $\tilde{\phi}(t;x_o(\epsilon;t);u)$ converge uniformly in t, so that we may take weak limits in (3.24) to obtain the

† Here and elsewhere it is understood that we will be dealing with subsequences as required.

the maximum principle:

$$[\Psi(t), \tilde{f}(t;x_o(t);u_o(t))] - \tilde{g}(t;x_o(t);u_o(t))$$

$$= \text{Max}\, [\Psi(t), \tilde{f}(t;x_o(t);u)] + [\phi_o(t), \tilde{\phi}(t;x_o(t);u)] - \tilde{g}(t;x_o(t);u) \qquad (3.27)$$

Suppose now that

$$\int_0^T \| \phi_o(\epsilon_n;t)\|^2 \, dt$$

is unbounded for every sequence ϵ_n going to zero. Let us pick a subsequence that makes $x_o(\epsilon_n;t)$ converge uniformly in t to $x_o(t)$, and the controls $u_o(\epsilon;t)$, in weak-star topology, to $u_o(t)$ as before so that (3.25), (3.26) hold. Let

$$k_n^2 = \int_0^T \| \phi_o(\epsilon_n;t)\|^2 \, dt$$

Let

$$\Psi_n(t) = \Psi(\epsilon_n;t)/k_n$$

$$\phi_n(t) = \phi_o(\epsilon_n;t)/k_n$$

Dividing thru by k_n in (3.23) let us note that $\Psi_n(t)$ is bounded. Clearly

$$\sup \int_0^T \| \dot{\Psi}_n(t)\|^2 \, dt < \infty$$

so that we have equicontinuity and may take (suitable subsequence renumbered) $\Psi_n(t)$ to converge uniformly in t to $\Psi(t)$ say. Similarly let us take $\phi_n(t)$ to converge weakly to $\phi_o(t)$ say. Then by taking (weak) limits in (3.23) we have:

$$\dot{\Psi}(t) + \tilde{f}_1(t;x_o(t);u_o(t))^* \Psi(t) - \tilde{\phi}_1(t;x_o(t);u_o(t)^* \phi(t) = 0$$

$$\Psi(T) = 0$$

And correspondingly, taking limits in (3.24) we obtain: [a.e. in 0<t<T]

$$[\Psi(t), \tilde{f}(t;x_o(t);u_o(t))]$$
$$= \text{Max}\,[\Psi(t);\tilde{f}(t;x_o(t);u)] + [\phi(t), \tilde{\phi}(t;x_o(t);u)] \quad \ldots\ldots \qquad (3.28)$$

we note that (3.27) and (3.28) can be combined by introducing a multiplier λ in front of $\tilde{g}(t;x_o(t);u_o(t))$ and saying $\lambda \geq 0$. There is now the possibility that $\Psi(t)$ is zero. To avoid this one can introduce additional conditions involving derivatives with respect to u.

Computational Aspects

We shall now study some of the questions that arise in examining computational aspects more deeply, and at the same time indicate a particular scheme for solving the epsilon problem. In doing so we shall need to make some additional assumptions which are natural in the practical context.

We can appreciate in a general way that as epsilon is made smaller and smaller we will run into computational accuracy problems while too large an epsilon will have no relation to the control problem we wish to of it for the problem. Let $C[0, T]$ denote the Banach space of continuous state functions under the sup norm. Let $\{b_n(t)\}$ denote a sequence of basis functions in $C[0, T]$. For each n, let $\mathcal{A}_n$ denote the set of functions spanned by $b_k(\cdot)$, $k = 1, \ldots n$, such that

$$\int_0^T \| \sum_1^n a_k b_k(t) \|^2 \, dt \qquad dt < m \tag{3.29}$$

For each n, we now consider the epsilon problem over the class of state functions (denoted S_n) of the form:

$$x_1 + \sum_1^n a_k \int_0^t b_k(s) \, ds \tag{3.30}$$

where the $\{a_k\}$ must satisfy (3.29), corresponding to condition (F'), and over controls $u(t)$ as before. (The controls are not approximated by basis functions.) Let us denote the corresponding infimum by $h_n(\epsilon)$. Clearly any admissible state function can be approximated uniformly in t by functions in S_n as closely as desired for large enough n, and of course S_n is also conditionally compact.

Let

$$h_n(\epsilon) = \delta_n(\epsilon)/\epsilon + g_n(\epsilon)$$

where the quantities are defined the same way as in section 2, the subscript n denoting restriction to S_n. It is evident that $\delta_n(\epsilon)$

and $g_n(\epsilon)$ are again monotone in the same fashion as before. Let $\delta_n(0)$ and $g_n(0+)$ denote the limits as ϵ goes to zero. Since $h_n(\epsilon)$ (unlike $h(\epsilon)$) has no given upper bound, $\delta_n(0)$ need not be zero. In fact we have:

$$0 \leq (\delta_n(\epsilon) - \delta_n(0))/\epsilon \leq (g_n(0+) - g_n(\epsilon)) \tag{3.31}$$

so that

$$\lim_{\epsilon \to 0} (h_n(\epsilon) - \delta_n(0)/\epsilon) = g_n(0+) \tag{3.32}$$

Thus $h_n(\epsilon)$ eventually increases without bound $[0(1/\epsilon)]$ as we make epsilon smaller. Of course

$$h(\epsilon) \leq h_{n+1}(\epsilon) \leq h_n(\epsilon)$$

and

$$h(\epsilon) = \lim h_n(\epsilon), \text{ for each } \epsilon > 0.$$

Let us now indicate a method for obtaining $h(\epsilon)$. We begin with any element of S_n, say with all $\{a_k\}$ set to be zero, in the absence of any prior ideas concerning the optimal state function.

Call this $x_1(t)$. We now make the following assumption (U_1):

$$\underset{u \in U}{\text{Min}} \quad \frac{1}{2\epsilon}\left(\| y-f(t;x;u)\|^2 + \| \phi(t;x;u)\|^2\right) + g(t;x;u)$$

is attained at a unique point in U, for each t, y, x and ϵ. Let $u_1(t)$ denote the minimal point in U for $y = \dot{x}_1(t)$, $x = x_1(t)$, so that in our previous notation:

$$m(\epsilon;t;\dot{x}_1(t);x_1(t)) = r(\epsilon;t;\dot{x}_1(t);x_1(t);u_1(t)) \tag{3.33}$$

(The assumption (U_1) can clearly be weakened to hold in a suitable neighbourhood.) We now choose $x_2(t)$ so that

$$\underset{x(\cdot)\epsilon S_n}{\text{Inf}} \int_0^T r(\epsilon;t;\dot{x}(t);x(t));u_1(t) \; dt + \varphi(x(T)) \tag{3.33a}$$

(which is attained in general by an element in the closure of S_n) is attained by $x_2(t)$. Note that the function $m(\epsilon;t;\dots)$ is not required to be known. We next determine $u_2(t)$ so that (3.33) holds with $x_1(t)$ replaced by $x_2(t)$, and continue on this way to produce the sequence

$$x_n(t), \; u_n(t)$$

Now

$$\int_0^T m(\epsilon;t;\dot{x}_{n+1}(t);x_{n+1}(t))\,dt + \varphi(x_{n+1}(T))$$

$$= \int_0^T r(\epsilon;t;\dot{x}_{n+1}(t);x_{n+1}(t);u_{n+1}(t))\,dt + \varphi(x_{n+1}(T))$$

$$\leq \int_0^T r(\epsilon;t;\dot{x}_{n+1}(t);x_{n+1}(t);u_n(t))\,dt + \varphi(x_{n+1}(T))$$

$$\leq \int_0^T r(\epsilon;t;\dot{x}_n(t);x_n(t);u_n(t))\,dt + \varphi(x_n(T))$$

$$= \int_0^T m(\epsilon;t;x_n(t);x_n(t))\,dt + \varphi(x_n(T)) \qquad (3.34)$$

and we have thus a monotone decreasing sequence. From any subsequence we can choose a further subsequence such that $x_n(t)$, $\dot{x}_n(t)$ converge in $C[0, T]$ to $x_o(t)$, $\dot{x}_o(t)$ say, and now because of condition (U_1), the corresponding $u_n(t)$ must converge to $u_o(t)$ where

$$m(\epsilon;t;\dot{x}_o(t);x_o(t)) = r(\epsilon;t;\dot{x}_o(t);x_o(t);u_o(t)) \qquad (3.35)$$

Moreover we must have:

$$\int_0^T m(\epsilon;t;\dot{x}_o(t);x_o(t))\,dt + \varphi(x_o(T))$$

$$= \operatorname*{Lim}_n \int_0^T m(\epsilon;t;\dot{x}_n(t);x_n(t))\,dt + \varphi(x_n(T))$$

$$= \operatorname*{Inf}_{x(\cdot)\epsilon S_n} \int_0^T m(\epsilon;t;\dot{x}(t);x(t);u_o(t))\,dt + \varphi(x(T))$$

The last equality together with (3.35) means that $x_o(t), u_o(t)$ cannot be further improved by our procedure.

Next let us note that $x_o(t)$, $u_o(t)$ is a local minimum for the epsilon problem in the sense that

$$\int_0^T r(\epsilon;t;\dot{x}_o(t);x_o(t);u_o(t))\,dt \le \int_0^T r(\epsilon;t;\dot{x}_o(t);x_o(t);u(t))\,dt$$

for any admissible control u(t) while the first variation of

$$\int_0^T m(\epsilon;t;\dot{x}(t);x(t))\,dt + \varphi(x(T))$$

vanishes at $x(t) = x_o(t)$. For this purpose we assume that for any $h(\cdot)$ in $\mathcal{A}_n$,

$$x_o(t) + \theta \int_0^t h(s)ds$$

belongs to S_n for all sufficiently small $|\theta|$. Now because of (U_2), we see from (3.20) that we only need to show that

$$\frac{d}{d\theta} r(\epsilon;t;\dot{x}_o(t) + \theta\, h(t);x_o(t) + \theta \int_o^t h(s)ds\; ;\; u_o(t))\Big|_{\theta = 0}$$

is zero. But this follows from the fact that this is true for $x_n(t)$, $u_n(t)$ and we can take limits with respect to n. Clearly in any computational scheme we can only obtain a local minimum. In practice we assume that (at least for large enough order of approximation) there is only one local minimum, or, at least our search is confined to a region where there is only one minimum and it is the true minimum.

Note that under condition $[U_1]$, we have

$$h'_n(\epsilon) = - \delta_n(\epsilon)|\epsilon^2 \text{ for every } \epsilon > 0 .$$

$$\frac{\delta'_n(\epsilon)}{\epsilon} + g'_n(\epsilon) = 0 \quad \text{a.e. in } \epsilon > 0 .$$

To conclude the computational scheme, we need only now to describe how the infimum in (3.33a) is determined. Here again we only seek a local minimum. For this purpose, let us note that the functional

$$\int_o^T r(\epsilon;t;\dot{x}(t);x(t);u(t))\, dt \; + \; \varphi(x(T))$$

is now a function only of the coefficients $\{a_k\}$ and let us denote this function by

$$h(\epsilon;\alpha;u(\cdot)), \quad \alpha = \{a_k\}$$

Then we use the iteration:

$$\alpha_{m+1} = \alpha_m - H_m^{-1} G_m$$

where

$$G_m = \frac{\partial}{\partial a_k} h(\epsilon; \alpha_m; u(\cdot))$$

H_m is an n X n matrix with components:

$$\frac{1}{\epsilon} \int_0^T [\frac{\partial}{\partial a_i} (\dot{x}(t) - f(t;x(t);u(t)), \frac{\partial}{\partial a_j} (\dot{x}(t) - f(t;x(t);u(t))] \, dt$$

$$+ \quad \frac{1}{\epsilon} \int_0^T [\frac{\partial}{\partial a_i} \phi(t;x(t);u(t)), \frac{\partial}{\partial a_j} \phi(t;x;(t);u(t))] \, dt$$

$$+ \quad \frac{\partial}{\partial a_i} \frac{\partial}{\partial a_j} \varphi(x(T))$$

This is then a slight variation of the Newton-Raphson technique (in that no second derivatives of the integrand are used). The convergence of the scheme is proved by a minor modification on the usual proofs, as given in [14]for example. Of course other techniques can be used.

Finally, let $\delta > 0$ be given. Then, in theory, we can find an N and ϵ such that

$$| g(0+) - g_N(\epsilon) | < \delta$$

For this we only need to first find ϵ such that

$$g(0+) - g(\epsilon) < \delta/4 \quad \ldots\ldots \qquad (3.36)$$

Then since

$$\delta(\epsilon)/\epsilon \leq g(0+) - g(\epsilon)$$

we have that

$$g(0+) - h(\epsilon) \leq \delta/2$$

Next we choose N large enough so that

$$h_N(\epsilon) - h(\epsilon) \leq \delta/2$$

A. V. Balakrishnan

Computational Aspects

We shall now study some of the questions that arise in examining computational aspects more deeply, and at the same time indicate a particular scheme for solving the epsilon problem. In doing so we shall need to make some additional assumptions which are natural in the practical context.

We can appreciate in a general way that as epsilon is made smaller and smaller we will run into computational accuracy problems while too large an epsilon will have no relation to the control problem we wish to solve. This is best seen by examining a Ritz approximation, or our version of it for the problem. Let $C[0, T]$ denote the Banach space of continuous state functions under the sup norm. Let $\{b_n(t)\}$ denote a sequence of basis functions in $C[0, T]$. For each n, let $\mathcal{A}_n$ denote the set of functions spanned by $b_k(\cdot)$, $k = 1, \ldots n$, such that

$$\int_0^T \| \sum_1^n a_k b_k(t) \|^2 \, dt < m \tag{3.1}$$

For each n, we now consider the epsilon problem over the class of state functions (denoted S_n) of the form:

$$x_1 + \sum_1^n a_k \int_0^t b_k(s)\, ds \qquad (3.2)$$

where the $\{a_k\}$ must satisfy (3.1), corresponding to condition (F'), and over controls u(t) as before. (The controls are not approximated by basis functions.) Let us denote the corresponding infimum by $h_n(\varepsilon)$. Clearly any admissible state function can be approximated uniformly in t by functions in S_n as closely as desired for large enough n, and of course S_n is also conditionally compact.

Let

$$h_n(\varepsilon) = \delta_n(\varepsilon)/\varepsilon + g_n(\varepsilon)$$

where the quantities are defined the same way as in section 2, the subscript n denoting restriction to S_n. It is evident that $\delta_n(\varepsilon)$

and $g_n(\epsilon)$ are again monotone in the same fashion as before. Let $\delta_n(0)$ and $g_n(0+)$ denote the limits as ϵ goes to zero. Since $h_n(\epsilon)$ (unlike $h(\epsilon)$) has no given upper bound, $\delta_n(0)$ need not be zero. In fact we have:

$$0 \leq (\delta_n(\epsilon) - \delta_n(0))/\epsilon \leq (g_n(0+) - g_n(\epsilon)) \tag{3.3}$$

so that

$$\lim_{\epsilon \to 0} (h_n(\epsilon) - \delta_n(0)/\epsilon) = g_n(0+) \tag{3.4}$$

Thus $h_n(\epsilon)$ eventually increases without bound $[0(1/\epsilon)]$ as we make epsilon smaller. Of course

$$h(\epsilon) \leq h_{n+1}(\epsilon) \leq h_n(\epsilon)$$

and

$$h(\epsilon) = \lim h_n(\epsilon), \text{ for each } \epsilon > 0.$$

Let us now indicate a method for obtaining $h(\epsilon)$. We begin with any element of S_n, say with all $\{a_k\}$ set to be zero, in the absence of any prior ideas concerning the optimal state function.

Call this $x_1(t)$. We now make the following assumption (U_1):

$$\underset{u \in U}{\text{Min}} \quad \frac{1}{2\epsilon}\left(\| y - f(t;x;u)\|^2 + \| \phi(t;x;u)\|^2\right) + g(t;x;u)$$

is attained at a unique point in U, for each t, y, x and ϵ. Let $u_1(t)$ denote the minimal point in U for $y = \dot{x}_1(t)$, $x = x_1(t)$, so that in our previous notation:

$$m(\epsilon;t;\dot{x}_1(t);x_1(t)) = r(\epsilon;t;\dot{x}_1(t);x_1(t);u_1(t)) \tag{3.5}$$

(The assumption (U_1) can clearly be weakened to hold in a suitable neighbourhood.) We now choose $x_2(t)$ so that

$$\underset{x(\cdot) \in S_n}{\text{Inf}} \int_0^T r(\epsilon;t;\dot{x}(t);x(t));u_1(t)\ dt + \varphi(x(T)) \tag{3.6}$$

(which is attained in general by an element in the closure of S_n) is attained by $x_2(t)$. Note that the function $m(\epsilon;t;\ldots)$ is not required to be known. We next determine $u_2(t)$ so that (3.5) holds with $x_1(t)$ replaced by $x_2(t)$, and continue on this way to produce the sequence

$$x_n(t),\ u_n(t)$$

Now

$$\int_0^T m(\epsilon;t;\dot{x}_{n+1}(t);x_{n+1}(t))\,dt + \varphi(x_{n+1}(T))$$

$$= \int_0^T r(\epsilon;t;\dot{x}_{n+1}(t);x_{n+1}(t);u_{n+1}(t))\,dt + \varphi(x_{n+1}(T))$$

$$\leq \int_0^T r(\epsilon;t;\dot{x}_{n+1}(t);x_{n+1}(t);u_n(t))\,dt + \varphi(x_{n+1}(T))$$

$$\leq \int_0^T r(\epsilon;t;\dot{x}_n(t);x_n(t);u_n(t))\,dt + \varphi(x_n(T))$$

$$= \int_0^T m(\epsilon;t;\dot{x}_n(t);x_n(t))\,dt + \varphi(x_n(T)) \qquad (3.7)$$

and we have thus a monotone decreasing sequence. From any subsequence we can choose a further subsequence such that $x_n(t)$, $\dot{x}_n(t)$ converge in $C[0, T]$ to $x_o(t)$, $\dot{x}_o(t)$ say, and now because of condition (U_1), the corresponding $u_n(t)$ must converge to $u_o(t)$ where

$$m(\epsilon;t;\dot{x}_o(t);x_o(t)) = r(\epsilon;t;\dot{x}_o(t);x_o(t);u_o(t)) \qquad (3.8)$$

Moreover we must have:

$$\int_0^T m(\epsilon;t;\dot{x}_0(t);x_0(t))\,dt + \varphi(x_0(T))$$

$$= \operatorname*{Lim}_n \int_0^T m(\epsilon;t;\dot{x}_n(t);x_n(t))\,dt + \varphi(x_n(T))$$

$$= \operatorname*{Inf}_{x(\cdot)\in S_n} \int_0^T m(\epsilon;t;\dot{x}(t);x(t);u_0(t))\,dt + \varphi(x(T))$$

The last equality together with (3.8) means that $x_0(t), u_0(t)$ cannot be further improved by our procedure.

Next let us note that $x_0(t)$, $u_0(t)$ is a local minimum for the epsilon problem in the sense that

$$\int_0^T r(\epsilon;t;\dot{x}_0(t);x_0(t);u_0(t))\,dt \leq \int_0^T r(\epsilon;t;x_0(t);x_0(t);u(t))\,dt$$

for any admissible control u(t) while the first variation of

$$\int_0^T m(\epsilon;t;\dot{x}(t);x(t))\,dt + \varphi(x(T))$$

vanishes at $x(t) = x_0(t)$. For this purpose we assume that for any $h(\cdot)$ in $\mathcal{A}_n$,

$$x_0(t) + \theta \int_0^t h(s)ds$$

belongs to S_n for all sufficiently small $|\theta|$. Now because of (U_1), that we only need to show that

$$\frac{d}{d\theta} r(\epsilon;t;\dot{x}_o(t) + \theta h(t);x_o(t) + \theta \int_o^t h(s)ds \,;\, u_o(t))\Big|_{\theta = 0}$$

is zero. But this follows from the fact that this is true for $x_n(t)$, $u_n(t)$ and we can take limits with respect to n. Clearly in any computational scheme we can only obtain a local minimum. In practice we assume that (at least for large enough order of approximation) there is only one local minimum, or, at least our search is confined to a region where there is only one minimum and it is the true minimum.

Note that under condition $[U_1]$, we have

$$h'_n(\epsilon) = - \delta_n(\epsilon)|\epsilon^2 \text{ for every } \epsilon > 0 .$$

$$\frac{\delta'_n(\epsilon)}{\epsilon} + g'_n(\epsilon) = 0 \quad \text{a.e. in } \epsilon > 0 .$$

To conclude the computational scheme, we need only now to describe how the infimum in (3.6) is determined. Here again we only seek a local minimum. For this purpose, let us note that the functional

$$\int_o^T r(\epsilon;t;x(t);x(t);u(t))\, dt + \varphi(x(T))$$

is now a function only of the coefficients $\{a_k\}$ and let us denote this function by

$$h(\epsilon;\alpha;u(\cdot)), \quad \alpha = \{a_k\}$$

Then we use the iteration:

$$\alpha_{m+1} = \alpha_m - H_m^{-1} G_m$$

where

$$G_m = \frac{\partial}{\partial a_k} h(\epsilon; \alpha_m; u(\cdot))$$

H_m is an n X n matrix with components:

$$\frac{1}{\epsilon} \int_0^T [\frac{\partial}{\partial a_i} (\dot{x}(t) - f(t;x(t);u(t)), \frac{\partial}{\partial a_j} (\dot{x}(t) - f(t;x(t);u(t))] \, dt$$

$$+ \quad \frac{1}{\epsilon} \int_0^T [\frac{\partial}{\partial a_i} \phi(t;x(t);u(t)), \frac{\partial}{\partial a_j} \phi(t;x;(t);u(t))] \, dt$$

$$+ \quad \frac{\partial}{\partial a_i} \frac{\partial}{\partial a_j} \varphi(x(T))$$

This is then a slight variation of the Newton-Raphson technique (in that no second derivatives of the integrand are used). The convergence of the scheme is proved by a minor modification on the usual proofs. Of course other techniques can be used.

Finally, let $\delta > 0$ be given. Then, in theory, we can find an N and ϵ such that

$$|g(0+) - g_N(\epsilon)| < \delta$$

For this we only need to first find ϵ such that

$$g(0+) - g(\epsilon) < \delta/4 \quad \ldots\ldots \qquad (3.9)$$

Then since

$$\delta(\epsilon)/\epsilon \leq g(0+) - g(\epsilon)$$

we have that

$$g(0+) - h(\epsilon) \leq \delta/2$$

Next we choose N large enough so that

$$h_N(\epsilon) - h(\epsilon) \leq \delta/2$$

References

1. L. C. Young: "Calculus of Variations and Control Theory", W. B. Saunders, 1969.

2. A. V. Balakrishnan: "On a New Computing Technique in Optimal Control", SIAM Journal on Control, September 1968.

3. A. V. Balakrishnan: "A Computational Approach to the Maximum Principle", Journal of Computer and System Sciences, 1971.

4. E. J. McShane: "Relaxed Controls and Variational Problems", SIAM Journal on Control, 1967.

5. H. G. Eggleston: "Convexity", Cambridge University Press, 1968.

LECTURE NOTES II

A. V. Balakrishnan

Erice, July 1971

A. V. Balakrishnan

Stochastic Systems

1. Inducing Measures on C; the Wiener measure

Given a stochastic process, or equivalently, a consistent family of finite dimensional distributions, we can always construct a 'function-space' process with X as the sample space and a probability measure $p(\cdot)$ on the sigma-algebra $\mathscr{S}$ of subsets of X (that 'agrees' with the finite-dimensional distributions). Throughout we assume that for any two arbit ary time-points t_1, t_2, denoting the corresponding 'variables' by $x(t_1)$, $x(t_2)$, that we have:

$$E(|x(t_x)-x(t_1)|^r) \leq k \ |t_2 - t_1|^{1+\delta}$$

where

$r > 0$, $k > 0$, $\delta > 0$ are fixed constants independent of t_1, t_2; and $|\cdot|$ denotes the Euclidean norm.

Furthermore we shall assume that T is a compact interval. For simplicity of notation, we shall take it to be the unit interval $[0, 1]$ without loss of generality.

Let $C(0, 1)$ denote the class of continuous functions (with range in E) on the closed interval $[0, 1]$. Endowing it with the 'sup' norm, we know that it becomes a Banach space:

$$\|f\| = \sup |f(t)|, \quad 0 \leq t \leq 1$$

where $|\cdot|$ denotes the Euclidean norm. We note that it is a separable Banach space, and denote it by $\mathscr{C}$. By the Borel sets of $\mathscr{C}$ we mean the smallest sigma-algebra generated by all open sets. Let B be an m-dimensional Borel set in $E^{(m)}$. Then for arbitrary $t_1, t_2, \ldots t_n$, the sets in $\mathscr{C}$ defined by:

$$[f(\cdot) \in C \mid f(t_1), \ldots f(t_n) \in B]$$

are called 'cylinder sets' (and B is then referred to as the 'base').

<u>Lemma</u> The class of Borel sets in $\mathscr{C}$ coincides with the smallest sigma-algebra generated by the class of cylinder sets.

<u>Proof</u> It is clear that cylinder sets are Borel sets. Conversely, since $\mathscr{C}$ is separable, any open set in $\mathscr{C}$ can be expressed as the union of a countable number of closed spheres; and every closed sphere, say with center f_o and radius d can be expressed:

$$\bigcap_n [f(\cdot) \mid |f(r_n) - f_o(r_n)| \leq d]$$

where r_n denotes the countable collection of rational numbers in [0, 1]. Hence open sets are contained in the smallest sigma-algebra generated by cylinder sets.

Note in particular that the Borel sets are generated as the smallest sigma-algebra containing sets of the form, $f(.) \in \mathscr{C}$:

$$[f(\cdot) \mid f(t) \in I \ , \quad I \text{ an interval in } E]$$

Given any consistent set of distributions, we can induce a finitely-additive measure on the cylinder sets of $\mathscr{C}$. The main point of this chapter is to show that under condition (2.1), it can be extended to be a countably additive measure on the Borel sets of $\mathscr{C}$. We follow Parthasarathy [4]. First use Kolmogorov's theorem and go to $(X, S, p(\cdot))$. Then for each n, define a mapping $\varphi_n(\cdot)$, mapping X into $\mathscr{C}$ by:

$$\varphi_n(x(\cdot)) = f(\cdot)$$

$$f(t) = x(m/2^n) + 2^n(t-m/2^n)\,(x(\overline{m+1}/2^n) - x(m/2^n))$$

$$\text{for } m/2^n \leq t \leq (m+1)/2^n, \quad 0 \leq m < 2^n, \ m \text{ - integer.}$$

In other words we join the ordinates at the discrete timepoints $m/2^n$ by segments of straight lines. It is not difficult to see that $\varphi_n(\cdot)$ is measurable; that is to say, the inverse images of Borel sets:

$$\varphi_n^{-1}(B) = [x(\cdot) \in X;\ \varphi_n(x(\cdot)) \in B]$$

where B is a Borel set in $\mathscr{C}$, belong to $\mathscr{S}$. For this we have only to note that, if B is of the form:

$$B = [f(\cdot) \in \mathscr{C},\ f(t_1) \in I\ ,\ I \text{ Borel set in } E]$$

then taking

$$m/2^n \leq t_1 < (m+1)/2^n$$

we have that

$$\varphi_n^{-1}(B) = [x(\cdot) \in X,\ x(m/2^n) + 2^n(t_1 - m/2^n)(x(\overline{m+1}/2^n) - x(m/2^n)) \in I]$$

which is clearly in $\mathscr{S}$. Hence the smallest sigma-algebra generated by inverse images of this form are in $\mathscr{S}$. But the smallest sigma-algebra generated by sets of the form B is indeed the class of all Borel sets. Hence $\varphi_n(\cdot)$ is measurable. Next we come to the crucial part; and show that $\varphi_n(x)$ is a Cauchy sequence in $\mathscr{C}$ for all x in X, except for a fixed set of measure zero. For this we begin with the bound:

$$\|\varphi_n(x) - \varphi_{n-1}(x)\| \leq \sup_{1 \leq k < 2^n} |x(k/2^n) - x(\overline{k-1}/2^n)| \tag{2.2}$$

This is evident from the fact that the maximum 'deviation' occurs at the subdivision points $(2k+1)/2^n$, and here the deviation is

$$|x(\overline{2k+1}/2^n) - x(2k/2^n) - (\tfrac{1}{2})(x(\overline{2k+2}/2^n) - x(2k/2^n))|$$

$$= |(\tfrac{1}{2})\cdot(x(\overline{2k+1}/2^n) - x(\overline{2k+2}/2^n)) + (\tfrac{1}{2})(x(\overline{2k+1}/2^n) - x(2k/2^n))|$$

Next we invoke the Chebychev inequality; and obtain:

$$p(\|\varphi_n(x) - \varphi_{n-1}(x)\| \geq \epsilon)^{\dagger} \leq \frac{1}{\epsilon^r} \sum_{i=1}^{2^n} E(|x(i/2^n) - x(\overline{i-1}/2^n)|^r)$$

$$\leq \frac{k}{\epsilon^r} \sum_{i=1}^{2^n} (\frac{1}{2^n})^{1+\delta}$$

$$= \frac{K}{\epsilon^r} \frac{1}{2^{n\delta}}$$

Next let

$$0 < \theta < \delta/r$$

and let

$$A_n = [x \mid \|\varphi_n(x) - \varphi_{n-1}(x)\| \geq 2^{-n\theta}]$$

Then

$$p(A_n) \leq K\, 2^{nr(\theta - \delta/r)}$$

---

† The student should verify that the set considered is measurable $\mathscr{P}$!

and since

$$\sum_{n=1}^{\infty} K\, 2^{nr(\theta - \delta/r)} < \infty$$

it follows (Borel-Cantelli Lemma):

$$p(\bigcap_{n}^{\infty} \bigcup_{j=n}^{\infty} A_j) = 0$$

Let

$$F = \bigcap_{n=1}^{\infty} \bigcup_{j=n}^{\infty} A_j$$

For any x not in F we have that, denoting the complement of A_j by $\mathscr{C}$-A_j,

$$x \in \bigcup_{n=1}^{\infty} \bigcap_{j=n}^{\infty} (\mathscr{C} - A_j)$$

or, for each x not in F,

$$\| \varphi_n(x) - \varphi_{n-1}(x) \| < 2^{-n\theta}$$

for <u>all</u> n sufficiently large (depending on x, of course). Hence for such x, for all n sufficiently large and every p,

$$\|\varphi_n(x)-\varphi_{n+p}(x)\| \leq \sum_{n+1}^{n+p} 2^{-j\theta}$$

or,

$\varphi_n(x)$ is a Cauchy sequence in $\mathscr{C}$.

Denote the limit by $\varphi(x)$ for x not in F and define the limit to be zero (the zero function) on F. And define

$$\Psi(x) = \varphi(x) \text{ on the complement of } F$$

$$= 0 \text{ on } F$$

Then $\Psi(\cdot)$ is measurable.

Note that for any $x(\cdot)$ not in F, we have that, if

$$\Psi(x(\cdot)) = f(\cdot),$$

then at each subdivision point t_j (of the form $m/2^j$)

$$x(t_j) = f(t_j)$$

by construction. On the other hand, using (2.1) it follows that we have stochastic continuity; and hence if t_k converges to t, a subsequence $\{x(t_k)\}$ converges with probability 1 to $x(t)$. Hence for each t

$$x(t) = f(t)$$

with probability one where the exceptional set may depend on t. [In other words $(\Psi(x))(t)$ is equivalent to $x(t)$]. Next we define the probability measure on $\mathscr{C}$ by: for each Borel set B,

$$\hat{p}(B) = p(\Psi^{-1}(B))$$

Let us calculate the finite dimensional distributions corresponding to $\hat{p}$. Let I be an interval in E, and let

$$B = [f(\cdot) \in \mathscr{C}, \quad f(t) \in I]$$

Then

$$\hat{p}(B) = p(\Psi^{-1}(B))$$

$$= p(x \in X \mid (\Psi(x))(t) \in I)$$

$$= p(x \mid x(t) \in I)$$

by the equivalence just proved. Hence $\hat{p}$ 'agrees' with the given finite dimensional distributions.

Wiener Process

An important special case of this construction, for us the one central case, is Wiener measure.

Theorem 2.1 There exists a measure W on the Borel sets of $\mathscr{C}$ (which we shall denote henceforth by $\mathscr{B}_{\mathscr{C}}$) such that

$$W[f(\cdot) \mid f(0) - 0] = 1$$

and for any finite number of indices $t_1 < t_2 < \ldots < t_m$,

$$W[f(\cdot) \mid f(t_k) \in I_k \, , \quad k = 1, \ldots m]$$

where I_k are intervals in E, is given by

$$\int_{I_1} \cdots \int_{I_m} G(x_1, \ldots x_m) \, d|x_1| \ldots d|x_m|$$

where G(....) is mn-variate Gaussian with zero mean and

$$\int \cdots \int x_i x_j^* \, G(x_1, \ldots, x_m) \, d|x_1| \ldots d|x_m| = \min(t_i, t_j) I_n$$

Proof It is readily verified that the given joint distributions satisfy (2.1) with $r = 4$, $K = n(n+2)$, and $\delta = 1$, and satisfy the consistency requirements.

With ω denoting 'points' in $\mathscr{C}$ and defining

$$W(t;\omega) \quad 0 \leq t \leq 1$$

to be the 'function-value' at t corresponding to ω, the measure W being the Wiener measure above, we have a 'Wiener process' which is thus a Gaussian function-space process, with sample space $\mathscr{C}$. We note that for $t_1 < t_2 < t_3$,

$$W(t_2;\omega) - W(t_1;\omega)$$

is Gaussian with mean zero and covariance $(t_2-t_1)\, I_{(n)}$, $I_{(n)}$ being the $n \times n$ identity matrix and is independent of

$$W(t_3;\omega) - W(t_2;\omega).$$

Hence $W(t;\omega)$ is also a Gaussian process with independent increments such that

$$E((W(t_2;\omega) - W(t_1;\omega))(W(t_2;\omega) - W(t_1;\omega))^*) = (t_2-t_1)\, I_{(n)}$$

and has sample functions which are continuous. It is also readily verified that

$$R(t;s) = E(W(t;\omega)\, W(s;\omega)^*) = (\min\ s,t)\, I_{(n)} \tag{2.3}$$

Measurability For each t, $W(t;\omega)$ is actually continuous in ω, being a continuous linear functional on $\mathscr{C}$. Being thus continuous in both variables, t and ω, it is measurable jointly with respect to ω and t (Lebesgue measurable sets in $[0,1]$).

Let $\phi(t)$ be any function in $\mathscr{C}$. For each ω, $W(t;\omega)$ is a continuous function of t. Hence we can define the integral

$$\int_0^1 \phi(t)^* W(t;\omega)dt$$

as a Riemann-integral for every ω. Moreover, it defines actually a continuous linear functional on $\mathscr{C}$. In fact

$$\int_0^1 | \phi(t)^* (W(t;\omega_1) - W(t;\omega_2)) | dt$$

$$\leq \left(\sup_{0\leq t\leq 1} | \phi(t) |\right) \quad \|\omega_1-\omega_2\|$$

since

$$|W(t;\omega_1) - W(t;\omega_2)| \leq \|\omega_1-\omega_2\|$$

The integral defines a Gaussian of mean zero and with variance

$$\int_0^1 \int_0^1 \phi(t)^* R(t;s)\, \phi(s)\, ds \ dt$$

using (2.3) and the calculation can of course be carried further.

Stochastic Integrals: Linear Case

Let $\phi(t)$ be any function in $L_2(0,1)$ over E. We wish now to define the stochastic integral (with respect to the Wiener process)

$$\int_0^1 [\phi(t),\ dW(t;\omega)]$$

First let $\phi(t)$ be a function of the form:

$$\phi(t) = v_i \qquad t_i \leq t < t_{i+1}, \quad i = 0, \ldots n-1;$$

$$t_0 = 0 \leq t_1 \leq t_2 \ldots \leq t_n = 1$$

corresponding to an arbitrary subdivision of $[0, 1]$ into subintervals. Then we define

$$\int_0^1 [\phi(t),\ dW(t;\omega)] = \sum_{i=0}^{n-1} \left[v_i (W(t_{i+1};\omega) - W(t_i;\omega)) \right]$$

This defines a Gaussian random variable with mean zero and variance:

$$\sum_{i=0}^{n-1} \|v_i\|^2 (t_{i+1} - t_i) = \int_0^1 \|\phi(t)\|^2 \, dt = \|\phi(\cdot)\|^2$$

[The integral of course defines a continuous linear functional on C, and the norm of the functional is

$$\leq 2 \sum_{0}^{n-1} \|v_i\| \Big]$$

Consider now the L_2 space of all random variables (measurable $\mathcal{B}_{\mathcal{C}}$) with finite second moment with respect to Wiener measure. Denote this $L_2(\mathcal{C})$. The mapping takes step functions into $L_2(\mathcal{C})$. It is readily verified to be linear on the linear subspace of step functions of the type considered. But this subspace is dense in $L_2(0, 1)$. If we denote the mapping by $\mathcal{L}$, we have

$$E \, |(\mathcal{L}(\phi(\cdot)))|^2 = \| \phi(\cdot) \|^2 \quad \cdots \cdot \tag{2.4}$$

and hence can be extended to be continuous on $L_2(0, 1)$. In other words if $\phi(\cdot)$ is any element in $L_2(0, 1)$ we know that we can find a sequence $\phi_n(\cdot)$ of step functions of the type considered which converge to $\phi(\cdot)$, and since $\{\phi_n(\cdot)\}$ is then Cauchy sequence in $L_2(0, 1)$, so is

$$\mathcal{L}(\phi_n(\cdot))$$

in $L_2(\mathcal{C})$. Since $L_2(\mathcal{C})$ is complete it follows that we can define

$$\mathcal{L}(\phi(\cdot)) = \lim_n \text{ (in } L_2(\mathcal{C})) \; \mathcal{L}(\phi_n(\cdot))$$

The limit is of course independent of the particular sequence chosen, thanks again to (2.4), and moreover a subsequence of $\mathcal{L}(\phi_n(\cdot))$ will converge to the limit a.e. Also, of course, we will continue to have:

$$E \,|\, \mathscr{L}(\phi(\cdot)) \,|^2 = \| \, \phi(\cdot) \, \|^2$$

and further, if $\phi(\cdot)$, $\Psi(\cdot)$ are any two elements in $L_2(0,1)$ we have

$$E([\mathscr{L}(\phi(\cdot)), \mathscr{L}(\Psi(\cdot))]) = [\phi(\cdot), \Psi(\cdot)] \quad \cdots \tag{2.5}$$

where the right side indicates inner product in $L_2(0,1)$. It is hardly necessary to add that $\mathscr{L}$ is a linear continuous (isometric) transformation of $L_2[0,1]$ into $L_2(\mathscr{C})$ and

$$\mathscr{L}(\phi(\cdot))$$

is Gaussian with zero mean and variance

$$= \| \, \phi(\cdot) \, \|^2$$

If $\phi(\cdot)$ is actually continuous then we can approximate $\mathscr{L}(\phi(\cdot))$ by sums of the form, with $\{ t_i \}$ as before,

$$\sum_{i=0}^{n-1} [\phi(\tau_i), (W(t_{i+1};\omega) - W(t_i;\omega))]$$

where $t_i \le \tau_i \le t_{i+1}$, since the difference (between this sum and the integral)

$$= \sum_{i=0}^{n-1} \int_{t_i}^{t_{i+1}} [\phi(\tau_i)-\phi(t), \; dW(t;\omega)]$$

If $\phi(\cdot)$ is absolutely continuous with derivative in $L_2[0,1]$, we can 'integrate by parts':

$$\int_0^1 [\phi(t), dW(t;\omega)] = [\phi(1), W(1)] - \int_0^1 [\phi'(t), \; W(t;\omega)] \, dt \text{ with pr. one}$$

This follows from the calculation:

$$\sum_{i=1}^{n-1} [\phi(t_i), \; W(t_{i+1};\omega) - W(t_i;\omega)]$$

$$= -[\phi(0), \; W(0;\omega)] - \sum_1^n [\phi(t_i)-\phi(t_{i-1}), \; W(t_i;\omega)]$$

$$+ [\phi(1), \; W(1;\omega)]$$

and

$$\sum_1^n [\phi(t_i) - \phi(t_{i-1}), \; W(t_i;\omega)] = \sum_1^n \int_{t_{i-1}}^{t_i} [\phi'(\sigma) \; , \; W(t_i;\omega)] \, d\sigma$$

$$\longrightarrow \int_0^1 [\phi'(t), \; W(t;\omega)] \, dt$$

Also for any complete orthonormal sequence $\{\phi_n(\cdot)\}$ in $L_2[0,1]$, we have

$$\int_0^1 [\phi(t), d\,W(t;\omega)] = \sum_1^\infty [\phi, \phi_n] \int_0^1 [\phi_n(t), dW(t;\omega)] \quad \text{with pr. 1}$$

where

$$\int_0^1 [\phi_n(t), dW(t;\omega)]$$

are zero mean, unit variance, mutually independent Gaussian variables. We can pursue this a bit further, to obtain an expansion for the process.

Shepp Expansion: (Cf. Shepp [5]) Let $\phi_n(\cdot)$ be a complete orthonormal sequence in $L_2(0,1)$. Let

$$\zeta_n(\omega) = \int_0^1 [\phi_n(t), \; dW(t;\omega)]$$

Then the ζ_n are zero mean, unit variance, mutually independent Gaussian variables (one-dimensional). Let

$$W_n(t;\omega) = \sum_1^n \zeta_j(\omega) \int_0^t \phi_j(s)\, ds \qquad 0 \le t \le 1$$

Now let us fix t, and take any arbitrary element v in E, and define the following function $\Psi(\cdot)$ in $L_2(0,1)$:

$$\Psi(s) = v\,,\ 0 \leq s \leq t$$

$$= 0,\ \text{otherwise.}$$

Then first of all,

$$[\,v,\ W(t;\omega)] = \int_0^1 [\,\Psi(s),\ dW(s;\omega)]$$

for every ω by definition. Let

$$a_j = [\,\Psi, \phi_j] = [\,\int_0^t \phi_j(s)ds,\ v]$$

be the Fourier coefficients of $\Psi(\cdot)$ with respect to the basis $\phi_j(\cdot)$.

Let

$$\Psi_n(\cdot) = \sum_1^n a_j\ \phi_j(\cdot)$$

Then

$$\int_0^1 [\,\Psi_n(t),\ dW(t;\omega)]$$

$$= \sum_1^n a_j\ \zeta_j(\omega)$$

$$= [v,\ W_n(t;\omega)]$$

Hence (since $\mathscr{L}$ is a continuous mapping):

$$E[v, \dot{W}_n(t;\omega) - W(t;\omega)]^2 \longrightarrow 0 \quad \text{as } n \longrightarrow \infty$$

for every v in E, or

$$E \, \| W_n(t;\omega) - W(t;\omega) \|^2 \longrightarrow 0 \quad \text{as } n \longrightarrow \infty$$

Or, we have the expansion (in the mean square sense), for each t:

$$W(t;\omega) = \sum_1^\infty \zeta_n(\omega) \int_0^t \phi_n(s)\, ds \tag{2.6}$$

Actually, the convergence is with probability one also, since

$$[v, W_n(t;\omega)] = \sum_1^n a_j \, \zeta_j(\omega)$$

and by virtue of Kolmogorov's inequality this converges with probability one, since

$$\sum_1^\infty |a_j|^2 \, E[|\zeta_j(\omega)|^2] = \sum_1^\infty |a_j|^2 = t \, \|v\|^2$$

Problem for any $f(\cdot)$ in $L_2[0, 1]$, show that

$$\int_0^1 [f(t), dW(t;\omega)] = \sum_1^\infty \zeta_n(\omega) \, [f, \phi_n]$$

Problem

$$\int_0^1 E(\| W(t;\omega) - \sum_1^n \zeta_k(\omega) \int_0^t \phi_k(s)\, ds \|^2)\, dt \longrightarrow 0$$

Hint:

$$E \| W(t;\omega) - \sum_1^n \zeta_k(\omega) \int_0^t \phi_k(s) ds \|^2 = (nt - \sum_1^n \| \int_0^t \phi_k(s)\, ds \|^2) \geq 0,$$

and so,

$$\sum_1^n \| \int_0^t \phi_k(s) ds \|^2 \text{ converges boundedly to } (nt) \text{ , in } L_1[0, 1].$$

Linear Stochastic Equations

Let $F(s)$ be an m-by-n matrix function, Lebesgue measurable on $[0, 1]$ and such that

$$\int_0^1 \|F(s)\|^2 \, ds < \infty$$

Let

$$S(t;\omega) = \int_0^t F(s) \, dW(s;\omega) \qquad 0 \le t \le 1 \tag{2.7}$$

which we know defined for each t, for almost every ω. However it is not defined for every ω, and in particular, the exceptional set of points ω on which it is not defined may well depend on t. We wish now to rectify this situation. If F(s) is absolutely continuous with a square-integrable (on [0, 1]) derivative we can do this very simply. For then,

$$S(t;\omega) = \int_0^t F(s) \, dW(s;\omega) = [F(t), W(t;\omega)] - \int_0^t F'(s) \, W(s;\omega) \, ds$$

with probability one, and since the right side is defined for every ω, we may just define the left side to be that for every ω. Note that $S(t;\omega)$ is then continuous in t, $0 \le t \le 1$, for every ω, and hence a separable process. Moreover, we have the following basic bound:

Lemma Suppose $S(t;\omega)$ in (2.7) is determined as a continuous function of t for almost every ω. Then:

$$W \left[\sup_{0 \le t \le 1} \left\| \int_0^t F(s) \, dW(s;\omega) \right\| > \epsilon \right] \le \frac{1}{\epsilon^2} \int_0^1 \|F(s)\|^2 \, ds \tag{2.8}$$

Proof

Let

$$A = \left[\omega \,\middle|\, \sup_{0 \le t \le 1} \left\| \int_0^t F(s)\, dW(s;\omega) \right\| > \epsilon \right]$$

Then by the assumed continuity of $S(t;\omega)$ in t, we observe that A is measurable. In fact if

$$A_n(k) = \left[\omega \,\middle|\, \sup_{0 \le m \le 2^n} \left\| \int_0^{m/2^n} F(s)\, dW(s;\omega) \right\| \ge \epsilon - \frac{1}{k} \right], \text{ m-integer}$$

then $A_n(k)$ is clearly monotone in n for each k

$$A = \bigcap_k \bigcup_n A_n(k)$$

But the integral over non-overlapping intervals being independent, a simple application of the Kolmogorov inequality shows that

$$W(A_n(k)) \le \frac{1}{(\epsilon - \frac{1}{k})^2} \int_0^1 \| F(s) \|^2 \, ds$$

from which the Lemma follows.

Next let us recall that if $F(\cdot)$ is any element in $L_2(0,1)$, we can find a sequence $F_n(\cdot)$ of functions which are absolutely continuous with derivative in $L_2(0,1)$ such that

$$\int_0^1 \| F(s) - F_n(s) \|^2 \, ds \longrightarrow 0$$

Using the lemma we obtain that

$$W\left[\sup_{0\le t\le 1}\left\|\int_0^t (F_n(s)dW(s;\omega) - F_{n+p}(s)dW(s;\omega))\right\| > \epsilon\right]$$

$$\le \frac{1}{\epsilon^2}\int_0^1 \|F_n(s) - F_{n+p}(s)\|^2 \, ds$$

Let θ , $0<\theta<1$, be fixed, and let n_k be such that

$$\|F_{n_k} - F_{n_k+p}\|^2 \le 2^{-k\theta} \quad \text{for } \underline{\text{all}} \ n, m > n_k$$

Clearly $\{n_k\}$ can be taken to be increasing.
Choose

$$\epsilon^2 = 2^{-k\gamma} \ , \quad 0<\gamma<\theta<1$$

and let

$$B_k = \left[\omega \,\middle|\, \sup_{0\le t\le 1}\left\|\int_0^t (F_n(s)-F_m(s))\, dW(s;\omega)\right\| > 2^{-k\gamma|2}\right]$$

for $n, m \ge n_k$. Let

$$\Lambda = \bigcap_{m=1}^{\infty} \bigcup_{k=m}^{\infty} B_k$$

Then since

$$W(B_k) \le 2^{-k(\theta-\gamma)} \tag{2. }$$

we have (Borel Cantelli Lemma):

$$W(\Lambda) = 0$$

Hence for ω not in Λ we have that:

$$\sup_{0 \leq t \leq} \left\| \int_0^t (F_n(s) - F_m(s))dW(s;\omega) \right\| \leq 2^{-k\gamma/2}, \quad n, m \geq n_k$$

for all k > some N. In other words the sequence

$$\int_0^t F_{n_k}(s)\, dW(s;\omega)$$

converges uniformly in t with probability one, and hence the limit must be continuous in t. Denote this limit by $\hat{S}(t;\omega)$. Then

$$\hat{S}(t;\omega) = S(t;\omega) \text{ with pr. one}$$

and we have produced a continuous equivalent version of $S(t;\omega)$

Problem

Suppose $F(\cdot)$ is essentially bounded on $(0,1)$. Then show that (2.1) is satisfied and hence use Parathasarathy's construction to obtain a continuous version of $S(t;\omega)$.

Answer

$$E[\,|S(t_1;\omega) - S(t_2;\omega)|^4] = 3\,E[\,|\int_{t_1}^{t_2} F(s)d\,W(s;\omega)|^2]^2$$

$$= 3\left[\int_{t_1}^{t_2} \|F(s)\|^2\,ds\right]^2$$

$$\leq 3\ m^4(t_2-t_1)^2 \qquad (2.9)$$

where

$$m = \text{ess.} \sup_{0\leq s\leq 1} \|F(s)\|$$

Define $S(t;\omega)$ arbitrarily for each t for ω not in the defining set. Then map $\mathscr{C}$ into $\underline{X}$ by

$$\varphi(\omega) = x$$

$$x(t) = S(t;\omega)$$

This is a measurable map and then we can use the fact that (2.1) is satisfied. Hence we can obtain a new measurable map $\Psi(\cdot)$, mapping $\underline{X}$ into $\mathscr{C}$. Note that if

$$\Psi[x] = f$$

then for each t;

$f(t) = x(t)$ omitting a set of measure zero (depending on t) in X.

or,

$\Psi[\varphi(\omega)](t) = S(t;\omega)$, omitting a set of measure zero (depending on t) in $\mathscr{C}$.

and

$$\hat{S}(t;\omega) = \Psi[\varphi(\omega)](t)$$

is continuous in t with probability one.

Let us now consider the problem of solving the following linear stochastic integral equation:

$$x(t;\omega) = \xi(\omega) + \int_0^t A(s)\, x(s;\omega)ds + \int_0^t B(s)dW(s;\omega), \quad 0 \le t \le L < \infty \tag{2.10}$$

where $\xi(\omega)$ is a given random variable, independent of $W(\cdot;\omega)$.

$A(s)$ is a Lebesgue measurable m-by-m matrix function

$B(s)$ is a Lebesgue measurable m-by-n matrix function

and

$$\int_0^L \|A(s)\|^2 ds + \int_0^L \|B(s)\|^2 ds < \infty \tag{2.11}$$

Sometimes (2.9) is written in the differential form:

$$dx(t;\omega) = A(s)\, x(t;\omega)\, dt + B(t)\, dW(t;\omega)$$

but this is to be looked upon only as a shorthand notation for (2.10). What shall we mean by a solution of (2.10)? Minimally, any stochastic process on $[\mathscr{C}, \mathscr{B}_{\mathscr{C}}, W]$, for which the integrals in (2.10) can be defined. We can actually do a little better. By a 'continuous solution' of (2.10) we shall mean a stochastic process such that $x(t;\omega)$ is continuous in $0 \leq t \leq L$, and satisfies (2.10) for every t in $[0, L]$, omitting a fixed set of measure zero.

Lemma Suppose $x_1(t;\omega)$, $x_2(t;\omega)$ are two continuous solutions of (2.9). Then

$$x_1(t;\omega) = x_2(t;\omega)$$

for every t except for a fixed ω-set of measure zero. In fact

$$W\left[\sup_{0 \leq t \leq L} \|x_1(t;\omega) - x_2(t;\omega)\| = 0\right] = 1$$

Proof

Let

$$y(t;\omega) = x_1(t;\omega) - x_2(t;\omega)$$

Then $y(t;\omega)$ is a continuous solution of:

$$y(t;\omega) = \int_0^t A(s)\, y(s;\omega)\, ds \; \ldots\ldots\ldots \tag{2.12}$$

for ω not in Λ say, where Λ is a set of measure zero. But for each ω not in Λ, (2.12) is a 'deterministic' equation, and readily yields that

$$y(t;\omega) = 0 \quad \text{for every } t \text{ and } \omega \,\bar{\epsilon}\, \Lambda.$$

To produce a continuous solution, let $\Phi(t)$ be the fundamental matrix solution of

$$\dot{\Phi}(t) = A(t)\, \Phi(t) \; ; \; \Phi(0) = I.$$

Then let

$$S(t;\omega)$$

be a continuous version of

$$\int_0^t \Phi(s)^{-1} B(s)\, dW(s;\omega), \quad 0 \leq t \leq L$$

and finally let

$$x(t;\omega) = \Phi(t)(S(t;\omega) + \xi(\omega))$$

which is then a continuous function of t excepting for a set Λ of measure zero.

Let us next note that

$$\int_0^t B(s)\, dW(s;\omega)$$

$$= \int_0^t \phi(s)\, \phi(s)^{-1}\, B(s)\, dW(s;\omega)$$

and integrating by parts, we get:

$$= \Big[\, \phi(s)\, S(s;\omega) \,\Big]_0^t - \int_0^t \left(\frac{d}{ds}\, \phi(s)\right) S(s;\omega)\, ds$$

This equality can clearly be defined to hold for every ω not in Λ. But for ω not in Λ,

$$\int_0^t A(s)\, x(s;\omega)\, ds = \int_0^t \left(\frac{d}{ds}\, \phi(s)\right)\Big(S(s;\omega) + \xi(\omega) \Big)\, ds$$

$$= \phi(t)(S(t;\omega) + \xi(\omega)) - \int_0^t B(s) dW(s;\omega) - \xi(\omega)$$

$$= x(t;\omega) - \int_0^t B(s)\, dW(s;\omega) - \xi(\omega)$$

as required.

Problem

Let $x(t;\omega)$ be a (continuous) solution of (2.10) and let

$$R(t) = E\,(x(t;\omega)\, x(t;\omega)^*]$$

show that $R(t)$ is absolutely continuous with

$$\dot{R}(t) = A(t)\, R(t) + R(t)\, A(t)^* + B(t)\, B(t)^* \qquad \text{a.e.} \tag{2.13}$$

Answer

We note that

$$R(t) = \Phi(t)\, E\,[\xi(\omega)\, \xi(\omega)^*]\; \Phi(t)^*$$

$$+ \Phi(t) \int_0^t \Phi(s)^{-1}\, B(s)\, B(s)^*\, \Phi(s)^{-1*}\, ds\, \Phi(t)^*$$

Next let $\xi(\omega) = 0$ in (2.10) and define:

$$Y\,(t;\omega) = \int_0^t C(s)\, x(s;\omega)\, ds + \int_0^t D(s)\, dW(s;\omega), \quad 0 \le t \le L \tag{2.14}$$

where

$C(s)$ is q-by-m and continuous on $[0, L]$

$D(s)$ is q-by-n and continuous on $[0, L]$

Note that $Y(t;\omega)$ can be defined so that it is continuous in t, $0 \leq t \leq L$, excepting an ω-set of (Wiener) measure zero. Furthermore, we can define a stochastic integral with respect to the $Y(t;\omega)$ process. Let $f(\cdot)$ be a function in $L_2(0,L)^q$; we first define the integral for a step function taking on a finite number of values each on a subinterval, as before. We note that for a step function the following equality holds:

$$\int_0^L [f(t), dY(t;\omega)] = \int_0^L [f(t), c(t)x(t;\omega)]\, dt + \int_0^L [f(t), D(t)\, dW(t;\omega)] \tag{2.15}$$

Now if we take a Cauchy sequence of step functions, the right-side converges, and hence so does the left-side. And the stochastic integral is then defined as this limit. Hence finally (2.15) holds for every $f(\cdot)$ in $L_2[0,L]^q$. This may be indicated in differential notation as:

$$dY(t;\omega) = C(t)\, x(t;\omega)\, dt + D(t)\, dW(t;\omega) \tag{2.16}$$

We can carry this further. Define the following operators mapping $L_2[0,L]^q$ into $L_2[0,L]^n$:

$$Kf = g\ ;\ g(t) = B(t)^* \Phi(t)^{*-1} \int_t^L \Phi(s)^* C(s)^* f(s)\, ds$$

$$Df = g;\ g(t) = D^*(t)\, f(t) \tag{2.18}$$

Note that K is a Volterra operator. Let R denote the mapping $L_2[0, L]^q$ into $L_2[0, L]^q$:

$$R = (D + K)^* (D + K) \tag{2.19}$$

Then we can verify that for $f(\cdot)$, $h(\cdot)$ in $L_2[0, L]^q$:

$$E \left(\int_0^L [f(t), dY(t;\omega)] \right) \left(\int_0^L [h(t), dY(t;\omega)] \right) = [Rf, h] \tag{2.20}$$

where on the right side we use the innerproduct notation in $L_2[0, L]^q$. For this we have only to note that

$$\int_0^L [f(t), C(t)\, x(t;\omega)]\, dt$$

$$= \int_0^L [f(t), C(t) \int_0^t \Phi(t)\, \Phi(s)^{-1} B(s)\, dW(s;\omega)]\, dt$$

$$= \int_0^L [\Phi(t)^* C(t)^* f(t), \int_0^t \Phi(s)^{-1} B(s)\, dW(s;\omega)]\, dt$$

$$= \int_0^L \left[\int_s^L \Phi(t)^* C(t)^* f(t)\, dt\, ,\ \Phi(s)^{-1} B(s)\, dW(s;\omega) \right]$$

$$= \int_0^L \left[B(t)^* \Phi(t)^{*-1} \int_t^L \Phi(s)^* C(s)^* f(s)\, ds,\ dW(t;\omega) \right]$$

so that

$$\int_0^L [f(t),\ dY(t;\omega)]$$

$$= \int_0^L [D(t)^* f(t) + B(t)^* \Phi(t)^{*-1} \int_t^L \Phi(s)^* C(s)^* f(s)\, ds,\ dW(t;\omega)]$$

from which the left side of (2.20) is seen to be

$$[(D+K)f,\ (D+K)h]$$

Problem Let $J = (D+K)^*$

Show that if $\{\phi_n(\cdot)\}$ is any complete orthonormal sequence in $L_2[0, L]^n$, we have

$$Y(t;\omega) = \sum_1^\infty \varphi_n(\omega) \int_0^t \psi_n(s)\, ds \tag{2.21}$$

where

$$\psi_n = J\phi_n,\quad J = (D+K)^*$$

and

$$\varphi_n(\omega) = \int_0^L [\phi_n(t),\ dW(t;\omega)]$$

These considerations can be generalized as follows, illustrating our point of view on stochastic processes. Let $y(t;\omega)$ be a Gaussian process [with sample space not necessarily $\mathscr{C}$] of dimension n with continuous covariance function $R(t;s)$. Then let (taking a separable extension)

$$Y(t;\omega) = \int_0^t y(s;\omega)\, ds$$

Note that this is a (Gaussian) process providing a consistent family of finite dimensional distributions, and

$$E(\|Y(t_2;\omega) - Y(t_1;\omega)\|^2) = \mathrm{Tr} \int_{t_1}^{t_2} \int_{t_1}^{t_2} R(s;t)\, ds\, dt$$

$$\leq M\,(t_1 - t_2)^2$$

Hence (2.1) holds and hence we can define a stochastic process with sample space $\mathscr{C}$. We shall denote this by $Y(t;\omega)$. Then we can also define a stochastic integral, by the same procedure as before. And for $f(.)$, $h(\cdot)$ in $L_2[0,L]^n$:

$$E\left(\int_0^L [f(t),\, dY(t;\omega)]\right)\left(\int_0^L [h(t),\, dY(t;\omega)]\right) = [Rf, h]$$

where

$$Rf = g; \quad g(t) = \int_0^L R(t;s)\, f(s)\, ds$$

If $\{\phi_n(t)\}$ are the orthonormalized eigen-functions of R, we have:

$$Y(t;\omega) = \sum_1^\infty \varphi_n(\omega) \int_0^t \phi_n(s)\, ds \tag{2.22}$$

where

$$\varphi_n(\omega) = \int_0^L [\, \phi_n(t),\ dY(t;\omega)]$$

and

$$E[(\varphi_n(\omega))^2] = \lambda_n$$

and

$\{\lambda_n\}$ eigen-values of R corresponding to $\phi_n(\cdot)$

The series again converges with probability one.

Conditional Expectation and Martingale Theory

Let $(\Omega, \mathscr{B}, p)$ denote a probability triple and let ζ be any (n-dimensional) random variable such that

$$E[\,|\zeta|\,] < \infty$$

Let $\mathcal{B}_s$ denote a sub sigma-algebra of $\mathcal{B}$. Then ζ need not be measurable $\mathcal{B}_s$. We can construct a (sort of) projection which is. Define the set function

$$\nu(B) = \int_B \zeta \, dp,$$

for every B in $\mathcal{B}_s$. $\nu(\cdot)$ is absolutely continuous with respect to the measure $p(.)$ on $\mathcal{B}_s$. Hence by the Radon Nikodym theorem, there is a function $f(\omega)$, measurable $\mathcal{B}_s$, such that

$$\int_B \zeta \, dp = \int_B f(\omega) \, dp, \quad B \in \mathcal{B}_s \tag{2.23}$$

Let $\mathcal{B}_s'$ denote the sigma algebra of sets which are either in $\mathcal{B}_s$ or differ from such sets by null sets. Any function measurable with respect to $\mathcal{B}_s$ and satisfying (2.23) will be called the conditional expectation of ζ with respect to $\mathcal{B}_s$ and denoted:

$E[\zeta \mid \mathcal{B}_s]$

Let us now itemize some of the properties of conditional expectations:

i) $E[\zeta \mid \mathcal{B}_s]$ is a random variable $(\Omega, \mathcal{B}_s', p)$ and $E[\zeta \mid \mathcal{B}_s] = E[\zeta \mid \mathcal{B}_s']$ with probability one.

ii) since $\Omega \in \mathcal{B}_s$, $\int_\Omega \zeta \, dp = \int_\Omega E[\zeta \mid \mathcal{B}_s] \, dp$

or,

$$E[\zeta] = E(E[\zeta \mid \mathcal{B}_s])$$

iii) Let $\mathscr{B}_1, \mathscr{B}_2$ be two sub sigma-algebras of $\mathscr{B}$, such that

$$\mathscr{B}_1 \subset \mathscr{B}_2$$

[This means that every set in $\mathscr{B}_1$ is also in $\mathscr{B}_2$]

Then

$$E[\zeta \mid \mathscr{B}_1] = E[E[\zeta \mid \mathscr{B}_2] \mid \mathscr{B}_1] \tag{2.24}$$

<u>Proof</u>

Let $f_2(\omega) = E[\zeta \mid \mathscr{B}_2]$

Let $f_1(\omega) = E[\zeta \mid \mathscr{B}_1]$

Let B be any set in $\mathscr{B}_1$. Then

$$\int_B \zeta \, dp = \int_B f_1(\omega) \, dp$$

But since B belongs to $\mathscr{B}_2$,

$$\int_B \zeta \, dp = \int_B f_2(\omega) \, dp$$

Hence for every B in $\mathcal{B}_1$,

$$\int_B f_1(\omega)\, dp = \int_B f_2(\omega)\, dp$$

or, the rèsult follows.

iv) Let $\mathcal{B}_s$ be a sub-sigma algebra of $\mathcal{B}$ and let $h(\omega)$ be a pxn matrix valued function measurable $\mathcal{B}_s$. Suppose

$$E[\,|h(\omega)\zeta\,|\,] < \infty$$

Then

$$E[h(\omega)\zeta \mid \mathcal{B}_s] = h(\omega)\, E[\zeta \mid \mathcal{B}_s] \tag{2.25}$$

Proof

Let $h(\omega)$ be the characteristic function of a set B_o in $\mathcal{B}_s$. Then for any B in $\mathcal{B}_s$

$$\int_B h(\omega)\zeta\, dp = \int_{B\cap B_o} \zeta\, dp$$

$$= \int_{B\cap B_o} f(\omega)\, dp$$

$$= \int_B h(\omega)\, f(\omega)\, dp$$

where

$$f(\omega) = E[\zeta|\mathscr{B}_s]$$

By linearity, this result is extended to simple functions, and by the usual limiting arguments to any function measurable $\mathscr{B}_s$.

Definition: Let $\{x_\alpha(\omega)\}$ be any collection of random variables. By the smallest sigma algebra generated by this collection we shall mean the sm smallest sigma algebra containing all sets of the form:

$$[\omega|x_\alpha(\omega) \in B]$$

where B is a Borel set in E, or differs from such a set by ω-sets of measure zero. We shall denote the sigma algebra by:

$$\mathscr{B}\{x_\alpha(\cdot)\}$$

Lemma (Doob): Let $x_1(\omega), \ldots x_m(\omega)$ be m random variables, and let $\mathscr{B}_m$ denote the smallest sigma algebra generated by them. Let $f(\omega)$ be measurable with respect to $\mathscr{B}_m$. Then there exists a Borel function $g(x_1, \ldots x_m)$ defined on $E^{(m)}$ such that

$$f(\omega) = g(x_1(\omega), \ldots x_m(\omega)) \text{ a.e.} \tag{2.26}$$

Proof: The main thing to note is that the smallest sigma algebra generated by the m variables is precisely the class of sets of the form:

$$[\omega \mid \{x_1(\omega), \ldots, x_m(\omega)\} \in U_{mn}] \cup N$$

where U_{mn} is a Borel set in mxn dimensions and N is any set of measure zero. Now suppose $f(\omega)$ is a characteristic function of a set in $\mathcal{B}_m$ of the above type. Define:

$$g(x_1, \ldots x_m) = 1 \text{ on } U_{mn}$$

$$= 0 \text{ otherwise.}$$

Then $g(\ldots)$ is a Borel function and (2.26) holds. Next note that the class of functions for which (2.26) holds is a linear class. Hence (2.26) holds for simple functions $f(\cdot)$. Now given any function $f(\cdot)$ measurable $\mathcal{B}_m$, we can find a sequence of simple functions converging to it with probability one. Let $f_n(\omega)$ denote a sequence of such simple functions and let Λ denote the set of convergence. We know that

$$f_n(\omega) = g_n(x_1(\omega), \ldots x_m(\omega))$$

for Borel functions $g_n(\ldots.)$, and that Λ must be of the form:

$$[\omega \mid x_1(\omega), \ldots x_m(\omega) \in I] \cup N$$

where I is a Borel set in mxn dimensions and N is a set of measure zero. Then $g_n(x_1, \ldots x_m)$ converges on I. Define:

$$g(x_1, \ldots x_m) = \text{limit } g_n(x_1, \ldots x_m) \text{ on } I$$

$$= 0 \text{ otherwise}$$

Then $g(\ldots.)$ is a Borel function, and of course (2.26) holds. Note in particular that

$$E[\zeta | \mathscr{B}_m] = g(x_1(\omega), \ldots x_m(\omega))$$

for some Borel function $g(\ldots.)$. Because of this we shall on occasion use the notation:

$$E[\zeta | x_1(\omega), \ldots x_m(\omega)]$$

for the conditional expectation with respect to $\mathscr{B}_m$.

Independence

Let ζ, η denote random variables of dimension n and m respectively which are independent. Recall that this means that:

$$pr.(\zeta \in A, \eta \in B) = pr.(\zeta \in A) \cdot pr.(\eta \in B)$$

Let us calculate the conditional expectation:

$$E[\zeta | \eta]$$

Let B be any set in $\mathcal{B}(\eta)$. Then

$$\int_B \zeta \, dp = \int_\Omega f(\omega) \, \zeta \, dp$$

where

$f(\omega)$ = Identity matrix on B, and zero otherwise.

Then $f(\omega)$ and ζ are independent, and hence we have that:

$$\int_\Omega f(\omega) \, \zeta \, dp = E(\zeta) \, p(B)$$

Hence

$$\int_B \zeta \, dp = \int_B E(\zeta) \, dp$$

Or,

$E(\zeta | \eta) = E(\zeta)$ with probability one

Martingales-Discrete Parameter

A sequence of random variables $\{\zeta_n\}$ such that:

$$E[|\zeta_n|] < \infty$$

is called a Martingale if

$$E[\zeta_n | \mathscr{B}_{n-1}] = \zeta_{n-1} \tag{2.27}$$

where $\mathscr{B}_n$ is the smallest sigma algebra generated by the first n variables. Note in particular that:

$$E[\zeta_n | \zeta_{n-1}] = E[E[\zeta_n | \mathscr{B}_{n-1}] | \zeta_{n-1}]$$

$$= \zeta_{n-1}$$

Suppose $\{y_n\}$ is a sequence of independent, random variables with finite expectation, but with zero mean. Then

$$\zeta_n = \sum_1^n y_k$$

is a Martingale.

Moreover we have the following generalization of the Kolmogorov inequality to Martingales: (Doob):

Lemma: Let $\{x_n\}$ be a Martingale. Then:

$$\text{pr.}[\underset{j \le n}{\text{Max}} |x_j(\omega)| \ge \epsilon] \le (1/\epsilon^2)\, E[|x_n|^2] \tag{2.28}$$

Proof:

Let

$$A_k = [\omega \mid |x_k(\omega)| \geq \epsilon]$$

$$B_k = A_k - \bigcup_{j=1}^{k-1} A_j$$

Define

$$z_k(\omega) = x_k(\omega) \text{ for } \omega \text{ in } B_k$$

$$= 0 \text{ otherwise}$$

Then z_k is measurable $\mathscr{B}(x_1, \ldots x_k)$, and hence:

$$E([z_k, x_n]) = E(E([z_k, x_n] \mid x_k)) = E([z_k, E(x_n | x_k)])$$

$$= E([z_k, x_k])$$

where the Martingale property is invoked in the last equality. We can now readily proceed as in the proof of the original Kolmogorov inequality:

$$E(|x_n|^2) \geq \sum_1^n \int_{B_k} |x_n|^2 \, dp = \sum_1^n \int_{B_k} \left\{ |x_n - x_k|^2 + |x_k|^2 + 2[x_k, x_n - x_k] \right\} dp$$

and

$$\int_{B_k} [x_k, x_n - x_k] \, dp = E([z_k, x_n - x_k]) = E[z_k, x_k - x_k] = 0$$

so that :

$$E(|x_n|^2) \geq \sum_1^n \int_{B_k} (|x_k|^2) \, dp \geq (\epsilon^2) \, pr.(\bigcup_k B_k) = (\epsilon^2) \, pr.(\bigcup_k A_k)$$

which yields the inequality sought.

Doob's Martingale Convergence Theorem:

The following special case of Doob's theorem will be useful to us in the sequel:

Theorem 2.2 Suppose $\{\zeta_n\}$ is a Martingale such that

$$E(|\zeta_n|) \leq c$$

Then the sequence ζ_n converges with probability one to a variable ζ such that

$$E(|\zeta|) \leq c$$

Proof See Doob ([2] p. 319)

As an example of how this result is used, let us consider the following canonical situation. Let $x(\omega)$ be a variable with finite expectation, and let $y_n(\omega)$ be a sequence of random variables, and let $\mathcal{B}_n$ denote the sigma algebra generated by the first n variables $y_1, \ldots y_n$. Let $\mathcal{B}_\infty$ denote the sigma algebra generated by the whole sequence. Then

$\zeta_n = E[x(\omega)|\mathscr{B}_n]$ is a Martingale which converges with

is a Martingale which converges with probability one to:

$$\zeta = E(x(\omega)|\mathscr{B}_\infty)$$

To see this we have only to note that

$$E[\zeta_n|\mathscr{B}_{n-1}] = E[E[x(\omega)|\mathscr{B}_n]|\mathscr{B}_{n-1}] = E[x(\omega)|\mathscr{B}_{n-1}] = \zeta_{n-1};$$

$$E(|\zeta_n|) = E(|E(\zeta_{n+1}|\zeta_n)|) \leq E(|\zeta_{n+1}|)$$

and of course:

$$E(|\zeta_n| \leq E(|x(\omega)|)$$

Hence ζ_n converges with probability one. Let $\hat{\zeta}$ denote the limit. Then for any set B in $\mathscr{B}_n$, we observe that

$$\int_B \zeta_n \, dp = \int_B \zeta_m dp \quad , \quad m \geq n$$

and hence

$$E(\hat{\zeta}|\mathscr{B}_n) = \zeta_n$$

Hence

$$E(\zeta - \hat{\zeta}|\mathscr{B}_n) = 0 \text{ for every } n$$

Or, really

$$\int_B (\zeta - \hat{\zeta}) dp = 0 \text{ for every B in } \mathscr{B}_n \text{ for every n.} \tag{2.29}$$

But the class of sets B such that B belongs to $\mathcal{B}_n$ for some n is a field, and it generates $\mathcal{B}_\infty$. Hence (2.29) holds for every B in $\mathcal{B}_\infty$, or,

$$\zeta = \hat{\zeta} \text{ with probability one}$$

since both variables are measurable with respect to $\mathcal{B}_\infty$.

Continuous Parameter Martingales:

Let us now turn to continuous parameter Martingales which form the central part of our study. Let $Z(t;\omega)$ be a stochastic process, $t \in T$ where T is assumed to be an interval of the real line. Let F(t) be a sigma algebra of measurable sets such that $Z(t;\omega)$ is measurable F(t) and let

$$F(t_1) \subset F(t_2) \text{ for } t_1 < t_2$$

Then $Z(t;\omega)$ is said to be a Martingale with respect to $F(\cdot)$, or simply a Martingale if

(i) $E(|Z(t;\omega)|) < \infty, \quad t \in T$

(ii) $E(Z(t;\omega) \mid F(s)) = Z(s;\omega), \; s < t$

Let $\mathcal{B}(t)$ denote the sigma algebra generated by $Z(s;\omega)$ for $s \le t$. Then it is clear that $Z(t;\omega)$ continues to be a Martingale with respect to $\mathcal{B}(t)$.

The Wiener process $W(t;\omega)$ is a Martingale with respect to $\mathcal{B}(t)$. In fact:

$$E(W(t;\omega) - W(s;\omega) \mid W(\sigma_1;\omega), \ldots . W(\sigma_n;\omega)) = 0$$

for any finite number of indices $\sigma_1, \ldots, \sigma_n, \leq s$. The class of sets B such that

$$B \in \mathcal{B}(W(\sigma_1;\omega), \ldots . W(\sigma_n;\omega)), \quad \sigma_i \leq s$$

for some finite number of indices, forms a field which generates $\mathcal{B}(s)$, and hence

$$E(W(t;\omega) \mid \mathcal{B}(s)) = W(s;\omega) \text{ for } s \leq t.$$

Actually more is true because non-overlapping increments are actually independent:

$$E[W(t;\omega)-W(s;\omega))(W(t;\omega)-W(s;\omega))^* \mid \mathcal{B}s)] = (t-s)\, I_n, \; s < t$$

where I_n is the indentity matrix. Next let $F(s)$ be any rectangular m-by-n matrix function, Lebesgue measurable. For any rectangular matrix A, we define:

$$\|A\|^2 = \text{Trace } AA^* = \text{Trace } A^*A$$

Assume now that

$$\int_T \|F(s)\|^2 \, ds < \infty$$

Then

$$Z(t;\omega) = \int_0^t F(s) \, dW(s;\omega)$$

is also a Martingale with respect to $\mathscr{B}(t)$. In fact:

$$Z(t;\omega) - Z(s;\omega) = \int_s^t F(\sigma) \, dW(\sigma;\omega)$$

and for any simple function $f(\cdot)$, we know that

$$E\left[\int_s^t f(\sigma) \, dW(\sigma;\omega) \,\middle|\, \mathscr{B}(s)\right] = 0$$

and hence since $F(.)$ is the limit of such simple functions,

$$E\left[\int_s^t F(\sigma) \, dW(\sigma;\omega) \,\middle|\, \mathscr{B}(s)\right] = 0$$

Moreover we know that

$$E[(Z(t;\omega) - Z(s;\omega))\,(Z(t;\omega) - Z(s;\omega))^{*}) \mid \beta(s)] =$$

$$\int_s^t F(\sigma)\,F(\sigma)^{*}d\sigma$$

We shall now generalize the stochastic integral (stil in the linear set-up) to Martingales, specifically to a class of Martingales which, following Nelson [8], we shall call R_2 Martingales. Since we shall only be concerned with such Martingales in the sequel, we shall not insert the qualification unless another kind is involved. By an R_2 Martingale, we shall mean a Martingale with the additional property that:

$$E[[(Z(t;\omega) - Z(s;\omega))(Z(t;\omega) - Z(s;\omega))^{*} \mid F(s)] = \int_s^t P(\sigma)\,d\sigma$$

where $P(s)$ is a non-nagative definite matrix function, Lebesgue measurable, and

$$\text{Tr.} \int_T P(s)\,ds < \infty$$

To define the stochastic integral we follow the same procedure as before. We wish to define:

$$\int_T f(t)\,dZ(t;\omega)$$

for a class or rectangular (m-by-n say to be specific) matrix functions. The class will be the Hilbert space of Lebesgue measurable functions with inner-product defined by:

$$[f, g] = \int_T [f(s), g(s)P(s)] \, ds \quad ; \text{(here } [a,b] = \text{Tr. } ab^*\text{)}.$$

It will be convenient to keep using the generic notation $\mathscr{H}$ to denote the space. For most purposes we may assume that P(s) is continuous; in any case we shall always assume that we can avoid the trivial case where it is zero almost everywhere. To define the integral we begin with simple functions which we know are dense in $\mathscr{H}$. For simple functions it is readily verified that

$$E([\int_T f(t)dZ(t;\omega) \, , \quad \int_T g(t)dZ(t;\omega)]) = [f, g]$$

and the mapping being linear into $L_2(\Omega)$ of m-dimensional random variables, we can complete the definition as before. It should be noted that the basic Martingale property implies that

$$E[(Z(t_3;\omega) - Z(t_2;\omega))(Z(t_2;\omega) - Z(t_1;\omega)^*) \mid F(t_2)] = 0, \ t_1 < t_2 < t_3$$

An Example:

Let $W(t;\omega)$ denote the Wiener process, and let F(s) denote an m-by-n matrix function, Lebesgue measurable such that

$$F(s)F(s)^* > 0 \quad \text{a.e.}$$

and

$$\int_T \|F(s)\|^2 \, ds < \infty$$

where we shall now take $T = [0, 1]$.

Define

$$Z(t;\omega) = \int_0^t F(s) \, dW(s;\omega)$$

Then $Z(t;\omega)$ is an R_2 Martingale. Let

$$a(s) = \sqrt{F(s)F(s)^*}$$

which is then non-singular a.e. Let

$$b(s) = a(s)^{-1}$$

Then the function $b(s)$ has the property

$$\int_0^1 [b(s), b(s) F(s)F(s)^*] \, ds = m < \infty$$

Hence we can define

$$K(t;\omega) = \int_0^t b(s) \, dZ(s;\omega), \; 0 \leq t \leq 1$$

Thus defined, we obtain a Wiener process in m dimensions. To verify this, we note that for any two m dimensional vectors (m-by-1 matrices) we have that:

$$E([\,(u^* \int_s^t b(\sigma) dZ(\sigma;\omega))\,(v^* \int_s^t b(\sigma)\, dZ(\sigma;\omega))] =$$

$$= \int_s^t [u^* b(\sigma), \qquad v^* b(\sigma) F(\sigma) F(\sigma)^*]\, d\sigma$$

$$= u^* \int_s^t b(\sigma) F(\sigma) F(\sigma)^*\, b(\sigma)^* v\, d\sigma$$

so that for $s < t$

$$E[(K(t;\omega) - K(s;\omega))(K(t;\omega) - K(s;\omega))^*] = \int_s^t b(\sigma) F(\sigma) F(\sigma)^*\, b(\sigma)^*\, d\sigma$$

$$= (t-s)\, I_m$$

where I_m is the m-by-m unit matrix. Of course $K(t;\omega)$ is Gaussian.

Moreover, we have the representation (of Doob, Nelson):

$$Z(t;\omega) = \int_0^t a(s)\, dK(s;\omega)$$

For this we have only to note that the partial sums (we assume that a(t) is continuous from now on):

$$\sum_{0}^{n-1} a(t_i)(K(t_{i+1};\omega) - K(t_i;\omega)) , \qquad 0 \leq t_i < t_{i+1} \leq t ;$$

$$0 = t_0 \; ; \; t_n = t$$

$$= \sum_{0}^{n-1} a(t_i) \int_{t_i}^{t_{i+1}} b(s)\, dZ(s;\omega)$$

$$= \sum_{0}^{n-1} a(t_i)(b(t_i)(Z(t_{i+1};\omega) - Z(t_i;\omega))) - \sum_{0}^{n-1} a(t_i) \int_{t_i}^{t_{i+1}} (b(s)-b(t_i))dZ(s;\omega)$$

The first term

$$= Z(t;\omega)$$

while the second term goes to zero as the maximal subdivision length goes to zero. For

$$E\left(\sum_{0}^{n-1} a(t_i) \int_{t_i}^{t_{i+1}} (b(s)-b(t_i))dZ(s;\omega)\right)^2 =$$

$$\sum_{0}^{n-1} \int_{t_i}^{t_{i+1}} \| a(t_i)(b(s) - b(t_i))F(s) \|^2 \, ds$$

and

$$\| a(t_i)b(s) - b(t_i))F(s) \|^2 = \| a(t_i)(b(s) - b(t_i)) a(s) \|^2$$

$$= \| a(t_i) - a(s) \|^2$$

and $a(s)$ is uniformly continuous on $[0,1]$.

Finally, we note that by using Doob's Martingale inequality, we can obtain

$$\int_0^t F(s)\, dZ(s;\omega)$$

as a continuous function of t, for almost all ω.

Problem Show that $Y(t;\omega)$ defined by (2.16) is such that

$$Y(t;\omega) - \int_0^t C(s)x(s;\omega)\, ds = Z(t;\omega)$$

is an R_2 Martingale. Show that if

$$a(t) = D(t)D(t)^* > 0 \quad \text{a.e.}$$

and we assume say,

$$b(t) = (a(t))^{-1/2}$$

is essentially bounded on $[0, 1]$, then

$$\int_0^t b(s)\, dY(s;\omega) = \int_0^t b(s)\, C(s)\, x(s;\omega)\, ds + K(t;\omega)$$

where $K(t;\omega)$ is a Wiener process.

Radon-Nikodym Derivatives with Respect to Wiener Measure

Let $W(t;\omega)$ denote the n-dimensional Wiener process, $0 \le t \le 1$, inducing the Wiener measure on $\mathscr{C}=C(0,1)$ as we have indicated. Let $\varphi(\omega)$ denote a (Borel) measurable function mapping $\mathscr{C}$ into $\mathscr{C}$. Then $\varphi(\cdot)$ induces a measure on the Borel sets of $\mathscr{C}$ given by:

$$p_\varphi(B) = p_W(\varphi^{-1}(B))$$

where p_W denotes the Wiener measure. Very often we are interested in the case where $\varphi(\cdot)$ is defined by means of a stochastic integral:

$$x(t;\omega) = L(t) \int_0^t M(s)\, dW(s;\omega)$$

where say the functions $L(\cdot)$, $M(\cdot)$ are continuous. Of particular importance is to determine when the induced measure is absolutely continuous with respect to Wiener measure, and then to evaluate the Radon-Nikodym derivative. In the present section we shall study the case where $\varphi(\cdot)$ is a linear transformation (or affine transformation).

Theorem 2.3

Suppose $\varphi(\cdot)$ is a measurable map of $\mathscr{C}$ into $\mathscr{C}$. Let $p_\varphi(\cdot)$ denote the induced measure. Let $\{\phi_n\}$ be any complete orthonormal system in $L_2(0,1)^{(n)}$. Let

$$\zeta_n = \int_0^1 [\phi_n(t), dW(t;\omega)]$$

Let $\mathscr{B}_n$ denote the sigma algebra $\mathscr{B}(\zeta_1, \ldots\ldots, \zeta_n)$. Suppose for every B in $\mathscr{B}_n$, we have

$$p_\varphi(B) = \int_B H_n(\omega)\, dp_W$$

In other words, p_φ is absolutely continuous with respect to p_W on $\mathscr{B}_n$, for every n, and $H_n(\omega)$ denotes the corresponding derivative. Then p_φ is absolutely continuous with respect to p_W, and the derivative is given by

$$H(\omega) = \lim H_n(\omega)$$

omitting a set of Wiener measure zero.

Proof We observe first of all that the sequence of random variables $H_n(\omega)$ is a Martingale. For, for any set B in $\mathscr{B}_n$, we have that

$$p_\varphi(B) = \int_B H_{n+1}(\omega)\, dp_W = \int_B H_n(\omega)\, dp_W$$

since $\mathscr{B}_n$ is increasing with n. Also

$$E(H_n(\omega)) = 1$$

Hence by Doob Martingale convergence theorem we know that $H_n(\omega)$ converges with probability one. Denote the limit by $H(\omega)$. Then of course

$$E(H(\omega)) = 1$$

and if we denote by $\mathscr{B}_\infty$ the smallest sigma algebra containing every set in every $\mathscr{B}_n$, we know that

$$p_\varphi(B) = \int_B H(\omega)\, dp_W, \quad B \in \mathscr{B}_\infty$$

Next let $f(\cdot)$ be any element in $\mathscr{H} = L_2(0,1)^{(n)}$. Let

$$\zeta = \int_0^1 [f(t), \; dW(t;\omega)]$$

Let U be an open set in the real line, and let

$$B = [\omega \mid \zeta(\omega) \in U]$$

It is clearly enough to show that

$$p_\varphi(B) = \int_B H(\omega)\, dp_W$$

Unfortunately however B need not belong to $\mathscr{B}_\infty$. On the other hand,

$$\zeta(\omega) = \underset{n}{\text{limit}}\ \xi_n(\omega) \text{ with probability one}$$

where

$$\xi_n(\omega) = \sum_1^n [f, \phi_m]\, \zeta_m(\omega)$$

since $\{\phi_n\}$ is a complete orthonormal system. Let Λ denote the set of convergence of the variables ξ_n. Then Λ is in $\mathscr{B}_\infty$, and hence

$$p_\varphi(\Lambda) = \int_\Lambda H(\omega)\, dp_W = 1$$

since Λ has Wiener measure one. Hence

$$p_\varphi(B) = p_\varphi(B \cap \Lambda).$$

Now let

$$A_k = [\omega | \xi_k(\omega) \varepsilon U]$$

and let

$$A = \bigcup_{n=1}^{\infty} \bigcap_{n=k}^{\infty} A_k$$

Then $A \varepsilon \mathscr{B}_\infty$, and

$$A \supset B \cap \Lambda$$

Hence

$$p_\varphi(A) = p_\varphi(B) = \int_A H(\omega) dp_W = \int_B H(\omega) dp_W \; .$$

as we started to show.

<u>Corollary</u> Suppose $\varphi(\omega)$ is a linear transformation such that for any $f(\cdot)$ in $\mathscr{H}$, and any linear Borel set U,

$$\varphi(\cdot)^{-1} [\omega | \int_0^1 [f(t), dW(t;\omega)] \varepsilon U] = [\omega | \int_0^1 [h(t), dW(t;\omega)]] \varepsilon U$$

where

$$h = Mf;$$

and M is a linear bounded transformation mapping $\mathscr{H}$ into itself.

Suppose

$$M^*M = (I + J)^{-1}$$

where J is Hilbert-Schmidt. Then p_φ is absolutely continuous with respect to Wiener measure.

<u>Proof</u> Since J is compact and self-adjoint by definition, let $\{\phi_n\}$ denote the orthonormalized eigenfunctions, including those corresponding to zero eigenvalues, so that $\{\phi_n\}$ is a complete orthonormal system. Let γ_n denote the corresponding eigenvalues. Then we observe that the infinite product

$$\prod_1^\infty (1 + \gamma_n) \exp(-\gamma_n)$$

converges. Let γ denote its value. Let

$$\zeta_n(\omega) = \int_0^1 [\phi_n(t), dW(t;\omega)]$$

Define

$$H_n(\omega) = \prod_1^n \sqrt{1+\gamma_i} \quad \exp -\frac{1}{2} \sum_1^n \gamma_n(\zeta_i(\omega))^2 \tag{2.30}$$

Then it is not difficult to verify that for B in $\beta(\zeta_1, \ldots, \zeta_n)$,

$$p_\varphi(B) = \int_B H_n(\omega)\, dp_W$$

Hence the theorem applies. We can actually evaluate the limit. For this observe that

$$H_n(\omega) = \left(\prod_1^n \sqrt{(1+\gamma_i)}\, \exp -\frac{1}{2}\gamma_i)\right) \exp -\frac{1}{2}\left(\sum_{i=1}^{n} \gamma_i((\zeta_i)^2 - 1)\right) \tag{2.31}$$

The numerical factor in front converges to $\sqrt{\gamma}$. While in the random exponent:

$$\sum_1^n \gamma_m(\zeta_m(\omega)^2 - 1)$$

is a sum of independent random variables, each of mean zero, while

$$\sum_1^n E[\gamma_m^2\ (\zeta_m(\omega)^2 - 1)^2]$$

$$= \sum_1^n \gamma_m^2\ (3 + 1 - 2)$$

and hence converges. Hence the sum converges with probability one. Hence

$$H(\omega) = \sqrt{\gamma}\quad \exp -\frac{1}{2}\left(\sum_1^\infty \gamma_i(\zeta_i(\omega)^2 - 1)\right) \tag{2.32}$$

<u>Example</u> Let $x(t;\omega)$ denote a continuous solution of the stochastic equation:

$$x(t;\omega) = \int_0^t A(t)\, x(t;\omega)\, dt + W(t;\omega) \qquad 0 \le t \le 1$$

where let us assume that $A(t)$ is continuous in $[0,1]$. Then the measure induced by the process $x(t;\omega)$ on C, is absolutely continuous with respect to Wiener measure. In this case we can explicitly write the solution as:

$$x(t;\omega) = \Phi(t)\int_0^t \Phi(s)^{-1} dW(s;\omega)$$

which then defines the mapping $\varphi(\cdot)$ explicitly. This shows that the mapping M is given by

$$M = I + K$$

where

$$K f = g; \quad g(t) = \Phi(t)^{*-1} \int_t^1 \Phi(s)^* A(s)^* f(s)ds$$

Since K is thus a Volterra operator, we note that $(I + K)$ has a bounded inverse. In fact a simple calculation shows that

$$(I + K^*)^{-1} = (I - L)$$

where L is again a Volterra operator defined by:

$$Lf = g ; \quad g(t) = A(t)\int_0^t f(s)ds$$

Hence we have

$$(M^*M)^{-1} = (I-L^*)(I-L) \tag{2.33}$$

Thus

$$J = L^*L - (L + L^*) \tag{2.34}$$

and is obviously Hilbert-Schmidt.

It is pertinent to ask where (2.32) can be expressed as a functional on the (Wiener) process without having to go through the eigenfunctions. Before we do this in general, it is interesting to consider the special case where

$$A(t) = A(t)^* = A$$

Then the main simplification is that $(L + L^*)$ is trace-class. Indeed

$$(L + L^*)\, f = g; \quad g(t) = A \int_0^1 f(s)ds$$

so that the operator has finite-dimensional range, and

$$\mathrm{Tr}(L + L^*) = \mathrm{Tr}\ A$$

Hence J is trace-class, since L^*L being the product of Hilbert-Schmidt operators is clearly trace-class. The implication of this is that now,

$$\sum_1^\infty |\gamma_n| < \infty \tag{2.35}$$

so that the infinite product

$$\prod_{i=1}^{\infty} (1+\gamma_i)$$

converges. Indeed we can actually evaluate this product by the general formula that

$$\sum_1^{\infty} \text{Log}\,(1+\gamma_i) = \text{Tr}\,(\text{Log}(I-L^*)(I-L)) = -\text{Tr}\,(L+L^*) \qquad (2.36)$$

which is valid whenever $(L+L^*)$ is trace-class. A proof is given in the Appendix. Since the exponent in (2.32) is convergent, and (2.35) holds,

$$-\frac{1}{2}\sum_1^{\infty} \gamma_k \zeta_k(\omega)^2 \qquad (2.37)$$

converges also, with probability one. We can go on to evaluate the limit. For this let us note that

$$\gamma_k \zeta_k(\omega) = \sum_1^{\infty} [J\phi_k, \phi_m]\, \zeta_m(\omega)$$

Since

$$[J\phi_k, \phi_m] = [L\phi_k, L\phi_m] - [\phi_k, (L+L^*)\phi_m]$$

and

$$[\phi_k, (L+L^*)\phi_m] = [\int_0^1 \phi_k(t)dt,\ A\int_0^1 \phi_m(t)dt]$$

$$\sum_1^\infty A \int_0^1 \phi_m(t)dt \;\; \zeta_m(\omega) = A\ W(1;\omega)$$

and letting

$$L\phi_k = \Psi_k$$

we have:

$$\sum_1^\infty \Psi_k(t)\, \zeta_k(\omega) = A\ W(t;\omega)$$

Hence we obtain finally that

$$\nu_k \zeta_k(\omega) = \int_0^1 [\Psi_k(t),\ A\ W(t;\omega)]\ dt - [\int_0^1 \phi_k(t)dt,\ AW(1;\omega)]$$

Multiplying by $\zeta_k(\omega)$ and summing on k, the second term clearly yields

$$-[W(1;\omega),\ AW(1;\omega)]$$

while the first term

$$\int_0^1 [\sum_1^\infty \zeta_k(\omega)\ \Psi_k(t),\ A\ W(t;\omega)]\ dt$$

$$= \int_0^1 [A\ W(t;\omega),\ A\ W(t;\omega)]\ dt$$

For this we only need to note that

$$\int_0^1 [A\,W(t;\omega) - \sum_1^n \zeta_k(\omega)\Psi_k(t),\ A\,W(t;\omega)]\,dt$$

$$= \int_0^1 [W(t;\omega) - \sum_1^n \zeta_k(\omega) \int_0^t \phi_k(s)ds\ ,\ \ A^*A\,W(t;\omega)]\,dt$$

and

$$\left\{E(\left|\int_0^1 [W(t;\omega) - \sum_1^n \zeta_k(\omega) \int_0^t \phi_k(s)ds\ ,\ \ A^*A\,W(t;\omega)]\,dt\ \right|)\right\}^2$$

$$\leq \int_0^1 E(\| W(t;\omega) - \sum_1^n \zeta_k(\omega) \int_0^t \phi_k(s)ds \|^2 dt) \int_0^1 E(\| A^*AW(t;\omega) \|^2)dt$$

which goes to zero. Hence finally

$$\sum_1^\infty -\frac{1}{2}\,\gamma_n\,\zeta_n(\omega)^2 = -\frac{1}{2}\int_0^1 [AW(t;\omega),\ AW(t;\omega)]\,dt + \frac{1}{2}[AW(1;\omega),\ \ W(1;\omega)]$$

Hence we have:

$$H(\omega) = \exp(-\frac{1}{2}\ \mathrm{Tr}\,A - \frac{1}{2}\int_0^1 [AW(t;\omega),\ AW(t;\omega)]\,dt$$

$$+ \frac{1}{2}[AW(1;\omega),\ W(1;\omega)]) \qquad (2.38)$$

We shall verify this answer in another way eventually.

So far we have been concerned with mapping $\mathscr{C}$ into $\mathscr{C}$. This is of course not essential. Thus let $Y(t;\omega)$ be any stochastic process with continuous sample functions for almost all ω. Then for $0 \le t \le 1$,

$$\varphi(\omega) = Y(\cdot;\omega)$$

is a measurable map into $\mathscr{C}$, measureable with respect to Borel sets in $\mathscr{C}$. Hence we can define an induced measure p_φ as before. Of particular interest is the case where $Y(t;\omega)$ is a Gaussian process and the stochastical integral

$$\int_0^1 f(t)\, dY(t;\omega)$$

can be defined. We can then state the following theorem.

Theorem 2.4 Let $Y(t;\omega)$ be a Gaussian process with continuous sample paths. Let the process dimension be n. Suppose the stochastic integral can be defined for every $f(\cdot)$ in $L_2(0,1)^{(n)}$ such that

$$E\left(\int_0^1 [f(t), dY(t;\omega)]\right)\left(\int_0^1 [g(t), dY(t;\omega)]\right) = [Rf, g] \ldots \tag{2.39}$$

where R is a linear transformation of $L_2(0,1)^{(n)}$ into itself, and has the form:

$$R = (I + J)^{-1}$$

where J is self-adjoint Hilbert-Schmidt operator. Then the measure induced by the process is absolutely continuous with respect to Wiener measure.

<u>Proof</u> Let c denote the generic element in $\mathscr{C}$, and let $W(t;c)$ denote the Wiener process,

$$W(t;c) = c(t)$$

We need only note that for any $f(\cdot)$ in $L_2(0,1)^{(n)}$, and any linear Borel set U,

$$\varphi(\cdot)^{-1}\left([c| \int_0^1 [f(t), dW(t;c)] \in U\right) = [\omega| \int_0^1 [f(t), dY(t;\omega)] \in U]$$

The proof can clearly proceed exactly as in Theorem 2.3, using (2.39).

<u>Example</u> Let us consider the process $Y(t;\omega)$ defined by (2.16), and take the case where $L = 1$ and

$$D(t)D(t)^* > 0 \qquad \text{a.e.}$$

and assume that

$$b(t) = \sqrt{((D(t)D(t)^*)^{-1})}$$

is essentially bounded. Let

$$\tilde{Y}(t;\omega) = \int_0^t b(s)\, dY(s;\omega)$$

Then the measure induced by the $\tilde{Y}(\cdot;\ )$ process is absolutely continuous with respect to Wiener measure. Indeed

$$\int_0^1 [f(t), d\tilde{Y}(t;\omega)] = \int_0^1 [b(t)f(t), dY(t;\omega)]$$

so that, introducing the operator

$$b\, f = g; \quad g(t) = b(t)\, f(t)$$

we have that

$$E\left(\left(\int_0^1 [f(t), d\tilde{Y}(t;\omega)]\right)\left(\int_0^1 [h(t), d\tilde{Y}(t;\omega)]\right)\right) = [Rf, h]$$

where

$$R = b^*(D+K)^* (D + K)\, b = I + b^* K^* K\, b + b^* D^* Kb + b^*K^*D\, b \tag{2.40}$$

Now because $(R - I)$ is compact and self-adjoint, it is clear that R has a bounded inverse unless for some non-zero $f(\cdot)$

$$[Rf, f] = 0$$

Or, equivalently,

$$[(D + K)b\, f, (D + K)\, b\, f] = 0$$

or,

$$(D + K)\, b\, f = 0$$

Or, with slight abuse of notation:

$$D(t)^*\, b(t)\, f(t) + K\, b\, f = 0$$

But, multiplying by $D(t)$, and noting that

$$(D(t)D(t)^*)^{-1} = b(t)\, b(t)$$

we obtain that

$$b\, f + (b(t)^2\, D(t))\, K\, b\, f = 0$$

But this implies that $(b\, f)$ must be zero, since the second term involves a Volterra operator. Since $b(t)$ is non-singular, f must be zero. Hence $R f$ is zero only if f is zero, or R has a bounded inverse. Further it is clear that

$$R^{-1} = I + J$$

where J is self-adjoint, and Hilbert- Schmidt.

As a final example of Radon-Nikodym derivatives, let us consider the case of a non-zero mean. Thus let

$$Y(t;\omega) = W(t;\omega) + \int_0^t m(s)ds \tag{2.41}$$

where $m(\cdot)$ is in $L_2[0,1]^{(n)}$. Then the measure induced by the process $Y(t;\omega)$ is absolutely continuous with respect to Wiener process. For this, let $\{\phi_n\}$ be a complete orthonormal system and it is readily seen that for B in $\beta(\zeta_1, \ldots, \zeta_n)$, where as usual:

$$\zeta_n = \int_0^1 [\phi_n(t), \; dW(t;\omega)]$$

we have, denoting the new measure by p_Y,

$$p_Y(B) = \int_B H_n(\omega) \, dp_W$$

where

$$H_n(\omega) = \exp -\frac{1}{2}\Big(\sum_1^n [m, \phi_k]^2 - \sum_1^n 2 \, \zeta_k(\omega) \, [m, \phi_k]\Big)$$

from which it follows that the Radon-Nikodym derivative is given by:

$$H(\omega) = \exp -\frac{1}{2}\Big(\int_0^1 |m(t)|^2 \, dt - 2\int_0^1 [m(t), \; dW(t;\omega)]\Big) \tag{2.42}$$

Ito Integral

We shall now study one of the main tools of much of our theory: namely the Ito integral. The Ito integral is the non-linear version of the stochastic integrals we have considered, in that the integrand will now be a random process also. Let $Z(t;\omega)$ denote an R_2 Martingale, with $F(t)$ the increasing sigma-algebra and

$$E((Z(t;\omega) - Z(s;\omega))(Z(t;\omega) - Z(s;\omega))^*) | F(s)) = \int_s^t P(\sigma)d\sigma \tag{2.43}$$

We wish to define the integral

$$\int_0^1 f(t;\omega)\, dZ(t;\omega)$$

where $f(t;\omega)$ is an m-by-n matrix valued random process with the following properties:

(i) $f(t;\omega)$ is measurable jointly in t and ω, in t with respect to Lebesgue measure; $f(t;\omega)$ is measurable $F(t)$ for each fixed t. The significance of this is that $f(t;\omega)$ depends only on the 'past' of the process $Z(t;\omega)$. Sometimes this is indicated by saying that $f(t;\omega)$ is 'non-anticipatory', or 'physically realizable'.

(ii) $$\int_0^1 E([f(t;\omega), f(t;\omega)P(t)])\, dt < \infty \tag{2.44}$$

Let $\mathcal{H}$ denote the class of such functions; it is clearly a linear space.

Introduce an inner-product in $\mathcal{H}$ by:

$$[f, g] = \int_0^1 E[f(t;\omega), g(t;\omega) P(t)]\, dt \tag{2.45}$$

(Here let us recall that in the integrand:

$$[a, b] = \text{Tr. } ab^* = \text{Tr. } a^*b)$$

By a "simple" function in $\mathcal{H}$ we shall mean a function of the form:

$$f(t;\omega) = v_i(\omega), \quad t_i \leq t < t_{i+1}, \quad 0 \leq i \leq n-1,$$

$$t_0 = 0, \quad t_n = 1$$

It is implicit of course that $v_i(\omega)$ is measurable $F(t_i)$. As in the linear case, we define the integral for such a simple function by:

$$\sum_{i=0}^{n-1} v_i(Z(t_{i+1};\omega) - Z(t_i;\omega))$$

Note that

$$E([v_i(Z(t_{i+1};\omega) - Z(t_i;\omega)), v_j(Z(t_{j+1};\omega) - Z(t_j;\omega))])$$

$$= E(E(\text{ditto} \mid F(t_{j+1})) \text{ if } t_{j+1} \leq t_i, \text{ and hence zero.}$$

$$= \text{Tr. } E(v_i(Z(t_{i+1};\omega) - Z(t_i;\omega))(Z(t_{i+1};\omega) - Z(t_i;\omega))^* v_i^* =$$

$$\int_{t_i}^{t_{i+1}} E[v_i, v_i P(s)]\, ds, \quad \text{for } i = j$$

Hence, by a similar calculation as in the linear case we see that for any two simple functions, $f(t;\omega)$, $g(t;\omega)$ in $\mathscr{H}$, we have:

$$E([\int_0^1 f(t;\omega)\, dZ(t;\omega)\, ,\ \int_0^1 g(t)\, dZ(t;\omega)]) = [f, g]$$

In other words we have an inner-product preserving linear transformation from the class of simple functions in $\mathscr{H}$, into the space $L_2(\Omega)$ of matrix valued (m-by-n) ω-random variables with finite second moment.

It only remains to show that $\mathscr{H}$ is a Hilbert space, and that the simple functions are dense in it. Let us note first that given any function $f(t;\omega)$ in $\mathscr{H}$, we can find a sequence of functions $f_n(t;\omega)$ in $\mathscr{H}$ such that $f_n(t;\omega)$ is uniformly continuous in t in the mean square sense, and such that:

$$\| f - f_n \|^2 = [f_n - f,\ f_n - f] \longrightarrow 0$$

For this we only need to define

$$f_n(t;\omega) = \int_{t-1/n}^{t} f(s;\omega)\, \sqrt{P(s)} \quad ds \ (\int_{t-1/n}^{t} \sqrt{P(s)} \quad ds)^{-1}$$

Then when $P(s)$ is the Identity matrix for example, we have

$$f_n(t+\Delta;\omega) - f_n(t;\omega) = \int_{t-1/n}^{t-1/n+\Delta} f(s;\omega) ds + \int_t^{t+\Delta} f(s;\omega)\, ds$$

so that

$$E(\| f_n(t+\Delta;\omega) - f_n(t;\omega)\|^2) \leq 2\ E\left\|\int_{t-1/n}^{t-1/n+\Delta} f(s;\omega)ds\right\|^2 + 2\ E\left\|\int_t^{t+\Delta} f(s;\omega)ds\right\|^2$$

$$\longrightarrow 0 \text{ as } |\Delta| \longrightarrow 0, \text{ uniformly in } t.$$

The more general case is handled similarly.

Next let $g(t;\omega)$ be in $\mathscr{H}$, and uniformly continuous in the mean. Define

$$g_n(t;\omega) = g(k/2^n;\omega), \quad k/2^n \leq t < (k+1)/2^n;\ 0 \leq k \leq 2^n-1$$

Then it is easy to see that

$$\|g_n - g\|^2 \longrightarrow 0$$

Since the closure of the class of simple functions is already a Hilbert space, this is enough to show that $\mathscr{H}$ is a Hilbert space.

Here is an elementary canonical example which illustrates the difference between the ordinary integral and the Ito integral. Let $W(t;\omega)$ be the Wiener process. The the Ito integral

$$\int_0^1 [W(t;\omega)\ ,\ dW(t;\omega)]$$

is clearly definable. Let us attempt what is essentially an integration by parts. For this we begin with an approximating finite sum:

$$\sum_{i=0}^{m-1} [W(t_i;\omega),\ W(t_{i+1};\omega) - W(t_i;\omega)],\ t_0 = 0, \ldots . t_i < t_{i+1},\ t_m = 1$$

which we can rewrite as:

$$= - \sum_{0}^{m-1} [W(t_{i+1};\omega) - W(t_i;\omega)\ ,\ W(t_{i+1},\omega) - W(t_i;\omega)]$$

$$+ \sum_{0}^{m-1} [W(t_{i+1};\omega)\ , W(t_{i+1};\omega) - W(t_i;\omega)]$$

Here the second term can be expressed:

$$\sum_{i=0}^{m-1} \sum_{j=0}^{i} [W(t_{j+1};\omega) - W(t_j;\omega),\ W(t_{i+1};\omega) - W(t_i;\omega)]$$

and by interchanging the order of summation, we have that this is the same as:

$$\sum_{j=0}^{m-1} \sum_{i=j}^{m-1} [W(t_{j+1};\omega) - W(t_j;\omega)\ ,\ W(t_{i+1};\omega) - W(t_i;\omega)]$$

$$= \sum_{j=0}^{m-1} [W(t_{j+1};\omega) - W(t_j;\omega)\ ,\ W(1;\omega) - W(t_j;\omega)]$$

$$= \sum_{j=0}^{m-1} [W(t_{j+1};\omega) - W(t_j;\omega),\ W(1;\omega)] - \sum_{j=0}^{m-1} [W(t_{j+1};\omega) - W(t_j;\omega),\ W(t_j;\omega)]$$

and since the first term in this

$$= [W(1;\omega),\ W(1;\omega)]$$

and the second term is the sum we originally started with, we have

$$\sum_{i=0}^{m-1} [W(t_i;\omega), W(t_{i+1};\omega) - W(t_i;\omega)] = \frac{1}{2}[W(1;\omega), W(1;\omega)]$$

$$- \frac{1}{2} \sum_0^{m-1} \| W(t_{i+1};\omega) - W(t_i;\omega) \|^2$$

Now since the sum on the left converges as the subdivision size shrinks, the second term on the right converges, in the mean square sense also. But

$$E(\| W(t_{i+1};\omega) - W(t_i;\omega) \|^2) = n(t_{i+1} - t_i)$$

and hence

$$E(\sum_0^{m-1} (\| W(t_{i+1};\omega) - W(t_i;\omega) \|^2 - n(t_{i+1} - t_i))^2$$

$$= E \sum_0^{m-1} \| W(t_{i+1};\omega) - W(t_i;\omega) \|^4 - n^2 \sum_0^{m-1} (t_{i+1} - t_i)^2$$

$$= 0 \, (\max | t_{i+1} - t_i |) \longrightarrow 0$$

Hence

$$\int_0^1 [W(t;\omega), dW(t;\omega)] = + \frac{1}{2}[W(1;\omega), W(1;\omega)] - \frac{1}{2} (Tr.I)$$

The significant point is the appearance of the second term, which is a characteristic feature of the Ito integral.

Problem: Show that if we define the partial sum slightly differently:

$$\sum_{i=0}^{m-1} [W((t_i+t_{i+1})/2;\omega) , W(t_{i+1};\omega) - W(t_i;\omega)]$$

we get a completely different answer in the limit. In fact the difference

$$\sum_{i=0}^{m-1} [W((t_i+t_{i+1})/2;\omega) - W(t_i;\omega), W(t_{i+1};\omega) - W(t_i;\omega)]$$

$$\longrightarrow \frac{1}{2} (\mathrm{Tr.}\, I)$$

More generally, taking

$$\tau_i = \gamma(t_{i+1} - t_i) \qquad 0 < \gamma < 1$$

show that the limit of the partial sums:

$$\sum_{i=0}^{m-1} [W(t_i+\tau_i;\omega) - W(t_i;\omega) , W(t_{i+1};\omega) - W(t_i;\omega)]$$

$$\longrightarrow (\gamma)\ (\mathrm{Tr.}\, I)$$

Hence the partial sums:

$$\sum_{0}^{m-1} [W(t_i+\tau_i;\omega) , W(t_{i+1};\omega) - W(t_i;\omega)]$$

converge to

$$\frac{1}{2}[W(1;\omega),\ W(1;\omega)] + (\gamma - \frac{1}{2})(\text{Tr. I})$$

Let us now prove a useful generalization of this result. Thus let $W(t;\omega)$ denote the Wiener process, and let $H(t;\omega)$ be an n-by-1 process defined by:

$$H(t;\omega) = \int_0^t L(s)\, dW(s;\omega) \qquad 0 \leq t \leq 1$$

where $L(\cdot)$ is continuous. Let us calculate the Ito integral:

$$\int_0^1 [H(t;\omega),\ dW(t;\omega)]$$

Let us begin with an approximating sum:

$$\sum_{i=0}^{n-1} [H(t_i;\omega),\ (W(t_{i+1};\omega) - W(t_i;\omega))]$$

$$= (-1) \sum_{i=0}^{n-1} [(H(t_{i+1};\omega) - H(t_i;\omega)), ((W(t_{i+1};\omega) - W(t_i;\omega)))]$$

$$+ \sum_{i=0}^{n-1} [H(t_{i+1};\omega), (W(t_{i+1};\omega) - W(t_i;\omega))]$$

As before,

$$\sum_{i=0}^{n-1} [H(t_{i+1};\omega), (W(t_{i+1};\omega)-W(t_i;\omega))] = \sum_{j=0}^{n-1} [(H(t_{j+1};\omega)$$

$$-H(t_j;\omega)), (W(1;\omega) - W(t_j;\omega))] = [H(1;\omega), W(1;\omega)]$$

$$- \sum_{j=0}^{n-1} [(H(t_{j+1};\omega) - H(t_j;\omega)), W(t_j;\omega)]$$

Transposing, and taking limits, and noticing that

$$\sum_{j=0}^{n-1} [(H(t_{j+1};\omega) - H(t_j;\omega)), W(t_j;\omega)]$$

defines in the limit the Ito integral:

$$\int_0^1 [dH(t;\omega), W(t;\omega)]$$

we have:

$$\int_0^1 [H(t;\omega), dW(t;\omega)] + \int_0^1 [W(t;\omega), dH(t;\omega)]$$

$$= [H(1;\omega), W(1;\omega)] - \lim \sum_0^{n-1} [H(t_{i+1};\omega) - H(t_i;\omega)), (W(t_{i+1};\omega)$$

$$-W(t_i;\omega))]$$

As before the limit exists in the mean square sense, and we shall now show that the limit is actually equal to:

$$\int_0^1 \mathrm{Tr.}(L(t))\, dt$$

But this follows readily from the fact that:

$$E\Big(\sum_0^{n-1} [H(t_{i+1};\omega) - H(t_i;\omega)),\ W(t_{i+1};\omega) - W(t_i;\omega)]\Big)$$

$$= \sum_0^{n-1} E([\int_{t_i}^{t_{i+1}} L(s)dW(s;\omega),\ \int_{t_i}^{t_{i+1}} dW(s;\omega)])$$

$$= \int_a^b \mathrm{Tr.}\ L(t)\, dt$$

Hence

$$E\Big(\Big(\sum_0^{n-1} \Big\{[\int_{t_i}^{t_{i+1}} L(s)\, dW(s;\omega)\ ,\ \int_{t_i}^{t_{i+1}} dW(s;\omega)] - \int_{t_i}^{t_{i+1}} \mathrm{Tr.}\ L(t)dt\Big\}\Big)^2\Big) \quad (2.46)$$

$$= \sum_0^{n-1} E([\int_{t_i}^{t_{i+1}} L(s)dW(s;\omega),\ \int_{t_i}^{t_{i+1}} dW(s;\omega)] - \int_{t_i}^{t_{i+1}} \mathrm{Tr.}\ L(s)ds)^2$$

$$= \sum_0^{n-1} E([\int_{t_i}^{t_{i+1}} L(s)dW(s;\omega),\ \int_{t_i}^{t_{i+1}} dW(s;\omega)])^2 - \sum_0^{n-1} \Big(\int_{t_i}^{t_{i+1}} \mathrm{Tr.}L(s)ds\Big)^2$$

and

$$E\left(\left[\int_{t_i}^{t_{i+1}} L(s)dW(s;\omega), \int_{t_i}^{t_{i+1}} dW(s;\omega)\right]^2\right)$$

$$\leq E\left(\left\| \int_{t_i}^{t_{i+1}} L(s)dW(s;\omega)\right\|^2\right) n(t_{i+1}-t_i)$$

$$\leq \left(\int_{t_i}^{t_{i+1}} \text{Tr. } L(s)L(s)^* ds\right) n(t_{i+1}-t_i)$$

so that (2.46) clearly goes to zero as maximal length of subdivision goes to zero. Hence we finally have the result:

$$\int_0^1 \left[\int_0^t L(s)dW(s;\omega), dW(t;\omega)\right] = \left[\int_0^1 L(s)dW(s;\omega), \int_0^1 dW(s;\omega)\right]$$

$$- \int_0^1 \left[\int_0^t dW(s;\omega), L(t)dW(t;\omega)\right] - \int_0^1 \text{Tr. } L(t)\, dt \ldots\ldots \quad (2.47)$$

and of course again, the unusual thing to note is the appearance of the third term on the right.

This can clearly be generalized into a more symmetric form:

$$\int_0^1 \left[\int_0^t L(s)dW(s;\omega), M(t)dW(t;\omega)\right] = \left[\int_0^1 L(s)dW(s;\omega), \int_0^1 M(s)dW(s;\omega)\right]$$

$$- \int_0^1 \left[L(t)dW(t;\omega), \int_0^t M(s)dW(s;\omega)\right]$$

$$- \int_0^1 \text{Tr. } L(t)M(t)^* dt \quad (2.48)$$

Note that the right side can be put in the form:

$$\int_0^1 [M(t)\int_0^t L(s)dW(s;\omega)\ ,\ dW(t;\omega)] = \int_0^1 [dW(t;\omega),\ L(t)^* \int_t^1 M(s)^* dW(s;\omega)]$$

$$- \text{Tr.} \int_0^1 L(t)M(t)dt\ldots \tag{2.49}$$

but the integral (first term) on the right has to be interpreted just in the way we got it by adding the first terms on the right in (2.48).

<u>R-N Derivatives using Ito Integrals</u>

We shall now see how to express the Radon-Nikodym derivatives in terms of Ito integrals. First of all, let us consider the special result (2.38). Using (2.48), we have (by setting $L(t) = A$; $M(t) =$ Identity; and $A = A^*$):

$$\int_0^1 [AW(t;\omega),\ dW(t;\omega)] = \frac{1}{2}[AW(1;\omega),\ W(1;\omega)] - \frac{1}{2}\int_0^1 \text{Tr. } A\ dt$$

and substituting into (2.38) we have:

$$H(\omega) = \exp -\frac{1}{2}\int_0^1 [AW(t;\omega),\ AW(t;\omega)]\ dt + \int_0^1 [AW(t;\omega),\ dW(t;\omega)] \tag{2.50}$$

and we have the advantage that the Trace term disappears. This expression for the Radon-Nikodym derivative is actually valid for the general case of (2.33). But here let us consider the case where J in Theorem 2.4 is trace-class. We begin with a theorem of interest in itself.

<u>Theorem 2.5</u> Let L denote a Volterra operator, mapping $L_2(0,1)^{(n)}$ into itself.

$$Lf = g; \quad g(t) = \int_0^t L(t;s)\, f(s)ds; \quad L(t;s) \text{ continuous}$$

where L is also <u>trace-class.</u> Then for any orthonormal system $\{\phi_k\}$, we have:

$$\int_0^1 \left[\int_0^t L(t;s)dW(s;\omega),\ dW(t;\omega) \right] = \sum_1^\infty \sum_1^\infty \zeta_k\, \zeta_m\, [L\phi_k, \phi_m] \tag{2.51}$$

where

$$\zeta_k = \int_0^1 [\phi_k(t),\ dW(t;\omega)]$$

and the convergence of the infinite series is at least in the mean of order two.

<u>Proof</u> First let us consider the convergence of the series;

using the fact that the $\{\varphi_k\}$ are independent, zero mean Gaussian, and the fact that L is trace-class we can readily verify that

$$\varphi_N = \sum_1^N \sum_1^N \varphi_k \varphi_e [L\phi_k, \phi_e]$$

converges in the mean of order two. Moreover

$$E[(\sum_1^\infty \sum_1^\infty \varphi_k \varphi_e [L\phi_k, \phi_e])^2] \leq 2\sum_1^\infty \|L\phi_k\|^2 + 3\sum_1^\infty [L\phi_k, \phi_k]^2$$
$$\leq 5\sum_1^\infty \|L\phi_k\|^2 \tag{2.52}$$

Note also that

$$E(\sum_1^\infty \sum_1^\infty \zeta_k \zeta_m [L\phi_k, \phi_m]) = \sum_1^\infty [L\phi_k, \phi_k] = 0$$

since L being Volterra, its trace must be zero. Let us next consider the Ito integral on the left of (2.51). For an approximating finite sum:

$$\eta_n = \sum_{i=0}^{n-1} \left[\int_0^{t_i} L(t_i;s)\, dW(s;\omega),\ W(t_{i+1};\omega) - W(t_i;\omega)\right],$$

$$t_i < t_{i+1};\ t_0 = 0;\ t_n = 1,$$

we have, by using the Shepp expansion:

$$\eta_n = \sum_{m=1}^\infty \left(\sum_{k=1}^\infty \zeta_k \zeta_m \sum_{i=0}^{n-1}\left[\int_0^{t_i} L(t_i;s)\phi_k(s)ds, \int_{t_i}^{t_{i+1}} \phi_m(s)ds\right]\right) \tag{2.53}$$

Define the operator L_n by

$$L_n f = g;\ g(t) = \int_0^{t_i} L(t_i;s)f(s)ds,\quad t_i \leq t < t_{i+1}$$

Then L_n is clearly finite dimensional, and hence trace-class. But the sum in (2.53) can be written

$$\sum_m \sum_k \zeta_k \zeta_m [L_n \phi_k, \phi_m] = \eta_n$$

Let

$$\eta = \sum_1^\infty \sum_1^\infty \zeta_k \zeta_m [L\phi_k, \phi_m]$$

Then

$$E(\zeta_n - \zeta) = \text{Tr.}(L_n - L)$$

and more important

$$E[(\zeta_n - \zeta)^2] = E(\sum_1^\infty \sum_1^\infty \zeta_k \zeta_m [(L_n - L)\phi_k, \phi_m])^2$$

$$= 0(\| L_n - L \|_{H.S.}) \qquad (2.53a)$$

using (2.52).

But the Hilbert-Schmidt norm of $(L_n - L)$ clearly goes to zero, and hence (2.53a) goes to zero.

Theorem 2.6 Let $Y(t;\omega)$, $0 \leq t \leq 1$, be a Gaussian process with continuous sample paths such that the stochastic integral

$$\int_0^1 [f(t), \, dY(t;\omega)]$$

can be defined for each $f(\cdot)$ in $L_2(0,1)^n$ where

$$E\left(\left(\int_0^1 [f(t), dY(t;\omega)]\right)\left(\int_0^1 [g(t), \, dY(t;\omega)]\right)\right) = [Rf, g]$$

for $f(\cdot)$, $g(\cdot)$ in $L_2(0,1)^n$, where

$$R = (I + J)^{-1}$$

$$J = L^*L - (L + L^*)$$

and L has the form:

$$L f = h \; ; \; h(t) = M(t) \int_0^t L(s) f(s) \, ds$$

$M(\cdot)$, $L(\cdot)$ being continuous square matrices. Further it is assumed that J is trace-class. Then the measure induced on $\mathscr{C}$ by the mapping:

$$\varphi(\omega) = Y(\cdot;\omega)$$

is absolutely continuous with respect to Wiener measure, and the derivative is given by:

$$H(v) = \exp -\frac{1}{2} \int_0^1 [W_1(t;v), W_1(t;v)] \, dt + \int_0^1 [W_1(t;v), dW(t;v)] \qquad (2.54)$$

where

$$W_1(t;v) = M(t) \int_0^t L(s) dW(s;v)$$

and $W(t;v)$ is the Wiener process with

$$W(t;v) = v(t) \, , \; v(\cdot) \in \mathscr{C}$$

<u>Proof:</u> Let $\{\phi_k\}$ denote the complete orthonormal system of eigenvectors of J with corresponding eigenvalues γ_k. Then as we know from Theorems 2.3 and 2.4, the Radon-Nikodym derivative is given by:

$$H(v) = (\prod_k \sqrt{(1+\gamma_k)})\,(\exp -\frac{1}{2}\sum_1^\infty \gamma_k \zeta_k^2)$$

where

$$\zeta_k = \int_0^1 [\phi_k(t), dW(t;v)]$$

Also from (2.36), we have, since if J is trace-class, so is $(L + L^*)$:

$$\Pi\sqrt{(1+\gamma_k)} = \exp -\frac{1}{2}\ \mathrm{Tr.}\ (L + L^*)$$

Now we can write:

$$\sum_1^\infty \gamma_k \zeta_k^2 = \sum_1^\infty \zeta_k \int_0^1 [\Psi_k(t), dW(t;v)]$$

where

$$J\,\phi_k = \Psi_k$$

and

$$\int_0^1 [\Psi_k(t), dW(t;v)] = \sum_1^\infty [J\,\phi_k, \phi_m]\,\zeta_m$$

$$= \sum_1^\infty [L\phi_k, L\phi_m]\,\zeta_m - \sum_1^\infty [(L+L^*)\,\phi_k, \phi_m]\,\zeta_m$$

It is readily seen that

$$\sum_{m=1}^{\infty} [L\phi_k, L\phi_m]\, \zeta_m = \int_0^1 [L\phi_k\,,\; W_1(t;v)]dt$$

while writing:

$$L + L^* = Q + P$$

where

$$Qf = g; \quad g(t) = M(t) \int_0^t L(s)\, f(s)\, ds - L(t)^* \int_0^t M(s)^*\, f(s) ds$$

$$Pf = g; \quad g(t) = L(t)^* \int_0^1 M(s)^*\, f(s) ds$$

we note that P is finite dimensional, and hence trace-class, and hence Q must be trace-class, and Q being Volterra, its trace must be zero. Hence we obtain:

$$\text{Tr.}\ (L + L^*) = \text{Tr.}\ P = \int_0^1 \text{Tr.}\ L(t)\, M(t)\, dt = \int_0^1 \text{Tr.}\ M(t) L(t) dt \qquad \ldots (2.55)$$

We thus obtain, Q being now Volterra and trace-class,

$$\sum_1^\infty \gamma_k \zeta_k^2 = \int_0^1 [W_1(t;v), W_1(t;v)]\, dt - \sum\sum \zeta_k \zeta_m [Q\phi_k, \phi_m]$$

$$- \left[\int_0^1 M(s)^* dW(s;v), \int_0^1 L(s) dW(s;v)\right]$$

and using Theorem 2.6 we have that

$$\sum\sum \zeta_k \zeta_m [Q\phi_k, \phi_m] = \int_0^1 [M(t) \int_0^t L(s) dW(s;v), dW(t;v)]$$

$$- \int_0^1 [L(t)^* \int_0^t M(s)^* dW(s;v), dW(t;v)]$$

But from (2.48) we have that

$$\int_0^t [L(t)^* \int_0^t M(s)^* dW(s;v), dW(t;v)]$$

$$= \left[\int_0^1 M(s)^* dW(s;v), \int_0^1 L(s) dW(s;v)\right] - \int_0^1 \left[M(t) \int_0^t L(s) dW(s;v), dW(t;v)\right]$$

$$- \text{Tr.} \int_0^1 M(t)^* L(t)^* dt$$

Hence, substituting, we finally obtain:

$$\sum_1^\infty \gamma_n \zeta_k^2 = \int_0^1 [W_1(t;v), W_1(t;v)]\, dt - 2\int_0^1 [M(t) \int_0^t L(s) dW(s;v), dW(t;v)]$$

$$- \int_0^1 \text{Tr. } M(t)^* \, L(t) dt$$

from which (2.54) readily follows, upon using (2.55).

Example Suppose in (2.14), we take the special case where:

(i) $D(s)D(s)^* =$ Identity matrix.

(ii) $B(s)D(s)^* = 0 \ , \quad 0 \leq s \leq 1$

Then R in (2.19) becomes

$$R = I + K^*K$$

This implies that

$$R^{-1} = I + J$$

where J is trace-class, since K^*K is. By a theorem of Krein [11], this implies that

$$I + J = (I-L)^*(I-L)$$

where L is Volterra. After we do Kalman filtering, we shall see an alternate method for actually obtaining L. Once we can represent L as:

$$Lf = g; \quad g(t) = M(t) \int_0^t L(s)f(s)ds$$

we can apply Theorem 2.6 directly to obtain the R-N derivative in this case.

Example: Kalman Filtering

Perhaps the most significant application of stochastic differential systems at a useful level is the theory originally due to Kalman and since associated with his name. The approach here is quite different from the original versions based on Wiener-Hopf equation analysis, and considerably simpler.

We begin with a fundamental property of Martingales. (Cf. Nelson, Doob).

Lemma Let $Z_i(t;\omega)$, $i = 1, 2$ denote two Martingales with respect to the growing sigma algebra $\mathscr{F}(t)$, and let:

$$E(\| Z_i(1;\omega) - Z_i(0;\omega) \|^2) < \infty \quad , \quad i = 1, 2 \ldots . \tag{2.56}$$

Suppose that for i, j fixed, : $0 \leq t < 1$, :

$$\lim_{\Delta \to 0} \frac{1}{\Delta} E((Z_i(t+\Delta;\omega) - Z_i(t;\omega))(Z_j(t+\Delta;\omega) - Z_j(t;\omega))^* | \mathscr{F}(t)) = P_{ij}(t) \tag{2.57}$$

where the convergence of the random variable on the left is in the mean order one (L_1), and it is assumed that $P_{ij}(t)$ is continuous in t, $0 \leq t \leq 1$. Then for $0 \leq s < t \leq 1$,

$$E\left(\left(\int_s^t dZ_i(\sigma;\omega)\right)\left(\int_s^t dZ_j(\sigma;\omega)\right)^* \middle| \mathscr{F}(s)\right) = \int_s^t P_{ij}(\sigma)d\sigma \tag{2.58}$$

Proof Let $0 \leq a < 1$, and define

$$\Lambda_{ij}^a(t) = \left(\int_a^t dZ_i(s;\omega)\right)\left(\int_a^t dZ_j(s;\omega)\right)^*, \quad a \leq t < 1$$

Observe that for any $\Delta > 0$, $t + \Delta < 1$, we have:

$$E(\Lambda_{ij}(t+\Delta) - \Lambda_{ij}(t)) | \mathscr{F}(a)) = E[E[(\int_t^{t+\Delta} dZ_i(s;\omega))(\int_t^{t+\Delta} dZ_j(s;\omega))^* | \mathscr{F}(t)] \mathscr{F}(a)]$$

where we have used the fact that:

$$E[[(Z_i(t+\Delta;\omega) - Z_i(t;\omega))(Z_j(t;\omega) - Z_j(a;\omega))^*] | \mathscr{F}(t)] = 0$$

We can now follow the argument of Nelson (8) ; Let $\epsilon > 0$ be given. Let J denote the set of points in $[a, 1]$ such that for t in J:

$$E(|E(\Lambda_{ij}(t)|\mathscr{F}(a)) - \int_a^t P_{ij}(s)ds|) \leq \epsilon\ (t-a) \tag{2.58a}$$

Clearly J contains the point a. Also J is closed. Now we shall show that if t is any point for which (2.58a) holds, then it will hold for $t + \delta$, $0 < \delta < \Delta$, for some $\Delta > 0$. For this, we note that

$$E(\Lambda_{ij}(t+\delta)|\mathscr{F}(a)) = E[\Lambda_{ij}(t+\delta) - \Lambda_{ij}(t)|\mathscr{F}(t)|\mathscr{F}(a)) + E(\Lambda_{ij}(t)|\mathscr{F}(a))$$

and let us choose Δ_1 such that for all $\delta < \Delta_1$,

$$E(|(E(\Lambda_{ij}(t+\delta) - \Lambda_{ij}(t) | \mathscr{F}(t)) - \delta P_{ij}(t))|) < (\epsilon/2)\,\delta$$

Next let us choose Δ_2 such that for all $\delta < \Delta_2$,

$$\left|\int_t^{t+} P_{ij}(s)ds - \delta P_{ij}(t)\right| < (\epsilon/2)\delta$$

Choosing Δ to be the minimum of Δ_1 and Δ_2, we have: for $0 \leq \delta \leq \Delta$

$$E\left|(E(\Lambda_{ij}(t+\delta) - \Lambda_{ij}(t) | \mathscr{F}(a)) - \int_t^{t+\delta} P_{ij}(s)ds)\right|$$

$$= \quad E\left|(E(\Lambda_{ij}(t+\delta) - \Lambda_{ij}(t) | \mathscr{F}(t) - \int_t^{t+\delta} P_{ij}(s)ds \mid \mathscr{F}(a)))\right|$$

$$\leq \quad (\epsilon)\,\delta .$$

Hence (2.58a) follows for $t + \delta$, $0 < \delta < \Delta$. In particular this is true for $t = a$. Suppose now that the upperbound of t such that (2.58a) holds for $[a, t]$ is t_o, say. Then t_o must belong to J since J is closed. But if t_o is not equal to 1, we will have a contradiction because it can be extended by a non-zero amount. Since ϵ is arbitrary, it is clear that:

$$E[\Lambda_{ij}(t) | \mathscr{F}(a)] = \int_a^t P_{ij}(s)ds$$

which is clearly enough to prove the Lemma.

Corollary: Assume now that the Martingales $Z_i(s;\omega)$ are Gaussian, $0 \leq s \leq 1$, and that

$$Z_i(0;\omega) = 0, \; i = 1, 2$$

Assume further that (2.56) holds, and that (2.57) holds for $i = j = 2$, and for $i = 1$, $j = 2$. Let $\beta_2(t)$ the smallest sigma algebra generated by $Z_2(s;\omega)$, $s \leq t$. Then

$$E(Z_1(t;\omega) \mid \beta_2(t)) = \int_0^t r_{12}(s) dZ_2(s;\omega) \quad 0 < t \leq 1 \tag{2.59}$$

where $r_{12}(s)$ is defined as the limit:

$$r_{12}(s) = \lim_{\epsilon \to 0} P_{12}(s)(P_{22}(s) + \epsilon I)^{-1} \quad \text{a.e.} \; 0 < s < 1 \tag{2.60}$$

Proof It should be emphasized that it is not assumed that (2.57) holds for $i = j = 1$. First let us prove that for any Lebesgue measurable matrix function $f(.)$, having the same dimension as $P_{12}(s)$, such that

$$\int_0^1 \text{Tr. } f(s)^* f(s) P_{22}(s) ds < \infty \tag{2.61}$$

we have that:

$$E\left(\left(\int_0^1 dZ_1(s;\omega)\right)\left(\int_0^1 f(s)dZ_2(s;\omega)\right)^*\right) = \int_0^1 P_{12}(s)f(s)^*ds \tag{2.62}$$

For this it is enough to note that for $0 < s < t < 1$,

$$E\left(\int_0^1 dZ_1(s;\omega)\right)(Z_2(t;\omega) - Z_2(s;\omega))^*)$$

$$= E(Z_1(t;\omega) - Z_1(s;\omega))(Z_2(t;\omega) - Z_2(s;\omega))^*) = \int_s^t P_{12}(\sigma)d\sigma$$

from (2.58) ; hence it is immediate that (2.62) will hold for simple functions, and hence by the usual limiting arguments to any $f(.)$ satisfying (2.61). Next if

$$L(\epsilon;t) = P_{12}(t)(P_{22}(t) + \epsilon I)^{-1}$$

we have that: $L(\epsilon;t)$ satisfies (2.61), and that:

$$0 \leq E(\| Z_1(1;\omega) - \int_0^1 L(\epsilon;t)\, dZ_2(t;\omega)\|^2)$$

$$= E\| Z_1(1;\omega)\|^2 - \text{Tr.} \int_0^1 P_{12}(t)(P_{22}(t) + \epsilon I)^{-1}(P_{22}(t) + 2\epsilon I)$$

$$(P_{22}(t) + \epsilon I)^{-1}P_{12}(t)^*dt$$

By Fatou's Lemma, it follows that: r_{12} satisfies (2.61). Hence the integral:

$$\int_0^t r_{12}(s)dZ_2(s;\omega) \qquad 0 < t < 1$$

is well defined. Let f(.) satisfy (2.61). Then

$$E(((Z_1(t;\omega) - \int_0^t r_{12}(s)dZ_2(s;\omega)) \left(\int_0^t f(s)dZ_2(s;\omega))^* \right)$$

$$= \int_0^t (P_{12}(s)f(s)^* - r_{12}(s)P_{22}(s)f(s)^*)ds$$

But

$$r_{12}(s)P_{22}(s) = (P_{12}(s) \lim_{\epsilon \to 0} (P_{22}(s) + \epsilon I)^{-1} P_{22}(s)$$

$$= P_{12}(s)$$

which is trivially true at a point where

$$P_{22}(s) > 0$$

and otherwise because if

$$P_{22}(s)\, x = 0$$

so is

$$P_{12}(s)x = 0$$

as can be verified from (2.57). Hence

$$Z_1(t;\omega) - \int_0^t r_{12}(s)dZ_2(s;\omega)$$

is uncorrelated with, and being Gaussian, independent of,

$$\int_0^t f(s)dZ_2(s;\omega)$$

for every $f(\cdot)$ satisfying (2.61). But since the latter generate $\beta_2(t)$, (2.59) follows.

Let us consider now the linear stochastic equation (cf.(2.16)):

$$x(t;\omega) = \int_0^t A(s)\, x(s;\omega)\, ds + \int_0^t B(s)dW(s;\omega) \tag{2.63}$$

$$Y(t;\omega) = \int_0^t C(s)x(s;\omega)ds + \int_0^t D(s)dW(s;\omega) \tag{2.64}$$

where we shall for simplicity assume that all coefficients are continuous. Further we shall assume:

$$D(s)D(s)^* > 0 \quad \text{on} \quad [0,1]$$

Note that this implies that

$$b(s) = \sqrt{(D(s)D(s)^*)^{-1}}$$

is continuous. Then as we have seen, defining:

$$\hat{Y}(t;\omega) = \int_0^t b(s)dY(s;\omega)$$

we have:

$$\hat{Y}(t;\omega) = \int_0^t b(s)C(s)x(s;\omega)ds + \hat{W}(t;\omega)$$

where $\hat{W}(s;\omega)$ is now a Wiener process. Let $\hat{W}(t;\omega)$ be measurable the growing sigma-algebra $\mathscr{F}(t)$. Then of course $x(t;\omega)$, $Y(t;\omega)$, $\hat{Y}(t;\omega)$, $\hat{W}(t;\omega)$ are all measurable $\mathscr{F}(t)$. Let

$$\mathscr{B}_Y(t) = \text{sigma algebra generated by } Y(s;\omega),\ s \leq t$$

and similarly define $\mathscr{B}_{\hat{Y}}(t)$. Then since

$$Y(t;\omega) = \int_0^t (b(s))^{-1} d\hat{Y}(s;\omega)$$

we have that

$$\mathscr{B}_Y(t) = \mathscr{B}_{\hat{Y}}(t)$$

Since we have that

$$E(|x(t;\omega)|^2) < \infty$$

we can define the conditional expectation:

$$\hat{x}(t;\omega) = E(x(t;\omega) \,|\mathscr{B}(t))$$

where from now on we write

$$\mathscr{B}(t) \text{ for } \mathscr{B}_Y(t) = \mathscr{B}_{\hat{Y}}(t)$$

Observe now that:

$$E(x(t;\omega)|\mathscr{B}(s)) = E(E(\phi(t)\,\phi(s)^{-1}x(s;\omega) + \phi(t)\int_s^t \phi(\sigma)^{-1}B(\sigma)dW(\sigma;\omega))|\mathscr{F}(s))|\mathscr{B}(s))$$

$$= \phi(t)\,\phi(s)^{-1}\hat{x}(s;\omega)$$

But the leftside is a Martingale in s, $s < t$, for fixed t, and converges with probability one as s goes to t. But $Y(t;\omega)$ being continuous with probability one, we note that

$\mathscr{B}(t)$ = smallest sigma-algebra containing $\mathscr{B}(s)$ for every $s < t$

and hence the limit must be $\hat{x}(t;\omega)$. Thus $\hat{x}(t;\omega)$ is continuous from below with probability one. Hence by a theorem of Doob ([2], Theorem 2.6, p. 61, and remark on p. 62), there is a process equivalent to $\hat{x}(t;\omega)$ which is jointly measurable in t and ω, and we may clearly redefine $\hat{x}(t;\omega)$ to be this process. And since

$$E(|\hat{x}(t;\omega)|^2) \leq E(|x(t;\omega)|^2) < \infty$$

it follows that the integral

$$\int_0^t A(s)\,\hat{x}(s;\omega)ds\ , \quad t \leq 1$$

is well-defined (converges a.s.).

<u>Lemma:</u> The process:

$$\hat{x}(t;\omega) - \int_0^t A(s)\,\hat{x}(s;\omega)\,ds = Z_s(t;\omega) \quad 0 \leq t \leq 1$$

is a Gaussian Martingale. Moreover

$$E(|Z_s(t;\omega)|^2) < \infty$$

<u>Proof</u> Now for $t > s$, we have:

$$E(\hat{x}(t;\omega)|\mathscr{B}(s)) = E(x(t;\omega)\,|\,\mathscr{B}(t))|\mathscr{B}(s))$$

$$= E(x(t;\omega)\,|\mathscr{B}(s)) \ldots\ldots \qquad (2.65)$$

Hence

$$E(Z_s(t;\omega) - Z_s(s;\omega)\,|\mathscr{B}(s)) = E((x(t;\omega) - x(s;\omega))|\mathscr{B}(s))$$

$$- E(\int_s^t A(\sigma)\,\hat{x}(\sigma;\omega)d\sigma\,|\mathscr{B}(s))$$

$$= E(\int_s^t B(\sigma)dW(\sigma;\omega)|\mathscr{B}(s)) + E(\int_s^t A(\sigma)(x(\sigma;\omega)-\hat{x}(\sigma;\omega))d\sigma\,|\mathscr{B}(s))$$

But the first term is zero because $\mathscr{F}(s) \supset \mathscr{B}(s)$, and $\int_0^t B(\sigma)dW(\sigma,u)$ is a Martingale; the second term is zero, using (2.65).

<u>Lemma</u>: Let

$$Z_0(t;\omega) = Y(t;\omega) - \int_0^t C(s)\hat{x}(s;\omega)ds$$

Then $Z_0(t,u)$ is a Gaussian martingale. Moreover:

$$\lim_{\Delta \to 0} (1/\Delta)E(\int_t^{t+\Delta} dZ_0(s;\omega)\,(\int_t^{t+\Delta} dZ_0(s;\omega)^*)\,|\mathscr{B}(t)) = D(t)D(t)^* \qquad (2.66)$$

Proof We have only to note that:

$$Z_0(t;\omega) = Y(t;\omega) - \int_0^t C(s)x(s;\omega)\,ds + \int_0^t C(s)(x(s;\omega)-\hat{x}(s;\omega))ds$$

$$= \int_0^t D(s)dW(s;\omega) + \int_0^t C(s)(x(s;\omega) - \hat{x}(s,\omega))ds$$

so that

$$Z_0(t;\omega) - Z_0(s;\omega) = \int_s^t D(\sigma)dW(\sigma;\omega) + \int_s^t C(\sigma)(x(\sigma;\omega) - \hat{x}(\sigma;\omega))d\sigma$$

and just as in the previous lemma,

$$E((Z_0(t;\omega) - Z_0(s;\omega)) \,|\mathscr{B}(s)) = 0$$

Let us use the notation:

$$e(s;\omega) = x(s;\omega) - \hat{x}(s;\omega) \qquad (2.67)$$

and observe that

$$E(\| \int_t^{t+\Delta} C(s)\, e(s;\omega)ds \|^2 \,|\mathscr{B}(t)\,) = 0\,(\Delta^2)$$

since

$$E(|e(s;\omega)|^2) \leq E(|x(s;\omega)|^2)$$

and is bounded in $0 < s < 1$. Again:

$$E(\| \int_t^{t+\Delta} D(s)dW(s;\omega)\|^2 | \mathscr{B}(t)) = 0(\Delta)$$

Hence clearly:

$$\frac{1}{\Delta} E((\int_t^{t+\Delta} dZ_0(s;\omega))(\int_t^{t+\Delta} dZ_0(s;\omega))^* | \mathscr{B}(t)) = \frac{1}{\Delta}\int_t^{t+\Delta} D(s)D(s)^* ds + 0(\Delta^{1/2})$$

Hence (2.66) follows.

<u>Lemma:</u> Let

$$E(e(t;\omega)\, e(t;\omega)^*) = P(t)$$

Then

$$\operatorname*{Lim}_{\Delta \to 0} (1/\Delta) E(\int_t^{t+\Delta} dZ_s(\sigma;\omega))(\int_t^{t+\Delta} dZ_0(s;\omega))^*) | \mathscr{B}(t))$$

$$= P(t)C(t)^* + B(t)D(t)^* \quad \text{a.e.} \tag{2.68}$$

<u>Proof</u> Since $Z_0(s;\omega)$ satisfies (2.57) and $Z_s(t;\omega)$ satisfies (2.56), it only remains to calculate (2.68). We have seen that:

$$Z_s(t+\Delta;\omega) - Z_s(t;\omega) = \int_t^{t+\Delta} A(s)e(s;\omega)ds + \int_t^{t+\Delta} B(s)dW(s;\omega)$$

$$- e(t+\Delta;\omega) + e(t;\omega)$$

It is immediate that

$$|E((\int_t^{t+\Delta} A(s)\, e(s;\omega)ds\ (\int_t^{t+\Delta} dZ_0(s;\omega))^* | \mathcal{B}(t))| = 0(|\Delta|^{3/2})$$

Next for any $\Delta > 0$, $e(t+\Delta;\omega)$ is uncorrelated with $Y(s;\omega)$, $s \leq (t+\Delta)$, and hence with $Z_0(s;\omega)$, $s \leq (t+\Delta)$. It is also uncorrelated with (and hence also independent of) the random variables generating $\mathcal{B}(t)$. Hence

$$E(e(t+\Delta;\omega)\ (\int_t^{t+\Delta} dZ_0(s;\omega))^* | \mathcal{B}(t)) = E(e(t+\Delta;\omega))E(\int_t^{t+\Delta} dZ_0(s;\omega))^* = 0$$

Since

$$Z_0(s;\omega) = \int_0^s C(\sigma)e(\sigma;\omega)d\sigma + \int_0^s D(\sigma)dW(\sigma;\omega)$$

we have:

$$E(e(t;\omega)\ (\int_t^{t+\Delta} dZ_0(s;\omega))^* | \mathcal{B}(t)) = \int_t^{t+\Delta} E(e(t;\omega)e(s;\omega)^* | \mathcal{B}(t))C(s)^* ds$$

$$= \int_t^{t+\Delta} E(e(t;\omega)\ e(s;\omega)^*)\ C(s)^* ds$$

and

$$E(\int_t^{t+\Delta} B(s)dW(s;\omega)\ (\int_t^{t+\Delta} dZ_0(s;\omega))^* | \mathcal{B}(t)) = \int_t^{t+\Delta} B(s)D(s)^* ds + 0(\Delta^{3/2})$$

since

$$E\left(\int_t^{t+\Delta} B(s)dW(s;\omega)\left(\int_t^{t+\Delta} C(s)e(s;\omega)ds\right)^* \Big| \mathcal{B}(t)\right) = 0(\Delta^{3/2})$$

Hence (2.68) follows by taking the limit as Δ goes to zero. Note that a.e. in (2.68) is necessary since we cannot (at this stage) assert that $e(t;\omega)$ is continuous in t. Eventually we shall show that it is. (Cf. (2.83)).

Lemma: Under the assumption that

$$D(s)D(s)^* > 0, \quad 0 \leq s \leq 1$$

we can write:

$$\hat{x}(t;\omega) = \int_0^t k(t;s)\, dY(s;\omega)\ ; \quad 0 \leq t \leq 1 \tag{2.69}$$

where

$$\int_0^1 \int_0^t |k(t;s)|^2 ds\, dt < \infty \tag{2.69a}$$

Proof Let us first note that

$$\int_0^t f(s)dY(s;\omega) = \int_0^t (f(s)D(s) + h(s))\, dW(s;\omega)$$

where

$$h(s) = \int_s^t f(\sigma)C(\sigma)\phi(\sigma)d\sigma \;\; \phi(s)^{-1} \;B(s)$$

Define the operator L by

$$Lf = g \; ; \;\; g(s) = f(s)D(s) + \int_s^t f(\sigma)C(\sigma)\phi(\sigma)d\sigma \;\phi(s)^{-1}\; B(s)$$

Then it follows that

$$E((\int_0^t f(s)dY(s;\omega)) \; (\int_0^t q(s)\, dY(s;\omega))^*) \; = \int_0^t p(s)\, q(s)^* ds \qquad (2.70)$$

where

$$L^*Lf = p$$

Here f(.) is an m-by-q matrix function, and p(.) is m-by-q consistent with the choice of dimensions on p. 37 . Specifically:

$$L^*h = p \;\; ; \; p(s) = h(s)D(s)^* + \int_0^s h(\sigma)B(\sigma)^*\phi(\sigma)^{*-1}d\sigma \;\; \phi(s)^* \, C(s)^* \qquad (2.71)$$

where h(.) is m-by-n and p(.) is m-by-q.

Next

$$E(x(t;\omega)\ (\int_0^t q(s)dY(s;\omega))^*)$$

$$= E((\phi(t)\int_0^t \phi(s)^{-1}B(s)dW(s;\omega))\ (\int_0^t q(s)dY(s;\omega))^*)$$

$$= \int_0^t r(s)\ q(s)^*\ ds$$

where

$$r = L^* v \quad \text{and} \quad v(s) = \phi(t)\ \phi(s)^{-1}\ B(s), \quad 0 \leq s \leq t \tag{2.72}$$

Now

$$E(x(t;\omega) - \int_0^t k(t;s)dY(s;\omega))\ (\int_0^t q(s)dY(s;\omega))^* = \int_0^t z(s)\ q(s)^*\ ds$$

where

$$L^*L\,k - L^*v = z$$

where k stands for the function $k(t;\cdot)$. Hence for (2.69) to hold it is necessary and sufficient that

$$z(s) = 0\ ; \quad \text{or,} \quad L^*L\,k = L^*v \tag{2.73}$$

But because $D(s)D(s)^*$ is positive, L^*L has a bounded inverse, so that the first part of our result that there exists a function $k(t;s)$

satisfying (2.69) and such that

$$\int_0^t |k(t;s)|^2 ds < \infty$$

for each $0 < t < 1$, is immediate. We need however to show that the double integral in (2.69a) makes sense and that it is finite. For this we proceed to a closer examination of (2.73). Thus we note that if $Lf = g$, we can write:

$$g(s) = f(s)D(s) - \int_0^s f(\sigma)\, C(\sigma)\, \phi(\sigma) d\sigma\, \phi(s)^{-1} B(s)$$
$$+ \int_0^t f(\sigma)\, C(\sigma)\, \phi(\sigma) d\sigma\, \phi(s)^{-1} B(s)$$

The point in doing this is that the first two terms are independent of t. Again if we denote by $\tilde{L}$ the operator yielding the first two terms:

$$\tilde{L} f = g\ ; \quad f(s)D(s) - \int_0^s f(\sigma)\, C(\sigma)\, \phi(\sigma)\, d\sigma\, \phi(s)^{-1} B(s) = g(s) \qquad (2.74)$$

we have

$$L^*L\, f = L^*\tilde{L}f + \left(\int_0^t f(\sigma)C(\sigma)\, \phi(\sigma)\, d\sigma\right). (\phi(s)^{-1}B(s)D(s)^*$$
$$+ R(s)\, \phi(s)^*\, C(s)^*) \qquad (2.75)$$

where

$$R(s) = \int_0^s \phi(\sigma)^{-1} B(\sigma) B(\sigma)^* \phi(\sigma)^{*-1} d\sigma$$

Also, the function $r(\cdot)$ in (2.72) can be expressed:

$$r(s) = \phi(t) \phi(s)^{-1} B(s)D(s)^* + \phi(t)R(s)\phi(s)^*C(s)^* \tag{2.76}$$

Hence (2.73) can be written:

$$h(t;s) = -\int_0^t k(t;\sigma)C(\sigma)\phi(\sigma) d\sigma \; u(s) + \phi(t) u(s)$$

where

$$h(t;\cdot) = L^*\tilde{L} (k(t;\cdot))$$

$$u(s) = \phi(s)^{-1} B(s) D(s)^* + R(s) \phi(s)^* C(s)^*$$

We now exploit the fact that $h(t;s)$ factors into a function of time and a function of s:

$$h(t;s) = (-\int_0^t k(t;\sigma)C(\sigma)\phi(\sigma) d\sigma + \phi(t)) u(s)$$

But since $D(s)D(s)^*$ is assumed positive, we note that L^* has a bounded inverse in $L_2(0, 1)$, and so does $\tilde{L}$; moreover if

$$L^* h = p$$

we have

$$h(s) = h_1(s)D(s)$$

$$p(s)(D(s)D(s)^*)^{-1} = h_1(s) + \int_0^s h_1(\sigma)D(\sigma)\ B(\sigma)^*\phi(\sigma)^{*-1}d\sigma\ \phi(s)^*C(s)^*$$

Since the right side is 'identity plus Volterra operator', we can use a Neumann expansion to find the inverse; and similarly for $\tilde{L}$. But since each of these operations does not involve t, it is readily seen that this implies that it is possible to express $k(t;s)$ as

$$k(t;s) = k_1(t)k_2(s) \text{ (where } k_1(\cdot) \text{ is m-by-m)} \tag{2.77}$$

Hence substituting this into (2.73), and taking advantage of the forms in (2.75) and (2.76) we must have:

$$L^*\tilde{L}k_2 = u(\cdot);\quad u(s) = \phi(s)^{-1}B(s)D(s)^* + R(s)\phi(s)^*C(s)^*,\ 0 < s < 1$$

And $k_1(t)$ must satisfy:

$$\phi(t) - \int_0^t k_1(t)k_2(s)C(s)\phi(s)ds = k_1(t)$$

But since $\phi(t)$ is nonsingular, it follows that both $k_1(t)$ and

$$(I + \int_0^t k_2(s)C(s)\phi(s)ds)$$

are nonsingular, and hence

$$k_1(t) = \phi(t)\left(I + \int_0^t k_2(s)C(s)\phi(s)ds\right)^{-1} \quad (2.78)$$

from which (2.69a) follows. Of course we have obtained more than we sought to prove.

Now we can prove one of the main theorems (7, 11).

Theorem 2.7 Under the assumption that

$$D(s)D(s)^* > 0$$

for every s, $0 \leq s \leq 1$, we have for every t, $0 \leq t \leq 1$,

$$\mathscr{B}(t) = \text{smallest sigma algebra generated by } \{Z_0(s;\omega), s \leq t\} \quad (2.79)$$

Proof Let us recall that

$$Z_0(s;\omega) = Y(s;\omega) - \int_0^s C(\sigma)\hat{x}(\sigma;\omega)\, d\sigma \quad (2.80)$$

Hence for any q-by-q matrix function $f(\cdot)$ in (appropriate dimensional) $L_2(0,t)$ space, we have:

$$\int_0^t f(s)dZ_0(s;\omega) = \int_0^t f(s)dY(s;\omega) - \int_0^t f(s)C(s)\hat{x}(s;\omega)ds$$

But using (2.69)

$$\int_0^t f(s)C(s)\hat{x}(s;\omega)ds = \int_0^t f(s)C(s)\int_0^s k(s;\sigma)\, dY(\sigma;\omega)ds$$

$$= \int_0^t \left(\int_\sigma^t f(s)C(s)k(s;\sigma)ds\right)dY(\sigma;\omega)$$

Introduce now the operators:

$$H\, f = g; \quad g(\sigma) = f(\sigma) - \int_\sigma^t f(s)C(s)k(s;\sigma)ds\, , \quad 0 \leq \sigma \leq t$$

which maps $L_2(0,t)$ into itself. More importantly, in view of (2.69a) it differs from the identity operator by a Volterra operator with a square integrable kernel. Hence H has a bounded inverse.

Hence for any $g(\cdot)$ in $L_2(0,t)$,

$$\int_0^t g(s)dY(s;\omega) = \int_0^t f(s)dZ_0(s;\omega)\, , \quad f = H^{-1}\, g$$

Hence the random variables

$$\int_0^t g(s)dY(s;\omega)$$

are measurable with respect to the smallest sigma algebra generated by $\{Z_0(s;\omega), s \leq t\}$, and hence $B(t)$ is contained in that algebra. This proves (2.79) since obviously the rightside of (2.79) is contained in $B(t)$.

Theorem 2.8

$$Z_s(t;\omega) = \int_0^t (P(s)C^*(s) + B(s)\dot{D}(s)^*)(D(s)D(s)^*)^{-1}dZ_0(s;\omega) \tag{2.81}$$

Proof First of all, using (2.59), taking $Z_1(t;\omega)$ therein to be $Z_s(t;\omega)$, and $Z_2(t;\omega)$ to be $Z_0(t;\omega)$, and making use of (2.66), (2.68), we have that the conditional expectation of $Z_s(t;\omega)$ with respect to the smallest sigma algebra generated by $\{Z_0(s;\omega), s \leq t\}$, is given by the right side of (2.81). But this algebra, by Theorem 2.7 is the same as $B(t)$, and $Z_s(t;\omega)$ is of course measurable with respect to $B(t)$. Hence (2.81) follows.

Finally we note that:

Corollary 1. Let $\hat{R}(t) = E(\hat{x}(t;\omega)\hat{x}(t;\omega)^*)$

Then $\hat{R}(t)$ is absolutely continuous and,

$$\dot{\hat{R}}(t) = A(t)\hat{R}(t) + \hat{R}(t)A^*(t) + (P(t)C(t)^* + B(t)D(t)^*)(D(t)D(t)^*)^{-1}(C(t)P(t)+D(t)B(t)^*) \tag{2.82}$$

Proof We have only to note that by the theorem,

$$\hat{x}(t;\omega) = \int_0^t A(s)\hat{x}(s;\omega) + \int_0^t (P(s)C(s)^* + B(s)D(s)^*)(D(s)D(s)^*)^{-1} dZ_0(s;\omega) \tag{2.83}$$

and the result follows by specializing (2.13), and using (2.66).

Corollary 2. P(t) is absolutely continuous and $P(0) = 0$, and

$$\dot{P}(t) = A(t)P(t)+P(t)A(t)^*+B(t)B(t)^* - (P(t)C(t)^*+B(t)D(t)^*)(D(t)D(t)^*)^{-1} (C(t)P(t)+D(t)B(t)^*) \tag{2.84}$$

Proof We have only to note that

$$P(t) = E(x(t;\omega)x(t;\omega)^*) - E(\hat{x}(t;\omega)\hat{x}(t;\omega)^*)$$

and using (2.13) and (2.82), the result follows.

The equations (2.80), (2.83), (2.84) are the Kalman filtering equations.

Problem Let $\Psi(t)$ be a fundamental solution of the matrix equation:

$$\dot{\Psi}(t) = (A(t) - K(t)C(t))\,\Psi(t)$$

where

$$K(t) = (P(t)C(t)^* + B(t)D(t)^*)\ (D(t)D(t)^*)^{-1} \tag{2.85}$$

Then the function $k(t;s)$ in (2.69) is given by

$$k(t;s) = \Psi(t)\ \Psi(s)^{-1}\ K(s) \tag{2.86}$$

Hint: Substitute (2.80) into (2.83).

Problem In the notation of (2.86) we have

$$P(t) = \int_0^t \Psi(t)\ \Psi(s)^{-1}[B(s)B(s)^* - B(s)D(s)^*\ (D(s)D(s)^*)^{-1}D(s)$$

$$+K(s)C(s)P(s)]\ \Psi(s)^{*-1}\Psi(t)^*ds$$

Hint Use (2.84)

Time-Invariant Systems: Asymptotic Behavior:

Let us now specialize to the case where the system is 'time-invariant': that is, where the matrices $A(t)$, $B(t)$, $C(t)$, $D(t)$ are all constant, independent of t, and let us denote them by A, B, C, D. Of particular interest then is what happens to $P(t)$, $R(t)$ as t goes to infinity - the situation in regard to

$$R(t) = E[x(t;\omega)\, x(t;\omega)^*]$$

is straight-forward; and in fact, motivates the question regarding the others. First of all, note that we have (equation (2.13)):

$$\dot{R}(t) = A\,R(t) + R(t)\,A^* + BB^*; \quad R(0) = 0 \tag{2.13}$$

Also, this can be 'solved' explicitly, (being a linear equation), as:

$$R(t) = \int_0^t e^{A(t-s)}\, BB^*\, e^{A^*(t-s)}\, ds = \int_0^t e^{As}\, BB^*\, e^{A^*s} ds \tag{2.87}$$

Note that from this we also have:

$$\dot{R}(t) = e^{At}\, BB^*\, e^{A^*t} \geq 0$$

Now (2.87) converges as $t \longrightarrow \infty$ if the eigen-values of $(A + A^*)$ are all negative, strictly less than zero. For

$$\frac{d}{dt}[e^{At}x,\ e^{At}x] = [(A + A^*)\, e^{At}x, e^{At}x] \leq \lambda[e^{At}x, e^{At}x]$$

where λ is the largest eigen-value of $(A+A^*)$ and hence clearly

$$\|e^{At}x\|^2 \leq e^{\lambda t}\|x\|^2$$

or, $\|\cdot\|$, denoting operator norm of a matrix,

$$\|e^{At}\|^2 \leq e^{\lambda t}$$

and hence

$$\|e^{As}BB^* e^{A^*s}\| \leq \|BB^*\| \ \|e^{As}\|^2 \leq \|BB^*\| \ e^{\lambda s},$$

so that (2.87) converges as t goes to infinity. Actually, we can replace this requirement [of 'stability' of the A-matrix] by a weaker condition that

$$[(A+A^*)x, x] \leq -|\lambda|[x,x] \quad \text{for } x \in \bigcup_t (\text{Range } e^{At}B)$$

but since this is quite transparent and only adds to the notation, we shall forego this, and stick to requiring that A be stable. A related question is whether the limit, denoted $R(\infty)$ is non-singular; or, equivalently, for stable A, whether (A, B) is 'controllable', that is to say

$$B, AB, \ldots A^{m-1}B$$

are linearly independent (as linear operators), m being the dimension of A. Suppose now we assume (A, B) is controllable so that $R(\infty)$ is non-singular. Then from (2.13) we have

$$0 = A R(\infty) + R(\infty) A^* + B B^*$$

so that also:

$$0 = KA + A^* K + K BB^* K \tag{2.88}$$

where

$$K^{-1} = R(\infty) \tag{2.89}$$

Next, we shall confine ourselves to the case where 'signal and noise' are independent, namely:

$$B D^* = 0 \tag{2.90}$$

Now from

$$R(t) = P(t) + \hat{R}(t)$$

it follows that $R(t)$ is non-singular as soon as $P(t)$ is. Conversely, suppose P(t) is singular. Say:

$$v^* P(t) v = 0 \qquad \text{for some } m \times 1 \text{ vector } v.$$

This implies that

$$v^* x(t;\omega) = v^* \hat{x}(t;\omega)$$

But left side

$$= \int_0^t v^* e^{A(t-s)} B \, dW(s;\omega)$$

and right side

$$= \int_0^t v^* k(t,s)\, C(s)\, x(s)ds + \int_0^t v^* k(t,s)\, D\, dW(s;\omega) \tag{2.91}$$

where the first term:

$$= \int_0^t \int_\sigma^t v^* k(t,s)\, C(s)\, e^{As}\, ds\, e^{-A\sigma} B\, dW(\sigma;\omega)$$

Since

$$B D^* = 0$$

it follows that we must have

$$v^* k(t, s) D = 0 \qquad 0 < s < t$$

or,

$$v^* k(t, s) D D^* = 0$$

or since DD^* is non-singular,

$$v^* k(t, s) = 0 \qquad 0 < s < t$$

Hence both terms in (2.91) are zero. Hence so must

$$v^* e^{A(t-s)}B = 0 \qquad 0 < s < t \tag{2.92}$$

But this implies that (A, B) is not controllable. Also from (2.92) it follows that

$$v^* R(t) = 0$$

But since $R(s)$ is monotone increasing, we must have

$$v^* R(s) = 0 , \qquad 0 < s < t$$

Hence also

$$v^* P(s) = 0 \qquad 0 < s < t.$$

Hence R(t) and P(t) are singular or non-singular together. Note also that R(t) is non-singular for any $t > 0$, implies that (A, B) is controllable. Next let us note that P(t) satisfies:

$$\dot{P}(t) = A\,P(t) + P(t)A^* + BB^* - P(t)\,C^*(DD^*)^{-1}C\,P(t)$$

In what follows we may set

$$DD^* = I$$

without loss of generality (simply redefine C by $(\sqrt{DD^*})^{-1}C$)

Hence we write

$$\dot{P}(t) = A\,P(t) + P(t)\,A^* + BB^* - P(t)\,C^*C\,P(t) \tag{2.93}$$

Note that we can rewrite this as

$$\dot{P}(t) = (A-P(t)C^*C)\,P(t) + P(t)\,(A-P(t)C^*C) + BB^* + P(t)\,C^*C\,P(t) \tag{2.94}$$

Before we consider the asymptotic properties, we shall indicate a constructive method for solving (2.92), based essentially on Wonham [9].

An obvious iteration based on (2.94) would be

$$\dot{P}_{n+1}(t) = (A-P_n(t)C^*C)P_{n+1}(t) + P_{n+1}(t)\,(A-P_n(t)C^*C)$$

$$+ BB^* + P_n(t)\,C^*C\,P_n(t) \qquad (2.95)$$

[If we based the iteration on (2.93), we would have

$$P_{n+1}(t) = R(t) - \int_0^t e^{A(t-s)}(P_n(s)\,C^*C\,P_n(s))\,e^{A^*(t-s)}ds \qquad (2.96)$$

and here it does not follow that $P_{n+1}(t)$ is non-negative.]

We shall show that (2.95) is actually the Newton-Raphson equation - solving algorithm of

$$P(t) = R(t) - \int_0^t e^{A(t-\sigma)}\,P(\sigma)\,C^*C\,P(\sigma)\,e^{A^*(t-\sigma)}d\sigma \qquad (2.97)$$

For our purposes it seems most convenient to work with (2.97). Let C denote the Banach space of symmetric-matrix functions P(·) such that they are continuous on the closed interval $[0,\infty]$ [i.e., approach limits at $+\infty$], and vanishing at the origin. Then

$$Q(P) = q \qquad q(t) = \int_0^t e^{A(t-\sigma)} P(\sigma)\ C^*C\ P(\sigma) e^{A^*(t-\sigma)} d\sigma$$

clearly maps C into C. Moreover if we denote the Frechet derivative 'at' P(·) by $\mathscr{L}(P)$, we have

$$\mathscr{L}(P)\, P = 2\, Q(P)$$

$$\mathscr{L}(P)\, f = g\ ; \quad g(t) = \int_0^t e^{A(t-\sigma)} [f(\sigma) C^*C\ P(\sigma) + P(\sigma)\ C^*C\ f(\sigma)] e^{A^*(t-\sigma)} d\sigma$$

The Newton-Raphson iteration would then become:

$$\mathscr{L} P_{n+1} = P_n - (I + \mathscr{L}(P_n))^{-1} (P_n + Q(P_n) - R) \tag{2.98}$$

where P_n stands for the function $P_n(t)$, R for the function R(t). We shall show that for a suitable choice of the initial function $P_0(t)$, $(I + \mathscr{L}(P_n))^{-1}$ does have a bounded inverse. We shall now actually compute the inverse. Our main aim is of course to prove that P_n converges to P in C, for suitable initial choice. Let

$$(I + \mathscr{L}(P))\, g = f$$

Then we have that $(f(t) - g(t))$ is absolutely continuous and

$$\frac{d}{dt}(f(t) - g(t)) = A(f(t) - g(t) + (f(t) - g(t))A^* + g(t)\, C^*C\, P(t) + P(t) C^*C\, g(t)$$

or, assuming $f(t)$ is differentiable, so is $g(t)$, and

$$\dot{g}(t) = \dot{f}(t) - A\, f(t) - f(t)\, A^* - g(t)\, C^*C\, P(t) - P(t)\, C^*C\, g(t)$$

$$+ A\, g(t) + g(t)\, A^*$$

Let us now note that for our iteration (2.98), we have:

$$P_n - P_{n+1} + 2\, Q(P_n) - \mathscr{L}(P_n)\, P_{n+1} = P_n + Q(P_n) - R$$

or

$$(I + \mathscr{L}(P_n))\, P_{n+1} = Q(P_n) + R \tag{2.99}$$

or

$$\dot{P}_{n+1} = \dot{Q}(P_n) + \dot{R} - A(Q(P_n) + R) - (Q(P_n) + R)A^* - P_{n+1}\, C^*C\, P_n$$

$$- P_n\, C^*C\, P_{n+1} + A\, P_{n+1} + P_{n+1}\, A^*$$

But

$$\dot{Q}(P_n) = A\,Q(P_n) + Q(P_n)A^* + P_nC^*CP_n$$

$$\dot{R}(t) = AR(t) + R(t)A^* + BB^*$$

Hence

$$\dot{P}_{n+1} = BB^* + P_nC^*CP_n - P_{n+1}C^*CP_n - P_nC^*CP_{n+1} + AP_{n+1} + P_{n+1}A^*$$

$$= (A - P_nC^*C)P_{n+1} + P_{n+1}(A - P_nC^*C)^* + BB^* + P_nC^*CP_n \qquad (2.100)$$

We can then state the following

Theorem: Suppose A is stable. Suppose further that C, A is observable that is to say that the matrix:

$$\int_0^{\infty} e^{A^*t}\, C^*C\, e^{At}\, dt$$

is nonsingular. Denote its inverse by Λ. Then the iteration (2.100) with $P_0(t)$ defined by:

$$P_0(t) = \int_0^t e^{As} \Lambda C^* C \Lambda e^{A^* s} ds$$

yields a sequence of (real) symmetric non-negative definite matrix functions in the space $C[0, \infty]$, vanishing at the origin. This sequence is moreover monotone:

$$P_{n+1}(t) \leq P_n(t)$$

and converges to $P(t)$, the unique solution of (2.92) in any finite interval of the form $[0, T]$. Moreover $P(t)$ converges as t goes to infinity, and denoting the limit by $P(\infty)$, we have:

$$0 = A P(\infty) + P(\infty)A^* + BB^* - P(\infty)C^*CP(\infty) \tag{2.101}$$

Further

$$A - P(\infty)C^*C$$

is stable (has all eigenvalues with strictly negative real parts). Finally, $P(\infty)$ is non-singular if and only if $R(\infty)$ is (or A-B is controllable).

Proof First let us note that

$$0 = A^*(\Lambda^{-1}) + (\Lambda^{-1}) A + C^*C$$

and hence:

$$0 = A\Lambda + \Lambda A^* + \Lambda C^*C \Lambda \tag{2.102}$$

and hence: $P_0(t)$ converges as t goes to infinity, and

$$P_0(\infty) = \int_0^\infty e^{At} \Lambda C^*C \Lambda e^{A^*t} \, dt = \Lambda$$

Next we need

Lemma 1: Let K be a non-negative definite real symmetric matrix such that

$$0 \geq (A-KC^*C)K + K(A-KC^*C)^* + \gamma BB^* + \mu KC^*CK \tag{2.103}$$

where

$$\gamma \geq 0; \quad \mu > 0 \text{ (strictly positive!)}$$

Then the eigenvalues of $(A-KC^*C)$ have strictly negative real parts.

Proof Let H denote $(A-KC^*C)^*$. Suppose for some nonzero vector z,

$$Hz = \lambda z \;; \quad \text{Real part } \lambda = \sigma \geq 0$$

Then substituting in (2.103), we have

$$0 \geq 2\sigma \quad [Kz, z] + \gamma \| B^* z\|^2 + \mu \| CKz\|^2$$

and hence

$$CKz = 0$$

which implies that

$$Hz = A^*z$$

and hence

$$[(A + A^*) z, z] = [(H + H^*)z, z] = 2\sigma[z, z]$$

so that σ must be strictly negative.

<u>Lemma 2.</u> (Wonham [9]): Let K, P denote two real symmetric, non-negative definite matrices and let

$$\Psi(K;P) = (A-KC^*C)P + P(A-KC^*C)^* + BB^* + KC^*CK$$

Then

$$\Psi(K;P) \geq \Psi(P;P) \tag{2.104}$$

Proof We have only to note that we can write:

$$\begin{aligned}\Psi(K;P) &= \Psi(P;P) + (P-K)C^* CP + PC^*C(P-K) - PC^*CP + KC^*CP \\ &= \Psi(P;P) + (K-P)C^*C(K-P) \\ &\geq \Psi(P;P)\end{aligned}$$

as required.

Lemma 3. Suppose $P_n(t)$ is a real symmetric non-negative definite matrix function uniformly continuous on $[0,\infty]$, with $P(0)$ equal to zero. Further suppose

$$(A-P_n(\infty)C^*C)$$

is stable.

Define $P_{n+1}(t)$ by:

$$\begin{aligned}\dot{P}_{n+1}(t) &= (A-P_n(t)C^*C)P_{n+1}(t) + P_{n+1}(t)(A-P_n(t)C^*C)^* + BB^* \\ &\quad +P_n(t)C^*CP_n(t)\, P_{n+1}\end{aligned} \tag{2.105}$$

Then $P_{n+1}(t)$ has the same properties as $P_n(t)$.

Proof Since we are given that $P_n(t)$ converges as t goes to infinity, and that

$$A - P_n(\infty)C^*C$$

is stable, it follows that for all $t > T$, T sufficiently large, the eigenvalues of

$$A - P_n(t)C^*C$$

have also all strictly negative real parts, say all less than equal to σ, where σ is negative. Hence if $\phi(t)$ denotes a fundamental matrix solution of:

$$\dot{\phi}(t) = (A - P_n(t)C^*C)\,\phi(t)$$

we have that $t > s > T$:

$$\|\phi(t)\phi(s)^{-1}x\| \leq \|x\| \exp \sigma(t-s) \tag{2.106}$$

Next let us note that we can express the solution of (2.105) as:

$$P_{n+1}(t) = \int_0^t \phi(t)\phi(s)^{-1}\,\theta(s)\,\phi(s)^{*-1}\phi(t)^* ds \tag{2.107}$$

where

$$\theta(s) = BB^* + P_n(s)C^*CP_n(s)$$

and the main thing to note is that $\theta(s)$ is convergent at infinity. From (2.107) it is immediate that $P_{n+1}(t)$ is non-negative definite. Let $\epsilon > 0$ be given. Then we can find T large enough so that (2.106) holds and in addition

$$\|(\theta(t_2)-\theta(t_1))x\| \leq \epsilon \|x\|, \quad t_1, t_2 \geq T$$

Next let us note that for Δ sufficiently large so that, $\Delta \geq T$ and

$$\|(P_n(t)C^*C - P_n(\infty)C^*C)x\| \leq \epsilon\|x\|, \quad t \geq \Delta$$

we have

$$\|\phi(s+\Delta)\,\phi(\Delta)^{-1}x - e^{(A-P_n(\infty)C^*C)s}x\| \leq \epsilon\,\|x\|$$

For, setting

$$\dot{x}(t) = (A-P_n(t)C^*C)x(t); \; x(\Delta) = x$$

we have on the one hand

$$x(\Delta+s) = \phi(\Delta+s)\phi(\Delta)^{-1}x$$

while also:

$$x(\Delta+s) = \int_{\Delta}^{\Delta+s} e^{(A-P_n(\infty)C^*C)(\Delta+s-t)} (P_n(\infty)C^*C - P_n(t)C^*C)x(t)dt$$

$$+ e^{(A-P_n(\infty)C^*C)s} x$$

A simple estimation of the integral using (2.106) verifies (2.108); then we can write:

$$P_{n+1}(t) = \int_0^t \phi(t)\phi(t-s)\,\theta(s)\,\phi(t-s)^{*-1}\,\phi(t)^* ds$$

(2.108) and hence:

$$P_{n+1}(t_2) - P_{n+1}(t_1) = \int_0^T e^{(A-P_n(\infty)C^*C)s} (\theta(t_2-s)$$

$$- \theta(t_1-s))e^{(A-P_n(\infty)C^*C)^*s} ds$$

+ terms which go to zero with $T \to \infty$ by virtue of our estimates as can be directly verified.

The first term, for $t_2 \geq t_1 \geq 2\Delta$, is less than (in norm)

$$\varepsilon / 2 |\sigma|$$

Hence $P_{n+1}(t)$ converges as t goes to infinity. Hence from (2.105), $\dot{P}_{n+1}(t)$ also converges, and hence must have zero for its limit. Hence we have (in the notation of Lemma 2):

$$0 = \Psi(P_n(\infty); P_{n+1}(\infty)) \geq \Psi(P_{n+1}(\infty); P_{n+1}(\infty)) \tag{2.109}$$

by Lemma 2; and the last inequality, by Lemma 1 implies that

$$A - P_{n+1}(\infty) C^* C$$

is stable. This completes proof of the Lemma.

Next let us note that $P_0(t)$ satisfies the conditions of Lemma 3. For, from (2.102) we have:

$$0 = (A - \Lambda C^* C)\Lambda + \Lambda(A - \Lambda C^* C)^* + 3\Lambda C^* C C$$

and by Lemma 1, this implies the stability of $(A - \Lambda C^* C)$. Of course $P_0(t)$ converges as $t \to \infty$. Hence (2.100) yields the kind of sequence asserted. The monotonicity follows from Lemma 3. Thus following

Wonham as in [9], we have:

$$\dot{P}_n(t) - \Psi(P_n(t);P_n(t)) \geq \dot{P}_n(t) - \Psi(P_{n-1}(t);P_n(t))$$

$$= 0$$

$$= \dot{P}_{n+1}(t) - \Psi(P_n(t);P_{n+1}(t))$$

so that

$$\dot{P}_n(t) - \dot{P}_{n+1}(t) \geq \Psi(P_n(t);P_n(t)) - \Psi(P_n(t);P_{n+1}(t))$$

$$= H_n(t)(P_n(t)-P_{n+1}(t)) + (P_n(t) - P_{n+1}(t))H_n(t)^*$$

But if

$$\dot{Z}(t) \geq H_n(t)Z(t) + Z(t)H_n(t)^*$$

we must have:

$$\frac{d}{dt}\left(\phi(t)^{-1}Z(t)\,\phi(t)^{*-1}\right) \geq 0$$

as can be verified by calculation. Since $Z(0)$ is zero, it follows that

$$Z(t) \geq 0$$

Hence $P_n(t)$ is monotone non-increasing. In particular we have

$$P_{n+1}(\infty) \leq P_n(\infty)$$

Because $P_n(t)$ is non-negative definite, and the sequence $P_n(t)$ is monotone, we have that $P_n(t)$ converges for every t in the closed interval $[0, \infty]$. From (2.105), $\dot{P}_n(t)$ also converges, and it is evident that, denoting the limit of $P_n(t)$ by $P(t)$, $\dot{P}(t)$, $\dot{P}_n(t)$ converges to $\dot{P}(t)$, and $P(t)$ is of course the unique solution of (2.92). We have thus obtained a constructive method for solving (2.92); particularly noteworthy is the monotonic nature of the approximating sequence. Let us now examine the asymptotic properties.

Lemma 4: Under the condition that A is stable, (2.101) has a unique solution, in the class of real symmetric matrices.

Proof Let P, Q, denote two real symmetric matrix solutions of (2.101). First of all, by Lemma 1, both $(A-PC^*C)$ and $(A-QC^*C)$ are stable. Substituting P, Q into equation (2.101) we have, upon subtraction:

$$A(P-Q) + (P-Q)A^* + QC^*CQ - PC^*CP = 0$$

Let z be an eigen-vector of (P-Q):

$$(P-Q)z = \lambda z, \quad \lambda \text{ must of course be real.}$$

Then we have:

$$\lambda\ [(A+A^*)z, z] + \|CQz\|^2 - \|CPz\|^2 = 0$$

But

$$\|CQz\|^2 - \|CPz\|^2 = [C(P+Q)z,\ C(Q-P)z]$$

$$= (-\lambda)\ [C^*C(P+Q)z, z]$$

Hence if λ is not zero,

$$[(A+A^*)z, z] - [C^*C(P+Q)z, z] = 0 = [(A-C^*CP)z, z] + [(A^*-QC^*C)z, z]$$

which contradicts the stability condition. Hence all eigen-values must be zero, or P must equal Q.

Next let us note that we have also a constructive method for solving (2.101). Thus we have (see (2.109)):

$$0 = (A-P_n(\infty)C^*C)P_{n+1}(\infty) + P_{n+1}(\infty)(A-P_n(\infty)C^*C)^* + BB^* + P_n(\infty)C^*CP_n(\infty)$$

which is a linear equation for determining $P_{n+1}(\infty)$ from $P_n(\infty)$, and taking

$$P_0(\infty) = \Lambda$$

we have further that

$$P_{n+1}(\infty) \leq P_n(\infty)$$

so that $P_n(\infty)$ converges monotonically to the solution of (2.101).

Finally, that $P(t)$ converges to $P(\infty)$ as t goes to infinity follows from:

Lemma 5. The solution $P(t)$ of (2.95) with $P(0) = 0$, is actually monotonic non-decreasing as t increases.

Proof We follow Wonham [7]. Thus let (in the notation of Lemma 2)

$$\dot{\tilde{P}}(t) = \Psi(P(t+\tau); \tilde{P}(t)) \; ; \; \tilde{P}(0) = 0 \; ; \; \tau > 0 \text{ and fixed.}$$

Then from Lemma 2, we have:

$$\dot{P}(t) \leq \Psi(P(t+\tau);P(t))$$

and hence

$$\dot{\tilde{P}}(t) - \dot{P}(t) \geq \Psi(P(t+\tau); \tilde{P}(t)) - \Psi(P(t+\tau); P(t)))$$

which, just as in the proof of the monotonicity of the sequence $P_n(\cdot)$ in Lemma 3, implies that

$$\tilde{P}(t) \geq P(t)$$

But

$$\tilde{P}(t) = \int_0^t \phi(t+\tau)\,\phi(s+\tau)^{-1}(BB^* + P(s+\tau)C^*CP(s+\tau))\phi(s+\tau)^{*-1}\phi(t+\tau)^* ds$$

and by an obvious change of variable in the integrand, this is

$$= \int_\tau^{t+\tau} \phi(t+\tau)\,\phi(s)^{-1}(BB^* + P(s)C^*CP(s))\phi(s)^{*-1}\phi(t+\tau)^* ds$$

$$= P(t+\tau) - \int_0^\tau \phi(t+\tau)\phi(s)^{-1}(BB^* + P(s)C^*CP(s))\phi(s)^{*-1}\phi(t+\tau)^* ds$$

$$\leq P(t+\tau)$$

Hence

$P(t) \leq P(t+\tau)$ as required.

Hence $P(t)$ converges as t goes to infinity to the unique solution of (2.101). Finally suppose $P(\infty)$ is singular. Then by Lemma 5, so is $P(t)$ for every t, and as we have seen, this implies that (A-B) is not controllable.

A. V. Balakrishnan

References

1. A. N. Kolmogorov: Foundations of the Theory of Probability, Chelsea, 1950.

2. J. L. Doob: Stochastic Processes, John Wiley and Sons, 1953.

3. L. I. Gikhman and A. V. Skorokhod: Introduction to the Theory of Random Processes, W. B. Saunders, 1969.

4. K. Parthasarathy: "Probability Measures on Metric Spaces", Academic Press, 1967.

5. L. Shepp: Radon-Nikodym Derivatives of Gaussian Measures: Annals of Mathematical Statistics, 1966.

6. H. McKean: Stochastic Integrals, Academic Press.

7. W. M. Wonham: "Random Differential Equations,in Control Theory", in Probabilistic Methods in Applied Mathematics, Volume 2, Academic Press, 1970.

8. E. Nelson: "Dynamical Theories of Brownian Motion", Princeton University Press, 1967.

9. W. M. Wonham: "On a Matrix Riccati Equation of Stochastic Control", SIAM Journal on Control, Volume 6, No. 4, 1968.

10. M. Coeve: Probability Theory, Van Nostrand, 1954.

11. I. Gohberg and M. G. Krein: Volterra Operators, AMS Translations, 1970.

Example: Linear Stochastic Control

Let us next consider stochastic control problems for the linear system:

$$x(t;\omega) = \int_0^t A(s)\, x(s;\omega)ds + \int_0^t B(s)\, u(s)ds + \int_0^t F(s)dW(s;\omega),$$

$$0 \le t \le 1$$

$$Y(t;\omega) = \int_0^t C(s)\, x(s;\omega)ds + \int_0^t G(s)dW(s;\omega), \quad 0 \le t \le 1$$

where we assume that all the coefficients are continuous on $[0, 1]$, and

$$G(s)G(s)^* > 0$$

and of course $W(s;\omega)$ is a Wiener process as before. The control problem is that of finding an optimal control function $u(t)$, $u(t)$ being measurable $\beta_Y(t)$, so as to minimize:

$$\int_0^1 E([Q(s)x(s;\omega),\ x(s;\omega)])ds + \lambda \int_0^1 E[u(t),\ u(t)]\ dt \tag{2.110}$$

where $Q(s)$ is continuous in s and is non-negative definite, and λ is a fixed positive constant. Since $u(t)$ is now also a random process, let us denote it by $u(t;\omega)$. It is implicit that $u(t;\omega)$ is jointly measurable in t and ω.

Whatever the choice of $u(t;\omega)$, let

$$\hat{x}(t;\omega) = E(x(t;\omega) \mid \beta_Y(t))$$

Then we have the Kalman filter equations characterizing $\hat{x}(t;\omega)$:

$$\hat{x}(t;\omega) = \int_0^t A(s)\hat{x}(s;\omega)ds + \int_0^t (P(s)C(s)^* + F(s)G(s)^*)(G(s)G(s)^*)^{-1} dZ_0(s;\omega)$$

$$+ \int_0^t B(s)u(s;\omega)ds \qquad (2.111)$$

where

$$Z_0(s;\omega) = Y(s;\omega) - \int_0^s C(\sigma)\hat{x}(\sigma;\omega)d\sigma \qquad (2.112)$$

and is a Wiener process with covariance $G(s)G(s)^*$. (Cf equations (2.83), (2.66)).

The matrix $P(s)$ is determined by: (Cf. (2.84)):

$$\dot{P}(t) = A(t)P(t) + P(t)A(t)^* + F(t)F(t)^* - (P(t)C(t)^* + F(t)G(t)^*)$$

$$(G(t)G(t)^*)^{-1} (C(t)P(t) + G(t)F(t)^*) \qquad (2.113)$$

and in particular does not depend on the control. Next in (2.110) we can write

$$E([Q(s)x(s;\omega), x(s;\omega)]) = Tr.\ Q(s)\ C((\hat{x}(s;\omega) + e(s;\omega))(\hat{x}(s;\omega) + d(s;\omega))^*)$$

$$= Tr.\ Q(s)\ E(\hat{x}(s;\omega)\ \hat{x}(s;\omega)^*) + Tr.\ Q(s)\ P(s) \tag{2.114}$$

where

$$e(s;\omega) = x(s;\omega) - \hat{x}(s;\omega)$$

The point in doing this is that the problem is thus reduced to that of choosing $u(t;\omega)$ so as to minimize:

$$\int_0^1 E([Q(s)\hat{x}(s;\omega), \hat{x}(s;\omega)]\ ds + \lambda \int_0^1 E([u(t;\omega),\ u(t;\omega)]\ dt \tag{2.115}$$

where $x(t;\omega)$ satisfies (2.111). Here we shall exploit our knowledge of deterministic control problems. Thus, let us fix the sample point ω, and consider the problem of minimizing:

$$\int_0^1 [Q(s)\hat{x}(s;\omega),\ \hat{x}(s;\omega)]ds + \lambda \int_0^1 [u(s;\omega),\ u(s;\omega)]\ ds \tag{2.116}$$

for each ω. For this purpose it is convenient to write

$$v(t;\omega) = \int_0^t (P(s)C(s)^* + F(s)G(s)^*)(G(s)G(s)^*)^{-1}\ dZ_0(s;\omega)$$

$$w(t;\omega = \int_0^t \phi(t)\ \phi(s)^{-1}\ d\ v(s;\omega)$$

where $\phi(t)$ is a fundamental matrix solution of

$$\dot{\phi}(t) = A(t)\,\phi(t)$$

and finally:

$$\hat{x}(t;\omega) = \int_0^t \phi(t)\phi(s)^{-1}B(s)\,u(s;\omega)\,ds + w(t;\omega) \tag{2.117}$$

Note that $w(t;\omega)$ is continuous in t (as we have seen). Since ω is fixed, we consider the control functions as functions of t alone for the moment. Let $L_2(0,1)$ denote the usual L_2 space for control functions $u(t)$. Introduce the linear bounded operator on this space:

$$Lf = g\ ;\quad g(t) = \int_0^t \phi(t)\phi(s)^{-1}B(s)\,f(s)ds, \quad 0 \leq t \leq 1$$

Then, we can rewrite (2.116) as:

$$[Q(Lu + w),\ (Lu + w)] + \lambda\,[u, u] \tag{2.118}$$

where Q stands for the operator corresponding to multiplication by $Q(s)$, u stands for $u(t)$, w for $w(t;\omega)$, and inner-products in two spaces have been used. Being a quadratic form with λ positive, it is obvious by a routine first variation (gradient with respect to u) that the unique minimum is given by

$$\lambda u_0 + L^*Q L u_0 = - L^*Q w \tag{2.119}$$

where L^* denotes the adjoint of L, and is given by:

$$L^*f = h \; ; \; h(t) = \int_t^1 B(t)^* \phi(t)^{*-1} \phi(s)^* f(s)ds \quad 0 \leq t \leq 1$$

From (2.119) we have:

$$u_0(t;\omega) = -B(t)^* z(t;\omega)/\lambda \tag{2.120}$$

where

$$z(t;\omega) = \int_t^1 \phi(t)^{*-1} \phi(s)^* Q(s)\hat{x}(s;\omega)ds \tag{2.121}$$

and is the unique solution of:

$$\dot{z}(t;\omega) = -A(t)^* z(t;\omega) - Q(t) \hat{x}(t;\omega) \; ; \; z(1;\omega) = 0 \tag{2.122}$$

Here (2.120) is the best 'open loop' solution. It is unfortunately NOT measurable $\beta_Y(t)$, since, as is evident from (2.121), $z(t;\omega)$ is not measurable $\beta_Y(t)$ (being independent to the sigma-algebra generated by $Z_0(t;\omega)$, the latter being the same as $\beta_Y(t)$).

Next we observe that

$$\dot{P}(t) + P(t)A(t) + A(t)^* P(t) + Q(t) - P(t)B(t)B(t)^*P(t)/\lambda = 0 \tag{2.123}$$

subject to $P(1) = 0$

has a unique non-negative solution which we shall denote by $P_c(t)$. We shall now show that we have the decomposition:

$$u_0(t;\omega) = u_1(t;\omega) + u_2(t;\omega)$$

where

$$u_1(t;\omega) = -B(t)^* P_c(t) \hat{x}(t;\omega)/\lambda \tag{2.124}$$

$$u_2(t;\omega) = \frac{-B(t)^*}{\lambda} \int_t^1 \phi(t)^{*-1} \phi(s)^* P_c(s) dv(s;\omega) \tag{2.125}$$

This follows from the directly verifiable relation (using (2.111), (2.120), (2.122) and (2.123) in differential form:

$$d(z(t;\omega) - P_c(t) \hat{x}(t;\omega)) = -\left(A(t)^* - \frac{P_c(t)B(t)B(t)^*}{\lambda}\right)(z(t;\omega) - P_c(t)\hat{x}(t;\omega))$$
$$-P_c(t)dv(t;\omega)$$

and since

$$z(1;\omega) = 0; \quad P_c(1) = 0$$

(2.125) follows. Note now that $u_1(t;\omega)$, and $u_2(t;\omega)$ are independent of each other; $u_1(t;\omega)$ is measureble $\beta_Y(t)$, while $u(t;\omega)$ is independent of it. Hence

$$u_1(t;\omega) = E[u_0(t;\omega) | \beta_Y(t)] \tag{2.126}$$

It is apparent that $u_1(t;\omega)$ should be the optimal Stochastic Control. We shall now prove this more formally. Let us consider the pre-Hilbert space H_u of control functions $u(t;\omega)$ such that it is jointly measurable in t and ω and further $u(t;\omega)$ is measurable $\beta_Y(t)$ for each t, with inner product defined by

$$[u, v] = \int_0^T E[u(t;\omega), v(t;\omega)] dt \tag{2.127}$$

Then the main thing to note is that if we define the linear transformation L by:

$$Lu = v; \quad v(t;\omega) = \int_0^t \phi(t) \phi(s)^{-1} B(s) u(s;\omega) ds$$

the adjoint is given by:

$$L^*y = q; \quad q(t;\omega) = E [\int_t^1 \phi(t)^{*-1} \phi(s)^* y(s;\omega) ds \mid \beta_Y(t)] \tag{2.128}$$

It is clear also that the functional (2.115) rewritten in the form

$$[Q(Lu + w), Lu + w] + \lambda [u, u]$$

yields a quadratic form over Hu with the unique solution:

$$\frac{-B(t)^*}{\lambda} E(z(t;\omega) \mid \beta_Y(t))$$

where $z(t;\omega)$ is the unique solution of:

$$\dot{z}(t;\omega) = -A(t)^* z(t;\omega) - Q(t)\hat{x}(t;\omega); \quad z(1;\omega) = 0$$

Let

$$\tilde{z}(t;\omega) = E(z(t;\omega) \mid \beta_Y(t))$$

Then we have, in differential notation:

$$d(z(t;\omega) - P_c(t)\,\hat{x}(t;\omega)) = -(A^* - P_c BB^*/\lambda)\,(z(t;\omega) - P_c(t)\hat{x}(t;\omega))$$

$$-(P_c BB^*/\lambda)\,(z(t;\omega) - \tilde{z}(t;\omega)) - P_c(t)dv(t;\omega)$$

Because $z(1;\omega)$ and $P_c(1)$ vanish and

$$E((z(s;\omega) - \tilde{z}(s;\omega)) \mid \beta_Y(t)) = 0 \quad \text{for} \quad s \geq t$$

it follows that:

$$E(z(t;\omega) - P_c(t)\hat{x}(t;\omega))/\beta_Y(t)) = 0$$

thus proving the optimality of

$$\frac{-B(t)^*}{\lambda} P_c(t) \hat{x}(t;\omega)$$

Note that the filtering and control can thus be treated separately - this is referred to as the 'separation' principle - see Wonham [7]. It was first derived by Joseph and Tou [12]. Our treatment is quite different from both of these.

A. V. Balakrishnan

References

1. A. N. Kolmogorov: Foundations of the Theory of Probability, Chelsea, 1950.

2. J. L. Doob: Stochastic Processes, John Wiley and Sons, 1953.

3. L. I. Gikhman and A. V. Skorokhod: Introduction to the Theory of Random Processes, W. B. Saunders, 1969.

4. K. Parthasarathy: "Probability Measures on Metric Spaces", Academic Press, 1967.

5. L. Shepp: Radon-Nikodym Derivatives of Gaussian Measures: Annals of Mathematical Statistics, 1966.

6. H. McKean: Stochastic Integrals, Academic Press.

7. W. M. Wonham: "Random Differential Equations,in Control Theory", in Probabilistic Methods in Applied Mathematics, Volume 2, Academic Press, 1970.

8. E. Nelson: "Dynamical Theories of Brownian Motion", Princeton University Press, 1967.

9. W. M. Wonham: "On a Matrix Riccati Equation of Stochastic Control", SIAM Journal on Control, Volume 6, No. 4, 1968.

10. M. Coeve: Probability Theory, Van Nostrand, 1954.

11. I. Gohberg and M. G. Krein: Volterra Operators, AMS Translations, 1970.

12. P. D. Joseph and J. T. Tou: "On Linear Control Theory", AIEE Transactions, Applications and industry, Vol. 80, p. 193-196, 1961.

A. V. Balakrishnan

Identification and Adaptive Control: An Example

As an example of an adaptive control problem we shall take the flight control problem outlined by Taylor and Rediess [1]. The problem here is the design of an automatic control system for a hypothetical aerospace vehicle which operates throughout a wide range of flight conditions that alter its dynamic characteristics.

A functional block diagram of the basic aircraft, control servoactuator, and measurement dynamics is shown in figure A-1. The description of these elements given here is necessarily simplified, but it represents many of the important characteristics that would have to be considered in the preliminary design of a stability augmentation system for this type of aircraft. Only longitudinal modes of response are considered, and the phugoid mode is assumed to be negligible.

1. Basic Dynamics and Measurements

The basic dynamical equations for command and windgust inputs are:

$$\dot{\alpha} = Z_\alpha \alpha + \dot{\theta} + Z_{\delta e}\delta_e + Z_\alpha \frac{w_g}{\bar{V}}$$

$$\ddot{\theta} = M_\alpha \alpha + M_{\dot{\theta}}\dot{\theta} + M_{\delta_e}\delta_e + M_\alpha \frac{w_g}{\bar{V}}$$

$$n_Z = \frac{\bar{V}}{g}(\dot{\theta}-\dot{\alpha})$$

The quantity w_g is the random vertical gust velocity and is assumed to have zero mean with spectral density given by

$$G_w(\omega) = \frac{2\sigma^2\omega_c}{\omega^2+\omega_c^2}$$

where ω is angular frequency

$$\sigma = 5.0$$

$$\omega_c = \bar{v}/1000 \text{ rad/sec}$$

The control servoactuator is taken approximately as a first-order lag plus an output rate limit, with dynamic equations:

$$\dot{\delta}_e = S(-k\,\delta_e + k\cdot\delta_{e_c})$$

$$k = 20 \text{ rad/sec}$$

$$\begin{aligned} S(x) &= x \quad |x| \leq 0.5 \\ &= 0.5 \quad x > 0.5 \\ &= -0.5 \quad x < -0.5 \end{aligned}$$

and δ_{e_c} is the total servoactuator command signal including the pilot's command.

Observations

The following quantities are observed: the pitch rate $[\dot{\theta}]$, the pitch attitude $[\theta]$, the normal acceleration $[n_z]$ and the angle of attack $[\alpha]$. Denoting the observations by the subscript i, we have:

$$\theta_i = \dot{\theta} - \lambda_{\theta_1} \dot{\nu}_1 - \lambda_{\theta_2} \dot{\nu}_2 + \theta_n$$

where

$$\lambda_{\theta_1} = -0.025 \text{ rad/ft}$$

$$\lambda_{\theta_2} = +0.040 \text{ rad/ft}$$

$\dot{\theta}_n$ is white Gaussian with spectral density:

$$0.0005 \text{ (rad/sec)}^2$$

ν_1 and ν_2 are the first and second bending mode deflections (at a specific reference station) and are governed by the equations:

$$\ddot{\nu}_1 + (2\xi_1\omega_1)\,\dot{\nu}_1 + (\omega_1)^2\nu_1 = \Psi\,\delta_e$$

$$\ddot{\nu}_2 + (2\zeta_2\omega_2)\,\dot{\nu}_2 + (\omega_2)^2\nu_2 = \Psi\,\delta_e$$

where

$$\Psi = 4\,\bar{v}\,Z_{\delta_e}$$

The frequencies ω_1 and ω_2 and the damping ratios ζ_1 and ζ_2 are assumed normally distributed with means and variances listed in Table 2, in [1].

Indicated pitch attitude

$$\theta_i = \theta - \lambda_{\theta_1}\,\nu_1 - \lambda_{\theta_2}\,\nu_2 + \theta_n$$

where θ_n is white Gaussian with spectral density $(.0001)^2$.

Indicated normal acceleration:

$$n_{Z_i} = n_Z + \frac{\ell_z}{g}\,\ddot{\theta} + \frac{\phi_1}{g}\,\ddot{\nu}_1 + \frac{\phi_2}{g}\,\ddot{\nu}_2 + n_{z_n}$$

n_{Z_n} is white Gaussian with spectral density $(0.01)^2$ and

$$\ell_z = 10$$

$$\phi_1 = -\,0.15$$

$$\phi_2 = +\,0.45$$

Indicated Angle of Attack

$$\alpha_i = k_\alpha\left[\dot{\alpha} + \frac{w_g}{\bar{v}} + \frac{\ell_\alpha}{\bar{v}}\,\dot{\theta}\right] - \lambda_{\alpha_1}\nu_1 - \lambda_{\alpha_2}\nu_2$$

where

$$\lambda_{\alpha_1} = -0.075$$

$$\lambda_{\alpha_2} = +0.19$$

$$\ell_\alpha = 32 \text{ ft}$$

Note that angle of attack measurement may be taken to be noise free. It is assumed that servoactuator position and rate can be measured with no error and that the measurements are available.

State Space Formulation

We now turn to a state space formulation of both the dynamics and the observations. We do this as follows:

$$\alpha = x_1$$

$$\theta = x_2$$

$$\theta = x_3$$

$$\delta_e = x_4$$

$$y = n_Z$$

$$M_\alpha = M_1$$

$$M_\theta = M_3$$

$$M_{\delta e} = M_4$$

$$Z_{\delta e} = Z_4$$

$$Z_\alpha = Z_1$$

$$\xi = w_g/\bar{v}$$

$$u = \text{control}$$

$$\dot{x}_1 = Z_1 x_1 + x_3 + Z_4 x_4 + Z_1 \xi$$

$$\dot{x}_2 = x_3$$

$$\dot{x}_3 = M_1 x_1 + M_3 x_3 + M_4 x_4 + M_1 \xi$$

$$\dot{x}_4 = 20\, S(u + \delta_p - x_4)$$

$$S(x) = x \quad -.5 < x < .5$$

$$= .5, \qquad x > .5$$

$$= -.5, \qquad x < -.5$$

$$y = \frac{\bar{v}}{g}(x_3 - \dot{x}_1)$$

$$= \frac{\bar{v}}{g}[x_3 - Z_1 x_1 - x_3 - Z_4 x_4 - Z_1 \xi]$$

Observations (Measurements)

Given

$$\nu_1(t) = \frac{4\bar{v}\, Z_4}{\omega_1\sqrt{1-\zeta_1^2}} \int_0^t e^{-\zeta_1\omega_1(t-\sigma)} \sin \omega_1\sqrt{1-\zeta_1^2}(t-\sigma)\, x_4(\sigma)\, d\sigma$$

$$\nu_2(t) = \frac{4\bar{v}\, Z_4}{\omega_2\sqrt{1-\zeta_2^2}} \int_0^t e^{-\zeta_2\omega_2(t-\sigma)} \sin \omega_2\sqrt{1-\zeta_2^2}\,(t-\sigma)\, x_4(\sigma)\, d\sigma$$

Ind. Pitch Rate

$$v_1 = x_3 + (0.025)\,\dot{\nu}_1 - (0.040)\,\dot{\nu}_2 + N_1(.0005)$$

Ind. Pitch Attitude

$$v_2 = x_2 + (0.025)\,\nu_1 - (0.040)\,\nu_2 + (0.0001)\,N_2$$

Ind. Normal Accel.

$$v_3 = y + \frac{10}{g}\dot{x}_3 + \frac{(-.15)}{g}\ddot{\nu}_1 + \frac{(.45)}{g}\ddot{\nu}_2 + (0.01)\,N_3$$

$$= y + \frac{10}{g}\,[M_1x_1 + M_3x_3 + M_4x_4 + M_1\xi] - \frac{(0.15)}{g}[\ddot{\nu}_1] + \frac{(0.45)}{g}\ddot{\nu}_2 + (0.01)\,N_3$$

Ind. Angle of Attack

$$v_4 = k_1\left[x_1 + \xi + \frac{32}{\bar{v}}x_3\right] + (0.075)\,\nu_1 - (0.19)\,\nu_2 + (0.00005)N_4$$

$$k_1 = k_\alpha$$

Let

$$4\bar{v}\,Z_4\tilde{\nu}_1 = \nu_1 \qquad 4\bar{v}\,Z_4\tilde{\nu}_2 = \nu_2$$

$$v_1 = x_3 + (4\bar{v}\,Z_4)\,(0.025\,\dot{\tilde{\nu}}_1 - 0.040\,\dot{\tilde{\nu}}_2) + (.005)\,N_1$$

$$v_2 = x_2 + (4\bar{v}\,Z_4)\,(0.025\,\tilde{\nu}_1 - 0.040\,\tilde{\nu}_2) + (0.001)\,N_2$$

$$v_3 = \frac{\bar{v}}{g}\,[x/3 - Z_1x_1 - x/3 - z_4x_4 - Z_1x_5]$$

$$+ \frac{10}{g}\,[M_1x_1 + M_3x_3 + M_4x_4 + M_1x_5]$$

$$+ \left[\frac{(-0.15)}{g}\ddot{\tilde{\nu}}_1 + \frac{(0.45)}{g}\ddot{\tilde{\nu}}_2\right](4\bar{v}\,Z_4) + (.01)\,N_3$$

$$= x_1\left(\frac{10M_1}{g} - \frac{\bar{v}z_1}{g}\right) + x_3\left[\frac{10M_3}{g}\right]$$

$$+ x_4\left(\frac{10M_4}{g} - z_4\frac{\bar{v}}{g}\right)$$

$$+ x_5\left[\frac{10M_1}{g} - \frac{z_1\bar{v}}{g}\right]$$

$$+ (.01)\, N_3 + (4\bar{v}\, Z_4) \left(\frac{0.15}{g}\ddot{\tilde{\nu}}_1 + \frac{.45}{g}\ddot{\tilde{\nu}}_2\right)$$

$$v. = k_1 x_1 + k_1 x_5 + k_1 \frac{32}{\bar{v}} x_3 + 0.05\, \ddot{\tilde{\nu}}_1 - 0.19\, \ddot{\tilde{\nu}}_2$$

Now ξ is a random process with zero mean and spectral density:

$$p_\xi(f) = \frac{50}{} \frac{1}{\bar{v}^2} \frac{\bar{v}}{(1000)} \frac{1}{4\ ^2 f^2 + \bar{v}^2 10^{-6}}$$

$$= \left(\frac{5}{(100)v}\right) \frac{1}{4\pi^2 f^2 + \bar{v}^2/10^6}$$

$$= \left(\frac{25}{\bar{v}^2}\right) \frac{2\bar{v}\, 10^{-3}}{4\pi^2 f^2 + \bar{v}^2/10^6}$$

The corresponding correlation function is

$$\left(\frac{25}{\bar{v}^2}\right) \exp - \frac{\bar{v}t}{1000}$$

Writing

$$\xi = x_5$$

we have the following state space representation for the windgust:

$$\dot{x}_5 = -\frac{\bar{v}}{1000} x_5 + \frac{1}{\sqrt{20\bar{v}}} N_0$$

N_0, N_1, N_2, N_3 are all unit power white Gaussian.

We have then the following state-space representation for the system and the observations:

$$x = \begin{bmatrix} x_1 \\ x_2 \\ x_3 \\ x_4 \\ x_5 \end{bmatrix}$$

$$\dot{x}_1 = Z_1 x_1 + 0\,x_2 + x_3 + Z_4 x_4 + Z_1 x_5$$

$$\dot{x}_2 = 0\,x_1 + 0\,x_2 + x_3 + 0\,x_4 + 0\,x_5$$

$$\dot{x}_3 = M_1 x_1 + 0\,x_2 + M_3 x_3 + M_4 x_4 + M_1 x_5$$

$$\dot{x}_4 = 0\,x_1 + 0\,x_2 + 0\,x_3 + 20\,x(u + \sigma_p - x_4) + 0\,x_5$$

$$\dot{x}_5 = 0\,x_1 + 0\,x_2 + 0\,x_3 + 0\,x_4 + \left(\frac{-\bar{v}}{1000}\right) x_5$$

The observation vector v

$$v = \begin{bmatrix} v_1 \\ v_2 \\ v_3 \\ v_4 \end{bmatrix}$$

is given by

$$v = C\,x + (4\,\bar{v}\,Z_4)\,\tilde{\nu} + G\,N$$

where

$$C = \begin{bmatrix} 0 & 0 & 1 & 0 & 0 \\ 0 & 1 & 0 & 0 & 0 \\ \dfrac{10\,M_1 - \bar{v} Z_1}{g} & 0 & \dfrac{10\,M_3}{g} & \dfrac{10\,M_4 - \bar{v}\,Z_4}{g} & \dfrac{10\,M_1 - \bar{v}\,Z_1}{g} \\ k_1 & 0 & \dfrac{32 k_1}{\bar{v}} & 0 & k_1 \end{bmatrix}$$

$$\tilde{\nu} = \begin{bmatrix} .025\,\tilde{\nu}_1 - 0.040\,\tilde{\nu}_2 \\ .025\,\tilde{\nu}_1 - 0.040\,\tilde{\nu}_2 \\ \dfrac{-0.15}{g}\,\nu_1 + \dfrac{0.45}{g}\,\tilde{\nu}_2 \\ .075\,\tilde{\nu}_1 - 0.19\,\tilde{\nu}_2 \end{bmatrix}$$

$$G = \text{diagonal}\,(.0005,\ .0001,\ .01,\ d_4)$$

$$N = \begin{bmatrix} N_1 \\ N_2 \\ N_3 \\ N_4 \end{bmatrix}$$

where N_1, N_2, N_3, N_4 are unit power white Gaussian.

Note: The dimensions of the various matrices and vectors will change depending on whether it is the Identification Phase, State Filtering Phase or Control Phase although we shall use the same generic notation as A, B, etc. In the main text a subscript (i-for identification, s-for state filtering, c-for control) will serve to distinguish the phases. The notation in the Appendices where the general theory is given will also use the same letters A,B, etc. and it is intended that appropriate specialization will be made to apply to particular cases in the text.

2. System Identification

The first (and less standard) problem is that of estimating system parameters (System Identification) from observed data. Here we can distinguish between closed loop identification and open loop identification. In closed loop identification, the feedback control input alone is used in contrast to open loop identification where it is assumed that we can have a separate input or probe signal. For the present this distinction is unnecessary; we assume that the servo-actuator output can always be measured - in other words, that $x_4(t)$ is known [or can be calculated from the dynamic equations knowing $u(t)$ and $\dot{\delta}_p(t)$]. The main point is that then in the Identification mode we have the representation:

$$x = \begin{bmatrix} x_1 \\ x_2 \\ x_3 \\ x_5 \end{bmatrix}$$

$$\dot{x} = A_i x + B_i x_4(t) + F_i N_0 \qquad (2.1)$$

where

$$A_i = \begin{bmatrix} Z_1 & 0 & 1 & Z_1 \\ 0 & 0 & 1 & 0 \\ M_1 & 0 & M_3 & M_1 \\ 0 & 0 & 0 & \frac{-\bar{v}}{1000} \end{bmatrix}$$

$$B_i = \begin{bmatrix} Z_4 \\ 0 \\ M_4 \\ 0 \end{bmatrix}$$

$$F_i = \begin{bmatrix} 0 \\ 0 \\ 0 \\ \frac{1}{\sqrt{20\bar{v}}} \end{bmatrix}$$

$$v = \begin{bmatrix} v_1 \\ v_2 \\ v_3 \\ v_4 \end{bmatrix} \qquad N = \begin{bmatrix} N_1 \\ N_2 \\ N_3 \\ N_4 \end{bmatrix}$$

$$v = C_i x + (4\bar{v} Z_4)\,\tilde{\nu} + GN + H_i x_4 \qquad (2.1)$$

$$C_i = \begin{bmatrix} 0 & 0 & 1 & .0 \\ 0 & 1 & 0 & 0 \\ \dfrac{10M_1 - \bar{v} Z_1}{g} & 0 & \dfrac{10M_3}{g} & \dfrac{10M_1 - \bar{v} Z_1}{g} \\ k_1 & 0 & \dfrac{32k_1}{\bar{v}} & k_1 \end{bmatrix}$$

$$\tilde{\nu} = \begin{bmatrix} .025\,\tilde{\nu}_1 - 0.040\,\tilde{\nu}_2 \\ .025\,\tilde{\nu}_1 - 0.040\,\tilde{\nu}_2 \\ \dfrac{-0.15}{g}\tilde{\nu}_1 + \dfrac{0.45}{g}\tilde{\nu}_2 \\ .075\,\tilde{\nu}_1 - 0.19\,\tilde{\nu}_2 \end{bmatrix}$$

$$H_i = \begin{bmatrix} 0 \\ 0 \\ \dfrac{10M_4 - \bar{v} Z_4}{g} \\ 0 \end{bmatrix} ; \quad G = \text{Diag} \begin{bmatrix} .0005 & & & \\ & .0001 & & \\ & & .01 & \\ & & & d_4 \end{bmatrix}$$

$$N = \begin{bmatrix} N_1 \\ N_2 \\ N_3 \\ N_4 \end{bmatrix} \quad \text{Ind. white Gaussians with unit power}$$

A general theory of identification for such a system is given in Appendix I. The theory splits into two distinct cases according as to whether d_4 is non-zero ["non-singular case"] or d_4 is zero ['singular case']. Of primary interest (because of an essential simplification) is the 'singular case' where the windgust is high so that in the angle of attack measurement ($v_4(t)$) the error due to measurement is "small in comparison with the windgust component". [Note that this is the only measurement where the windgust enters directly.] Specializing the general treatment therein, we have for our particular problem

$$v_4(t) = k_1x_1 + \frac{32k_1}{\bar{v}} x_3 + k_1x_5 + ((0.075)\tilde{v}_1 - (0.19)\tilde{v}_2)(4\bar{v}Z_4)$$

Hence, if we assume that all the coefficients are known, [as we would in calculating conditional distribution of the observations given the coefficients2 we can actually 'solve' for the windgust component as:

$$x_5 = \frac{v_4 + 4\bar{v}Z_4(0.19\tilde{v}_2 - 0.075\tilde{v}_1)}{k_1} - x_1 - \left(\frac{32}{\bar{v}}\right) x_3 \qquad (2.2)$$

Substituting this into the state equations (2.1), we can 'eliminate' x_5 from the equations and (since $x_4(t)$ is known) thereby obtain a set of equations in x_1, x_2, x_3 only. We can solve the latter, noting that $v_4(t)$ plays the role of a 'forcing term'. We can thus obtain the state vector $x(t)$ exactly; the problem thus reduces to that of identification without state noise, which has been extensively studied [3]. Thus we seek to find the coefficients to minimize:

$$\frac{1}{T}\int_0^T \left\{ \frac{1}{d_1^2}[v_1-(4\bar{v}Z_4)(.025\tilde{v}_1-.040\tilde{v}_2)-x_3]^2 \right.$$

$$+\frac{1}{d_2^2}[v_2-(4\bar{v}Z_4)(.025\tilde{v}_1-.040\tilde{v}_2)-x_2]^2$$

$$+\frac{1}{d_3^2}[v_3-(4\bar{v}Z_4)(\frac{0.45\tilde{v}_2}{g}-\frac{0.15\tilde{v}_1}{g}) - (\frac{10M_1-\bar{v}Z_1}{g})x_1$$

$$\left. -\frac{10M_3}{g}x_3 - (\frac{10M_1-\bar{v}Z_1}{g})(x_5) - (\frac{10M_4-\bar{v}Z_4}{g})\ x_4]^2 \right\} dt$$

where x_5 is determined by (2.2).

This minimization is carried out by a Newton-Raphson technique as indicated in [3], by actually seeking a root of the gradient with respect to the unknown parameters. Note that $x_1(t)$, $x_2(t)$, $x_3(t)$ now contain the unknown parameters, and that at the end of the computation of unknown parameters, we obtain an estimate of the state variables as well. Of course, strictly speaking, the integrals should be written as Ito integrals as indicated in the Appendix I. Note that the equations determining $x_1(t)$, $x_2(t)$, $x_3(t)$ are:

$$\dot{x}_1(t) = (1 - \frac{32Z_1}{\bar{v}})\, x_3(t) + \frac{Z_1}{k_1}(v_4 + 4\bar{v}Z_4(0.19\tilde{v}_2 - 0.075\tilde{v}_1))$$

$$\dot{x}_2(t) = x_3(t)$$

$$\dot{x}_3(t) = (M_3 - \frac{32M_1}{\bar{v}})\, x_3(t) + \frac{M_1}{k_1}(v_4 + 4\bar{v}Z_4(0.19\tilde{v}_2 - 0.075\tilde{v}_1))$$

and the initial conditions at $t = 0$ being taken to be zero.

<u>Non-Singular Case</u>

Let us next consider the case where

$$d_4 > 0$$

so that G is non-singular. By writing

$$G^{-1} v$$

in place of v, we can then clearly assume G to be the Identity matrix, and thereby simplify the notation. We shall now specialize the material in Appendix I to our specific problem. Thus let $m(t)$ denote the system response in the absence of windgust noise and observation noise. In other words,

$$m(t) = C_i \int_0^t e^{A_i(t-\sigma)} B_i x_4(\sigma) d\sigma + H_i x_4(t)$$

Let the matrix P_i be the solution of the steady-state Riccati Equation (since our estimates are good only 'asymptotically' we simplify the calculation in this manner):

$$0 = A_i P_i + P_i A_i^* + FF^* - P_i C_i^* C_i P_i$$

Appendix II indicates the iterative method used for solving this equation. Let

$$h(t) = m(t) - C_i \int_0^t e^{(A_i - P_i C_i^* C)(t-s)} P_i C_i^* m(s) ds$$

$$y(t) = -C \oint_0^t e^{(A_i - P_i C^* C)(t-s)} P_i C_i^* v(s) ds$$

where the 0 indicates Ito integral.

Let

$$\nabla m(t)$$

denote the gradient with respect to all the unknown parameters, and let θ denote the unknown parameter vector, and let

$$\hat{m}(t) = \int_0^t e^{(A_i - P_i C^* C)(t-s)} P_i C_i^* \nabla m(s) ds - \nabla m(t)$$

Then we seek a root of:

$$\frac{1}{T}\int_0^T [y(t)-h(t), \hat{m}(t)]dt + \frac{1}{T}\int_0^T [\hat{m}(t), v(t)]dt = G(\theta) = 0$$

using the Newton-Raphson algorithm:

$$\theta_{n+1} = \theta_n - R^{-1} G(\theta_n)$$

where

$$R = \frac{1}{T}\int_0^T \hat{m}(t)\, \hat{m}(t)^* \, dt$$

and the identifiability conditions insure that R is non-singular for large enough T. We use the given mean values to start the iteration.

3. Control Theory

At the first level, we can assume that we use the estimated system parameters to devise a feedback control to meet the performance requirements [see [1] for the latter]. Here we can make use of standard optimal control theory. Since the given performance requirements cannot be handled directly in the theory, we proceed in the usual way by using appropriate 'soft' criteria. Moreover, we shall assume a linear control system - that in particular the servo-actuator does NOT saturate, so that in reference to the basic state equations (1.1) (1.2), we can write

$$\dot{x}_4(t) = 20 \ (u + \delta_p - x_4(t)) \tag{3.1}$$

This simplification is reasonable; and is necessary in any case if we wish to obtain any useful answers from stochastic control theory.

We then have the following canical linear control problem:

$$x(t) = \int_0^t A_c \, x(s)ds + \int_0^t B_c \, u(s)ds + F_c \, W_0(t) + \int_0^t B_c \, \delta p(s)ds$$

(3.2)

$$\int_0^t v(s)ds = \int_0^t C \ x(s)ds + \int_0^t 4 \, vZ_4 \tilde{v}(s)ds + G \, W(t)$$

where

$$A_c = \begin{bmatrix} Z_1 & 0 & 1 & Z_4 & Z_1 \\ 0 & 0 & 1 & 0 & 0 \\ M_1 & 0 & M_3 & M_4 & M_1 \\ 0 & 0 & 0 & -20 & 0 \\ 0 & 0 & 0 & 0 & \frac{-v}{1000} \end{bmatrix}$$

$$B_c = \begin{bmatrix} 0 \\ 0 \\ 0 \\ 20 \\ 0 \end{bmatrix} \qquad F_c = \begin{bmatrix} 0 \\ 0 \\ 0 \\ 0 \\ \frac{1}{\sqrt{20\bar{v}}} \end{bmatrix}$$

and C and G are as before. We again have two cases to distinguish depending on whether d_4 is zero or not. But first let us formulate the optimization criterion. We wish to choose u(t) such that in the absence of pilot input (that is $\delta_p(t) \equiv 0$) we minimize

$$\lim_{T\to\infty} \left[\frac{1}{T} \int (n_Z(t)^2 + \lambda u(t)^2)dt\right]$$

where λ is some fixed positive constant to allow for the saturation constraint. Now

$$n_Z(t) = \frac{-\bar{v}}{g}(Z_1 x_1 + Z_4 x_4 + Z_1 x_5)$$

so that

$$n_Z(t)^2 = [Q\, x(t),\; x(t)] \tag{3.3}$$

where

$$Q = \frac{\bar{v}^2}{g^2} \begin{bmatrix} Z_1 \\ 0 \\ 0 \\ Z_4 \\ Z_1 \end{bmatrix} \begin{bmatrix} Z_1 & 0 & 0 & Z_4 & Z_1 \end{bmatrix}$$

[For the control analysis we are of course assuming that the system parameters are known.] It is well-known that the optimal control is given by

$$u(t) = -B_c^* P_c \hat{x}(t)/\lambda \tag{3.4}$$

where P_c is the solution of

$$A_c^* P_c + P_c A_c + Q - \frac{P_c B_c B_c^* P_c}{\lambda} = 0 \tag{3.5}$$

This equation is again solved by the iteration technique described in Appendix II.

Note that P_c is independent of the state noise variance. By the separation theorem, $\hat{x}(t)$ is given by the Kalman Filter (for the case where $d_4 > 0$ so that G is non-singular):

$$\hat{x}(t) = \int_0^t (A - P_s C^*(GG^*)^{-1} C)\, \hat{x}(s)ds$$

$$+ P_s C^*(GG^*)^{-1} \oint_0^t v(s)ds + \int_0^t B_c\, u(s)ds \tag{3.6}$$

where (in the steady state) P_s is determined from

$$0 = A_c P_s + P_s A_c^* + FF^* - P_s C^*(GG^*)^{-1} C P_s \tag{3.7}$$

In the case where $d_4 = 0$ (corresponding to high windgust) we note that

$$\hat{x}(t) = x(t) \tag{3.8}$$

as explained in section 2 and in Appendix I. In fact in our current notation letting

$$C = \begin{bmatrix} c_1 \\ c_2 \\ c_3 \\ c_4 \end{bmatrix}$$

where c_1, c_2, c_3, c_4 are 1×5 matrices, and noting that

$$c_4 F_c = k \neq 0$$

we have

$$\dot{x}(t) = (A - F_c\, k^{-1}\, c_4\, A)\, x(t) + (B - F_c\, k^{-1}\, c_4\, B)\, u(t)$$

$$+ F_c\, k^{-1}\, [v_4(t) - 4 \bar{v}\, Z_4 (0.075\, \tilde{v}_1 - 0.19\, \tilde{v}_2)] \tag{3.9}$$

Using the optimal feedback control, the actual value of the normal acceleration component

$$\lim_{T \to \infty} \frac{1}{T} \int_0^T n_Z(t)^2 dt$$

$$= \lim_{T \to \infty} \frac{1}{T} \int_0^T [Q\, x(t),\, x(t)]\, dt$$

$$= \operatorname{Tr} Q J$$

where in the non-singular case

$$J = P_s + J_a \tag{3.10}$$

where J_a is the solution of

$$\left(A - \frac{B_c B_c^* P_c}{\lambda}\right) J_a + J_a \left(A - \frac{B_c B_c^* P_c}{\lambda}\right) + P_s C^* (GG^*)^{-1} C P_s = 0 \tag{3.11}$$

This follows from the fact that in the steady state

$$E[(x(t)x(t)^*] = E[(x(t)-\hat{x}(t))(x(t)-\hat{x}(t)^*)] + E[\hat{x}(t)\hat{x}(t)^*)] = P_s + J_a$$

and

$$\hat{x}(t) = \int_0^t \left(A - \frac{B_c B_c^* P_c}{\lambda}\right) \hat{x}(s)ds + P_s C^* (GG^*)^{-1} (Z_0(t))$$

where $Z_0(t)$ is a Wiener process with

$$E[Z_0(t)\ Z_0(t)^*] = \int_0^t GG^*\ ds$$

In the singular case $(d_4 = 0)$ we have

$$J = J_a$$

and J_a is the solution of

$$\left(A - \frac{B_c B_c^* P_c}{\lambda}\right) J_a + J_a \left(A - \frac{B_c B_c^* P_c}{\lambda}\right)^* + FF^* = 0 \tag{3.12}$$

Note that in the absence of feedback, the normal acceleration due to windgust is given by J_b the solution of

$$A J_b + J_b A^* + FF^* = 0$$

The reduction in decibels is given by

$$10 \text{ Log } (\text{Tr } Q J_a) - 10 \text{ Log } (\text{Tr } Q J_b)$$

and this remains the same even if FF* increases by any multiplicative factor.

Optimization of Step Response

Let us next consider the problem of meeting the requirements on desired response to step input (that is, $\delta_p(t)$ is a step function). Here again we can follow standard optimization theory. Thus let the state dynamics be (setting windgust noise to be zero):

$$\dot{x} = A_c x + B_c \, u(t)$$

we consider the 'deterministic' case where the state is known completely. Our optimization criterion is: Minimize

$$\lim_{T\to\infty} \frac{1}{T}\int_0^T (n_Z(t) - \delta_p(t))^2 dt + \frac{\lambda}{T}\int_0^T |u(t)|^2 dt \tag{3.13}$$

where λ is a positive constant to be chosen appropriately later. The corresponding theory is new with this paper; and is given in Appendix III. The optimal control $u_o(t)$ is given by:

$$u_o(t) = -\frac{B_i^*}{\lambda}[P_c(t)\, x(t) + r(t)]$$

where

$$P_c A_c + A_c^* P_c - \frac{P_c B_c B_c^* P_c}{\lambda} + L^*L = 0 \tag{3.14}$$

L being defined by:

$$n_Z(t) = L\, x(t)$$

$$\left(A^* - \frac{P_c B_i B_i^*}{\lambda}\right) r = L^* \delta_0(t) \tag{3.15}$$

Remembering that

$$Q = L^*L$$

the optimal feedback control is the same as that obtained in the stochastic case. Thus the pilot input $\delta_p(t)$ is shaped by the rule:

$$+ \frac{B_i^*}{\lambda} (A^* - P_c B_c B_c^*/\lambda)^{-1} L^* \delta_p(t) \tag{3.16}$$

and the feedback control remains the same as in the stochastic case. For a unit step function, the corresponding average mean-square 'tracking-error' is

$$\lim_{T \to \infty} \frac{1}{T} \int_0^T (n_Z(t) - \delta_p(t))^2 dt$$

$$= \left(L \left(A_c - \frac{B_c B_c^* P_c}{\lambda} \right)^{-1} \left(\frac{B_c B_c^*}{\lambda} \right) \left(A_c^* - \frac{P_c B_c B_c^*}{\lambda} \right)^{-1} L^* - 1 \right)^2$$

of course this average mean-square error is only the square of the difference between the steady-state output corresponding to the unit step, and the unit step and is not very significant, since the precise steady state value is not important and can be scaled up as desired. However the criterion does yield the same feedback gain as in the stochastic case. Stability is of course guaranteed.

This brings up then the possibility of using a 'shaping filter' with memory. Ideally the frequency transform of this shaping filter should be the inverse of the frequency transform of the feedback system: Denoting the former by K(f), we must have, ideally,

$$K(f) = [L\,[2\pi if - (A_c - B_c B_c^* P_c/\lambda)]^{-1}\left(\frac{B_c B_c^*}{\lambda}\right)(A_c^* - P_c B_c B_c^*/\lambda)^{-1} L^*]^{-1}$$

and this involves differentiators. The problem is thus solved completely, in theory. In practice a suitable approximation can be chosen. It must be noted that the input is no longer a constant so that the considerations of Appendix III for Z(t) now hold, strictly speaking, only asymptotically.

APPENDIX I

A. V. Balakrishnan

IDENTIFICATION THEORY

The identification problem can be formulated as: Given

$$x(t) = \int_0^t A\, x(s)ds + \int_0^t B\, u(s)\, ds + F\, W_0(t)$$

$$v(t) = \int_0^t C\, x(s)ds + \int_0^t D\, u(s)ds + G\, W(t) \qquad A(1)$$

where $W_0(\cdot)$ and $W(\cdot)$ are Wiener processes in the appropriate dimensions, perhaps correlated with each other, $u(\cdot)$ is a given known input, and $v(\cdot)$ is the observed output. It is desired to identify some or all of the parameters in the matrices A, B, C, D, F. The matrix G is a known constant matrix. For most of the analysis we assume G is a non-singular square matrix; we treat the case where G is singular separately. In spite of the considerable literature on Identification [4], such a problem in this generality has not been studied hitherto.

The first question that we wish to answer is, when can we identify such a system? Of course the answer must depend on the notion of identifiability used. While the literature on Identification abounds with many 'recipes', there is often little by way of any measure of goodness of the estimates obtainable. There is not in any case much agreement on what constitutes 'identifiability'. In the present paper we take the following approach which has at least the virtue of being mathematically precise. First we assume there does exist a set of 'true' values for the unknown parameters. Let θ denote the set of unknown parameters; θ then takes its values in some finite-dimensional Euclidean space. Let θ_0 denote the true value. An estimate based on observed data for a time-interval T will be denoted θ_T. We shall say that a system is identifiable if we can find an estimate which is asymptotically unbiassed and is consistent. That is to say:

i) $\lim_{T \to \infty} E(\theta_T) = \theta_0$

ii) θ_T converges with probability one to θ_0 as T goes to infinity.

Such a definition requires a precise formulation of the problem involving 'noise' processes, of course. Fortunately, the practical Flight Control problem shows that this is not unrealistic. In the present paper we shall show that under certain conditions, which we term 'identifiability conditions', it is possible to find such a class of estimates for the problem under consideration. Moreover, we can develop a computational algorithm for generating such an estimate. Our estimate is a maximum-likelihood estimator; or, more correctly (and in order that the difference in approach can be emphasized) it is a root of the gradient of the likelihood functional. Because we are dealing with 'continuous' data, the term 'likelihood functional' will have to be clarified. We shall first consider the case where G is non-singular; without loss of generality, we can clearly take it to be the Identity; which we do. The main thing to note then is that measure induced by the process v(•) for any finite time-interval [0, T] on the [Banach] space C of continuous functions on [0, T] in the usual manner, is absolutely continuous with respect to the Wiener measure thereon. This is true for any assumed value for θ. By the 'likelihood functional' we mean the corresponding Radon-Nikodym derivative. We shall show that under the 'identifiability conditions', there is a non-zero neighborhood of θ_0 in which the gradient of the likelihood functional has a root for all T bigger than some T_0. We shall take such a root as the estimate θ_T, and show that it is asymptotically unbiassed and is consistent, provided the identifiability conditions are satisfied. We begin then with the calculation of the R-N derivative for the case where G is the identity matrix. For this we note that the system (1) can be rewritten with:

$$\tilde{v}(t) = v(t) - \int_0^t m(s)ds$$

where

$$m(t) = D\, u(t) + C \int_0^t e^{A(t-s)} B\, u(s)ds \qquad A(2)$$

in the form:

$$\begin{aligned} x(t) &= \int_0^t A\, x(s)ds + F\, W_0(t) \\ \tilde{v}(t) &= \int_0^t C\, x(s)ds + W(t) \end{aligned} \qquad A(3)$$

Mainly for notational simplicity we shall assume $W_0(.)$ and $W(.)$ are independent. The fact that G is the identity matrix will imply that the process $v(\cdot)$ is absolutely continuous with respect to Wiener measure. However this can be explicitly demonstrated by using the well-known Kalman filtering equations for the system (3). Thus let

$$\hat{x}(t) = E[x(t) \mid \tilde{v}(s), \quad s \leq t]$$

Then we have:

$$\hat{x}(t) = \int_0^t A\,\hat{x}(s)\,ds + \int_0^t P(s)\,c^*\,dW(s) \qquad (4)$$

$$\tilde{v}(t) = \int_0^t C\,\hat{x}(s)\,ds + W(s) \qquad (5)$$

where $W(\cdot)$ is the Wiener process and $P(\cdot)$ is the unique solution of:

$$\dot{P}(t) = AP(t) + P(t)A^* - P(t)C^*CP(t) + FF^*; \; P(0) = 0 \qquad A(6)$$

Multiplying the differential form of (5) by $P(t)C^*$, we have:

$$\int_0^t P(s)\,C^*\,d\tilde{v}(s) = \int_0^t P(s)C^*C\,x(s)ds + \int_0^t P(s)C^*\,dW(s) \qquad A(7)$$

and substituting for the last term on the eight using (4), we have finally:

$$\hat{x}(t) - \int_0^t (A-P(s)C^*C)\,\hat{x}(s)ds = \int_0^t P(s)C^*\,d\,\tilde{v}(s)$$

Letting $\phi(t)$ to be a fundamental matrix solution of

$$\dot{\phi}(t) = (A - P(t)C^*C)\,\phi(t)$$

we have:

$$\hat{x}(t) = \Phi(t)\int_0^t \Phi(s)^{-1}P(s)C^* \, d\tilde{v}(s)$$

and hence

$$W(t) = \tilde{v}(t) - \int_0^t K(t;s)d\tilde{v}(s) \qquad A(8)$$

where:

$$K(t;s) = \int_s^t C\,\Phi(\sigma)\,d\sigma\,\Phi(s)^{-1}\,P(s)C^*$$

In the differential form, (8) becomes:

$$dW(t) = d\,\tilde{v}(t) - \int_0^t L(t;s)\,d\,\tilde{v}(s)\,dt;\; L(t,s) = C\Phi(t)\Phi(s)^{-1}P(s)C^*$$

Let

$$y(t) = -\int_0^t L(t;s)dv(s)$$

$$h(t) = -\int_0^t L(t;s)m(s)ds + m(t)$$

<u>Theorem</u> The Radon-Nikodym derivative of the measure induced by the process $v(\cdot)$ on C with respect to Wiener measure is given by:

$$H(v) = -\frac{1}{2}\left\{\int_0^T \|y(t) - h(t)\|^2 dt + 2\int_0^T [y(t)-h(t), dv(t)]\right\} \qquad A(9)$$

where the second integral is an Ito integral, and $v(\cdot) \in C$.

<u>Remark</u> Note that the use of the Ito integral in (9) eliminates the determinant used in [2].

We proceed next to the gradient equation. For this, let θ denote the vector of unknown parameters.
We shall show that for large T it is enough to seek a root of

$$\frac{1}{T}\int_0^T [\hat{m}(t), (y(t) - h(t))]\, dt + \frac{1}{T}\int_0^T [\hat{m}(t), dv(t)] \qquad A(10)$$

where

$$\int_0^t L(t;s)\ \nabla_\theta\, m(s)ds - \nabla_\theta\, m(t) = \hat{m}(t)$$

and ∇_θ denotes the gradient with respect to θ. Let

$$m_i(t) = \int_0^t + L(t;s) \frac{\partial}{\partial\theta_i}\, m(s)ds - \frac{\partial}{\partial\theta_i}\, m(t)$$

where θ_i denotes the ith unknown parameter. Then the main 'identifiability condition' is that the matrix with components

$$\frac{1}{T}\int_0^T [m_i(t), m_j(t)]\, dt$$

be positive definite in the limit as T goes to infinity. Using this matrix a Newton-Raphson technique for finding the root can be readily developed. Further it can be shown, for example, that for our particular problem, the identifiability condition is satisfied under an appropriate almost-periodic character imposed on the input in open loop mode.

Singular Case: Let us now consider the case (corresponding to large windgust component)

$$v(t) = \begin{matrix} v_1(t) \\ v_2(t) \end{matrix}$$

$$D = \begin{matrix} D_1 \\ D_2 \end{matrix}$$

and

$$C = \begin{matrix} C_1 \\ C_2 \end{matrix}$$

with

$$v_1(t) = C_1 \int_0^t x(s)ds + D_1 \int_0^t u(s)ds$$

$$v_2(t) = C_2 \int_0^t x(s)ds + D_2 \int_0^t u(s)ds + W_2(t)$$

where $W_2(t)$ is a Wiener process. Then of course the measure induced by the process $v(t)$ cannot be absolutely continuous with respect to Wiener measure. But:

Theorem: Suppose C_1F is non-singular, that is to say:

$$(C_1F)^* \ C_1F > 0$$

Denote this (square) matrix by K. Then

$$\hat{x}(t) = x(t)$$

being the unique solution of

$$x(t) = \int_0^t (A - FK^{-1}(C_1F)^*C_1A)x(s)ds + \int_0^t (B - FK^{-1}(C_1F)^*C_1B)u(s)ds$$

$$+FK^{-1}(C_1F)^*(\dot{v}_1(t) - D_1u(t)) \quad ; \ x(0) = 0 \qquad A(11)$$

Thus $x(t)$ is known without error and the measure induced by $v_2(t)$, conditioned on $x(t)$ being given, with respect to Wiener measure is given by:

$$\exp - \frac{1}{2} \left\{ \int_0^T \| C_2x(s) + D_2u(s) \|^2 ds - 2 \int_0^T [C_2x(s)+D_2u(s), dv_2(s)] \right\} \qquad A(12)$$

where the second integral is an Ito integral.

The minimization can proceed as before; indeed, the calculations are simpler.

APPENDIX II

A. V. Balakrishnan

In this Appendix we indicate the iterative technique used to find the solution of the steady state Riccati equation

$$A P + P A^* + Q - P C^* CP = 0$$

where Q is non-negative definite and we assume A is stable. The iteration is:

$$A P_{n+1} + P_{n+1} A^* + Q - P_n C^*C P_n = 0$$

This is a linear equation for P_{n+1}; moreover let us assume that the system is observable so that

$$\int_0^\infty e^{A^*t} C^*C \, e^{At} \, dt$$

is actually non-singular, and denote the inverse by Λ, so that

$$A^*\Lambda^{-1} + \Lambda^{-1} A + C^*C = 0$$

Then choose

$$P_0 = \Lambda$$

with this choice the approximating sequence P_n is actually a monotone decreasing sequence of non-negative definite matrices, the limit being the solution sought. For a proof see [5].

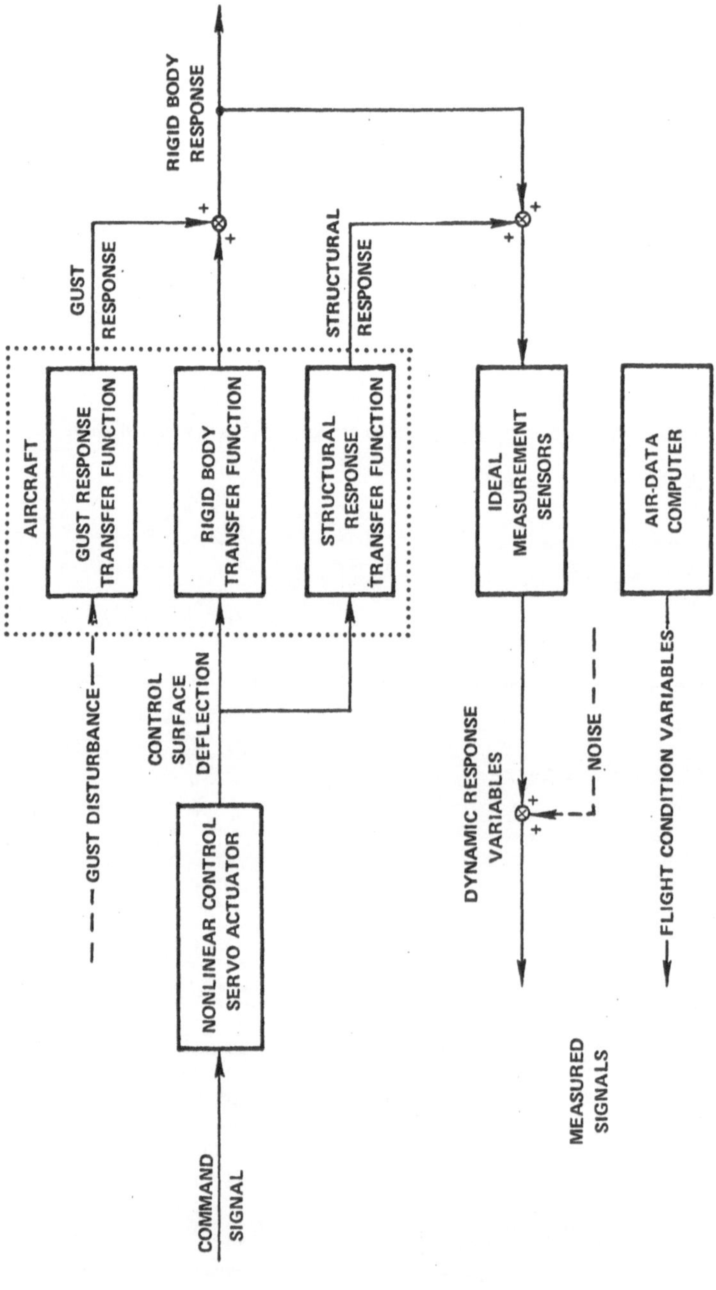

Figure A. Functional Block Diagram of the Aircraft, Control Servo Actuator and Measurement Dynamics.

NOMENCLATURE

F_s	pilot command input force, lbs.
g	acceleration due to gravity constant, ft/sec^2
G_w	gust power spectral density (ft/sec)^{2}sec.
K_α	angle-of-attack calibration coefficient
M	Mach number
M_α, M_{δ_e}, $M_{\dot\theta}$	Dimensional pitch moment coefficients
n_z	normal acceleration, "g's".
q	dynamic pressure, lbs/ft^2
$\bar{v}$	true velocity, ft/sec
w_g	incremental vertical velocity due to air gusts, ft/sec
z	distance from c.g. to accelerometer, ft.
Z_α, Z_{δ_e}	Dimensional normal force coefficients
α	angle-of-attack, rad
α (sub)	distance from c.g. to angle-of-attack vane, ft.
δ_e	elevator deflection angle, rad

NOMENCLATURE (Continued)

δ_{e_c}	commanded elevator deflection angle, rad
δ_s	pilot command input deflection, inches
ξ_j	damping ratio of the jth bending mode
θ	pitch attitude, rad
λ_{α_j}	slope of the jth bending mode at angle-of-attack vane, rad/ft
λ_{θ_j}	slope of the jth bending mode at pitch attitude and rate sensor, rad/ft
ν_j	displacement of the jth bending mode at reference station, ft
ϕ_j	relative displacement of the jth bending mode at the accelerometer
Ψ	forcing function coefficient for bending modes, ft/sec^2
ω_c	characteristic frequency of gust power spectral density, rad/sec natural
ω_j	frequency of the jth bending mode, rad/sec

APPENDIX III

Taking the state equation as:

$$\dot{x} = Ax + Bu \qquad x(0) = 0$$

we wish to find u(t) so as to minimize:

$$\lim_{T\to\infty} \left\{ \frac{1}{T} \int_0^T \| L\, x(t) - \delta_p(t) \|^2 dt + \frac{\lambda}{T} \int_0^T |u(t)|^2 dt \right\}$$

where $\delta_p(t)$ is a unit step function. By the usual analysis, the optimal u(t), denoted $u_o(t)$ is given by:

$$u_o(t) = -\frac{B^*}{\lambda} \int_t^\infty e^{A^*(\sigma - t)} [L^*L\, x(\sigma) - L^*\, \delta_p(\sigma)] d\sigma$$

[The main point to note is the appearance of ∞ as the upper limit in the integral.] We can express this solution alternately as

$$u_o(t) = -\frac{B^*}{\lambda} Y(t) \qquad 0 < t < \infty$$

where

$$\dot{Y}(t) = -A^* Y(t) - L^*L\, x(t) + L^*\, \delta_p(t)$$

$$\dot{x}(t) = A\, x(t) - \frac{BB^*}{\lambda} Y(t)$$

Now if A, B is controllable,

$$PA + A^*P - \frac{PBB^*P}{\lambda} + L^*L = 0$$

has a unique non-negative definite solution such that

$$A - \frac{BB^*P}{\lambda}$$

is stable.

Next let

$$Z(t) = P\,x(t) - Y(t)$$

AIII (1) Then we have:

$$\dot{Z}(t) = P\,\dot{x}(t) - \dot{Y}(t)$$

and substituting from AIII (2) and AIII (3), we have

$$Z(t) = -\left[A^* - \frac{PBB^*}{\lambda}\right] Z(t) - L^*\,\delta_p(t)$$

AIII (2) and hence letting D denote $(-A^* + \frac{PBB^*}{\lambda})$, we have

AIII (3)

$$Z(t) = e^{+Dt} Z(0) - \int_0^t e^{D(t-s)} L^* \, ds$$

$$= e^{Dt} [Z(0) + D^{-1} L^*] - D^{-1} L^*$$

Again from AIII (2), AIII (3) it is clear that $Z(t)$ must have a finite value for the time average and since D is now unstable (eigen-values have positive real parts) it follows that

$$Z(0) + D^{-1} L^* = 0$$

or

$$Z(t) = - D^{-1} L^*$$

and finally

$$u_o(t) = - \frac{B^*}{\lambda} [P \, x(t) + D^{-1} L^*]$$

We note that this result is already available in the literature [6] although derived on less firm grounds, and in particular not on the basis of the time-average error criterion.

References

1. L. Taylor and H. Rediess: "Flight Control Design Challenge", JACC, June 1970.

2. A. V. Balakrishnan: "Flight Control System Design I: Identification of Flight Parameters, Problem Formulation", UCLA Engineering Report No. 70-63, July 1970.

3. L. Taylor and K. Iliff: "A Modified Newton-Raphson Method for Determining Stability Derivatives from Flight Data," in 'Computing Methods in Optimization Problems', II, Academic Press, 1969.

4. A. V. Balakrishnan and V. Peterka: "Identification in Automatic Control Systems", Automatica, Vol. 5, 1969.

5. A. V. Balakrishnan: "Stochastic Differential Systems I", Lecture Notes, UCLA, January 1971.

6. M. Athans and P. Falb: "Optimal Control", p. 804, McGraw-Hill, 1966.

CENTRO INTERNAZIONALE MATEMATICO ESTIVO

(C.I.M.E.)

R. GLOWINSKI

METHODES ITERATIVES DUALES POUR LA MINIMISATION DE FUNCTIONNELLES CONVEXES

Corso tenuto ad Erice dal 27 giugno al 7 luglio 1971

METHODES ITERATIVES DUALES POUR LA MINIMISATION DE FONCTIONNELLES CONVEXES (*)

par R. Glowinski (**)

(Université de Paris)

1 Introduction

Il arrive qu'on puisse associer á un problème d'optimisation "difficile" (parce que la fonction á minimiser est non différentiable , ou parce que les contraintes sont non-triviales) un problème d'optimisation plus simple dont la solution fournit celle du problème initial ; on indiquera dans ce qui suit un procédé basé sur l'utilisation de la dualité , via le lagrangien , permettant une telle transformation dans un certain nombre de cas importants.

On en deduira , de façon trés naturelle, deux algorithmes d'optimisation associés á cette transformation et permettant , en fait , de résoudre silmultanément les deux problèmes.

L'exposé qui suit reprend en grande partie les considérations sur les méthodes itératives duales développées dans Glowinski-Lions-Tremolieres [1], chapitre 2 et systématiquement utilisées dans le meme ouvrage.

(*) Exposé fait au CIME á Erice , Sicile-Italia en Juin-Juillet 1971

(**) IRIA et Université Paris 6.

2. Cadre fonctionnel

On considère un espace de Hilbert V, de dimension finie ou infinie et la fonctionnelle :

(2.1) $$J_o(v) = \frac{1}{2} a(v,v) - (f , v)$$

où

$a(u,v) = a(v,u) \quad \forall u,v \in V$ est un forme bilinéaire continue sur V vérifiant

(2.2) $$a(v,v) \geq \alpha \|v\|^2 \quad ; \alpha > o , \forall \quad v \in V$$

où $\|v\|$ = norme de v dans V, et où dans (2.1) $v \rightarrow (f,v)$ est une forme linéaire continue sur V.

On se donne par ailleurs

(2.3) M = ensemble convexe fermé de V.

(2.4) $\begin{cases} L = \text{espace de Hilbert, et } \Phi \text{ fonction de } V \quad > L, \\ \text{ou de } M \rightarrow L \text{ , linéaire ou non.} \end{cases}$

(2.5) Λ = ensemble convexe fermé de L.

On suppose que, $(\quad , \quad)_L$ désignant le produit scalaire dans L, pour toute fonction $q \in L$, on a :

(2.6) $\begin{cases} v \longrightarrow (q, \Phi(v))_L \text{ est convexe et semi continue} \\ \text{inférieurement sur V faible.} \end{cases}$

On considère alors le problème:

(2.7) $$\inf_{v \in M} \left[J_o(v) + \sup_{q \in \Lambda} (q, \Phi(v))_L \right]$$

Remarque 2-1

On prend ainsi par définition le problème sous une forme adaptée à la dualité (cf. Glowinski-Lions-Tremolieres [1] , ch. I).

On verra dans la suite de l'exposé que la formulation (2. 7) contient des probléme importants.

Remarque 2-2:

La fonction $v \longrightarrow \sup_{q\in\Lambda} (q, \Phi(v))_L$ est convexe semi-continue inférieurement pour la topologie faible de v; il en est donc de même pour J définie par:

$$(2.8) \qquad J(v) = J_o(v) + \sup_{q\in\Lambda} (q, \Phi(v)).$$

Pour étudier la convergence des algorithmes ci-aprés on fera l'une ou l'autre des hypothèses suivantes :

$$(2.9) \qquad \Phi \in \mathcal{L}(V;L) \text{ et } \Lambda \text{ borné dans } L.$$

oú

(2.10) M est borné dans V et Φ est Lipschitzienne de $M \longrightarrow L$, i. e;

$$\|\Phi(u) - \Phi(v)\|_L \leq C_1 \|u-v\| \quad \forall u, v \in M$$

Notons que la fonction J (v) est strictement convexe, avec , dans le cas (2. 9), $\lim_{\|v\|\to\infty} J(v) = +\infty$. Donc sous les hypothèses(2. 9) ou (2. 10) il existe u unique dans M tel que:

$$J(u) \leq J(v), \quad \forall \quad v \in M$$

3 - Exemples

Exemple 3. 1 :

Soit Ω un ouvert borné de R^n, de frontière $\Gamma = \partial\Omega$, on prend

$$(3.1) \qquad V = H_o^1(\Omega) = \left[v \mid v_1 \ \frac{\partial v}{\partial x_i} \in L^2(\Omega),\ i = 1, n,\ v|_\Gamma = 0 \right]$$

$$(3.2) \qquad M = V$$

$$(3.3) \qquad a(u, v) = \int_\Omega \text{gradu} . \text{gradv} \ dx$$

(3.4) $(f, v) = \int_\Omega fvdx \quad , f \in L^2(\Omega)$

(3.5) $L = (L^2(\Omega))^n$

(3.6) $\Phi v = \text{grad } v$

donc $\Phi \in \mathcal{L}(V ; L)$

(3.7) $\Lambda = [q \mid q \in L , |q(x)| \leq g \text{ p.p. dans } \Omega] , g = c^{te} > 0$

avec $|q(x)| = \left(\sum_{i=1}^{u} q_i^2(x)\right)^{\frac{1}{2}}$.

On rappelle que $v \longrightarrow \int_\Omega |\text{gradv}|^2 dx$ définit une norme Hilbertienne dans $H'_o(\Omega)$, on est donc dans le cadre (2.9) et on montrerait facilment :

(3.8) $\sup_{q \in \Lambda} (q, \Phi(v)) = \max_{q \in \Lambda} (q, \Phi(v)) = g \int_\Omega |\text{gradv}| dx$:

ce qui correspond pour le problème (2.8) á l'écoulement d'un fluide rigide visco-plastique dans un cylindre de section Ωx (cf. Cea-Glowinski [1], Glowinski-Lions-Tremolieres [1], Lions-Duvaut [1] etc..)

Exemple 3.2 :

On considére dans $V = H'_o(\Omega)$ le problème d'élasto-plasticité :

(3.9) $\min_{v \in k} \left(\frac{1}{2} a(v,v) - \int_\Omega f v dx\right) = \min_{v \in k} J_o(v)$

avec

(3.10) $K = [v \mid v \in H^1_o(\Omega) , |\text{grad } v| \leq 1 \text{ p.p.}]$

et $a(u,v)$ et f comme dans (3.3), (3.4).

Le problème (3.9) admet une solution et une seule, la fonction J_o étant strictement connexe avec $\lim_{\|v\| \to \infty} J_o(v) = +\infty$ et K étant fermé dans $V = H^1_o(\Omega)$.

En pratique [(1)], on considérera le restriction de J_o á un sous espece

(1) Si on travaille en approximation interne.

V_h de $V = H'_o(\Omega)$, V_h étant de dimension finie de façon que lorsque v parcourt V_h la fonction :

(3. 11) $$\Phi(v) = |\operatorname{grad} v|^2 - 1$$

demeure dans un espace L de dimension finie.

On notera que K est borné dans $H'_o(\Omega)$, on peut donc restreindre le problème à un ensemble M borné de V_h.

On introduit :

(3. 12) $$\Lambda = \left\{\text{cône des vecteurs } q = \{q_i\} \in L \text{ avec } q_i \geq 0 \ \forall i\right\}$$

Alors :

(3. 13) $$\sup_{q\in\Lambda} \int_\Omega q \ \Phi(v) \ dx = \begin{cases} 0 & \text{si } |\operatorname{grad} v|^2 - 1 \leq 0 \\ +\infty & \text{sinon} \end{cases}$$

de sorte que (2. 7) dévient (la restriction à V_h de) :

$$\inf. \left(\frac{1}{2}\int_\Omega |\operatorname{grad} v|^2 dx - \int_\Omega fv \, dx\right), \ v \in K$$ donc du problème (3. 9).

Il est facile de voir que la fonction $\Phi : H'_o(\Omega) \rightarrow L^1(\Omega)$ est lipschitzienne sur les bornés de telle sorte que par restriction à V_h on a. (2. 10).

En fait la restriction à V_h n'est pas necessaire ici mais est utile pour la vérification de (4. 2) ci-après, car l'existence de $q \in \Lambda$, associée au problème (3. 9) restraint à V_h, résultera de la théorie des multiplicateurs de Lagrange en dimension finie, cf. R. T. Rockafellar [1]

4 - Un algorithme de recherche de point-selle

Introduisons le Langrangien :

(4. 1) $$\mathcal{L}(v, q) = J_o(v) + (q, \Phi(v))_L$$

et faisons l'hypothèse :

$$(4.2)\quad \begin{cases} \mathcal{L}(v,q) \text{ admet un point-selle sur } M \times \Lambda, \text{i.e. } \exists \text{ un point} \\ \{u,p\} \in M \times \Lambda \text{ tel que :} \\ \mathcal{L}(u,q) \leq \mathcal{L}(u,p) \leq \mathcal{L}(v,p), \quad v \in M, \quad q \in \Lambda \end{cases}$$

Remarque 4. 1

L'hypothèse (4. 2) a lieu dans le cas où (2. 9) est satisfaite : c'est le théorème classique de Ky-Fan [2], Sion [1] si M est borné; si M n'est pas borné, on "approche" M par $M_R = M \cap \{v / \|v\| \leq R\}$ et on considère la restriction de $\mathcal{L}$ à $M_R \times \Lambda$; il existe alors un point selle $\{u_R, p_R\}$. Tout se ramène à verifier que u_R et p_R sont bornés indépendamment de R lorsque $R \longrightarrow +\infty$; pour p_R cela résulte de l'hypothèse Λ borné, il reste donc à verifier que $\|u_R\| \leq C$.

Mais :

$$J_o(u_R) + (p_R, \Phi(u_R)) \leq J_o(v) + (p_R, \Phi(v)) \qquad \forall v \in M_R$$

Fixant v on en deduit :

$$J_o(u_R) + (p_R, \Phi(u_R)) \leq C$$

donc $J_o(u_R) \leq C + C \|u_R\|$ d'où le résultat puisque

$$J_o(u_R) \leq \frac{\alpha}{2} \|u_R\|^2 - C\|u_R\|$$

Définition de l'algorithme : (1)

On se donne p^o, on calcule u^o, puis p^1 ect. . . . La règle générale étant :

$$(4.3)\quad \begin{cases} p^n \text{ connu } (\in \Lambda)\, u^n \text{ est alors l'élément de M minimisant} \\ J_o(v) + (p^n, \Phi(v))_L \end{cases}$$

---

(1) C'est un algorithme de type UZAWA cf. H. UZAWA

On définit ensuite :

(4. 4) $$p^{n+1} = P_\Lambda (p^n + \rho_n \Phi(u^n))$$

où :

(4. 5) P_Λ = opérateur de projection de $L \longrightarrow \Lambda$ et où $\rho_n > 0$ est convenablement choisi.

Remarque 4. 2 :

La fonction $v \longrightarrow J_o (v) + (p^n , \Phi (v))_L$ est strictement convexe et est " infinié à l'infini " dans le cas (2. 9) ; (4. 3) définit donc bien u^n de façon unique.

Remarque 4. 3 : (Motivation de l'algorithme)

Si $\{u, p\}$ est un point-selle alors :

(4. 6) $$J_o (u) + (p , \Phi(u))_L \leq J_o (v) + (p , \Phi (v))_L \qquad \forall v \in M$$

ce qui conduit à (4. 3) et d'aprés (4. 2) :

(4. 7) $$(q - p , \Phi (u))_L \leq 0 \qquad \forall q \in \Lambda$$

ce qui équivant à :

(4. 8) $$p = P \Lambda (p + \rho \Phi(u)) \qquad \forall \rho > 0$$

et (4. 8) explique (4. 4).

Remarque 4. 4 : En fait l'algorithme n'est pas complétement determiné puisque dans (4. 3) il faut trouver une méthode de calcul pour u^n. On verra au N. 5 une variante où l'on précise ce choix.

Convergence de l'algorithme :

Théorème 4. 1 : On suppose que (2. 3) ,... (2. 6) ont lieu ainsi que l'une des hypothèses (2. 9) ou (2. 10). Alors l'algorithme défini par (4. 3), (4. 4) est convergent au sens suivant :

(4. 9) $u^n \longrightarrow u$ dans V fort

u étant la solution de (2. 7), lorsque :

(4. 10) $0 < \rho_0 \leq \rho_n \leq \rho_1$, ρ_1 suffisemment petit.

Démonstration : De (4. 4) et (4. 8) on déduit en posant

(4. 11) $r^n = p^n - p$

et en utilisant le fait que P_Λ est une contraction:

(4. 12) $\| r^{n+1} \|_L \leq \| r^n + \rho^n (\Phi (u^n) - \Phi (u))\|_L$

par ailleurs (4. 3) équivant à :

(4. 13) $(J'_0 (u^n), v - u^n) + (p^n, \Phi (v) - \Phi (u^n))_L \geq 0$ $\forall v \in M$

et (4. 6) à :

(4. 14) $(J'_0 (u), v - u) + (p, \Phi(v) - \Phi (u))_L \geq 0$ $\forall v \in M$

Prenant $v = u$ (resp. $v = u^n$) dans (4. 13) (resp. (4. 14)) on en dé= duit :

(4. 15) $a(u^n-u, u^n- u) + (p^n - p, \Phi(u^n) - \Phi(u))_L \leq 0$

Sous l'hypothèse (2. 10) on a :

$$\|u^n\| < \text{constante}$$

et sous l'hypothèse (2. 9) Λ est borné, donc Φ étant Liepschitzienne

$$|(p^n - p, \Phi(u^n) - \Phi(u))_L| \leq c \, \|u^n - u\|$$

donc dans tous les cas u^n est borné dans V et

(4.16) $(p^n - p, \Phi(u^n) - \Phi(u))_L \leq - a(u^n - u, u^n - u)$

On déduit de (4.12) que :

$$\|r^{n+1}\|^2_L \leq \|r^n\|^2_L + 2\rho_n (r^n, \Phi(u^n) - \Phi(u))_L +$$

$$+ \rho_n^2 \|\Phi(u^n) - \Phi(u)\|^2_L$$

$$\leq \text{(d'aprés (4.16))} \|r^n\|^2_L - 2\alpha\rho_u \|u^n - u\|^2 + \rho_u \|\Phi(u^n) -$$

$$- \Phi(u)\|^2_L$$

Mais dans tous les cas :

$$\|\Phi(u^n) - \Phi(u)\| \leq C \|u^n - u\| \quad \text{et donc:}$$

$$\|r^{n+1}\|^2 \leq \|r^n\|^2 + (C\rho_n^2 - 2\alpha\rho_u)\|u^n - u\|^2$$

Choisissons $[\rho_o, \rho_1]$ de façon que

$$2\alpha\rho_n - C\rho_n^2 \geq \beta > 0 \quad \text{si } \rho_n \in [\rho_o, \rho_1]$$

alors :

(4.17) $\|r^{n+1}\|^2_L + \beta\|u^n - u\|^2 \leq \|r^n\|^2$

Donc $\|r^n\|$ décroit avec u, donc $\|r^n\|^2_L \longrightarrow c$ lorsque $n > +\infty$ et alors d'aprés (4.17)

$$\beta \|u^n - u\|^2 > 0$$

d'où (4.9)

On verra au n. 6 l'application à l'exemple 3.1 de l'algorithme précédent ; l'application à l'exemple 3.2 sera considérée dans un exposé séparé consacré à l'Analyse Numérique du problème (3.9) : cf Glowinski [2].

5. - Un deuxième algorithme de recherche de point-selle [1]

On complète maintenant la Remarque 4. 4.

On se place dans le cadre de l'hypothèse (2. 9) avec

(5. 1) $M = V$

alors dans (4. 3) u^n est défini par :

(5 2) $J'_0 (u^n) + \Phi^* p^n = f$

ce qui s'écrit encore

(5. 3) $Au^n + \Phi^* p^n - f = 0$

où $A \in \mathcal{L}(V, V')$ est défini par $a(u,v) = (Au, v) \quad \forall u, v \in V$.

Si on introduit un algorithme itératif pour la résolution de (5. 3) on est naturellement conduit à l'algorithme suivant: supposant u^n et p^n connus ou calculès on définit u^{n+1} par:

(5. 4) $u^{n+1} = u^n - \rho_1 S^{-1} (Au^n + \Phi^* p^n - f)$

où :

S = identité si V' = V = espace de dimension finie, toute matrice symétrique définie positive répondant plus généralement à la question. S = opérateur de dualité de $V \longrightarrow V'$ si V est la dimension infinie [2]

On définit ensuite p^{n+1} par:

(5. 5) $p^{n+1} = P_\Lambda (p^n + \rho_2 \Phi u^{n+1})$

où dans (5. 4) et (5. 5) ρ_1 et ρ_2 sont deux paramètres > 0 à choisir convenablement. Ou a le :

---

(1) - L'algorithme considéré est de type ARROW-HURWICZ [1]

(2) - Si on sait que $u \in W$, W Hilbert inclu dans V avec densité et injection continue (d'où $W \subset V \subset V' \subset W'$), on peut prendre pour S un opérateur de dualité de $W \longrightarrow W'$

Théorème 5.1 : On suppose que (2.2), (2.4), (2.9), (5.1) ont lieu.

On peut alors choisir $\rho_1, \rho_2 > 0$ de façon que l'algorithme (5.4), (5.5) soit convergent au sens :

(5.6) $\qquad u^n \longrightarrow u$ dans V fort

u étant la solution de (2.7)

Démonstration :

Ou a, (u, p) désignant un point selle de $\mathcal{L}$:

(5.7) $\qquad u = u - \rho_1 S^{-1} (Au + \Phi^* p - f)$

(5.8) $\qquad p = P_{\Lambda} (p + \rho_2 \Phi u)$

Posant $w^n = u^n - u$, $r^n = p^n - p$ ou a :

(5.9) $\qquad w^{n+1} = w^n - \rho_1 S^{-1} (Aw^n + \Phi^* r^n)$

(5.10) $\qquad \| r^{n+1} \|_L^2 \leq \| r^n \|_L^2 + 2 \rho_2 (r^n, \Phi w^{n+1})_L + \rho_2^2 \| \Phi w^{n+1} \|_L^2$

On déduit de (5.9) que :

$$Sw^{n+1} = Sw^n - \rho_1 (Aw^n + \Phi^* r^n)$$

d'où en multipliant scalairement par w^{n+1} :

(5.11) $\qquad \|w^{n+1}\| = ((S - \rho_1 A) w^n, w^{n+1}) - \rho_1 (\Phi^* r^n, w^{n+1})$

comme $A = A^*$, $(Av, v) \geq \alpha \|v\|^2$, ou a :

$$\|(S - \rho_1 A) v\| \leq \beta \|v\|, \quad 0 < \beta < 1, \text{ pour } 0 < \rho_1 \leq \rho_o$$

Alors:

(5.12) $\qquad \| w^{n+1} \|^2 \leq \beta \|w^n\| \, \|w^{n+1}\| - \rho_1 (\Phi^* r^n, w^{n+1})$

On cherche ρ_1, ρ_2 sous la forme:

$$\rho_1 = \rho \quad , \rho_2 = \rho c$$

Utilisant (5. 12) dans (5. 11), il vient :

$$\|r^{n+1}\|_L^2 \leq \|r^n\|_L^2 + 2c\,(\beta\|w^n\|\,\|w^{n+1}\| - \|w^{n+1}\|^2)$$
$$+\rho^2 c^2 \|\Phi\|^2 \|w^{n+1}\|^2$$
$$\leq \|r^n\|_L^2 + 2c\beta\,\|w^n\|\,\|w^{n+1}\| - \|w^{n+1}\|^2) +$$
$$+[\rho^2 c^2\|\Phi\|^2 - 2C(1-\beta)]\,\|w^{n+1}\|^2$$
$$\leq \|r^n\|_L^2 + c\beta(\|w^n\|^2 - \|w^{n+1}\|^2) + [\rho^2 c^2\|\Phi\|^2 - 2C(1-\beta)\,\|w^{n+1}\|^2$$

Mais :

$2C(1-\beta) - \rho^2 c^2 \|\Phi\|^2 \geq \gamma > 0$ si $0 < \rho \leq \rho_o^*$, de sorte que si $0 < \rho \leq \min(\rho_o, \rho_o^*)$ on a :

$$\|r^n\|_L^2 + c\beta\|w^n\|^2 - (\|r^{n+1}\|^2 + c\beta\|w^{n+1}\|^2) \geq \gamma\|w^{n+1}\|^2 \tag{5. 13}$$

Donc :

$\|r^n\|_L^2 + c\beta\|w^n\|^2$ décroit avec n, donc converge et donc le premier member de (5. 13) tend vers 0, donc $\|w^{n+1}\| \longrightarrow 0$.

Remarque 5. 1 Le comportement de ρ^n dépend des caractéristiques du problème et en particulier des proprietés de Φ, par exemple

$\Phi \in \mathcal{L}(V,L)$, Φ surjectif conduit á Φ^* injectif et si $V = M$ on a unicité de p , et on peut montrer que toute la suite p^u converge vers p (faiblement si V est de dimension infinie), plus généralement , sous les hypotheses précédentes, la suite p^u étant bornée d'aprés (4. 17) ou (5. 13) , posséde des points adhérents (faibles, au moins, en dimension infinie) et si p est un tel point adhérent ou montre sans difficulté que (u, p) est point-selle de $\mathcal{L}$ sur $M \times \Lambda$.

<u>Remarque 5. 2</u> La première méthode fournit une estimation par valeurs inférieures de $\inf_{v \in M} [J_o(v) + \sup_{q \in \Lambda} (q, \Phi(v))_L]$

En effet , par définition de u^n

$$J_o(U^u) + (p^u, \Phi(v^u))_L \leq J_o(u) + (p^u, \Phi(u))_L$$

$$\leq J_o(U) + \sup_{q\in\Lambda}(q, \Phi(u))_L = \inf_{v\in M}[J_o(v) + \sup_{q\in\Lambda}(q, \Phi(v))_L]$$

<u>Remarque 5. 3</u> (Motivation de la terminologie <u>méthodes itératives duales</u>)

Si defini en (4. 1) admet un point selle (u,p) sur $M \times \Lambda$ on a :

$$(5.\ 14) \qquad \operatorname*{Max}_{q\in\Lambda} \operatorname*{Min}_{v\in M} \mathcal{L}(v,q) = \operatorname*{Min}_{v\in M} \operatorname*{Max}_{q\in\Lambda} \mathcal{L}(v,q) = \mathcal{L}(u,p)$$

L'existence du point selle (u,p) entraine donc que le problème (p) defini par (2. 7) admet une solution ; on appellera <u>problème primal</u> le problème (p) ; de même le probléme (p^*) défini par :

$$(5.\ 15) \qquad \max_{q\in\Lambda} \inf_{v\in M} (J_o(v) + (q, \Phi(v))_L$$

admet une solution ; si on note

$$(5.\ 16) \qquad J^*(q) = \inf_{v\in M} (J_o(v) + (q, \Phi(v))_L$$

alors (p^*) peut encore s'écrire

(5.17) $\underset{q \in \Lambda}{\text{Max}} \; J^*(q)$

On appellera (p^*) le problème dual de (p), cette terminologie étant compatible avec les notions de dualité developpées, entre autres, par R. T. ROCKAFELLAR [1].

On vérifiera par ailleurs que la méthode d'UZAWA i. e. l'algorithme étudié au N 4 n'est autre que la méthode du gradient avec projection (c f. MOSCO [1]) appliquée au problème dual.

A proprement parler, les deux algorithmes étudiés sont du type primal-dual puisque l'on détermine et on exprime le gradient de la fonctionnelle duale J^* en fonction de la variable primale v(q) associée à q et solution du problème :

(Pq) $\underset{v \in M}{\text{Min}} \; [J_o(v) + (q, \Phi(v))_L$

6. - Application du premier algorithme à l'exemple 3.1.

On a vu au N 3 (c.f. (3.8)) que dans le cas de l'exemple 3.1 la forme explicite du problème général (2.8) est donnée par :

(6.1) $\underset{v \in H'_o(\Omega)}{\text{Min}} \int_\Omega |\text{grad} v|^2 dx + 2g \int_\Omega |\text{grad} v| \, dx - 2 \int_\Omega fv dx]$

problème admettant une solution unique dans $H'_o(\Omega)$ et dont la difficulté essentielle réside dans la non-différentiabilité du terme $2g \int_\Omega |\text{grad} v| \, dx$.

L'application au problème (6.1), de l'algorithme (4.3), (4.4) conduit, compte tenu du N 3, (et de quelques modifications mineures) à :

$$(6.2)\quad \begin{cases} p^o \text{ donné} \\ -\Delta u^{n+1} - g \ \text{div}\ p^n = f \\ u^{n+1}|_\Gamma = 0 \\ p^{n+1} = P_\Lambda \ (p^n + \rho_n \ g \ g \ \text{v} \ \text{ad} u^{n+1}) \end{cases}$$

avec

$$(6.3)\quad P_\Lambda \ (q) = \frac{gq}{\sup (g, |q|)}$$

La condition de convergence est alors :

$$(6.4)\quad 0 < \alpha \leq \rho_n \leq \beta < 2$$

Mise en oeuvre numerique de (6. 2)

Pour une étude approfondie de l'aspect numerique que de (6. 2) on renvoie à CEA-GLOWINSKI [1], GOURSAT [1], GLOWINSKI-LIONS-TREMOLIERES [1].

On va utiliser une approximation par différences finies : on se donne donc $\Omega \subset R^2$ (n = 2 étant le cas physique réel pour ce problème), et ayant introduit un pas $h > 0$ (destiné à tendre vers zéro), on définit successivement :

$$(6.5)\quad Rh = [M_{ij} \mid M_{ij} \in R^2, \quad M_{ij} = (x_i, y_j), \ x_i = ih \ , \ y_j = jh \ , i, j \in Z]$$

$$(6.6)\quad \sigma_{ij} = \left] x_i - \frac{h}{2}, \ x_i + \frac{h}{2} \right[\times \left] y_j - \frac{h}{2}, \ y_j + \frac{h}{2} \right[$$

$\sigma_{i \pm \frac{1}{2} j}$ (resp. $\sigma_{ij \pm \frac{1}{2}}$) = translaté de $\pm \frac{h}{2}$ de σ_{ij}, parallélement à ox (resp. oy).

$$(6.7)\quad \rho_{ij} = \sigma_{ij} \cup \sigma_{i \pm \frac{1}{2} j} \cup \sigma_{ij \pm \frac{1}{2}}$$

(6. 8) $\quad \Omega_{oh} = [M_{ij} \mid M_{ij} \in R_h, \rho_{ij} \subset \Omega]$

et on notera $\tilde{u}_{ij}$, la valeur approchée (du moins on l'espère) de u en M_{ij} ; même rotation avec f_{ij} pour f. On posera:

(6. 9) $\quad u_h = [u_{ij} \mid M_{ij} \ \Omega_{oh}]$

Pour définir la variable duale p $(\in (L^2(\Omega))^2)$, il est commode d'introduire un réseau Q_h, de pas h également, décalé par rapport à R_h de la façon indiquée sur la figure 6. 1 :

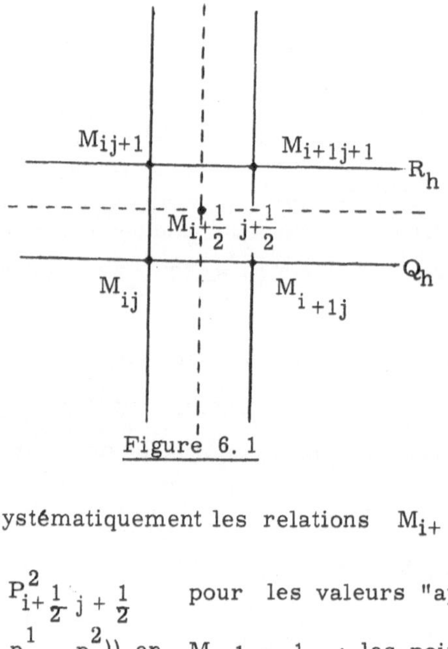

Figure 6. 1

On utilisera systématiquement les relations $M_{i+\frac{1}{2}j+\frac{1}{2}}$ et $P^1_{i+\frac{1}{2}\ j+\frac{1}{2}}$, $P^2_{i+\frac{1}{2}\ j+\frac{1}{2}}$ pour les valeurs "approchées" de p^1 et p^2($p = (p^1, p^2)$) en $M_{i+\frac{1}{2}\ j+\frac{1}{2}}$; les points $M_{i+\frac{1}{2}\ j+\frac{1}{2}}$ à prendre en compte sont ceux qui sont centre d'un carré de cote h (v. figure 6. 1) dont un sommet au moins appartient à Ω_{oh} et on notera Ω_{1h} l'ensemble de ces points; on posera également :

(6. 10) $\quad Ph = \left[P_{i+\frac{1}{2}\ j+\frac{1}{2}} \ \middle|\ M_{i+\frac{1}{2}\ j+\frac{1}{2}} \in \Omega_{1h}\right]$

Dans ces conditions l'algorithme (6. 2) est " approché" par :

(6. 11)
$$\begin{cases} p_h^{\circ} \text{ donné} \\ -\dfrac{u_{i+ij}^{n+1} + u_{i-ij}^{n+1} + u_{ij+1}^{n+1} + u_{ij-1}^{n+1} - 4u_{ij}^{n+1}}{h^2} - g\, D_{ij}\, p_h^n = f_{ij} \quad (M_{ij} \in \Omega_{oh}) \\ p^n_{i+\frac{1}{2}\, j+\frac{1}{2}} = p_h\,(\rho^n_{i+\frac{1}{2}\, j+\frac{1}{2}} + \rho_n g\, G_{i+\frac{1}{2}\, j+\frac{1}{2}}\, u_h^{n+1}) \\ (\rho_n > 0)\,,\ M_{i+\frac{1}{2}\, j+\frac{1}{2}} \in \Omega_{1h} \end{cases}$$

avec :

(6. 12)
$$\begin{cases} D_{ij}\, p_h^n = \dfrac{p^{1n}_{i+\frac{1}{2}\, j+\frac{1}{2}} - p^{1n}_{i-\frac{1}{2}\, j+\frac{1}{2}} + p^{1n}_{i+\frac{1}{2}\, j-\frac{1}{2}} - p^{1n}_{i+\frac{1}{2}\, j-\frac{1}{2}}}{2h} + \\ + \dfrac{p^{2n}_{i+\frac{1}{2}\, j+\frac{1}{2}} - p^{2n}_{i+\frac{1}{2}\, j-\frac{1}{2}} + p^{2n}_{i-\frac{1}{2}\, j+\frac{1}{2}} - p^{2n}_{i-\frac{1}{2}\, j-\frac{1}{2}}}{2h} \end{cases}$$

approchant div p en M_{ij},

(6. 13)
$$\begin{cases} G^1_{i+\frac{1}{2}\, j+\frac{1}{2}}\, u_h^{n+1} = \dfrac{u^{n+1}_{i+1j+1} - u^{n+1}_{i\, j+1} + u^{n+1}_{i+1j} - u^{n+1}_{ij}}{2h} \\ G^2_{i+\frac{1}{2}\, j+\frac{1}{2}}\, u_h^{n+1} = \dfrac{u^{n+1}_{i+1j+1} - u^{n+1}_{2+1j} + u^{n+1}_{ij+1} - u^{n+1}_{ij}}{2h} \end{cases}$$

approchant $\dfrac{\partial u}{\partial x}$, $\dfrac{\partial u}{\partial y}$ en $M_{i+\frac{1}{2}}, {}_{j+\frac{1}{2}}$,

(6.14)
$$\begin{cases} (P_h(q_h))_{i+\frac{1}{2}\,j+\frac{1}{2}} = \dfrac{g\, q_{i+\frac{1}{2}\,j+\frac{1}{2}}}{\max(g, |q_{i+\frac{1}{2}\,j+\frac{1}{2}}|)} \\ \text{avec} \\ |q_{i+\frac{1}{2}\,j+\frac{1}{2}}| = \sqrt{(q^1_{i+\frac{1}{2}\,j+\frac{1}{2}})^2 + (q^2_{i+\frac{1}{2}\,j+\frac{1}{2}})^2} \end{cases}$$

Dans (6.11), (6.13), on convient de prendre $u^{n+1}_{kl} = 0$ si $M_{kl} \notin \Omega_{oh}$

On démontre dans CEA-GLOWINSKI [1], GLOWINSKI-LIONS-TREMOLIERES [1] que (6.11), (6.12), (6.13), (6.14) correspond à l'application de l'algorithme du N. 4 (l'algorithme d'UZAWA) à la minimisation d'une fonctionnelle $J_h(v_h)$, convexe, non différentiable, en dimension finie ; J_h étant une approximation de la fonctionelle du problème (6.1), explicitée dans le deux réferences ci-avant ; on y démontre également la convergence de (6.11) pour $0 < \alpha_h \leq \rho_n \leq \beta_h$. En ce qui concerne la convergence de $u_h \longrightarrow u$ lorsque $h \longrightarrow 0$, on démontre cette propriété dans CEA-GLOWINSKI [1], GLOWINSKI-LIONS-TREMOLIERES [1], TREMOLIERES [1], pour une topologie appropriée (du type norme de SOBOLEV discretisée dans $H^1_o(\Omega)$).

Résolution numérique d'un exemple :

On a pris $\Omega =]0,1[\times]0,1[$, $f = 10$, $g = 1$, $h = \frac{1}{20}$

Le processus itératif (6.11) est initié avec $p^o_h = 0$, le test d'arrêt choisi étant :

$$\sum_{M_{ij}\in\Omega_{oh}} |u^{n+1}_{ij} - u^n_{ij}| \leq 10^{-3}$$

On remarque, d'après (6. 11), que u_h^{n+1} (à p_h^n donné) est solution d'un problème de DIRICHLET approché résolu par surrelaxation avec ω optimal, en initialisant à u_h^n ; en ce qui concerne ρ_n on a travaillé en fait avec $\rho_n = c^{te} = \rho$.

Le paramètre ρ ayant sa valeur optimale (constatée expérimentalement) soit $\rho_{opt.} \simeq 1{\cdot}1$, il y a convergence en 12 itérations, soit un temps d'éxécution de l'ordre de 1 s. sur IBM 360/91 ; plus généralement on a représenté figure 6. 2 la dépendance en ρ de la vitesse de convergence, et on remarque le caractère réaliste de l'estimation (6. 4) puisque si pour $\rho = 2$ il y a encore convergence (en 150 itérations environ), il y a divergence pour $\rho = 2.1$.

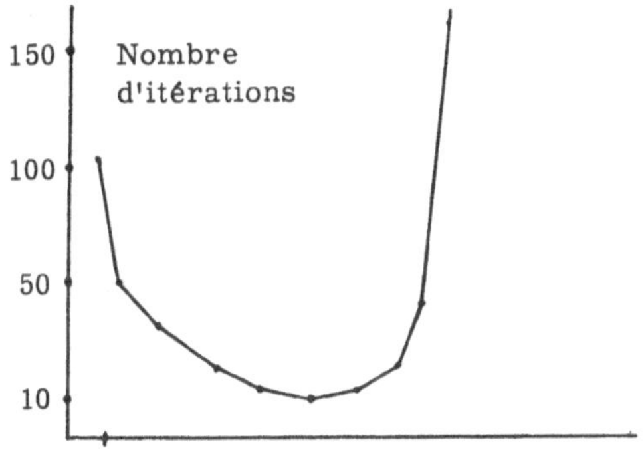

Figure 6. 2

Sur la figure 6. 3 on a représenté les zones de Ω où grad u = 0 et celles on grad u $\neq$ 0.

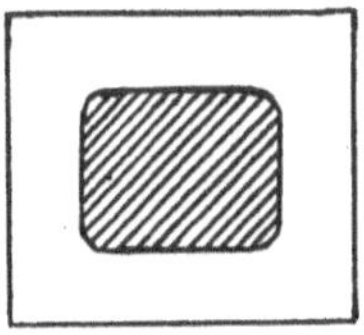

on a hachuré la zone où grad u = 0

Figure 6. 3

7. - Considérations sur la minimisation par dualité de fonctionelles convexes non différentiables

7. 1 : Généralités

On a vu dans les N. précédents et en particulier au N. 6 , que l'introduction d'un Lagrangien et l'utilisation de techniques de dualité, fournit une méthode élégante pour la minimisation de certaines fonctionnelles non différentiables et à ce sujet, on renvoie à CEA-GLOWINSKI-NEDELEC [1] où on trouvera un certain nombre d'exemples en dimension infinie (incluant l'exemple 3. 2, traité au N. 6) ; en fait la méthode est bien adaptée aux fonctionnelles dont la partie non différentiable J_1 est positivement homogène de degré 1 i. e.

$$J_1 (\lambda v) = \lambda J(v) \qquad \forall V \geq 0.$$

Nous voudrions dans ce N. indiquer un exemple trés simple de minimisation de fonctionelle non différentiables pour lequel , le problème dual étant également non différentiable, la méthode preécédente ne s'applique pas (tout au moins directement, c.f. Remarque 7. 1)

7. 2 : Position du problème

On suppose $V=R^N$, on se donne A matrice symétrique définie positive , $f \in R^n$ et J_o définie par :

$$J_o (v) = (Av, v) - 2 (f, v) \tag{7.1}$$

où (. , .) désigne le produit scalaire euclidien standard R^N.

On pose :

$$\left\{ \begin{aligned} \|v\|_1 &= \sum_{i=1}^{N} |v_i| \\ \|v\| &= (\sum_{i=1}^{N} v_i^2 \ \frac{1}{2}) \\ \|v\| &= \max_i |v_i| \end{aligned} \right. \tag{7.2}$$

On considère alors la fonctionnelle , visiblement non différentiable définie par :

(7. 3) $$J(v) = J_o(v) + \|v\|_1^2$$

J étant continue, strictement convexe (grâce à J_o) avec $\lim_{\|v\| \to +\infty} J(v) = +\infty$, admet sur V un optimum unique, soit u ; le problème:

(7. 4) $$\min_{v \in V} J(v)$$

est donc bien posé.

Avant de dégager , en 7. 3 , les propriétés de dualité du problème (7. 4) on va indiquer comment il est possible d'utiliser les résultats des N. précédents pour répondre (7. 4) ; ceci fuit l'objet de la :

Remarque 7. 1 : On paramètre (7. 4) de la façon suivante :

Soit $\lambda \geq 0$; à (7. 4) on associe la famille de problèmes (P_λ) définie par :

$$(P_\lambda) \quad \min_{v \in V} \left[J_o(v) + \lambda \|v\|_1 \right]$$

Alors :

Proposition 7. 1 :

i) (P_λ) admet une solution unique , u_λ , $\forall \lambda \geq 0$.

ii) $\lambda \longrightarrow \|u_\lambda\|_1$ est une fonction continue décroissante de $R^+ \longrightarrow R^+$

Démonstration : Le point (i) est évident.

Pour démontrer (ii) on remarquera (c. f. U. MOSCO [1]) que si u_λ, u_μ sont les solutions respectivement associées à λ et μ on a :

(7.5) $$\begin{cases} (J'_o(u_\lambda), u_\mu - u_\lambda) + (\|u_\mu\|_1 - \|u_\lambda\|_1) \geq 0 \\ (J'_o(u_\mu), u_\lambda - u_\mu) + (\|u_\lambda\|_1 - \|u_\mu\|_1) \geq 0 \end{cases}$$

d'où par addition :

(7.6) $$(J'_o(u_\mu) - J'_o(u_\lambda), u_\mu - u_\lambda) + (\mu - \lambda)(\|u_\mu\|_1 - \|u_\lambda\|_1) \leq 0$$

or :

(7.7) $$(J'_o(u_\mu) - J'_o(u_\lambda), u_\mu - u_\lambda) \geq 2c\|u_\mu - u_\lambda\|^2$$

où c ($c > 0$) est la plus petite valeur propre de A.

d'où $(\mu - \lambda)(\|u_\mu\|_1 - \|u_\lambda\|_1) \leq 0 \quad \forall \lambda, \mu \in R^+$ ce qui démontre la décroissance; pour démontrer la continuité on remarque que $\|u_\lambda\|_1 \leq \|u_o\|_1 \quad \forall \lambda \geq 0$ d'où, compte tenu de (7.6), (7.7) :

(7.8) $$2c\|u_\mu - u_\lambda\|^2 \leq |\mu - \lambda| \, \|u_o\|_1$$

ce qui démontre la continuité (uniforme) de l'application

$\lambda \to u_\lambda : R^+ \longrightarrow V$ donc celle de $\lambda \to \|u_\lambda\|_1 : R^+ \to R^+$

On en tire la :

Proposition 7.2: L'équation $\lambda = \|u_\lambda\|_1$ admet une solution et une seule qui est égale à $\|u\|_1$.

Démonstration : Evidente et illustrée par la figure 7.1 ($u_o = A^{-1}f$)

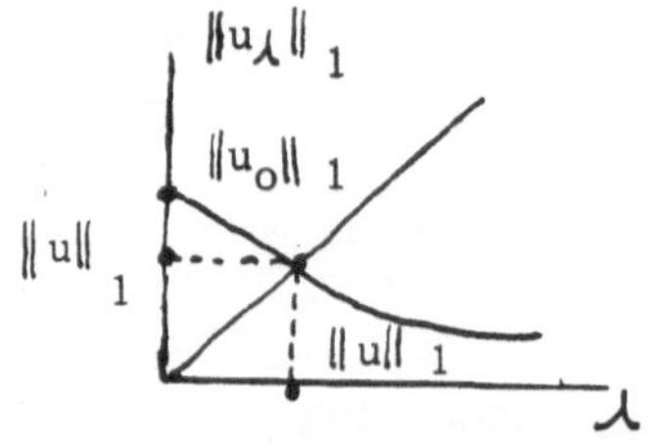

Figure 7.1

Donc si on sait résoudre (P_λ) on est ramené á la résolution du problème á une variable $\lambda - \|u_\lambda\|_1 = 0$; la fonction $\lambda \to \|u_\lambda\|_1$ étant continue et décroissante cela ne pose pas de difficultés.

En ce qui concerne (P_λ), on vérifiera qu'il entre dans le cadre des problèmes étudiés dans les N. précédents (avec $L = R^n$, $\Phi = I$, $\Lambda = -\left[\frac{\lambda}{2}, \frac{\lambda}{2}\right]^n$) et que l'algorithme étudié au N° 4 prend la forme :

$$\text{(7.9)} \qquad \begin{cases} p^o \text{ donné} \\ Au^n + p^n = f \\ p^{n+1} = P \quad (p^n + \rho^n u^n) \end{cases}$$

avec :

$$\text{(7.10)} \qquad (P_\Lambda(q))_i = \max\left(-\frac{\lambda}{2}, \min\left(\frac{\lambda}{2}, q_i\right)\right)$$

On démontrerait la convergence de (7. 11) sous la condition :

$$\text{(7.11)} \qquad 0 < \alpha \leq \rho_n \leq \beta < 2c$$

C étant (toujours) la plus petite valeur propre de A.

Ceci termine la remarque 7. 1.

<u>7.. 3 : Détermination d'un problème dual.</u>

1) On va ramener le problème non différentiable (7. 4) , problème sans contrainte à un problème avec contraintes , en effet (7. 4) est identique à :

$$\text{(7.12)} \qquad \begin{cases} \underset{v,p}{\text{Min}} \left[(Av, v) - 2\,(f, v) + p^2\right] \\ \text{sous la contrainte} \\ p - \|v\|_1 = 0 \end{cases}$$

Dans (7. 12) la contrainte est non convexe , mais un raisonnement trés simple montre qu'on peut la remplacer par $\|v_1\| - p \leq 0$ qui est de type convexe en (v, p) ; d'oú équivalence entre (7. 4), (7. 12) et (7. 13) défini par :

$$(7.13) \quad \begin{cases} \underset{v,p}{\text{Min}} \; [(Av, v) - 2(f, v) + p^2] \\ \text{sous la contrainte} \\ \|v\|_1 - p \leq 0 \end{cases}$$

2) Soit $\mathcal{L}: R^N \times R \times R \longrightarrow R$ définie par : (1)

$$(7.14) \quad \mathcal{L}(v, p; \lambda) = (Av, v) - 2\,(f, v) + p^2 + 2\,\lambda\,(\|v\|_1 - p)$$

A l'aide des théorèmes classiques de min-max (VON NEUMAN, SION ect. . .) ou directement, on montrerait que $(u, \|u\|_1 ; \|u\|_1)$ est point selle unique de $\mathcal{L}$ sur $R^n \times R \times R_+$ et que $\|u\|_1$ est multiplicateur de Lagrange des problèmes (7. 12), (7. 13).
On déduit de tout ceci :

$$(7.15) \quad \underset{v}{\text{Min}}\; J(v) = \underset{\lambda \geq 0}{\text{Max}}\; \underset{v,p}{\text{Min}}\; \mathcal{L}(v, p; \lambda)$$

mais :

$$(7.16) \quad \underset{v,p}{\text{Min}}\; \mathcal{L}(v, p; \lambda) = \underset{v}{\text{Min}}\; [(Av, v) - 2\,(f, v) + 2\,\lambda\,\|v\|_1 - \lambda^2]$$

d'oú :

---

(1) Ce λ est á distinguer de celui de la remarque 7. 1.

$$\underset{v}{\text{Min}}\ J(v) = \underset{\lambda \geq 0}{\text{Max}}\ \underset{v}{\text{Min}} \left[(Av,v) - 2(f,v) + 2\lambda \|v\|_1 - \lambda^2 \right]$$

$$= \underset{\lambda \geq 0}{\text{Max}}\ \underset{t,\|t\|_\infty \leq \lambda}{\text{Max}}\ \underset{v}{\text{Min}} \left[(Av,v) - 2(f,v) + 2(t,v) - \lambda^2 \right]$$

(7.17) $$= \underset{t}{\text{Max}}\ \underset{v}{\text{Min}}\ \underset{\lambda,\|t\|_\infty \leq \lambda}{\text{Max}} \left[(Av,v) - 2(f,v) + 2(t,v) - \lambda^2 \right]$$

$$= \underset{t}{\text{Max}}\ \underset{v}{\text{Min}} \left[(Av,v) - 2(f,v) + 2(t,v) - \|t\|_\infty^2 \right]$$

Mais $\underset{v}{\text{Min}} \left[(Av,v) - 2(f,v) + 2(t,v) \right] = -(A^{-1}(f-t), (f-t))$

d'oú posant $g = A^{-1} f$ la formulation du problème dual

(7.18) $$\underset{t}{\text{Min}} \left[(A^{-1}t,t) - 2(g,t) + \|t\|_\infty^2 \right]$$

la fonctionnelle $J^*(t) = (A^{-1}t,t) - 2(g,t) + \|t\|_\infty^2$ est non différentiable et l'utilisation d'une méthode de gradient sur le problème dual (7.18) conduit á des difficultés liées á la non différentiabilité de $t \to \|t\|_\infty^2$

8. Un problème de contrôle optimal avec fonction cout non différentiable

Soit Ω un ouvert borné de R^n de frontière suffisemment régulière et $Q = \Omega \times]0,T[$ (T fini), $\Sigma = \Gamma \times]0,T[$;

on se donne :

(8.1) $$u_{ad} = [v \mid v \in L^2(Q),\ v \geq 0 \text{ p.p.}]$$

" l'équation d'état " :

(8.2) $$\begin{cases} \dfrac{\partial y}{\partial t} - \Delta y = f + v & \text{sur } \Omega \times]0,T[\\ y(x,t) = 0 \text{ sur } \Sigma \\ y(x,o) = y_o(x) & \text{dans } \Omega \end{cases}$$

avec $f \in L^2(Q)$, $y_o \in L^2(Q)$.

Le contrôle v étant donné, (8. 2) admet une solution unique dans $H^1(Q)$, soit y (v) , et on peut définir la fonction coût :

(8. 3) $$J(v) = \int_Q |y(v) - z_d|^2 \, dx\, dt + \alpha \int_Q |v|^2 dx\, dt \quad 2\beta \int_Q |v| \, dx\, dt$$

avec $\alpha > o$, $\beta > o$ et $z_d \in L^2(Q)$.

Il résulte de LIONS [1] que le problème :

(8. 4) $$\underset{v \in \mathcal{U}_{ad}}{\text{Min}} \; J(v)$$

admet une solution unique , soit u , qui est le contrôle optimal. Compte tenu de LIONS , loc. cit., et des N. 2 et 3 , on montrerait facilement l'existence et l'unicité de (y, p, u, λ) avec $|\lambda(x\ t)| \leq 1$ p. p. sur Q solution de :

(8. 5) $$\begin{cases} \dfrac{\partial y}{\partial t} - \Delta y = f + u & \text{sur } \Omega \times]0, T[\\ y(x,t) = o & \text{sur } \Sigma \\ y(x,o) = y_o(x) & \text{sur } \Omega \end{cases}$$

(8. 6) $$\begin{cases} -\dfrac{\partial p}{\partial t} - \Delta p = y - z_d & \text{sur } \Omega \times]0, T[\\ p(x,t) = o & \text{sur } \Sigma \\ p(x,T) = o & \text{sur } \Omega \end{cases}$$

(8. 7) $$\begin{cases} \alpha u + \beta\lambda + p \geq 0 & \text{p. p. sur } Q \\ u \geq 0 & \text{"} \\ |\lambda| \leq 1 & \text{"} \\ (\alpha u + \beta\lambda + p)\, u = 0 & \text{"} \\ \lambda u = |u| & \text{"} \end{cases}$$

Réciproquement si (y, p, u, λ) est solution de (8. 5), (8. 6), (8. 7) u est contrôle optimal.

L'utilisation de (8. 5), (8. 6), (8. 7), associée á des algorithmes du type de ceux développés aux N. 4 et 5 , a donné des résultats numériques satisfaisants pour le problème (8. 4) , même pour des valeurs assez petites de α.

9.- Une remarque sur le comportement des multiplicateurs de Lagrange.

Il a souvent été constaté que, la fonctionnelle J étant donnée, l'efficacité des méthodes duales (en particulier celles des N. 4 et 5) était une fonction " décroissante" du convexe K, ceci n'a rien de trés surprenant puisqu'á la limite lorsque K se réduit á un point le problème d'optimisation considéré n'admet pas de multiplicateurs de Lagrange en général. On va mettre ce phénomème en évidence sur un exemple trés simple :

Soit $V = R$, J définie par :

(9. 1) $J(y) = y^2 - 2y$

On prend pour K_ε le convexe $[y \mid |y| \leq \varepsilon]$, l'appartenance á K_ε étant donnée par :

(9. 2) $y^2 - \varepsilon^2 \leq 0$

D'oú le Lagrangien :

(9. 3) $\mathcal{L}(y, \lambda) = y^2 - 2y + \lambda(y^2 - \varepsilon^2)$

La solution optimale de

(9. 4) $\operatorname*{Min}_{y \in K_\varepsilon} J(y)$

est donnée de façon évidente par :

(9.5) $$y_\varepsilon = \min (1, \varepsilon)$$

et le multiplicateur de Lagrange λ_ε correspondant s'obtient á partir de :

(9.6) $$\min_{y \in K_\varepsilon} J(y) = \min_{y \in R} \mathcal{L}(y, \lambda_\varepsilon)$$

d'oú :

(9.7) $$\begin{cases} \lambda_\varepsilon = 0 & \text{si } \varepsilon \geq 1 \\ \lambda_\varepsilon = \dfrac{1-\varepsilon}{\varepsilon} & \text{si } \varepsilon < 1 \end{cases}$$

et lorsque $\varepsilon \to 0$ i.e. $K_\varepsilon \to \{0\}$, $\lambda_\varepsilon \to +\infty$

R. Glowinski

BIBLIOGRAPHIE

ARROW K. J. - HURWICZ L. [1] : Dans Arrow-Hurwicz-Uzawa, Studies in linear and non linear programming. - Stanford University Press 1958.

CEA J. - GLOWINSKI R. [1] : Méthodes numériques pour l'écoulement laminaire d'un fluide rigide visco-plastique incompressible - A paraître - (Rapport IRIA disponible)

CEA J. - GLOWINSKI R. - NEDELEC J. C. - [1] : Méthodes duales pour la minimisation de fonctionnelles non différentiables - A paraitre dans les proceedings du Colloque d'analyse numérique de DUNDEE-1971, SPRINGER VERLAG.

DUVAUT G. - LIONS J. L. : Les inéquations en mécanique et en physique - DUNOD 1971

GLOWINSKI R. [1] : Méthodes Numérique pour l'écoulement stationnaire d'un fluide rigide visco-plastique incompressible - Proceedings of the 2. nd. Int. Conf. on Num. Methods in Fluid Dynamics- Lecture Notes in physics, 8 , Springer-Verlag 1971

[2] : Exposé à cette réunion CIME consacré à l'Analyse Numérique du Problème elasto-plastique.

GLOWINSKI-LIONS-TREMOLIERES. [1] : Livre sur l'Analyse Numérique des inéquations variationnelles, à paraitre an 1972 chez DUNOD

GOURSAT M. [1] : Analyse Numérique de problèmes d'élasto-plasticité et de visco-plasticité. - Thése de 3 cycle - PARIS-IRIA, 1971

KY-FAN [1] : Sur un théorème de Min Max C.R.A.S. - Paris, 259, 3925-3928, Paris.

LIONS J.L. [1] : Controle optimal de systèmes gouvernés par des équations aux dérivées particelles - DUNOD-GAUTHIER-VILLARS 1968

MOSCO U. [1] : Exposé á cette réunion CIME

ROCKAFELLAR R.T. [1] : Convex Analysis - Princeton Univ. Press 1970

SION M. [1] : On general min max theorems - Pacific J. Math. 8, 1958, 171-176.

UZAWA H. [1] : Dans ARROW- HURWICZ-UZAWA loc. cit.

CENTRO INTERNAZIONALE MATEMATICO ESTIVO

(C.I.M.E.)

J. L. LIONS

APPROXIMATION NUMÉRIQUE DES INÉQUATIONS D'ÉVOLUTION

Corso tenuto ad Erice dal 21 giugno al 7 luglio 1971

APPROXIMATION NUMERIQUE DES INÉQUATIONS D'ÉVOLUTION.

J.L. LIONS (Paris)

Introduction.- On donne dans ce cours les méthodes fondamentales pourla résolution numérique des inéquations d'évolution intervenant en Mécanique et en Physique.

Lex expériences numérique, faites à l'I.R.I.A. (Paris), seront présentés avec toues les détails dans un livre de R. Glowinski, R. Trémolières et l'A., à paraitre chez Dunod.

Plan détaillé.

CHAPITRE 1.- Inéquations d'evolutions parabolique. Type I.

1. Exemples.
2. Formulation générale.
3. Solutions fortes et faibles.
4. Généralités sur les méthodes constructives d'approximation.
   - 4.1 Réduction à un équation parabolique. Pénalisation.
   - 4.2 Réduction à un équation parabolique. Régularisation.
   - 4.3 Réduction à un inéquation elliptique. Régularisation elliptique.
   - 4.4 Réduction à un inéquation elliptique. Discrétisation.
   - 4.5 Inéquation d'évolution et points selles.

J.L. Lions

CHAPITRE I.- Inequations d'evolutions paraboliques Type I.

1.- Exemples.

Exemple 1.1.- La théorie de la diffusion en milieux poreux (cf. Duvaut-Lions [1]) conduit à des problèmes du type suivant:

Soit Ω ouvert borné de $\mathbb{R}^n$ (n=2 ou n=3 dans les applications) de frontière Γ "régulière". Soit $\vec{n}$ la normale à Γ dirigée vers l'extérieur de Ω. On cherche une fonction $u=u(x,t)$, $x \in \Omega$, $t > 0$, solution de

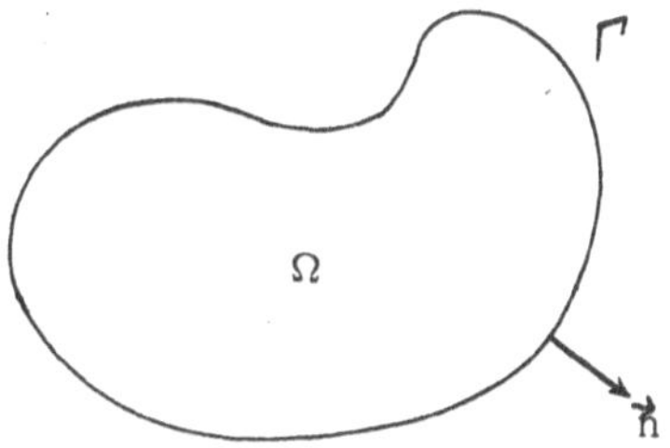

(1.1) $\frac{\partial u}{\partial t} - \Delta u = f$ dans $\Omega \times]0,T[$, $T>0$ fini quelconque,

(où f est donnée dans $\Omega \times]0,T[$), avec la condition initiale

(1.2) $u(x,0) = u_o(x)$, u_n donné dans Ω

et les conditions aux limites

$$(1.3) \quad \begin{cases} u \geq 0 & \text{sur } \Sigma = \Gamma \times]0,T[\,, \\ \frac{\partial u}{\partial n} \geq 0 & \text{sur } \Sigma, \\ u \frac{\partial u}{\partial n} = 0 & \text{sur } \Sigma \,. \end{cases}$$

Remarque 1.1.-

Le problème (1.1), (1.2),(1.3) est non linéaire à cause des conditions aux limites (1.3).

Remarque 1.2.-

D'après $u \frac{\partial u}{\partial n} = 0$ sur Σ, on voit que

$u = 0$ sur $\Sigma_o \subset \Sigma$

$\frac{\partial u}{\partial n} = 0$ sur $\Sigma - \Sigma_0$.

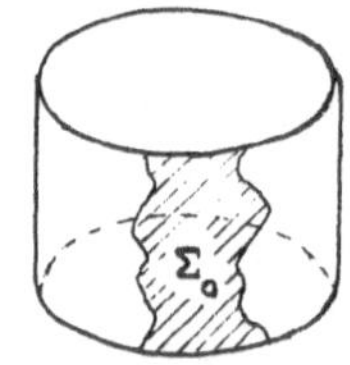

Mais Σ_0 n'est pas donné à priori.

Orientation.

Le but de ce premièr chapitre est: a) de montrer brièvement comment le problème (1.1),(1.2),(1.3) (et, a vrai dire, des problèmes beaucoup plus généraux) est bien posé ([1]);

b) de donner des méthodes d'approximation numérique de la solution du problème.

Donnons un 2ème exemple.

Exemple 1.2.-

On cherche u satisfaisant à (1.1),(1.2) et aux conditions aux limites

$$(1.4) \quad \begin{cases} \left|\frac{\partial u}{\partial n}\right| \leq g \quad \text{sur } \Sigma \text{ , g constante } > 0, \\ u \frac{\partial u}{\partial n} + g|u| = 0 \text{ sur } \Sigma \text{ .} \end{cases}$$

Autrement dit:

$$(1.5) \quad \begin{cases} \frac{\partial u}{\partial n} < g \Rightarrow u=0, \\ \frac{\partial u}{\partial n} = g \Rightarrow u \leq 0, \\ \frac{\partial u}{\partial n} = -g \Rightarrow u \geq 0. \end{cases}$$

On verra que ce problème est encore bien posé.

([1]) Pour une étude plus systématique de la théorie, cf. H BRÉZIS [1], J.L. LIONS [1][2] .

2. Formulation générale.

Nous donnous maintenant une formulation "abstraite" de problèmes d'inequations de type parabolique, puis nous montrons comment cette formulation contient, en particulier, les exemples du N.1.

Soient V et X deux espaces de Hilbert ([1]) sur $\mathbb{R}$, avec

(2.1) $V \subset H$, V dense dans H, l'injection de V dans H étant continue.

On désigne par:

(2.2) $|\ \ |$ la norme dans H, $(\ ,\ )$ le produit scalaire correspondant dans H, $\|\ \ \|$ la norme dans V

D'après (2.1), il existe une constante $c>0$ telle que

(2.3) $|v| \leqslant c\|v\| \qquad \forall v \in V.$

On se donne ensuite:

(2.4) $a(u,v)$=forme bilinéaire continue sur $V \times V$, coercive au sens: il existe λ tel que $a(v,v)+\lambda|v|^2 \geqslant \alpha\|v\|^2$, $\alpha>0$, $\forall v \in V$,

et on se donne encore:

(2.5) K = ensemble convexe fermé dans V;

(2.6) j = fonction convexe continue de $V \to \mathbb{R}$.

On identifie H à son dual et l'on introduit l'espace V' dual de V

([1]) On peut aller beaucoup plus loin, en prenant pour V un espace de Banach réflexif. Cf. Lions [2] et la bibliografie de ce livre.

de sorte que

(2.7) $$V \subset H \subset V'.$$

Si $f \in V'$, on désigne par (f,v) son produit scalaire avec $v \in V$; cette notation est compatible avec celle du produit scalaire dans H.

Le problème.

On cherche une fonction $t \to u(t)$ de $[0,T] \to V$ [1] telle que

(2.8) $u(t) \in K$,

(2.9) $$\left(\frac{\partial u(t)}{\partial t}, v-u(y)\right)+a(u(t),v-u(t)) \geq (f(t),v-u(t)) \quad \forall v \in K,$$

(2.10) $u(0)=u_o$.

Un autre problème est:

On cherche $u=u(t)$ de $[0,T] \to V$ telle que

(2.11) $$\left(\frac{\partial u(t)}{\partial t}, v-u(t)\right)+a(u(t),v-u(t))+j(u)-j(u(t)) \geq (f(t),v-u(t)) \quad \forall v \in V,$$

avec (2.10).

L'inéquation (2.9) ou (2.11) est ce qu'on appellera ici une inéquation parabolique de type I [2].

Remarque 2.1.

Si K=V ou si j=0, (2.9) et (2.11) se réduisant à l'equation:

(2.12) $$\left(\frac{\partial u(t)}{\partial t}, v\right)+a(u(t),v)=(f(t),v) \quad \forall v \in V.$$

---

[2] Cf. les inéquations du type II au Chap.3.

[1] Dont il faudra précisèr les propriétés.

Remarque 2.2.-

Si la fonction $v \to j(v)$ est différentiable sur V alors (2.11) équivaut à l'équation (en général non linéaire):

(2.13) $$\left(\frac{\partial u(t)}{\partial t}, v\right) + a(u(t),v) + (j'(u(t)),v) = (f(t),v), \qquad \forall v \in V.$$

Remarque 2.3.-

Introduisons $A \in \mathcal{L}(V;V')$ par

(2.14) $$a(u,v) = (Au,v) \; .$$

Alors (2.13) équivaut à

(2.15) $$\frac{\partial u(t)}{\partial t} + Au(t) + j'(u(t)) = f(t).$$

Remarque 2.4.-

Si l'on considère la fonction ψ_k indicatrice de K ([1]):

(2.16) $$\psi_k(v) = \begin{cases} 0 & \text{si} \quad v \in K, \\ +\infty & \text{si} \quad v \notin K \end{cases}$$

alors (2.8) (2.9) équivaut à :

(2.17) $$\left(\frac{\partial u(t)}{\partial t}, v-u(t)\right) + a(u(t),v-u(t)) + \psi_k(v) - \psi_k(u(t)) \geq$$
$$\geq (f(t),-v-u(t)) \qquad \forall v \in V.$$

Les inéquations (2.9) sont donc des cas particuliers de l'inéquation

---

([1]) La fonction ψ_k est convexe et semi continue inférieurement.

$$(2.18)\quad \left\{ \begin{array}{l} (\frac{\partial u(t)}{\partial t},v-u(t))+a(u(t),v-u(t))+\psi(v)-\psi_K(u(t)) \geq (f(v),v-u(t)) \\ \qquad\qquad\qquad\qquad\qquad\qquad\qquad\qquad \forall v \in V, \end{array} \right.$$

où ψ est une <u>fonction convexe propre</u> (cf. le cours de U. Mosco [1]). Utilisant la notion de <u>sous différential</u>, on voit que (2.18) équivaut à

$$(2.19)\qquad -(\frac{\partial u(t)}{\partial t} + Au(t)-f(t)) \in \partial\varphi(u(t)),$$

equation parabolique <u>multivoque</u>.

<u>Exemple</u> 2.1.-

Voyons comment l'enoncé général recouvre le problème de l'Exemple 1.1.

On introduit (notations des cours de R. Glowinski et U. Mosco):

$$(2.20)\qquad V = H^1(\Omega), \qquad H=L^2(\Omega),$$

$$(2.21)\qquad K= \{v \mid v \in H^1(\Omega), \quad v \geq 0 \text{ sur } \Gamma\},$$

$$(2.22)\qquad a(u,v)= \sum_{i=1}^{n} \int_\Omega \frac{\partial u}{\partial x_i}\frac{\partial v}{\partial x_i}\,dx$$

Alors le problème (2.8),(2.9),(2.10) <u>équivaut</u> au problème (1.1),(1.2),(1.3).

<u>Exemple</u> 2.2.-

On prend V,H,a(u,v) comme en (2.20),(2.22) et l'on introduit

$$(2.23)\qquad j(v) = g\int_\Gamma |v|d\Gamma .$$

Alors le problème (2.11),(2.10) <u>équivaut</u> au problème (1.1), (1.2),(1.4).

Oriéntation.

On va maintenant précisér à quel sens on entend les "solutions" des problèmès précédents.

3.- Solutions fortes et faibles.

Solutions fortes.

Par "solution forte" du problème (2.8),(2.9),(2.10) on entendra une fonction u telle que

(3.1) $u \in L^2(0,T;V)$, (1)

(3.2) $\frac{\partial u}{\partial t} \in L^2(0,T;V')$

(3.3) $u(t) \in K$ p.p. en t (ou pour tout t $\in [0,T]$)

(3.4) sauf peut être pour t dans un ensemble $Z \subset [0,T]$ de mesure nulle, on a:
$$\left(\frac{\partial u(t)}{\partial t}, v-u(t)\right)+a(u(t),v-u(t)) \geqslant (f(t),v-u(t)) \qquad \forall v \in K,$$

et naturellement (2.10) (2):

(3.5) $u(0)=u_0$.

Evidemment (3.3) impose (3)

---

(1) $L^2(0,T;X)$=espace des "fonctions" $t \to u(t)$ de $[0,T] \to X$ qui sont mesurables et telles que $\int_0^T ||u(t)||_X^2 \, dt < \infty$.

(2) Il résulte de (3.1) et (3.2) que $t \to u(t)$ est, après modification éventuelle sur un ensemble de mesure nulle, continue de $[0,T] \to H$. Alors u(0) a un sens.

(3) Tant que l'on travaille avec des solutions fortes où (3.3) a lieu pour tout t.

(3.6) $u_o \in K$.

Il est important pour les applications d'introduire une notion de solution faible (cf. Lions-Stampacchia [1], Brézis [2]). Pour simplifier l'exposé nous prenons

(3.7) $u_o = 0$ (donc $0 \in K$).

On observe alors que si u est solution "forte" de (3.4), on a:

$$(3.8) \quad \begin{cases} \int_0^T \left[\left(\frac{\partial v}{\partial t}, v-u \right) + a(u,v-u) - (f,v-u) \right] dt \geq 0 \\ \forall v \in L^2(0,T;V) \text{ tel que } \frac{\partial v}{\partial t} \in L^2(0,T,V') \text{ et } v(t) \in K \text{ p. p.} \\ \text{et } v(0)=0. \end{cases}$$

Mais comme $\frac{\partial u}{\partial t}$ n'intervient plus dans (3.8) on peut définir u comme solution faible si u satisfait à (3.1),(3.3) et (3.8).

Remarque 3.1.-

On a évidemment des notions analogues de solutions "fortes" et "faibles" relativement à l'inéquation (2.11).

Remarque 3.2.-

Seuil de régularité.

La solution u(t) des problèmes précédents n'est pas une fonction "très régulière" de t, quelle que soit la régularité des donnés f et u_o.

Prenons en effet

$V=H=\mathbb{R}$, $a(u,v)=0$ (qui vérifie (2.4) lorsque V=H),

$u_o>0$, $f=-1$,

$K = \{v \mid v \geq 0\}$

La solution est indiquée sur le graphe ci contre. On voit que, en particulier, $\frac{\partial^2 u}{\partial t^2} \notin L^2(O,T)$.

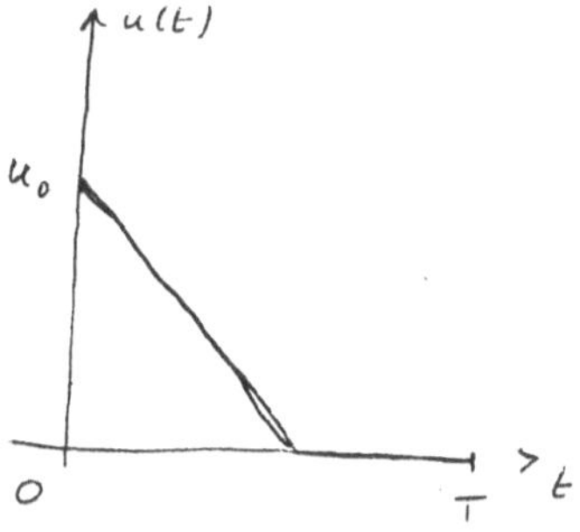

Resultats généraux.

On démontre ([1]) les résultats suivants (cf. par ex. Lions [2] et la bibliographie de ce travail):

Théorème 3.1.- On suppose $f \in L^2(O,T;V')$. On suppose que (2.4) a lieu. Il existe alors une fonction u et un seule satisfaisant à (3.1),(3.3) (3.8).

Théorème 3.2.- On suppose que (2.4) a lieu et que

$$f, \frac{\partial f}{\partial t} \in L^2(O,T;V'),$$

$$u_o \in K, \quad f(O)-Au_o \in H.$$

Il existe alors un solution forte et une seule de (3.1)...(3.5).

Remarque 3.3.-

On a des énoncés analogues pour l'inéquation (2.11).

Remarque 3.4.-

L'unicité des solutions fortes est immédiate. Pour l'unicité des solutions faibles, si u_1 et u_2 sont deux solutions éventuelles, on introduit:

([1]) Nous donnons ci après quelques indications sur les méthodes constructives possibles de démonstration et au Chap.2 nous donnons l'approximation numérique de la solution (qui peut, d'ail-

$$w = \frac{1}{2}(u_1+u_2) \, , \qquad \text{puis } w_\varepsilon \text{ solution de}$$

$$\varepsilon \, w'_\varepsilon + w_\varepsilon = w, \qquad w_\varepsilon(0)=0$$

et l'on prend $v=w_\varepsilon$ dans chacun des inéquations (3.8) relatives à u_1 et u_2 .

On additionne et on peut alors faire $\varepsilon \to 0$. Cf. H. Brézis [2].

4.- Généralités sur les méthodes constructives d'approximation.

4.1.- Réduction à une équation parabolique. Pénalisation. Soit β un opérateur de pénalisation attaché à K (cf. Lions [2], p.370 et les cours de R. Glowinski et U. Mosco). On "approche" (3.4) par l'équation pénalisée

$$(4.1) \quad (\frac{\partial u}{t}, v)+a(u_\varepsilon(t),v)+\frac{1}{\varepsilon}(\beta(u_\varepsilon(t)),v)=(f(t),v) \qquad \forall\, v \in V,$$

où $\varepsilon>0$ est déstiné à tendre vers 0, avec

$$(4.2) \qquad u_\varepsilon(0)=u_0.$$

Il s'agit d'un problème parabolique non linéaire monotone (car β est, par définition, monotone de $V \to V'$) dont on sait qu'il admet une solution unique.

On montre (Lions [2]) que, par ex. sous les conditions du Théorème 3.1, $u_\varepsilon \to u$ dans $L^2(0,T;V)$ faible lorsque $\varepsilon\to 0$ où u est solution faible.

On peut ensuite approcher u_ε par l'une des méthodes de reso

4.2.- Réduction à un équation parabolique. Régularisation.

Dans le cas des inéquations (2.11) on peut introduire $j_\varepsilon(v)$, fonctionnelle convexe différentiable approchant $j(v)$. Par ex. si $j(v)=g\int_\Gamma |v|d\Gamma$, on prendra

$$j_\varepsilon(v)=g\int_\Gamma \varphi_\varepsilon(v)d\Gamma$$

où φ_ε est convexe, differentiable, et par exemple $\varphi_\varepsilon(\lambda)=|\lambda|$ si $|\lambda|\geq\varepsilon$

On "approche" alors (2.11) par l'équation régularisée

$$(4.3)\qquad (\frac{\partial u_\varepsilon}{\partial t}, v)+a(u_\varepsilon,v)+(j'_\varepsilon(u_\varepsilon),v)=(f,v) \qquad \forall v\in V,$$

avec (4.2).

Il s'agit là encore d'une équation parabolique non linéaire monotone et on verifie encore que $u_\varepsilon \to u$ dans $L^2(O,T;V)$ faible, u solution faible.

4.3.- Réduction à un inéquation elliptique. Régularisation elliptique.

Pour réduire par ex. (3.8) à un situation dejà connue, nous nous sommes ramenés jusq'ici à des équations d'évolution. On peut essajer de se ramener à des inéquations stationnaires. Pour cela, on considère le probleme de nature elliptique: trouver $u_\varepsilon \in \mathcal{K}_o$ où

$$(4.4)\qquad \mathcal{K}_o = \{v \mid v\in L^2(O,T;V),\ \frac{\partial v}{\partial t}\in L^2(O,T;H),v(t)\in K \ \text{p.p.},\ v(O)=O\},$$

solution de

$$(4.5)\begin{cases} \varepsilon \int_0^T \left(\frac{\partial u_\varepsilon}{\partial t}, \frac{\partial(v-u_\varepsilon)}{\partial t}\right) dt + \int_0^T \left[\left(\frac{\partial u_\varepsilon}{\partial t}, v-u_\varepsilon\right)+a(u_\varepsilon, v-u_\varepsilon)- \right. \\ \left. \qquad -(f, v-u_\varepsilon)\right] dt \geq 0 \\ \forall v \in \mathfrak{K}_0. \end{cases}$$

Cette inéquation entre dans le cadre de inéquations variationnelles elliptiques étudiées dans le cours de R. Glowinski et U. Mosco (1).

On montre (cf. Lions-Stampacchia [1] par ex.) l'existence de u_ε solution de (4.5) et la convergence de u_ε vers la solution faible lorsque $\varepsilon \to 0$.

Remarque 4.1.-

On peut utiliser pour l'approximation de la solution u_ε de (4.5) les méthodes des cours Glowinski et Mosco. Il est possible (mais non encore verifié sur des exemples numériques) que l'usage de (4.5) soit utile pour des calculs sur de longs intervalles de temps. Cf. Carasso [1] pour le cas de équations.

Remarque 4.2.-

On peut utilisér simultanément les idées de 4.1 on 4.2 et 4.3. On peut donc se ramener à des équations stationnaires.

4.4.- Réduction à une inéquation elliptique. Discrétisation.

La méthode peut être la plus naturelle de réduction au cas

---

(1) On notera que le problème (4.5) est non symétrique même si l'on part d'une forme a(u,v) symétrique.

elliptique est d'utilisér la discrétisation de la derivée en t.

En raison de l'importance essentielle de ce procédé pour les applications numériques, nous étudions cela en détail au Chap. 2..

4.5.- Inéquations d'évolution et points selles (cf. Trémolières [1]).

On introduit

(4.6) $$L(u,w) = \int_0^T [(v',v-w)+a(v,v-w)-(f,v-w)]dt$$

et l'ensemble

(4.7) $$\mathcal{K} = \{v \mid v\in L^2(0,T;V),\ \frac{\partial v}{\partial t}\in L^2(0,T;V'),\ v(0)=0,\ v(t)\in K\}.$$

On vérifie que si u est solution forte alors

(4.8) $$L(u,w) \leqslant L(u,u)=0 \leqslant L(v,u) \qquad \forall v,w\in\mathcal{K}$$

autrement dit: $\{u,u\}$ est point selle de $L(v,w)$ sur $\mathcal{K}\times\mathcal{K}$.

Réciproquement, soit $\{u,\hat{u}\}$ point selle de $L(u,w)$ sur $\mathcal{K}\times\mathcal{K}$ Alors

(4.9) $$L(u,w) \leqslant L(u,\hat{u}) \leqslant L(v,\hat{u}) \qquad \forall v,w\in\mathcal{K}$$

d'où l'on déduit (en observant que $L(u,u)=L(\hat{u},\hat{u})=0$) que $L(u,\hat{u})=0$. Alors

$$L(v,\hat{u})\geqslant 0 \qquad \text{et} \qquad L(w,u)=-L(u,w)\geqslant 0$$

donc d'après l'unicité dans le Théorème 3.1 (ou 3.2) on a:

$$u = \hat{u}.$$

Donc si u est solution, alors $\{u,u\}$ est point selle de $L(v,w)$ sur $\mathcal{K}\times\mathcal{K}$ et réciproquement.

On peut déduire de là une méthode de démonstration de l'existence de solutions.

En effet si K est borné dans V, alors $\mathcal{K}$ est borné dans l'espace W des fonctions $v \in L^2(0,T;V)$ avec $\frac{\partial v}{\partial t} \in L^2(0,T;V')$ et l'existence d'un point selle (nécessairement de la forme $\{u,u\}$) est conséquence d'un résultat classique di Von Neumann.

Si K n'est pas borné, on introduit

$$K_R = \{v \mid v \in K, \ ||v|| \leq R\};$$

soit u_R la solution de

$$(4.10) \qquad \left(\frac{\partial u_R}{\partial t}, v-u_R\right) + a(u_R, v-u_R) \geq (f, v-u_R) \qquad \forall v \in K_R.$$

Prenant dans (4.10 v=0 on en déduit que

$$(4.11) \qquad \int_0^T ||u_R||^2 dt \leq c = \text{constante indépendante de } R$$

d'où l'on déduit l'existence d'une solution faible u et

$$(4.12) \qquad u_R \to u \text{ dans } L^2(0,T;V) \text{ faible lorsque } R \to +\infty \ .$$

CHAPITRE 2.- Approximation par discrétisation des inéquations paraboliques de type I.

1.- Approximation d'un couple d'espaces Constante de stabilité.

Soient V et H deux espaces de Hilbert ([1]) comme au Chap.1, N.2. Dans la théorie des approximations internes (cf. Aubin [1], Céa [1]) on considère une famille de sous espaces V_h de V ([2])

(1.1) $$V_h \subset V$$

avec

(1.2) V_h de dimension finie N(h), N(h)→∞ lorsque "h→0",

et les V_h réalisant une approximation de V (cf. les cours de Strang, Glowingki dans ce volume).

La norme de V - resp. de H - induit sur V_h une norme,

$$||v|| \quad (\text{resp } |v|)$$

et comme V_h est de dimension finie ces deux normes sont équivalentes sur V_h. Evidemment (2.3) Chap. 1 est valable sur V_h, i.e. $|v| \leq c||v|| \forall v \in V_h$, et d'après l'équivalence:

(1.3) $$\begin{cases} \forall h \text{, il existe une constante } S(h) \text{ telle que} \\ ||v|| \leq S(h)\,|v| \qquad \forall v \in V_h. \end{cases}$$

La constante S(h) est dite: constante de stabilité du couple {V,H} pour la famille V_h.

Remarque 1.1-

On a (sauf si V=H);

(1.4) $$S(h) \to +\infty \quad \text{si } h \to 0.$$

(1) Cela s'etend aux espaces de Banach -

(2) h est un paramètre scalair ou vectoriel representand soit le maillage dans la méthode des différences finies soit la "triangulation" en simplexes dans la méthode des éléments finis (cf le cours de Strang, ce volume et Ciarlet-Raviant [1]

En effet sinon les normes $|\quad|$ et $||\quad||$ seraient équivalentes sur V, d'où V=H puis que V est dense dans H.

Cette remarque fait toute la différence entre le cas des équations différentielles (où V=H) et le cas des équations aux dérivées partielles (où $V \subset H$ strictement).

Remarque 1.2.-

Evaluations de S(h).

Si $V=H^1(\Omega)$ ou $H_0^1(\Omega)$ et $H=L^2(\Omega)$ alors

(1.5) $$S(h) = O\left(\frac{1}{\sqrt{|h|}}\right)$$

où h désigne:

soit le pas du maillage

soit la longuer maximum du coté de la triangulation (dont les angles sont supposés uniformement $\geq \vartheta_0 > 0$).

Si $V=H^m(\Omega)$ ou $H_0^m(\Omega)$ et $H=L^2(\Omega)$ alors

(1.6) $$S(h) = O\left(\frac{1}{|h|^m}\right).$$

Cf. Aubin [1] Bramble-Schatz, Ciarlet-Raviart [1], Fix-Strong [1], Zlamal [1].

Remarque 1.3.-

Pour simplifier l'exposé, nous utiliserons seulement les approximation internes.

Mais, à des difficultés d'écriture près, tout ce qui suit s'étend aux approximations externes.

2.- Schémas d'approximation des inéquations paraboliques de type I.

Notations.-

On introduit un pas Δt de discrétisation en t [1].

On désigne par u_h^n ou plus simplement u^n l'approximation de u à l'instant $n\Delta t$.

Cela suppose choisis:

a) une approximation V_k de V;

b) une approximation K_h de K et j_h de j (cf. les cours de Glowinski et Mosco dans ce Volume).

Commençons par l'approximation de l'inéquation (2.9) Chap. 1.

Si l'on remplace $\frac{\partial u(t)}{\partial t}$ par $\frac{u^{n+1}-u^n}{\Delta t}$, et si l'on remplace dans (2.9) v-u(t) par v-u*, u* à choisir, on obtient:

$$(2.1) \qquad \left(\frac{u^{n+1}-u^n}{\Delta t}, v-u^*\right)+a(u^?,v-u^*) \geq (f^n,v-u^*) \qquad \forall v \in K_h,$$

où $u^?$ est également à choisir, et où $u^{n+1} \in K_n$.

Raisonnons par analogie formelle avec le cas des équations (ce qui correspondrait au cas $K_h=V_h$ dans (2.1)). Alors les schémas classiques sont:

$$(2.2) \qquad \left(\frac{u^{n+1}-u^n}{\Delta t}, v\right)+a(u^n,v) = (f^n,v), \text{ schéma "explicite" } [2],$$

---

[1] On peut étendre ce qui suit au cas de pas variables Δt^n.

[2] Bien que, dans le cas des éléments finis, on ait à résoudre un système non diagonal.

(2.3) $\left(\frac{u^{n+1}-u^n}{\Delta t}, v\right) + a(u^{n-1}, v) = (f^n, v)$, schéma <u>implicite</u>,

et plus généralement, si l'on introduit

(2.4) $u^{n+\vartheta} = \vartheta u^{n+1} + (1-\vartheta) u^n$, $0 \leqslant \vartheta \leqslant 1$,

on considère le schéma

(2.5) $\left(\frac{u^{n+1}-u^n}{\Delta t}, v\right) + a(u^{n+\vartheta}, v) = (f^n, v)$

Le cas $\vartheta = \frac{1}{2}$ correspond au schéma de Crank-Nicolson

Cela conduit raisonnablement à prendre dans (2.1) $u^? = u^{n+\vartheta}$ d'où:

(2.6) $\left(\frac{u^{n+1}-u^n}{\Delta t}, v-u^*\right) + a(u^{n+\vartheta}, v-u^*) \geqslant (f^n, v-u^*) \quad \forall v \in K_h$,

toujours avec $u^{n+1} \in K_h$ (1).

Reste à choisir u^* de façon que

(i) le schéma <u>ait un sens</u>:

(ii) on puisse analyser <u>la stabilité et la convergence du schéma</u>.

Relativement au point (i), notons que l'<u>on ne peut pas prendre</u> $u^* = u^n$, l'inéquation (2.6) correspondante <u>ne définit pas</u> u^{n+1} de manière unique. Il reste alors deux possibilités: $u^* = u^{n+1}$ et, plus généralement $u^* = u^{n+\sigma}$, $0 < \sigma \leqslant 1$. (Notons que u^{n+1} peut ne pas appartenir à K_h pour $\sigma < 1$).

Nous nous bornerons dans ce qui suit au cas $\sigma = 1$, d'où les schémas

(2.7) $\left(\frac{u^{n+1}-u^n}{\Delta t}, v-u^{n+1}\right) + a(u^{n+\vartheta}, v-u^{n+1}) \geqslant (f^n, u-u^{n+1}), \quad \forall v \in K_h$,

---

(1) Ou bien une "approximation" de u^{n+1} est dans K_h

avec

(2.8) $u^{n+1} \in K_h,$

et

(2.9) $u^o =$ approximation dans V_h de u_o.

L'inéquation (2.7) équivaut à

(2.10) $$(u^{n+1}, v-u^{n+1}) + \vartheta \Delta t\, a(u^{n+1}, v-u^{n+1}) \geqslant$$

$$\geqslant (u^n + \Delta t\, f^n, v-u^{n+1}) - (1-\vartheta)\Delta t\, a(u^n, v-u^{n+1})$$

$$\forall v \in K_h;$$

pour Δt assez petit on a <u>sur</u> V_h

$$\vartheta \Delta t\, a(v,v) + |v|^2 \geqslant \beta |v|^2, \qquad \beta > 0 \quad (^1)$$

de sorte que (2.10) definit u^{n+1} de façon unique dans K_h.

<u>Remarque 2.1</u>.-

Si $\vartheta = 0$, (2.10) se réduit à

(2.11) $$(u^{n+1}, v-u^{n+1}) \geqslant (u^n + \Delta t f^n - \Delta t\, A_h u^n, v-u^{n+1}) \qquad \forall v \in K_h$$

où

$$a(u,v) = (A_h u, v), \quad A_h \text{ opérateur de } V_h \to V_h' .$$

Introduisons $\bar{u}^{n+1}$ par

(2.12) $$\frac{\bar{u}^{n+1} - u^n}{\Delta t} + A_h\, u^n = f^n.$$

---

(1) Cela même sans l'hypothèse (2.4), Chap. 1; cette hypothèse interviendra dans l'analyse de la <u>stabilité</u> et de la <u>convergence</u>.

Alors (2.11) équivaut à

(2.13) $\qquad (u^{n+1}-\bar{u}^{n+1}, v-u^{n+1}) \geq 0 \qquad \forall v \in K_h$

d'où

(2.14) $\qquad u^{n+1} = P_{K_h}(\bar{u}^{n+1}),$

où P_{K_h} = projecteur de $V_h \to K_h$ pour la norme $|\ |$.

On dit que le schéma que correspond à $\vartheta=0$ est <u>semi-explicite</u> ([1]).

Pour $\vartheta=1$ on obtient un schéma <u>implicite</u>. Ces schémas ont été introduits dans Lions [3].

<u>Remarque</u> 2.2.-

Une variante de (2.12) (2.14) (Trémolières [1]) est:

(2.15) $$\frac{\bar{u}^{n+1}-u^n}{\Delta t} + A_h\,\bar{u}^{n+1} = f^n,$$

(2.16) $$u^{n+1} = P_{K_h}(\bar{u}^{n+1}).$$

C'est un schéma "semi-implicite".

<u>Remarque</u> 2.3.-

Si l'on prend dans (2.6) $u^*=u^{n+\vartheta}$ alors (2.6) équivaut à

(2.17) $$\begin{cases} \left(\dfrac{u^{n+\vartheta}-u^n}{\Delta t},\ v-u^{n+\vartheta}\right)+a(u^{n+\vartheta},\ v-u^{n+\vartheta}) \geq (f^n, v-u^{n+\vartheta}) & \forall v \in K_h, \\ u^{n+\vartheta} \notin K_h\ ; \end{cases}$$

noter qu'alors $u_n \in K_h$.

<u>Remarque</u> 2.4.-

Tout ce qu'est fait dans ce Chapitre s'adapte aussitôt à l'inequation (2.11), Chap. 1.

([1]) Pour les optimistes ! Les pessimistes préfèrent "semi-implicite".

Remarque 2.5.-

Pour le résolution de (2.10) on utilisera les méthodes indiquées dans les cours de Glowinski et Mosco. On utilisera en outre le fait que, en toute probabilité, u^{n+1} est "voisin" de u^n.

Orientation.

On va maintenant étudier le stabilité, puis la convergence, du schéma (2.7), (2.8).

3.- Analys de la stabilité

On effectue une translation de u^o ce qui revient à supposer que

(3.1) $\qquad 0 \in K , \qquad u^o = 0.$

Pour un peu simplifier l'exposé, on supposera que

(3.2) $\qquad a(v,v) \geq \alpha||v||^2 \qquad \forall v \in V.$

On va démontrer le:

Théorème 3.1.- On suppose que (3.1) et (3.2) ont lieu. On considère le schéma (2.7)(2.8)(2.9) (avec $u^o=0$). On suppose que $N\ \Delta t=T$ fixé.

On a:

(3.3) $\qquad |u^n| \leq C \qquad (0 \leq n \leq N),$

(3.4) $\qquad \sum_{n=o}^{N} \Delta t||u^n||^2 \leq C$

où les C désignent des constantes indépendantes de h et Δt, et cela:

(3.5) $\qquad$ quel que soit Δt lorsque $\theta=1$,

si la condition de stabilité suivante a lieu :

(3.6) $$(1-\vartheta)S(h)^2\Delta t \leqslant C \qquad \text{lorsque} \quad 0 \leqslant \vartheta \leqslant 1$$ [1]

Remarque 3.1.-

Il y a une différence avec le cas des équations où les conditions de stabilité n'apparaissent que si $\vartheta \leqslant 1/2$. Mais nous ignorons si (3.6) est le meilleur résultat possible. La difficulté vient de ce que l'on ne peut, en général, prendre $v=u^{n+1}-u^{n+\vartheta}$ dans (2.7).

Démonstration.

On fait v=O dans (2.7). Il vient:

$$\frac{1}{\Delta t}(u^{n+1}-u^n,u^{n+1})+a(u^{n+\vartheta},u^{n+1}) \leqslant (f^n,u^{n+1})$$

d'où

(3.7) $$\frac{1}{2\Delta t}(|u^{n+1}|^2-|u^n|^2+|u^{n+1}-u^n|^2)+a(u^{n+\vartheta},u^{n+1})\leqslant(f^n,u^{n+1}).$$

Mais

$$a(u^{n+\vartheta},u^{n+1})=\vartheta a(u^{n+1},u^{n+1})+(1-\vartheta)a(u^n,u^{n+1}) =$$

$$= \vartheta a(u^{n+1})+(1-\vartheta)a(u^n)+(1-\vartheta)a(u^n,u^{n+1}-u^n)$$ [2]

et

$$(f^n,u^{n+1})=(f^n,u^{n+\vartheta})+(f^n,u^{n+1}-u^{n+\vartheta})$$

$$=(1-\vartheta)(f^n,u^n)+\vartheta(f^n,u^{n+1})+(1-\vartheta)(f^n,u^{n+1}-u^n)$$

de sorte que (3.7) donne

$$\frac{1}{2\Delta t}(|u^{n+1}|^2-|u^n|^2+|u^{n+1}-u^n|^2)+\vartheta a(u^{n+1})+(1-\vartheta)a(u^n) \leqslant$$

---

[1] S(h) est la constante de stabilité introduite dans (1.3). On notera que si $\vartheta=1$ la condition (3.6) disparait et on retrouve (3.5). La démonstration qui suit donne une évaluation de la constan

$$\leqslant \vartheta(f^n,u^{n+1})+(1-\vartheta)(f^n,u^n)+(1-\vartheta)(f^n,u^{n+1}-u^n) -(1-\vartheta)a(u^n,u^{n+1}-u^n).$$

Avec les majorations

$$(f^n,u^{n+1})\leqslant||f^n||_* \, ||u^{n+1}|| \;(^1)\leqslant \frac{1}{2\alpha}\,||f^n||_*^2+\frac{1}{2}\,a(u^{n+1}),$$

$$(f^n,u^n)\leqslant ||f^n||_* \, ||u^n||\leqslant \frac{1}{\alpha}\,||f^n||_*^2+\frac{1}{4}\,a(u^n),$$

$$(f^n,u^{n+1}-u^n)\leqslant||f^n||_* \, ||u^{n+1}-u^n||$$

$$\leqslant \frac{1}{2\varepsilon}\,||f^n||_*^2+\frac{\varepsilon S(h)^2}{2}\,(u^{n+1}-u^n)^2, \qquad \varepsilon>0 \text{ quelconque},$$

$$|a(u^n,u^{n+1}-u^n)|\leqslant M||u^n||\,||u^{n+1}-u^n||$$

$$\leqslant \frac{1}{4}\,a(u^n)+\frac{M^2}{\alpha}\,S(h)^2|u^{n+1}-u^n)^2 \qquad (^2)$$

On obtient:

$$|u^{n+1}|^2-u^n|^2+(1-(\varepsilon+\frac{2}{\alpha}\,M^2)\Delta t\;S(h)^2(1-\vartheta))\;|u^{n+1}-u^n|^2+$$

$$+(1-\vartheta)\Delta t\;a(u^n)+\vartheta\Delta t\;a(u^{n+1})\leqslant C\Delta t||f^n||_*^2.$$

On fait donc l'hypothèse $(C=\varepsilon+\frac{2}{\alpha}\,M^2))$

$$(3.9) \qquad 1-C(1-\vartheta)\;\Delta t\;S(h)^2\geqslant\beta>0$$

(hypothèse qu'est sans objet si $\vartheta=1$); cf. (3.5) et (3.6).

Alors (3.8) donne

$$(3.10) \quad |u^{n+1}|^2-|u^n|^2+\beta|u^{n+1}-u^n|^2+\alpha\Delta\;t((1-\vartheta)\,||u^n||^2+$$

$$+\vartheta||u^{n+1}||^2)\leqslant c\;\Delta t||f^n||_*^2.$$

---

(1) $||\quad||_*$ = norme dans V' duale de $||\quad||$

(2) M est la constante de continuité de a(u,v).

On en déduit par sommation

$$(3.11)\qquad |u^{n+1}|^2+\beta \sum_{q=0}^{n} |u^{q+1}-u^q|^2+$$

$$+\alpha \sum_{q=0}^{n} \Delta t(1-\vartheta)||u^q||^2+\vartheta||u^{q+1}||^2) \leqslant$$

$$\leqslant c \sum_{q=0}^{n} \Delta t||f^q||_*^2 < c \qquad \text{lorsque } n \leqslant N.$$

De (3.11) on déduit (3.3) (3.4) et on obtient en outre l'estimation supplémentaire:

$$(3.12)\qquad \sum_{n=0}^{N} |u^{n+1}-u^n|^2 \leqslant C.$$

Remarque 3.2.-

Pour l'analyse des schémas du type précédent et d'autres schémas, cf. R. Trémolières [1].

4.- Etude de la convergence.

Définissons:

$$(4.1)\qquad u_{\Delta t}(t)=u^n \quad \text{dans l'intervalle} \quad [n\Delta t,\ (n+1)\Delta t[.$$

On va démontrer le:

Théorème 4.1.- On se place dans les conditions du Théorème 3.1.- Alors lorsque $h \to 0$ et, lorsque $\Delta t \to 0$

(4.2) sans condition supplémentaires si $\vartheta=1$,

et avec

$$(4.3)\qquad S(h)^2\Delta t \to 0 \qquad \text{si} \qquad 0 \leqslant \vartheta < 1,$$

on a:

(4.4) $u_{\Delta t} \to u$ dans $L^2(0,T;V)$ faible,

<u>où</u> u <u>est solution "faible" de</u> (3.8), <u>Chap</u>.1.

<u>Démonstration</u>.

Soit v une fonction régulière (par ex. C^1) de $[0,T] \to K$ avec v(0)=0 et soit

$$v_h^n = v^n = \text{approximation dans } K_h \text{ de } v(n\Delta t).$$

On prend $v=v^{n+1}$ dans (2.7):

$$(4.5) \quad \left(\frac{u^{n+1}-u^n}{\Delta t}, v^{n+1}-u^{n+1}\right) + a(u^{n+\vartheta}, v^{n+1}-u^{n+1}) \geq (f^n, v^{n+1}-u^{n+1}).$$

On en deduit que

$$(4.6) \quad \sum_{n=0}^{N} \Delta t\left(\frac{v^{n+1}-v^n}{\Delta t}, v^{n+1}-u^{n+1}\right) + \sum_{n=0}^{N} \Delta t\, a(u^{n+\vartheta}, v^{n+1}-u^{n+1}) - $$
$$- \sum_{n=0}^{N} \Delta t (f^n, v^{n+1}-u^{n+1}) \geq 0.$$

En effet le 1 membre de (4.6) vaut

$$\sum_{n=0}^{N} \left[\Delta t\left(\frac{u^{n+1}-u^n}{\Delta t}, v^{n+1}-u^{n+1}\right) + a(u^{n+\vartheta}, v^{n+1}-u^{n+1}) - (f^n, v^{n+1}-u^{n+1})\right] +$$
$$+ \sum_{n=0}^{N} \Delta t (v^{n+1}-v^n-(u^{n+1}-u^n), v^{n+1}-u^{n+1}) \geq \text{(d'après (4.5))} \geq$$
$$\geq \sum_{n=0}^{N} \frac{1}{2}\left[|v^{n+1}-u^{n+1}|^2 - |v^n-v^n|^2 + |v^{n+1}-u^{n+1}-(v^n-u^n)|^2\right] > 0 \text{ d'où (4.6)}$$

On écrit (4.6) sous la forme

$$(4.7) \quad \sum_{n=0}^{N} \Delta t\left(\frac{v^{n+1}-v^n}{\Delta t}, v^{n+1}-u^{n+1}\right) + \sum_{n=0}^{N} \Delta t\, a(u^{n+\ }, v^{n+1}) -$$
$$- \sum_{n=0}^{N} \Delta t (f^n, v^{n+1}-u^{n+1}) > \sum_{n=0}^{N} \Delta t\, a(u^{n+\ }, u^{n+1}).$$

D'après (3.4) on a:

$$(4.8) \qquad \int_{0}^{T} ||u_{\Delta t}||^2 dt \leq C$$

et l'on peut donc supposer que (par extraction)

$$(4.9) \qquad u_{\Delta t} \to w \text{ dans } L^2(0,T;V) \text{ faible.}$$

Alors (4.7) donne

$$(4.10) \qquad \int_{0}^{T} [(v',v-w)+a(w,v)-(f,v-w)]dt \geq \lim.\inf \sum_{n=0}^{N} \Delta t\ a(u^{n+\vartheta},u^{n+1}).$$

Mais comme $u^{n+1}=u^{n+\vartheta}+(1-\vartheta)(u^{n+1}-u^n)$, le 2eme membre de (4.10) vaut

$$(4.11) \qquad \lim.\inf. \left[\sum_{n=0}^{N} \Delta t\ a(u^{n+\vartheta})+(1-\ ) \sum_{n=0}^{N} \Delta t\ a(u^{n+\vartheta},\ u^{n+1}-u^n)\right].$$

Admettons un instant que sous la condition (3.6) on a ([1]):

$$(4.12) \qquad \sum_{n=0}^{N} \Delta t\ a(u^{n+\vartheta},u^{n+1}-u^n) \to 0.$$

Alors (4.11) donne

$$\lim.\ \inf. \sum_{n=0}^{N} \Delta t\ a(u^{n+\vartheta}) \geq \int_{0}^{T} a(w)dt$$

ce qui joint à (4.10) montre que

$$\int_{0}^{T} \left[(v',v-w)+a(w,v-w)-(f,v-w)\right] dt \geq 0$$

pour toute fonction v régulière de $[0,T] \to K$ avec $v(0)=0$, d'où par prolongement par continuité $\forall v \in L^2(0,T;V)$, $v' \in L^2(0,T;V')$,

---

([1]) Le cas $\vartheta=1$ est immédiat.

$v(0)=0$, $v(t) \in K$.

Donc $w=u$ et le théorème est démontré, sous réserve de la vérification de (4.12).

Pour cela, on note que

$$\left|\sum_{n=0}^{N} \Delta t\, a(u^{n+\vartheta}, u^{n+1}-u^n)\right| \leqslant c \sum_{n=0}^{N} \sqrt{\Delta t}\, \|u^{n+\vartheta}\|\, S(d)\sqrt{\Delta t}\,|u^{n+1}-u^n| \leqslant$$

$$\leqslant c\left(\sum_{n=0}^{N} \Delta t\,\|u^{n+\vartheta}\|^2\right)^{1/2} S(h)\sqrt{\Delta t}\left(\sum_{n=0}^{N} |u^{n+1}-u^n|^2\right)^{1/2}$$

$\leqslant$ (d'après (3.4), et (3.12)) $\leqslant C\, S(h)\, \sqrt{\Delta t}$ d'où

(4.12) d'après (4.3).

<u>Remarque 4.1</u>.-

La démonstration précédente montre l'existence d'un solution faible du problème (cf. Chap. 1, N.4.4).

CHAPITRE 3. - Inéquations d'évolution paraboliques de type II.

1.- Exemples.

Exemple 1.1.- La théorie de l'asservissement (cf. Duvaut-Lions [1]) conduit à des problèmes du type suivant; les notations sont celles du Chap .1, N.1.

On cherche une fonction u=u(x,t) solution de

$$\frac{\partial u}{\partial t} - \Delta u = f \qquad \text{dans } \Omega \times]0,T[, \tag{1.1}$$

$$u(x,0)=u_0(x), \qquad x \in \Omega, \tag{1.2}$$

$$\frac{\partial u}{\partial t} \geqslant 0,\ \frac{\partial u}{\partial n} \geqslant 0, \qquad \frac{\partial u}{\partial t}\frac{\partial u}{\partial n} = 0 \quad \text{su } \Sigma. \tag{1.3}$$

Remarque 1.1.-

L'équation (1.1) et la condition initiale (1.2) sont identiques à celles du Chap.1 , N.1. Par contre les conditions aux limites (non linéaires) dans (1.3) sont différentes des conditions (1.3), Chap.1.

Remarque 1.2.- (Analogue à la Remarque 1.2, Chap.1).

D'après la 3ème condition (1.3), on voit que $\frac{\partial u}{\partial t} = 0$ sur une partie Σ_0^* de Σ et $\frac{\partial u}{\partial n} = 0$ sur $\Sigma-\Sigma_0^*$. Mais Σ_0^* n'est pas donné a priori.

Exemple 1.2.-

Une variante de l'Exemple 1.1 est: on cherche u avec

$$\left\{\begin{array}{l} \frac{\partial u}{\partial t} \geqslant 0 \quad \text{dans } \Omega \times]0,T[, \\ \frac{\partial u}{\partial t} - \Delta u - f \geqslant 0, \\ \frac{\partial u}{\partial t}\left(\frac{\partial u}{\partial t} - \Delta u - f\right) = 0, \end{array}\right. \tag{1.4}$$

avec la condition initiale (1.2) et la condition aux limites

(1.5) $u = 0 \quad \text{sur} \quad \Sigma$

Exemple 1.3.-

On cherche u satisfaisant à (1.1)(1.2) et à

(1.6) $$\begin{cases} \left| \frac{\partial u}{\partial n} \right| \leqslant g, \\ \frac{\partial u}{\partial n} \frac{\partial u}{\partial t} + g \left|\frac{\partial u}{\partial t}\right| = 0 \quad \text{sur } \Sigma . \end{cases}$$

2.- Formulation générale.

Les notations sont celles du Chap.1, N.2.

On cherche une fonction $t \to u(t)$ de $[0,T] \to V$ telle que

(2.1) $$\frac{\partial u(t)}{\partial t} = u'(t) \in K ,$$

(2.2) $$(u'(t),v-u'(t))+a(u(t),v-u'(t)) \geqslant (f(t),v-u'(t)) \quad \forall v \in K,$$

(2.3) $$u(0) = v_o.$$

Variante.

On cherche u telle que

(2.4) $$(u'(t),v-u'(t))+a(u(t),v-u'(t))+j(v)-j(u'(t)) \geqslant (f(t),v-u'(t) \quad \forall v \in V,$$

avec (2.3).

Les inéquations (2.2) où (2.4) seront appelée inéquations paraboliques de type II.

Remarque 2.1.-

Si K=V ou si j=0, (2.2) et (2.4) se reduisant à l'équation

ordinaire (2.12) Chap.1. La distinction entre les types I et II n'a de sens que pour les inéquations.

Remarque 2.2.-

Pour les inéquations, les inéquations de type II sont effectivement distinctes du type I ! Prenons en effet la situation de la Remarque 3.2, Chap.1. Alors (2.2) équivaut à

$$u' \geqslant 0, \quad u'-f \geqslant 0, \qquad u'(u'-f) = 0$$

donc

$$u' = f^{+}. \tag{2.5}$$

Si $f=-1$ on trouve $u'=0$ qui donne $u(t)=u_o$ solution différente de celle de la Remarque 3.2, Chap.1.

Notons en outre que (2.5) montre qu'il y a en général un seuil de régularité.

Remarque 2.3.-

Pour la résolution des inéquations de type II nous devrons supposer a symétrique (il suffirant que la "partie principale" de a soit symétrique; il est probable, mais non démontré, qu'une condition de ce genre est nécessaire pour que les problèmes de type II soient bien posés).

Remarque 2.4.-

La formulation (2.2) suggère de nombreuses variantes. En voici quelques unes:

1) asservissement retardé (cf. Duvant-Lions [1]); soit $\tau>0$ donné; on cherche $u(t)$ solution de

$$(2.6)\quad \begin{cases} (u'(t), v-\frac{u(t)-u(t-\tau)}{\tau}) + a(u(t), v-\frac{u(t)-u(t-\tau)}{\tau}) \geq (f(t), v-\frac{u(t)-u(t-\tau)}{\tau}) \\ \qquad \forall v \in K, \\ \frac{u(t)-u(t-\tau)}{\tau} \in K \ ; \end{cases}$$

2) variante de l'asservissement retardé (cf. D. Viaud [1]); soit $\tau > 0$ donné; on définit

$$(2.7)\qquad B\,u(t) = \frac{u(t)-u(n\tau)}{t-n\tau} \quad \text{pour } t \in [n\tau, (n+1)\tau[;$$

on cherche u(t) solution de

$$(2.8)\qquad (u'(t), v-Bu(t)) + a(u(t), v-Bu(t)) \geq (f(t), v-Bu(t)) \qquad \forall v \in K,$$

$$(2.9)\qquad Bu(t) \in K \ ;$$

3) notons que, en utilisant $A \in \mathcal{L}(V;V')$ défini par $a(u,v) = (Au,v)$, l'inéquation (2.2) s'écrit

$$(2.10)\qquad (u'+Au-f, v-u') \geq 0 \qquad \forall v \in K;$$

on peut introduire les puissances fractionnaires A^{ϑ} de A et considérer l'inéquation

$$(2.11)\qquad (u'+Au-f, v-A^{\vartheta}u') \geq 0 \qquad \forall v \in K,$$

avec

$$(2.12)\qquad A^{\vartheta}u'(t) \in K \ .$$

Remarque 2.5.-

Tout ce qui a été dit dans le Chap. précédents et tout ce qui sera fait dans la suite vaut en remplaçant a(u,v) par a(t;u,v) dépendant convenablement de t.

Remarque 2.6.- Exemples.

Prenons d'abord

$$V=H^1(\Omega),\quad H=L^2(\Omega),\quad K=\{v \mid v\geqslant 0 \text{ sur } \Gamma\},$$

$$a(u,v) = \sum_{i=1}^{n}\int_\Omega \frac{\partial u}{\partial x_i}\frac{\partial v}{\partial x_i}\,dx .$$

Alors (2.1)(2.2)(2.3) équivalent au problème de l'Exemple 1.1.

Si l'on prend

$$K= \{v \mid v\geqslant 0 \text{ dans } \Omega,\ v\in H_0^1(\Omega)\},$$

on trouve le problème de l'Exemple 1.2.

Si l'on prend dans (2.4)

$$j(v) = g\int_\Gamma |v|\,d\Gamma,$$

et V,H, a comme ci dessus, on trouve le problème de l'Exemple 1.3.

Relativement au problème (2.2),(2.3) (ou (2.4) (2.3)) on a le résultat suivant:

Théorème 2.1.- On suppose que

(2.13) $$a(u,v) = a(v,u) \qquad \forall u,v\in V,$$

(2.14) il existe λ tel que $a(v,v)+\lambda|v|^2\geqslant\alpha\ ||v||^2,\ \alpha\geqslant 0,\ \forall v\in V.$

On suppose f donné avec

(2.15) $$f,f'\in L^2(0,T;H)$$

et u_0 donné avec

(2.16) $$f(0)-Au_0 = 0$$ (1)

(1) On peut, plus généralement, supposer qu'il existe $v_0\in K$ tel que $(v_0+Au_0-f(0),\ v-v_0)\geqslant 0\ \forall v\in K$ – ou bien qu'il existe $v_-\in V^0$ tel que

Il existe alors une fonction u et un seule, solution de (2.2),(2.3) (ou (2.4) (2.3)) telle que

(2.17) $$u, u' \in L^{\infty}(0,T;\ V),$$

(2.18) $$u'' \in L^2(0,T;H)$$

L'unicité est immediate.

Pour la démonstration de l'existence on a des méthodes analogues à celles évoquées au Chap. 1, N.4. Nous allons développer seulement, dans les deux N. qui suivent, la méthode de discrétisation.

3.- Schémas d'approximation.

Notations.

Las notations sont celles du Chap. 2.

Si φ^n est une suite (n=0,....) on posera

(3.1) $$\delta\varphi^n = \frac{1}{\Delta t}(\varphi^{n+1}-\varphi^n),$$

$$\delta^2\varphi^n = \delta(\delta\varphi^n).$$

On considère alors le schéma suivant:

(3.2) $$\begin{cases} (\delta u^n + Au^{n+\theta} - f^n, v - \delta u^n) \geq 0 & \forall v \in K_h, \\ \delta u^n \in K_h. \end{cases}$$

Comme $u^{n+\theta} = \theta u^{n+1} + (1-\theta) u^n = \theta \Delta t \delta u^n + u^n$, (3.2) équivaut à

(3.3) $((1+\theta\Delta tA)\ \delta u^n, v-\delta u^n) \geq (f^n - Au^n, v-\delta u^n) \quad \forall v \in K_h,\ \delta u^n \in K_h,$

ce qui montre que, dès que Δt est assez petit, δu^n est défini de façon unique par (3.3). Ensuite u^{n+1} est calculé par: $u^{n+1} =$

$=u^n+\Delta t\ \delta u^n$.

On démarre le calcul des δu^n par

(3.4) $$\delta u^o = 0$$ ([1])

et de

(3.5) $$u^o = \text{approximation de } u_o.$$

Cela définit donc de proche en proche la suite u^n.

Si $\vartheta=1$ le schéma est dit "implicite". En fait (cf. (3.3)) il est "implicite pour tout $\vartheta>0$ et "semi explicite" si $\vartheta=0$.

Remarque 3.1.-

Le schéma analogue à (3.2) relatif à l'inéquation (2.4) s'écrit:

(3.6) $$(\delta u^n+Au^{n+\vartheta}-f^n, v-\delta u^n)+j(v)-j(\delta u^n)\geq 0 \qquad \forall v\in V_h.$$

4.- Stabilité et convergence (cf. D. Viaud |1|).

Ecrivons l'analogue de (3.2) pour n+1:

(4.1) $$(\delta u^{n+1}+Au^{n+1+\vartheta}-f^{n+1}, v-\delta u^{n+1})\geq 0 \qquad \forall v\in K_h.$$

Prenons dans (3.2) (resp. (4.1)) $v=\delta u^{n+1}$ (resp. $v=\delta u^n$) et ajoutons:

(4.2) $$|\delta u^{n+1}-\delta u^n|^2+a(u^{n+1+\vartheta}-u^{n+\vartheta}, \delta u^{n+1}-\delta u^n)\leq (f^{n+1}-f^n, \delta u^{n+1}-\delta u^n).$$

Mais

$$\delta u^{n+1}-\delta^n=\Delta t\delta^2u^n,\ u^{n+1+\vartheta}-u^{n+\vartheta}=\Delta t\ \delta u^{n+\vartheta},\ f^{n+1}-f^n=\Delta t\ \delta f^n.$$

([1]) Cela si justifie à partir de (2.16).

On déduit donc de (4.2) après division par Δt que

$$\Delta t|\delta^2u^n|^2+a(\delta u^{n+\vartheta},\ \delta u^{n+1}-\delta u^n)\leqslant\Delta t(\delta f^n,\delta^2u^n)\leqslant$$

$$\leqslant \frac{\Delta t}{2}|\delta^2u^n|^2+ \frac{\Delta t}{2}|\delta f^n|^2.$$

donc

(4.3) $$\Delta t|\delta^2u^n|^2+2a(\delta u^{n+}\ ,\ \delta u^{n+1}-\delta u^n)<\Delta t\ |\delta f^n|^2.$$

On va en déduire le

Théorème 4.1.- On se place dans les hypothèses du Théorème 2.1. On a alors:

(4.4) $||u^n||,||\delta u^n|| \leqslant$ constante indépendante de Δt et de h,

(4.5) $$\sum_{n=0}^{N} \Delta t|\delta^2u^n|^2 \leqslant \text{constante}$$

lorsque h et $\Delta t\to 0$, et cela

(4.6) sans condition si $\frac{1}{2}\leqslant\vartheta\leqslant 1$,

et

(4.7) $\begin{cases}\text{sous une condition de stabilité de la forme}\\ \Delta t\ S(h)^2\leqslant c \quad \text{si}\quad 0\leqslant\vartheta\leqslant 1/2.\end{cases}$

Démonstration [1].

On vérifie (en utilisant le symétrie de a(u,v)) que

$$2a(\delta u^{n+\vartheta},\delta\ u^{n+1}-\delta u^n)=a(\delta u^{n+1})-a(\delta u^n)+(2\vartheta-1)a(\delta u^{n+1}-\delta u^n)).$$

Donc (4.3) donne:

---

[1] On supposera que $a(v,v)\geqslant\alpha\ ||\ v||^2$.

(4.8) $\Delta t\,|\delta^2 u^n|^2 + a(\delta u^{n+1}) - a(\delta u^n) + (2\vartheta-1)\,a(\delta u^{n+1}-\delta u^n) \leqslant \Delta t\,|\delta f^n|^2.$

Distinguons deux cas:

<u>1er cas:</u> $2\vartheta-1 \geqslant 0.$

On somme (4.8); il vient

$$\sum_{n=0}^{q} \Delta t\,|\delta^2 u^n|^2 + a(\delta u^{q+1}) + (2\vartheta-1) \sum_{n=0}^{q} a(\delta u^{n+1}-\delta u^n) \leqslant \tag{4.9}$$

$$\leqslant \Delta t \sum_{n=0}^{q} |\delta f^n|^2.$$

On en déduit (4.4)(4.5) et en outre

$$\begin{cases} \text{si } \frac{1}{2} < \vartheta \leqslant 1 \text{ on a:} \\ \sum_{n=0}^{N} a(\delta u^{n+1}-\delta u^n) \leqslant C. \end{cases} \tag{4.10}$$

<u>2^em cas:</u> $0 \leqslant \vartheta < 1/2.$

On note que

$a(\delta u^{n+1}-\delta u^n) \leqslant C\,S(h)^2\,|\delta u^{n+1}-\delta u^n|^2 = C\,S(h)^2\,\Delta t^2\,|\delta^2 u^n|^2$

donc $(2\vartheta-1)\,a(\delta u^{n+1}-\delta u^n) \geqslant -C(1-\vartheta)\,S(h)^2\Delta t^2\,|\delta^2 u^n|^2$ et (4.8) implique donc que

(4.11) $\Delta t\,[1-C(1-2\ )S(h)^2\Delta t]\ \delta^2 u^n|^2 + a(\delta u^{n+1}) - a(\delta u^n) \leqslant \Delta t\,|\delta f^n|^2.$

Si donc l'on suppose que

(4.12) $1-C(1-2\vartheta)S(h)^2\Delta t \geqslant \gamma > 0,$

on déduit de (4.11) que

(4.13) $\gamma\ \Delta t\,|\delta^2 u^n|^2 + a(\delta u^{n+1}) - a(\delta u^n) \leqslant \Delta t\,|\delta f^n|^2$

d'où l'on déduit (4.4) et (4.5).

Examinons maintenant la convergence.

Théorème 4.2.- On se place dans les hypothèses du Théorème 4.1. On définit $u_{\Delta t}$, $\delta u_{\Delta t}$, $\delta^2 u_{\Delta t}$ comme les fonctions égales à u^n, δu^n et $\delta^2 u^n$, δu^n et $\delta^2 u^n$ sur $[n\Delta t,(n+1)\Delta t[$. On a lorsque h et $\Delta t \to 0$:

$$(4.14) \qquad u_{\Delta t}, \ \delta u_{\Delta t} \ \to \ u, u' \ \text{dans } L^\infty(0,T;V) \text{ faible étoile,}$$

$$(4.15) \qquad \delta^2 u_{\Delta t} \to u'' \ \text{dans } L^\infty(0,T;H) \ \text{faible étoile,}$$

et cela sans conditions supplémentaire si $\frac{1}{2} \leqslant \vartheta \leqslant 1$ et sous la condition

$$(4.16) \qquad \Delta t \ S(h)^2 \to 0 \qquad \text{si} \qquad 0 \leqslant \vartheta < \frac{1}{2} .$$

Démonstration.

Soit v une fonction régulière de $[0,T] \to K$ et soit

$$v_h^n = v^n = \text{approximation dans } K_h \text{ de } v(n\ \Delta t).$$

Nous prenons dans (3.2) $v=v^n$ et nous sommons en n, après multiplication par Δt; il vient:

$$(4.17) \quad \sum_{n=0}^{N} [\Delta t(\delta u^n, v^n) + \Delta t\ a(u^{n+\vartheta}, v^n) - \Delta t(f^n, v^n - \delta u^n)] \geqslant \sum_{n=0}^{N} \Delta t |\delta u^n|^2 + X,$$

où

$$(4.18) \qquad X = \sum_{n=0}^{N} \Delta t\ a(u^{n+\vartheta}, \delta u^n).$$

Mais on vérifie que

$$(4.19) \qquad X = \sum_{n=0}^{N} a(u^{n+\vartheta}, u^{n+1} - u^n)$$

$$= \sum_{n=0}^{N} \frac{1}{2} [a(u^{n+1}) - a(u^n) + (2\vartheta - 1) a(u^{n+1} - u^n)]$$

$$= \frac{1}{2} a(u^{N+1}) - \frac{1}{2} a(u^o) + Y,$$

(4.20) $$Y = (2\vartheta-1) \sum_{n=0}^{N} (u^{n+1}-u^n)$$

D'après les estimations (4.4) et (4.5), $u_{\Delta t}$, $\delta u_{\Delta t}$ demeurent dans un borné de $L^\infty(0,T;V)$ et $\delta^2 u_{\Delta t}$ dans un borné de $L^2(0,T;H)$, de sorte que l'on peut extraire une sous suite, encore notée $u_{\Delta t}$, telle que

$$u_{\Delta t} \to w \quad \text{dans } L^\infty(0,T;V) \quad \text{faible étoile,}$$

$$\delta u_{\Delta t} \to w_1 \quad \text{dans } L^\infty(0,T;V) \quad \text{faible étoile,}$$

$$\delta^2 u_{\Delta t} \to w_2 \quad \text{dans } L^\infty(0,T;H) \quad \text{faible étoile.}$$

On vérifie sans peine que

$$w_1 = w', \qquad w_2 = w''$$

de sorte que

$$u^{N+1} \to w(T).$$

Distinguons alors deux cas.

1er cas: $\frac{1}{2} \leqslant \vartheta \leqslant 1$

On a alors $Y \geqslant 0$ donc $X \geqslant \frac{1}{2} a(u^{N+1}) - \frac{1}{2} a(u^o)$ d'où

(4.21) $$\liminf X \geqslant \frac{1}{2} a(w(T)) - \frac{1}{2} a(u_o).$$

Utilisant (4.21) dans (4.17) on en déduit que

(4.22) $$\int_0^T [(w'',v)+a(w,v)-(f,v-w')]dt \geqslant \int_0^T |w'|^2 dt + \frac{1}{2} [a(w(T))-a(u_o)].$$

Mais

$$\frac{1}{2} [a(w(T))-a(u_o)] = \int_0^T a(w,w')dt$$

de sorte que (4.22) équivaut à

(4.23) $$\begin{cases} \int_0^T [(w',v-w')+a(w,v-w')-(f,v-w')]dt \geq 0 \quad \forall v \text{ régulière} \\ \text{de } [0,T] \\ \to K \end{cases}$$

et donc par prolongement par continuité, $\forall v \in L^2(0,T;V)$ avec $v(t) \in K$ p.p. On déduit de là que w est solution du problème, d'où w=u.

<u>2ème cas</u>: $0 \leq \vartheta < \frac{1}{2}$.

Cette fois Y est ≤ 0, mais le résultat final est valable si l'on vérifie que

(4.24) $$Y \to 0 .$$

Or

$$a(u^{n+1}-u^n) \leq c\|u^{n+1}-u^n\|^2 \leq C\ S(h)^2 |u^{n+1}-u^n|^2 \leq$$

$$\leq C\ S(h)^2 \Delta t^2 |\ \delta u^n|^2 ;$$

mais (4.4) entraine que, en particulier, $|\delta u^n| \leq C$, donc

$$|Y| < C \sum_{n=0}^{N} \Delta t^2 S(h)^2 = c(N\Delta t)\Delta t\ S(h)^2 \leq c\ \Delta t \cdot S(h)^2$$

donc $\to 0$ d'après l'hypothèse (4.16).

<u>Remarque 4.1</u>.-

On a des résultats analogues relatifs à (3.6).

<u>Remarque 4.2</u>.-

On a démontré, par cette méthode, l'<u>existence</u> d'une solution du problème.

J.L. Lions

CHAPITRE 4 .- Inéquations d'évolution du 2eme ordre en t.

1.- Exemples.

De nombreux problèmes de Mécanique et de Physique ([1]) (cf. Duvaut-Lions [1]) conduisent à des problèmes du type suivant:

Exemple 1.1.-

Avec les notations du Chap. 1, N.1, on cherche une fonction u solution de

$$(1.1) \qquad \frac{\partial^2 u}{\partial t^2} - \Delta u = f \qquad \text{dans } \Omega \times]0,T[\, ,$$

$$(1.2) \qquad u(x,0)=u_0(x), \quad \frac{\partial u}{\partial t}(x,0) = u_1(x), \qquad x \in \Omega,$$

$$(1.3) \qquad \frac{\partial u}{\partial t} \geq 0, \quad \frac{\partial u}{\partial n} \geq 0, \quad \frac{\partial u}{\partial t}\frac{\partial u}{\partial n} = 0 \qquad \text{sur } \Sigma \quad .$$

Remarque 1.1.-

On notera que les conditions aux limites (1.3) sont identiques à celles de l'Exemple 1.1, Chap.3; on peut donc faire des observations analogues à celles de la Remarque 1.2, Chap.3.

Exemple 1.2.-

On cherche u solution de (1.1)(1.2) avec les conditions aux limites

$$(1.4) \qquad \left|\frac{\partial u}{\partial n}\right| \leq g, \quad \frac{\partial u}{\partial t}\frac{\partial u}{\partial n} + g\left|\frac{\partial u}{\partial t}\right| = 0 \qquad \text{sur } \Sigma \quad .$$

Remarque 1.2.-

Des problèmes de ce type se posent également pour le système de l'élasticité; cf. Duvaut-Lions, loc. cit.

([1]) Par ex. pour les operateurs de Maxwell. Cf. Duvaut-Lions, loc. cit., Chap.7.

Remarque 1.3.-

L'opérateur (1.1) est hyperbolique. Mais des problèmes de ce genre se posent également pour des opérateurs non hyperboliques. Par ex. trouver u solution de

(1.5) $$\frac{\partial^2 u}{\partial t^2} + \Delta^2 u = f \qquad \text{dans } \Omega \times]0,T[,$$

les conditions initiales (1.2) et les conditions aux limites

(1.6) $$\Delta u = 0 \qquad \text{sur } \Sigma,$$

(1.7) $$\frac{\partial u}{\partial t} \geq 0 , \qquad - \frac{\partial}{\partial n} \Delta u \geq 0, \qquad \frac{\partial u}{\partial t} (\frac{\partial}{\partial n} \Delta u) = 0 \quad \text{sur } \Sigma.$$

2.- Formulation générale.

Les notations sont celles du Chap.1, N.2.

On cherche une fonction $t \to u(t)$ de $[0,T] \to V$ telle que

(2.1) $$u'(t) \in K,$$

(2.2) $$(u''(t),v-u'(t))+a(u(t),v-u'(t)) \geq (f(t),v-u'(t)) \qquad \forall v \in K,$$

(2.3) $$u(0) = u_o, \qquad u'(0) = u_1 .$$

Variante.

On cherche u telle que

(2.4) $$(u''(t),v-u'(t))+a(u(t),v-u(t))+j(v)-j(u'(t)) \geq (f(t),v-u(t)) \qquad \forall v \in V,$$

et les conditions initiales (2.3).

Remarque 2.1.-

Si K=V on si j=0, ou retrouve les équations du 2eme ordre en t:

(2.5) $(u''(t),v)+a(u(t),v)=(f(t),v) \qquad \forall v\in V,$

avec (2.3).

Exemples.

Exemple 2.1.-

On prend:

$$V = H^1(\Omega), \quad H = L^2(\Omega),$$

$$a(u,v) = \sum_{i=1}^{n} \int_\Omega \frac{\partial u}{\partial x_i} \frac{\partial r}{\partial x_i} dx ,$$

$$K = \{v \mid v\geq 0 \text{ sur } \Gamma\} .$$

Le problème (2.1),(2.2) (2.3) correspond à l'Exemple 1.1.

Exemple 2.2.-

V,H et a étant choisis comme ci dessus, on prend

$$j(v) = \int_\Gamma g |v| d\Gamma .$$

Le problème (2.4) (2.3) correspond alors à l'Exemple 1.2.

Exemple 2.3.-

On prend:

$$V = \{v \mid v, \Delta v \in L^2(\Omega)\} ,$$

$$H = L^2(\Omega),$$

$$K = \{v \mid v\in V, v\geq 0 \text{ sur } \Gamma \text{ }(^1)\} ,$$

$$a(u,v) = \int_\Omega \Delta u \, \Delta v \, dx .$$

Alors le problème (2.1),(2.2),(2.3) correspond à l'Exemple de la Remarque 1.3.

(1) On peut définir $v\in H^{-1/2}(\Gamma)$ (cf. Lions-Magenes [1],Chap.2) de sorte que " $v\geq 0$" a un sens.

On a le résultat suivant:

<u>Théorème 2.1</u>.- <u>On suppose que</u>

(2.6) $$a(u,v) = a(v,u) \qquad \forall u,v \in V,$$

(2.7) il existe $\lambda \in \mathbb{R}$ tel que $a(v,v)+\lambda|v|^2 \geq \alpha||v||^2, \alpha>0, \forall v \in V.$

<u>On suppose que</u>

(2.8) $$f, f' \in L^2(0,T;H)$$

(2.9) $$u_0 \in D(A) \quad (\text{i.e. } Au_0 \in H),$$

(2.10) $$u_1 \in K,$$

<u>Il existe une fonction</u> u <u>et une seule telle que</u>

(2.11) $$u,u' \in L^\infty(0,T;V),$$

(2.12) $$u'' \in L^\infty(0,T;H),$$

<u>et vérifiant</u> (2.1),(2.2),(2.3).

<u>Remarque 2.2</u>.-

On a un résultat analogue pour le problème (2.4), (2.3).

<u>Remarque 2.3</u>.-

On peut étendre le résultat précédent au cas où seule la "partie principale" de a est symétrique.

<u>Remarque 2.4</u>.-

On a défini et donné un résultat d'existence pour des solutions <u>fortes</u>.

On peut aussi définir des solutions <u>faibles</u> de ces problèmes. Cf. Lions [4] et une étude plus générale et plus simple dans Brézis [1].

Remarque 2.5. Seuil de régularité.

On rencontré ici encore le phénomène de "seuil de régularité" déjà rencontré au Chapitre 1, Remarque 3.2 et au Chap.3, Remarque 2.2.

Prenons en effet $V=H=R$, $a=0$, $K=\{v \mid v\geq 0\}$. Alors (2.1),(2.2) (2.3) équivalent à

$$u'\geq 0, \quad u''-f\geq 0, \quad u'(u''-f)=0,$$

$$u(0)=u_0, \quad u'(0)=u_1.$$

Posant $u'=w$, on a donc:

$$w\geq 0, \quad w'-f\geq 0, \quad w(0)=u_1$$

ce qui ramène au Chap. 1, Remarque 3.2 ([1]).

L'unicité dans le Théorème 2.1 est immédiate

Pour la démonstration de l'existence, on a des méthodes analogues à celles évoquées au Chap.1, N.4.- Nous allons développer seulement, dans les deux numéros qui suivent, la méthode de discrétisation.

3.- Schémas d'approximation.

Notations

$$d^n = \frac{u^{n+1}-u^{n-1}}{2\Delta t} \quad . \tag{3.1}$$

([1]) Naturellement le cas général ne se réduit pas à la situation du Chap. 1 !

Schéma "semi-explicite".

Supposons u^q comme di 0 à n; on démarre avec

$$u^o = \text{approximation de } u_o,$$

$$u^1 = \quad " \quad " \quad u_1.$$

On détermine alors u^{n+1} par:

(3.2) $$d^n \in K_h,$$

(3.3) $$\left(\frac{u^{n+1} - u^n + u^{n-1}}{\Delta t^2}, v-d^n\right) + a(u^n, v-d^n) \geq (f^n, v-d^n) \quad \forall v \in K_h.$$

Cela définit bien u^{n+1}. En effet (3.3) équivaut à

(3.4) $$(d^n, v-d^n) \geq \left(\frac{u^n - u^{n-1}}{\Delta t} + \frac{\Delta t}{2} f^n - \frac{\Delta t}{2} A_h u^n, v-d^n\right) \quad \forall v \in K_h,$$

ce qui definit d^n de façon unique.

Remarque 3.1.-

Si l'on definit u^{n+1} par

(3.5) $$\frac{\bar{u}^{n+1} - 2u^n + \bar{u}^{n-1}}{\Delta t^2} + A_h u^n = f^n,$$

alors (3.4) equivaut à

$$(d^n, v-d^n) \geq \left(\frac{\bar{u}^{n+1} - u^{n-1}}{2\Delta t}, v-d^n\right) \quad \forall v \in K_h$$

i.e.

(3.6) $$d^n = P_{K_h}\left(\frac{\bar{u}^{n+1} - u^{n-1}}{2\Delta t}\right),$$

où P_{K_h} = projecteur sur K_h pour la norme $|\ |$.

Remarque 3.2.- Schéma implicite.

On détermine cette fois u^{n+1} par

$$(3.7)\quad \begin{cases} d^n \in K_h, \\ \left(\dfrac{u^{n+1}-2u^n+u^{n-1}}{\Delta t^2}, v-d^n\right)+a(u^{n+1},v-d^n) \geq (f^n,v-d^n) \quad \forall v \in K_h. \end{cases}$$

Le schéma (3.7) est <u>inconditionnellement stable.</u>

<u>Remarque 3.3</u>.-

Nous avons commencé par donner le schéma (3.3) à cause de sa "symétrie". Mais ce schéma a, du point de vue pratique, tendance à "découpler" les u^n selon la parité de l'indice " n " et donc à engendrer des oscillations.

Un autre schéma, évitant cet écueil, est le suivant; posant

$$(3.8)\quad \delta^n = \frac{u^{n+1}-u^n}{\Delta t},$$

on considère le schéma:

$$(3.9)\quad \begin{cases} \left(\dfrac{\delta^n-\delta^{n-1}}{t}, v-\delta^n\right)+a(u^n, v-\delta^n) \geq (f^n,v-\delta^n) \quad \forall v \in K_h, \\ \delta^n \in K_h. \end{cases}$$

Si u^n est connu, (3.9) définit δ^n, d'où u^{n+1} par (3.8).

Nous allons maintenant étudier la stabilité de (3.3) puis (Remarque 4.1 à la fin) celle du schéma (3.9) - les démonstrations étant d'ailleurs très voisines.

4.- <u>Stabilité et convergence</u>.

Nous allons démontrer les résultats suivants:

<u>Théorème 4.1</u>. (<u>Stabilité</u>).-<u>On se place dans les condition du Théorème 2.1. Alors, si la condition de stabilité suivante a lieu:</u>

(4.1) $\Delta t\ S(h) \leqslant C$ ([1])

on a:

(4.2) $||u^n||\ ,\quad ||\dfrac{u^{n+1}-u^n}{\Delta t}|| \leqslant C,$

(4.3) $\left|\ \dfrac{u^{n+1}-2u^n+u^{n-1}}{\Delta t^2}\ \right| \leqslant C\ .$

Théorème 4.2 (convergence).- *Les hypothèses sont celles du Théorème 2.1.- On introduit* $u_{\Delta t}$, $\delta u_{\Delta t}$, $\delta^2 u_{\Delta t}$ *égale à* u^{n+1}, $\dfrac{u^{n+1}-u^{n-1}}{2\Delta t}$ $\dfrac{u^{n+1}-2u^n+u^{n-1}}{\Delta t^2}$ *dans l'intervalle* $[n\Delta t, (n+1)\Delta t[$. *Alors, lorsque* h *et* $\Delta t \to 0$:

(4.4) $u_{\Delta t} \to u$, $\delta u_{\Delta t}$ u' dans $L^\infty(0,T;V)$ faible étoile,

(4.4) $\delta^2 u_{\Delta t} \to u''$ dans $L^\infty(0,T;H)$ faible étoile,

où u *est la solution du problème* (2.1),(2.2),(2.3),(2.11),(2.12), *si*

(4.5) $\Delta t\ S(h) \to 0.$

Démonstration du Théorème 4.1. ([2]).

Introduisons:

(4.6) $\delta^n = \dfrac{u^{n+1}-u^n}{\Delta t}\ .$

Alors (3.3) s'écrit

([1]) C = constante dont on obtient une estimation dans la Démonstration ci après.

([2]) Pour un peu simplifier l'exposé, on suppose que $0 \in K$ et que $a(v) \geqslant \alpha\ ||v||^2$.

$$(4.7) \qquad \left(\frac{\delta^n-\delta^{n-1}}{\Delta t}, v-d^n\right)+a(u^n,v-d^n)\geq(f^n,v-d^n) \qquad \forall v\in K_h.$$

Prenant v=0 et notant que

$$(4.8) \qquad d^n = \frac{1}{2}(\delta^n+\delta^{n-1}),$$

il vient :

$$(4.9) \qquad \frac{1}{2\Delta t}(\delta^n-\delta^{n-1},\delta^n+\delta^{n-1})+\frac{1}{2\Delta t}\, a(u^n,u^{n+1}-u^{n-}\ )\leq$$

$$\leq (f^n, \frac{u^{n+1}-u^{n-1}}{2\Delta t});$$

en multipliant par 2 t et en notant que a(u,v) est symétrique, on en déduit

$$(4.10) \qquad |\delta^n|^2-|\delta^{n-1}|^2+a(u^n,u^{n+1})-a(u^{n-1},u^n)\leq(f^n,u^{n+1}-u^{n-1}).$$

Par sommation, on en déduit que

$$(4.11) \qquad |\delta^n|^2+a(u^n,u^{n+1})\leq|\delta^o|^2+a(u^o,u^1)+\sum_{q=0}^{n}(f^q,u^{q+1}-u^{q-1})$$

d'où

$$(4.12) \qquad |\delta^n|^2+a(u^n)\leq-a(u^n,u^{n+1}-u^n)+c+\sum_{q=0}^{n}(f^q,u^{q+1}-u^{q-1}).$$

Mais

$$|a(u^n,u^{n+1}-u^n)|\leq c\|u^n\|\,S(h)\,|u^{n+1}-u^n|\leq c\|u^n\|\,S(h)\Delta t|\delta^n|$$

$$\leq \frac{1}{2}a(u^n)+c(\Delta t S(h))^2|\delta^n|^2$$

de sorte que (4.12) donne

$$(4.13) \qquad [1-c(\Delta t S(h))^2]\ |\delta^n|^2+\frac{1}{2}a(u^n)\leq c+c\sum_{q=0}^{n}|f^q|^2\Delta t +$$

$$+ c \sum_{q=0}^{n} \Delta t |\delta^q|^2.$$

On choisit donc Δt et h de façon que

(4.14) $$c(\Delta t\ S(h))^2 \leqslant 1-\gamma, \quad \gamma>0.$$

Alors (4.13) donne pour Δt assez petit

(4.15) $$\gamma|\delta^n|^2 + \frac{1}{2} a(u^n) \leqslant c + c \sum_{q=0}^{n-1} |\delta^q|^2.$$

De (4.15) et de l'inégalité de Gronwall discrète, on déduit que

(4.16) $$|\delta^n| + ||u^n|| \leqslant c.$$

On va maintenant obtenir d'autres estimations a priori . Ecrivons (4.7) pour n+1 au lieu de n:

(4.17) $$\left(\frac{\delta^{n+1}-\delta^n}{\Delta t}, v-d^{n+1}\right) + a(u^{n+1}, v-d^{n+1}) \geqslant (f^{n+1}, v-d^{n+1}) \quad \forall v \in K_h.$$

On prend $v=d^{n+1}$ (resp. $v=d^n$) dans (4.7) (resp. (4.17)). Additionant et divisant par Δt^2, il vient

(4.18) $$\left(\frac{\delta^{n+1}-2\delta^n+\delta^{n-1}}{\Delta t^2}, \frac{d^{n+1}-d^n}{\Delta t}\right) + a\left(\frac{u^{n+1}-u^n}{\Delta t}, \frac{d^{n+1}-d^n}{\Delta t}\right) \leqslant$$
$$\leqslant \left(\frac{f^{n+1}-f^n}{\Delta t}, \frac{d^{n+1}-d^n}{\Delta t}\right).$$

Mais

$$\frac{d^{n+1}-d^n}{\Delta t} = \frac{1}{2\Delta t}(\delta^{n+1}-\delta^{n-1})$$

de sorte que (4.18) s'écrit encore:

(4.19) $$\left(\frac{\delta^{n+1}-2\delta^n+\delta^{n-1}}{\Delta t^2}, \frac{\delta^{n+1}-\delta^{n-1}}{2\Delta t}\right) + a\left(\delta^n, \frac{\delta^{n+1}-\delta^{n-1}}{2\Delta t}\right) \leqslant$$

$$\leqslant (g^n, \frac{\delta^{n+1}-\delta^{n-1}}{2\Delta t})$$

où l'on a posé

(4.20) $$g^n = \frac{1}{\Delta t}(f^{n+1}-f^n)$$

Mais (4.19) équivaut à (4.9) où l'on remplace u^n par δ^n.

On a donc le résultat analogue à (4.16); sous l'hypothèse (4.14), on a

(4.21) $$\left|\frac{u^{n+1}-2u^n+u^{n-1}}{\Delta t^2}\right| + \|\delta n\| < c$$

d'où (4.2) et (4.3).

Démonstration du Théorème 4.2.-

D'après (4.2),(4.3), on a:

$u_{\Delta t}$, $\delta u_{\Delta t}$ demeurent dans un borné de $L^\infty(0,T;V)$,

$\delta^2 u_{\Delta t}$ demeure dans un borné de $L^\infty(0,T;H)$

et on peut donc supposer, par extraction d'un sous suite, que

(4.22) $u_{\Delta t}$, $\delta u_{\Delta t} \to$ w,w' dans $L^\infty(0,T;V)$ faible étoile,

(4.23) $\delta^2 u_{\Delta t} \to$ w" dans $L^\infty(0,T;H)$ faible étoile.

On a alors:

(4.24) $u^{N-1} \to w(T)$ dans V faible,

(4.25) $\delta^{N-1} \to w'(T)$ dans V faible.

Soit v=v(t) une fonction régulière de $[0,T] \to K$. Posons:

(4.26) $v_h^n = v^n$ = approximation dans K_h de $v(n\Delta t)$.

Faisons $v=v^n$ dans (4.7); on en déduit, après multiplication par Δt:

$$\Delta t\left(\frac{\delta^n-\delta^{n-1}}{\Delta t}, v^n\right) + \Delta t a(u^n,v^n) - \Delta t(f^n, v^n-d^n) \geq$$

$$\geq (\delta^n-\delta^{n-1}, d^n) + \Delta t\, a(u^n,d^n) =$$

$$= (\delta^n-\delta^{n-1}, \frac{\delta^n+\delta^{n-1}}{2}) + \frac{1}{2} a(u^n, u^{n+1}-u^{n-1})$$

$$= \frac{1}{2}[|\delta^n|^2 - |\delta^{n-1}|^2] + \frac{1}{2}[a(u^n,u^{n+1}) - a(u^{n-1},u^n)]$$

d'où; après sommation en n::

(4.27) $$\sum_{n=1}^{N-1} \left[\Delta t\left(\frac{\delta^n-\delta^{n-1}}{\Delta t}, v^n\right) + \Delta t\, a(u^n,v^n) - \Delta t(f^n, v^n-d^n)\right] \geq$$

$$\geq \frac{1}{2}|\delta^{N-1}|^2 - \frac{1}{2}|\delta^o|^2 + \frac{1}{2} a(a^{N-1}, u^N) - \frac{1}{2} a(u^o, u')$$

$$= \frac{1}{2}|\delta^{N-1}|^2 - \frac{1}{2}|\delta^o|^2 + \frac{1}{2} a(u^{N-1}) - \frac{1}{2} a(u^o) + r,$$

où

(4.28) $$r = \frac{1}{2} a(u^{N-1}, u^N - u^{N-1}) - \frac{1}{2} a(u^o, u^1 - u^o).$$

Admettons un instant que

(4.29) $\quad r \to 0$ lorsque $\Delta t \to 0$ (avec (4.5)).

On deduit alors de (4.27) que

(4.30) $$\int_0^T [(w'',v) + a(w,v) - (f, v-w')]dt \geq$$

$$> \lim\inf \left[\frac{1}{2}|\delta^{N-1}|^2 + \frac{1}{2} a(u^{N-1})\right] - \frac{1}{2}|w'(0)|^2 - \frac{1}{2} a(w(0))$$

$$\geq \text{(d'après (4.24) (4.25))}$$

$$\geq \frac{1}{2}\ |w'(T)|^2 + \frac{1}{2}\, a(w(T)) - \frac{1}{2}|w'(O)|^2 - \frac{1}{2}\, a(w(O)) =$$

$$= \int_0^T \left[(w'',w') + a(w,w')\right] dt$$

de sorte que

$$(4.31) \qquad \int_0^T \left[(w^v, v-w') + a(w,v-w') - (f,v-w')\right] dt \geq 0$$

pour toute fonction v par exemple continue de $[O,T] \to K$ - et par prolongement par continuité, $\forall v \in L^2(O,T;V)$ avec $v(t) \in K$ p.p.

On déduit de là que w=u=solution du problème, d'où le Théorème sous réserve de la vérification de (4.29).

Mais

$$|r| \leq c\|u^{N-1}\| \Delta t\ S(h)\, |\delta^{N-1}| + c\|u^O\|\ \Delta t\ S(h)\ |\delta^O| \leq c\ \Delta t\ S(h)$$

d'où le résultat d'après (4.5).

Remarque 4.1.

Les résultats pour le schéma (3.9) sont tout à fait analogues aux précédents.

Faisant v=O dans (3.9) on en déduit:

$$(4.32) \qquad \left(\frac{\delta^n - \delta^{n-1}}{\Delta t}, \delta^n\right) + a(u^n,\delta^n) \leq (f^n,\delta^n).$$

Mais

$$a(u^n,\delta^n) = a(u^{n+1},\ \delta^n) - \Delta t\ a(\delta^n)$$

d'où en portant dans (4.32) et en multipliant par Δt:

$$(\delta^n-\delta^{n-1},\delta^n)+a(u^{n+1},a^{n+1}-u^n)\leqslant\Delta t^2 a(\delta^n)+\Delta t(f^n,\delta^n)$$

soit

$$(4.33)\quad \begin{cases} |\delta^n|^2-|\delta^{n-1}|^2+|\delta^n-\delta^{n-1}|^2+a(u^{n+1})-a(u^n)+a(u^{n+1}-u^n)\leqslant \\ \leqslant 2\Delta t\ a(\delta^n)+2\Delta t(f^n,\delta^n) \\ \leqslant c\ \Delta t^2 S(h)^2|\delta^n|^2+2\Delta t\ |f^n||\delta^n|. \end{cases}$$

Par sommation on en déduit

$$(4.34)\quad |\delta^n|^2+a(u^{n+1})\leqslant c_1+c_2(\Delta t\ S(h))^2\sum_{q=0}^{n}|\delta^q|^2+2\sum_{q=0}^{n}\Delta t|f^q||\delta^q|$$

d'où l'on déduit, <u>si</u>

$$(4.35)\quad c_2(\Delta(t)S(h))^2<1$$

que

$$(4.36)\quad |\delta^n|^2+a(u^{n+1})\leqslant \text{ constante.}$$

On obtient une <u>estimation supplémentaire</u> en considerant (3.9) pour n+1, soit

$$(4.37)\quad \left(\frac{\delta^{n+1}-\delta^n}{\Delta t},\ v-\delta^{n+1}\right)+a(u^{n+1},v-\delta^{n+1})\geqslant(f^{n+1},v-\delta^{n+1}).$$

Prenant $v=\delta^{n+1}$ (resp. $v=\delta^n$) dans (3.9) (resp. (4.37)), il vient par addition et après division par Δt^2:

$$(4.38)\quad \left(\frac{\delta^{n+1}-2\delta^n+\delta^{n-1}}{\Delta t^2},\ \frac{\delta^{n+1}-\delta^n}{\Delta t}\right)+a\left(\frac{u^{n+1}-u^n}{\Delta t},\ \frac{\delta^{n+1}-\delta^n}{\Delta t}\right)\leqslant$$

$$\leqslant\left(\frac{f^{n+1}-f^n}{\Delta t},\ \frac{\delta^{n+1}-\delta^n}{\Delta t}\right).$$

Introduisant g^n comme en (4.20) et posant

(4.39) $$\delta_1^n = \frac{1}{\Delta t}(\delta^{n+1}-\delta^n),$$

(4.38) s'écrit:

(4.40) $$(\frac{\delta_1^n-\delta_1^{n-1}}{t},\delta_1^n|+a(\delta^n,\delta_1^n)\leqslant(g^n,\delta_1^n)$$

qui est l'analogue de (4.32) avec u^n (resp. δ^n) remplacé par δ^n (resp. δ_1^n).

On a donc l'analogue de (4.36) à savoir

(4.41) $$|\delta_1^n|^2+a(\delta^{n+1}) \leqslant \text{costante}$$

d'où les resultats analogues aux Théorèmes 4.1 et 4.2.-

CHAPITRE 5.- Compléments et problèmes.

1.- Ecoulement de fluides de Bingham bidimensionnels.

Décrivons un modèle - Pour la motivation physique, cfr. Duvaut -Lions [1], Chap.6.

On introduit:

$$V=\{v \mid v \in (H_0^1(\Omega))^2, \quad \text{Div } v=0\},$$

$$H=\{v \mid v \in (L^2(\Omega))^2, \quad \text{Div } v=0 \ , n\, v=0 \quad \text{sur } \Gamma\}, \ (^1)$$

(1.1) $$a(u,v) = \sum_{i,j=1}^{2} \int_\Omega \frac{\partial u_i}{\partial x_j} \frac{\partial v_i}{\partial x_j} dx \ ,$$

(1.2) $$b(u,v,w) = \sum_{i,j} \int_\Omega u_i \frac{\partial v_i}{\partial x_j} w_j \ dx \ ,$$

(1.3) $$j(v) = g \int_\Omega \Big(\sum_{i,j} \big(\frac{\partial v_i}{\partial x_j}\big)^2\Big)^{1/2} dx \ , \qquad g>0.$$

On cherche une fonction $u=\{u_1,\dots,u_n\}$ (la vitesse de l'écoulement) vérifiant

(1.4) $$(u'(t),v-u(t))+\mu a(u(t),v-u(t))+b(u(t),u(t),v-u(t))+$$
$$+g\ j(v)-g\ j(u(t))\geqslant(f(t),v-u(t)) \qquad \forall v\in V$$

et

(1.5) $$u(0)=u_0 \qquad \text{donné.}$$

Dans (1.4) μ est donné >0.

---

(1) n = normale à Γ dirigée vers l'extérieur de Ω.

Remarque 1.1.-

Si g=0, (1.4) équivaut à

$$(u'(t),v)+\mu\, a(u(t),v)+b(u(t),u(t),v)=(f(t),v) \quad \forall v\in V; \tag{1.6}$$

c'est le système classique des équations de Navier-Stokes.

Remarque 1.2.-

L'inéquation (1.4) est une inéquation parabolique, mais l'opérateur aux dérivées partielles est (à la différence des chapitres précédents) non linéaire (à cause du facteur $b(u,u,v-u)$).

On démontre (cf. Duvaut-Lions, loc. cit.) que, le problème (1.4) (1.5) admet une solution unique vérifiant

$$u\in L^2(0,T;V), \qquad u'\in L^2(0,T;V'). \tag{1.7}$$

Notons que, pour les écoulements tridimensionnels, on montre l'existence d'une solution faible, l'unicité éventuelle dans la classe des solutions faibles où l'on peut montrer l'existence étant un problème ouvert ([1]).

Schéma d'approximation.

On introduit encore

$$u^{n+\vartheta}=\vartheta u^{n+1}+(1-\vartheta)u^n . \tag{1.8}$$

Partant de u^o=approximation dans V_h ([2]) de u_o, on définit de proche en proche u^{n+1} par

---

([1]) Comme dans le cas des équations de Navier-Stokes.

([2]) La construction de V_h est l'une des difficultés du problème Cf. Fortin [1].

$$(1.9)\quad \begin{cases} \left(\dfrac{u^{n+1}-u^n}{\Delta t}, v-u^{n+1}\right)+ \mu a(u^{n+\vartheta}, v-u^{n+1})+b(u^{n+\vartheta}, u^{n+1}, v-u^{n+1})+ \\ \qquad +g\, j(v)-g\, j(u^{n+1}) \geq (f^n, v-u^{n+1}) \qquad \forall v \in V_h. \end{cases}$$

L'inéquation (1.9) équivaut à

$$\frac{1}{\Delta t}(u^{n+1}, v-u^{n+1})+\mu\vartheta a(u^{n+1}, v-u^{n+1})+b(\vartheta u^{n+1}+(1-\vartheta)u^n, u^{n+1}, v-u^{n+1})+$$

$$+g\, j(v)-g\, j(u^{n+1}) \geq \frac{1}{\Delta t}(u^n, v-u^{n+1})+(f^n, v-u^{n+1})$$

dont on montre qu'elle admet une solution unique pour Δt assez petit.

Etude de la stabilité. Estimations à priori.

Prenons v=0 dans (1.9).

Notons que $b(u^{n+\vartheta}, u^{n+1}, u^{n+1})=0$. Il vaut donc:

$$\left(\frac{u^{n+1}-u^n}{\Delta t}, u^{n+1}\right)+ \mu a(u^{n+\vartheta}, u^{n+1})+g\, j(u^{n+1}) \leq (f^n, u^{n+1})$$

on encore

$$(1.10)\qquad \left(\frac{u^{n+1}-u^n}{\Delta t}, u^{n+1}\right)+\mu\vartheta a(u^{n+1})-(1-\vartheta)\mu\, a(u^{n+1}-u^n, u^{n+1})+$$

$$+g\, j(u^{n+1}) \leq (f^n, u^{n+1}).$$

On en déduit:

$$(1.11)\qquad \frac{1}{2\Delta t}\left(|u^{n+1}|^2-|u^n|^2+|u^{n+1}-u^n|^2\right)+\mu\, a(u^{n+1}) \leq$$

$$\leq \|f^n\|_* \, \|u^{n+1}\|+c(1-\vartheta)\|u^{n+1}\|\,\|u^{n+1}-u^n\|$$

Mais

$$\|u^{n+1}-u^n\| \leq S(h)\, |u^{n+1}-u^n| \qquad (^1)$$

(1) $S(h)=O\left(\frac{1}{|h|}\right)$. Cf. Fortin [1].

et donc (1.11) entraine:

$$\frac{1}{2\Delta t}(|u^{n+1}|^2-|u^n|^2+|u^{n+1}-u^n|^2)+\mu\, a(u^{n+1})<$$

$$\leqslant \frac{\mu}{2}\, a(u^{n+1})+c||f^n||_*^2+c(1-\vartheta)^2 S(h)^2|u^{n+1}-u^n|^2$$

d'où

$$(1.12)\qquad |u^{n+1}|^2-|u^n|^2+(1-c(1-\vartheta)^2S(h)^2\Delta t)\ |u^{n+1}-u^n|^2+\mu\ \Delta t a(u^{n+1})\leqslant$$

$$\leqslant c||f^n||_*^2.$$

Par conséquent, si

$$(1.13)\qquad (1-\ )^2 S(h)^2 \Delta t \leqslant \text{constante convenable}$$

on a:

$$(1.14)\qquad |u^n|\leqslant c,\quad \sum_{n=0}^{N} ||u^n||^2\ \Delta t\leqslant c.$$

Ecrivons maintenant (1.9) pour n-1 au lieu de n:

$$(1.15)\qquad (\frac{u^n-u^{n-1}}{\Delta t},\ v-u^n)+\ \mu a(u^{n-1+\vartheta},v-u^n)+b(u^{n-1+\vartheta},u^n,v-u^n)+$$

$$+g\ j(v)-g\ j(u^n)\geqslant(f^n,v-u^n).$$

Prenant $v=u^n$ (resp. $v=u^{n+1}$) dans (1.9) (resp. (1.15)) et posant

$$(1.16)\qquad \frac{u^{n+1}-u^n}{\Delta t} = \delta^n$$

il vient, après addition et division par Δt:

$$(1.17)\qquad -(\delta^n-\delta^{n-1},\delta^n)-\mu\ a(u^{n+\vartheta}-u^{n-1+\vartheta},\delta^n)+b(u^{n+\vartheta},u^{n+1},\ \frac{u^n-u^{n+1}}{\Delta t})\ +$$

$$+\ b(u^{n-1+\vartheta},u^n,\ \frac{u^{n+1}-u^n}{\Delta t})\geqslant-(f^n-f^{n-1},\ \frac{u^{n+1}-u^n}{\Delta t})$$

Mais

$$b(u^{n+\vartheta},u^{n+1},u^n-u^{n+1})=b(u^{n+\vartheta},u^{n+1}-u^n,u^n-u^{n+1})+$$

$$+b(u^{n+\vartheta},u^n,u^n-u^{n+1})=b(u^{n+\vartheta},u^n,u^n-u^{n+1})$$

de sorte que (1.17) donne, après division par Δt:

$$(1.18)\begin{cases}(\frac{\delta^n-\delta^{n-1}}{\Delta t},\delta^n)+\mu\, a(\frac{u^{n+\vartheta}-u^{n-1+\vartheta}}{\Delta t},\delta^n)\leqslant(\frac{f^n-f^{n-1}}{\Delta t},\delta^n)+X,\\ X=-b(\frac{u^{n+\vartheta}-u^{n-1+\vartheta}}{\Delta t},u^n,\delta^n).\end{cases}$$

Mais

$$\frac{u^{n+}\;-u^{n-1+}}{\Delta t}=\vartheta\delta^n+(1-\vartheta)\;\delta^{n-1},$$

de sorte que (1.18) donne:

$$(1.19)\begin{cases}\frac{1}{2\Delta t}\left[|\delta^n|^2-|\delta^{n-1}|^2+|\delta^n-\delta^{n-1}|^2\right]+\mu a(\delta^n)-\mu(1-\vartheta)a(\delta^n-\delta^{n-1},\delta^n)\leqslant\\ \qquad\leqslant(\frac{f^n-f^{n-1}}{\Delta t},\delta^n)+X.\end{cases}$$

Mais

$$X=-b(\vartheta\delta^n+(1-\vartheta)\delta^{n-1},u^n,\delta^n)$$

$$=b(\vartheta\delta^n+(1-\vartheta)\delta^{n-1},\delta^n,u^n)$$

donc

$$(1.20)\qquad |X|\leqslant c\,||\vartheta\delta^n+(1-\vartheta)\delta^{n-1}||_{(L^4)^2}\,||\delta^n||\;||u^n||_{(L^4)^2}.$$

Mais comme la dimension d'espace vaut 2, on a:

$$||\varphi||_{(L^4)^2}\leqslant c\;|\varphi|^{1/2}||\;\varphi\;||^{1/2}\qquad\qquad \forall\varphi\in V$$

donc

$$|X|\leqslant c|\vartheta\delta^n+(1-\vartheta)\delta^{n-1}|^{1/2}||\vartheta\delta^n+(1-\vartheta)\delta^{n-1}||^{1/2}||\delta^n||\;|u^n|^{1/2}||u^n||^{1/2}$$

et comme d'après (1.14), $|u^n|\leqslant c$, on en déduit:

(1.21) $$|X| \leqslant c||\delta^n||^{3/2}|\delta^n|^{1/2}||u^n||^{1/2} + c||\delta^n||\,||\delta^{n-1}||^{1/2}|\delta^{n-1}|^{1/2}||u^n||^{1/2}$$

$$\leqslant \frac{\mu}{4}\, a(\delta^n) + c|\delta^n|^2||u^n||^2 + \frac{\mu}{4}\, a(\delta^{n-1}) + c|\delta^{n-1}|^2\, u^n \quad 2$$

Posant

(1.22) $$g^n = \frac{f^n - f^{n-1}}{\Delta t},$$

on déduit de (1.19) (1.21) que

(1.23) $$\frac{1}{2\Delta t}\left[|\delta^n|^2 - |\delta^{n-1}|^2 + |\delta^n - \delta^{n-1}|^2\right] + \frac{3\mu}{4}\, a(\delta^n) \leqslant$$

$$\leqslant \mu(1-\theta)\, a(\delta^n - \delta^{n-1}, \delta^n) + |g^n|\,|\delta^n| + \frac{\mu}{4}\, a(\delta^{n-1}) +$$

$$+ c(|\delta^n|^2 + |\delta^{n-1}|^2)\,||u^n||^2.$$

Mais

(1.24) $$|a(\delta^n - \delta^{n-1}, \delta^n)| \leqslant c||\delta^n||\, S(h)\, |\delta^n - \delta^{n-1}|$$

$$\leqslant \frac{\mu}{4}\, a(\delta^n) + c\; S(h)^2 |\delta^n - \delta^{n-1}|^2$$

de sorte que (1.23) et (1.24) entrainent

(1.25) $$\frac{1}{2\Delta t}\,(|\delta^n|^2 - |\delta^{n-1}|^2) + \frac{1}{2\Delta t}\,(1 - (1-\theta)\, c\Delta t\; S(h)^2)\, |\delta^n - \delta^{n-1}|^2 +$$

$$+ \frac{\mu}{2}\, a(\delta^n) \leqslant \frac{\mu}{4}\, a(\delta^{n-1}) + |g^n|\,|\delta^n| + c(|\delta^n|^2 + |\delta^{n-1}|^2)\,||u^n||^2.$$

On en déduit par sommation (et multiplication par Δt) [1]

---

[1] En supposant qu'une condition du type (1.13) a lieu, pour que

$$|\delta^n|^2+\sum_{q=1}^{n}|\delta^q-\delta^{q-1}|^2+\frac{\mu}{2}\sum_{q=1}^{n}\Delta t\ a(\delta^q)\leqslant$$

$$\leqslant\frac{\mu}{4}\sum_{q=}^{n}\Delta t a(\delta^q)+c+\sum_{q=1}^{n}\Delta t|g^q||\delta^q|+c\sum_{q=}^{n}\Delta t|\delta^q|^2||u^q||^2+$$

$$+c\sum_{q=}^{n}\Delta t|\delta^{q-1}|^2||u^q||^2$$

On en déduit, en supposant que $\sum_{q=0}^{N}\Delta t|g^q|^2\leqslant C$:

$$(1.26)\quad |\delta^n|^2+\sum_{q=}^{n}|\delta^q-\delta^{q-1}|^2+\frac{\mu}{4}\sum_{q=}^{n}\Delta t\ a(\delta^q)\leqslant$$

$$\leqslant c+c\sum_{q=1}^{n}\Delta t|\delta^q|^2+c\sum_{q=1}^{n}\Delta t(|\delta^q|^2+|\delta^{q-1}|^2)||u^q||^2.$$

D'après la 2eme inegalité (1.14) et l'inegalité de Gronwall discrète, on en tire:

$$(1.27)\qquad |\delta^n|\leqslant c,\qquad \sum_{n=0}^{N}\Delta t||\delta^n||^2\leqslant C$$

Donc:

Théorème 1.1.- On suppose que

$$(1.28)\qquad f,\ f'\in L^2(O,T;H),\qquad u_o\in D(A).$$

On a alors, u^{n+1} étant donné par le schéma (1.9):

$$(1.29)\qquad \sum_{n=0}^{N}\Delta t(||u^n||^2+||\frac{u^{n+1}-u^n}{t}||^2)\leqslant C,$$

$$(1.30)\qquad \left|\frac{u^{n+1}-u^n}{\Delta t}\right|\leqslant C$$

et cela, sous condition sur Δt si $\theta=1$ et sous la condition

(1.31) $\Delta t\ S(h)^2 \leq$ constante convenable si $0 \leq \vartheta < 1$.

On en déduit le théorème de convergence suivant:

Théorème 1.2.- On se place dans les conditions de Théorème 1.1.- On définit $u_{\Delta t}$, $\delta u_{\Delta t}$ comme les fonctions égales à u^n, δu^n sur $[u\Delta t, (n+1)\Delta t[$.

On a, lorsque h et $\Delta t \to 0$:

(1.32) $u_{\Delta t}, \delta u_{\Delta t} \to u, u'$ dans $L^2(O,T,V)$ faible,

(1.33) $\delta u_{\Delta t} \to u'$ dans $L^\infty(O,T;H)$ faible étoile,

où u est la solution de (1.4) (1.5) appartenant à $L^2(O,T;V)$ avec $u' \in L^2(O,T;V) \cap L^\infty(O,T;H)$, et cela sans condition supplémentaire si $\vartheta=1$ et sous la condition

(1.34) $\Delta t\ S(h)^2 \to 0$ si $0 \leq \vartheta < 1$.

La démonstration se fait selon les même principes que ci dessus. Cf. Fortin [1].

Remarque 1.3.-

On trouvera d'autres schémas et des applications numériques dans Fortin [1].

2.- Problèmes ouverts.

Nous ne mentionnons par ici les problèmes ouverts relatifs à la théorie des inéquations d'evolution (cf. Duvaut-Lions [1], Lions [5] pour un certain nombre de questions ouvertes) - mais plutôt des problèmes d'analyse numérique liés aux inequations d'evolution .

2.1.- Obtention d'estimations d'erreur.

La question est dejà ouverte dans le cas stationnaire.

Il serait, en particulier, très interessant d'etendre - dans la mesure du possible ([1]) - les résultats de Bramble - Schatz [1], Ciarlet-Raviart [1], Fix-Strong [1], Strong [1], Zlamal [1], aux inéquations stationnaires puis d'adapter l'analyse aux problèmes d'évolution.

2.2.- Obtention de schémas stables (et efficaces...) lorsque les convexes K=K(t) dépendent de t.

2.3.- Adaptation au cas des inéquations d'évolution des méthodes de Douglas-Du Pont [1].

([1]) Pour les schémas à haute précision, on se heurte à la difficulté du "seuil de regularité".

J. L. Lions

B i b l i o g r a p h i e.

J.P. AUBIN [1] Livre à paraitre.

J.H. BRAMBLE et A.H. SCHATZ [1] Least square methods for $2m^{th}$ order elliptic boundary value problems. A paraitre

[2] Rayleigh-Ritz-Galerkin Methods for Dirichlet's problem using subspaces without boundary conditions. Comm. Pure Applied Math. 23 (1970), 653-675.

H. BREZIS [1] Les inéquations variationnelles. Journal de Mathematiques. Paris 1972.

[2] Equations et inequations non linéaires dans les espaces vectoriels en dualité. Annales Inst. Fourier, XVIII, 1968, 115-175.

CARASSO [1] Thèse, Madison, 1970.

J. CEA [1] Approximation variationnelle des problèmes aux limites. Annales Inst .Fourier, XIV (1964), 345-444.

Ph. CIARLET, P.A. RAVIART [1] Livre à paraitre chez Dunod,1972.

J. DOUGLAS Jr et T. DU PONT [1] A paraitre.

G. DUVAUT et J.L. LIONS [1] Les inéquations en Mécanique et en Physique. Dunod Paris, 1971.

G. FIX et G. STRANG [1] On The Analysis of the Finite Element method. A paraitre.

M. FORTIN [1] Thèse Paris 1972.

R. GLOWINSKI [1] Cours CIME, ce Livre.

R. GLOWINSKI, J.L. LIONS et R. TRÉMOLIERÈS [1] Résolution numérique des inéquations de la Mécanique et de la Physique .Dunod 1972.

J.L. LIONS [1] Quelques problèmes de la théorie des inéquations variationnelles d'évolution. Cours CIME "Problems in Non Linear Analysis", 1970.- Ed. Cremonese, 1971.

[2] Quelques méthodes de résolution des problèmes aux limites non linéaires. Paris, Dunod-Gauthier Villars, 1969.

[3] Sur l'approximation de la solution d'inéquations d'evolution - C.R. Acad. Sc. Paris, t.269 (1966), pp.55-57.

[4] Sur un nouveau type de problème non linéaire pour opérateurs hyperbolique du 2eme ordre. Sém. J. Leray, Collège de France, 1965-66, vol.II, 17-33.

[5] Sur les inéquations aux dérivées partielles. U-spechi Mat. Nauk. (1971) (en Russe).

J.L. LIONS, E. MAGENES [1] Problèmes aux limites non homogènes et applications, vol.1. Dunod, 1968.

J.L. LIONS, G. STAMPACCHIA [1] Variational Inequalities. Comm. Pure Applied Math. XX (1967), 493-519.

U. MOSCO [1] Cours CIME, ce Livre

G. STRANG [1] Cours CIME, ce Livre.

R. TRÉMOLIÈRES [1] Thèse, Paris, à paraitre.

D. VIAND [1] , Thèse, Paris, à paraitre.

M. ZLAMAL [1] On the finite element method. Numer Math. 12 (1968) 394-409.

CENTRO INTERNAZIONALE MATEMATICO ESTIVO

(C. I. M. E.)

G. I. MARCHUK

INTRODUCTION INTO THE METHODS OF NUMERICAL ANALYSIS

Corso tenuto ad Erice dal 27 giugno al 7 luglio 1971

INTRODUCTION INTO THE METHODS OF NUMERICAL ANALYSIS

Introduction

Numerical mathematics being part of mathematics has currently at its disposal powerful techniques for solving problems of science and engineering.

Large - capacity electronic computers gave rise to algorithmic constructions and mathematical experimentation over a wide area of science and engineering. This attracted new research personnel to the problems of numerical mathematics. The valuable experience we had in solving applied problems, was later used to devise effective methods and algorithms in numerical mathematics.

The methods of numerical mathematics are closely related to the state of computer art. New concepts and methods are formed in numerical mathematics and its numerous applications influenced essentially by every new stage of computer technology. As a rule, with the perfection of existing methods and ideas new tendencies begin to appear in numerical mathematics which eventually become important scientific trends.

The standard of research in numerical mathematics is largely dependent on the actual connection with fundamental areas of mathematics. First of all I should like to mention functional analysis , differential equations , algebra and logic theory of probability, calculus of variations, ect. An interchange of the ideas between different branches of mathematics has been intensified in the recent decade. This is true in the first place for numerical mathematics which has used the results of fundamental mathematical areas to develop new and more sofisticated methods and to improve the old ones.

At the same time it should be emphasized that applications have an important influence on numerical mathematics. Thus, for instance, mathematical simulation often stimulated a discovery of new approaches which are now a most valuable possession of numerical mathematics. Such applied areas as hydrodynamics, atomic physics, mathematical economics and control theory are most important examples.

The studies of approximation, stability and convergence have provided the necessary basis for a wide research of effective difference schemes applied to the problems of mathematical physics. The algorithms of finite difference methods combine as a rule the aspect of a construction of a difference equation-analog as well as the aspect of the solution. Therefore the advance of constructive theory of finite difference methods depends on an inter coordinated development of the two aspects mentioned above.

One of such trends is connected with a systematic development of conservative difference schemes based on conservation laws inherent in most physical phenomena. For this purpose one uses equations of balance written for a separate mesh element of the domain, and then quadrature and interpolational formulas are applied. The resulting difference equations adequately transformed and summed over all mesh points satisfy integral conservation laws. These methods were of great importance in forming a general point of view on a construction of difference schemes for linear and quasilinear equations.

The second trend is concerned with finding efficient algorithms for multi-dimensional stationary problems of mathematical physics. As a result of the success achieved in a solution of simultaneous linear algebraic equations with Jacobi and block-tridiagonal matrices there have emerged a few excellent algorithms in which factorization of the difference operator is used.

Early sixties were marked by a major contribution to numerical mathematic associated with the names of Douglas , Peaceman and Rachford who suggested an alternating direction method. The success of the method was ensured by use of a simple reduction of a multi-dimensional problem to a sequence of onedimensional problems with Jacobi matrices which are convenient to handle. Ultimately the alternating direction method can be treated as the iterative method, where optimization of computation is carried out by a special selection of the compression operator. The latter is a product of simpler operators. Besides the selection of the compression operator there is a set of free relaxation parameters to the selected too.

Later Soviet mathematicians Yanenko, Diakonov, Samarsky and others suggested a so-called splitting-up method. The point is that the approximation of the initial operator by each auxiliary operator is not necessary but on the whole such an approximation in special norms.

A series of investigations has been devoted to a choice of optimization parameters of splitting-up by means of spectral and variational techniques.

The experience we have in a solution of one - dimensional problems represents a solid base when we come to the development of algorithms for more complex problems in mathematical physics. An important role in the development of new approaches to a solution of non-stationary two-dimensional problems belongs to the alternating direction method. However from the outset it turned our that the method had some drawbacks.

Further advance of the methods for multi-dimensional non-stationary problems is connected with splitting-up techniques based as a rule on inhomogeneous difference approximations of the initial differential operators. The mathematical technique is related with splitting of a compound operator to simple ones. If this approach is used the given equation can be solved by means of integration of simpler equations.

In this case the intermediate schemes have to satisfy the approximation and stability conditions only as a whole which permits flexible schemes to be constructed for practically all problems in mathematical physics.

Splitting-up schemes for implicit approximations have been suggested by Yanenko, Diakonov, Samarskii et al. and applied in various problems. Such schemes have stimulated a more general numerical approach to the problems of mathematical physics which has been called a weak approximation method. It turned out that the splitting-up method can be treated as a method of weak approximation of the initial equation by another , simpler one. The solution of the latter, under certain conditions, comes in a norm to that of the initial problem. It is natural that the method of weak approximation is applicable to hydrodynamics, meteorology, oceanology, radiation transfer theory and so on. Recently there has been found a class of splitting-up schemes equivalent in their accuracy to the Crank-Nicolson difference scheme and applied to non-stationary operators. These schemes are absolutely stable for the systems of equations with positive semi-definite operators depending explicity on coordinates and time. This method is easily extended to quasi-linear equations.

French scientists Lions, Teman, et al. have made an important contribution to the splitting-up methods and theoretically substantiated a number of new approaches. These investigations are especially important for fluid dynamics, theory of plasticity and control theory. The method of decomposition and decentralization formulated by these scientists should be specially mentioned. It is closely related to the method of weak approximation.

Lately there has been much interest in variational methods applied to problems in mathematical physics. The variational methods of Ritz, Galerkin, Trefz and others have long become classical in numerical mathematics. These methods are especially effective when one seeks functionals to a solution.

Besides not long ago there emerged a new trend in variational methods, a so-called method of finite elements or functions. The main idea of it was expressed by Courant in nineteen forties. The essence of this method is that one seeks an approximate solution in a form of a linear combination of functions with compact support of order of the mesh width h. In other words one takes as trial functions special functions in a polynomial form identically equal to zero outside of a fixed domain having a characteristic dimension of several h's. The main problem here is theory of approximation of the functions by a given system of finite elements.

An important contribution to the finite element method has been made by Oganesian, Ruhovtz, Rivkind and by Birkhoff, Shultz, Varga et al. The finite element method is closely associated with the application of a variational approach to constructing finite difference equations corresponding to differential equations in mathematical physics. French mathematicians have contributed to this area of research.

A solution of simultaneous algebraic equations and computing of eigenvalues and eigenvectors of matrices are important problems in numerical mathematics. Speaking about the numerical methods and problems in linear algebra of recent years it is necessary first of all to emphasize the growing interest in the solution of large systems of the corresponding equations, in the solution of ill-conditioned systems and in spectral problems for arbitrary matrices. Much attention has been paid to the use of a priori and a posteriori information in the process of a solution. Under the influence of computer development the old numerical methods in linear algebra have been reconsidered. The increasing use of computers has stimulated a creation of new algorithms well suited for automatic calculation.

Direct methods in linear algebra are especially important. A method is called direct if it provides a solution by a finite number of arithmetic operations. Direct methods play an important role when simultaneous linear algebraic equations are solved or inverse matrices and determinants are found. Using some elementary transformations one can represent the initial matrix as a product of two matrices, each being easily inverted. The Gauss elimination method is a classical example of direct methods.

As usual a difficult problem is a solution of simultaneous equations with ill-conditioned matrices. It is closely associated with a solution of conditionally properly posed problems of mathematical physics. One comes across this difficulty because the solution is sensitive to the accuracy of the matrix elements and vector components in the right-hand side of the equation. Though important results have been obtained here, it is only a beginning of extensive research which, evidently, will lead to general theory.

Iterative methods remain very important in linear algebra. An active progress of these methods has resulted in a number of powerful algorithms which are efficiently used on computers. This progress has been caused first of all by a need to solve problems in mathematical physics, economics and control theory involving large systems of equations with special matrices. Direct methods are in most cases ineffective for such problems though each new stage in the computer technology extends their applicability.

So far some trends have formed in a construction of iterative processes and methods. We shall focus our attention only on two of them. The first is associated with the use of spectral characteristics of the operators involved. The methods of this type can be described as follows. An iterative method is constructed with a matrix which depends on some set of parameters. There are two alternatives. First, all parameters of the set are the same for all iterative steps , the spectral radius of the transition matrix being minimized. Second, a sequence of parameter values is constructed so that the value of a parameter depends on the No. of iteration (provided that the error vector tends fast to zero over all initial approximations). Both methods use a priori information about spectra of the matrices involved. A choice of such parameters is part of optimization of the numerical algorithm. The major difficulty here is as a rule to determine boundaries of the spectra of the matrices.

Special optimization of iterative methods stimulates a formulation of a number of problems. Once again we shall discuss two of them. It will be noted that two spectral optimization methods are especially effective when we have a set of problems with the same operator but with different input data.

Much attention has been recently attracted to the Lanszos transform of arbitrary matrices which leads to an equivalent system of equations with a symmetric matrix whose spectrum occupies two segments symmetric with respect to zero. A possibility of such symmetrization as well as some new problems encourage the development of spectral methods accelerating convergence and using polynomials with the least deviation from zero on a set of segments. The second problem connected with spectral optimization is a search of effective methods intended to determine matrix eigenvalue with minimum modulus.

Let us discuss application of variational principles to iterative methods. Such methods allow a successive minimization of some functional (squared as a rule) which attains a minimum on a desired solution. There has been much interest in such problems.

When the variational approach to iterative methods is used one can select relaxation parameters on the basis of a posteriori information obtained at each step. This also is the case for the steepest descent method and the iterative method with minimal discrepancies. The above said is a merit of the variational approach. The rate of convergence seems to be not lower than the rate we get using Chebyshev polynomials. It is essential that such methods converge for both symmetric and non-symmetric matrices if these are positive definite. It has been possible to devise a number of effective methods like the method of minimal discrepancies for positive semi-definite matrices. A grave obstacle to developing nonstationary variational methods has been a necessity to store a more amount of intermediate data than in the case of using corresponding Chebyshev methods.

At present there are iterative methods which combine spectral and variational approaches. Lebedev has formulated conditions which must be satisfied by corresponding operators. Under these conditions unimprovable estimates of a number of arithmetic operations take place. There is also probabilistic technique intended to choose optimization parameters of iterative processes. A series of interesting results have been obtained by Vorobjev. The Young-Frankel overrelaxation method has not yet lost its importance. It has become classical and is generalized in a number of monographs.

We used to compare computational methods according to a number of arithmetic operations and memory requirements. Now we also ought to pay attention to their accuracy. It means that round-off error analysis has become an essential feature of the method itself. A systematic study of errors was first made by Wilkinson. His results were later systematized in his excellent monograph " An algebraic eigenvalue problem" where the method of equivalent perturbations was taken as a basic mathematical technique. As a result estimates of the norms of perturbations were obtained for all fundamental transformations in linear algebra.

In parallel with the method of equivalent parturbations there was an intensive development of statistical error theory. The results obtained by Bakhvalov, Voevodin et al. initiated an investigation of round-off errors. Certainly the statistical methods will play an important role in round-off error analysis.

The progress in computation technology has had an important influence on many branches of computer science which show a tendency to integration. The relations between : software, the methods in numerical and applied methematics, the theory of programming and languages - become so close that choice of a strategy for a solution of particular problems is now of paramount importance. Though optimization of individual components of computional process is as before a fundamental factor of the theory , the attention becomes more and more concentrated on optimization of the whole process.

Optimization of computation is obviously one of the central objects in numerical mathematics which stimulates exploration of new algorithms and new ways of their computer implementation.

The second trend is connected with a solution of classes of problems and with algorithm standardization. A large amount of computer-processed information must be systematized and put in order. The valuable experience which we have in the solution of the problems of science and technology allows us in many cases to set as an ultimate goal a creation of universal methods suitable to handle more or less wide classes of mathematical problems of the same type. At present a care must be taken to save the efforts of the society on a creation of numerous individual algorithms for individual and rare problems. It seems that a rational strategy for a solution of various rare problems is to construct universal algorithms self-adjusting to optimal operating conditions because they use a posteriori information. A rational strategy for a solution of frequently repeated problems is a careful implementation of specific algorithms. These two approaches combined will help to save social resources spent on a creation of efficient software. First steps have been made in the theory of universal algorithms which are self-adjusting to a kind of optimal operating conditions and a course of further research has been outlined.

Software is becoming a materialization of the society's intellect. The process of mathematization of sciences has given rise to an active development of the methods to simulate the phenomena occurring in nature and society. High-speed, large-capacity computers of new generations can store immediately available valuable information and multiaccess computers allow new forms of man-machine interaction using a conversational mode of operation. Therefore standardization of software in general and of numerical algorithms in particular is an urgent problem of scientific and technological progress.

The problem of software has stimulated a formulation of new problems in numerical mathematics, such as a construction of gride for complicated domains. For two-dimensional domains the above is close to its effective solution while for three - and multi - dimensional domains it is just being posed. This problem is closely connected with a construction of algorithms for large problems with high accuracy by difference, variational and other techniques or may be a combination of different methods. The solution of the problems with non-linear monotonous operators is especially important. The corresponding theory is at present intensively developed.

The success achieved in analytic transformations on a computer practically leads us to a solution of mathematical physics problems by the well-known technique of continuous function analysis. As the supply of visual aids for analytic computations grows, these methods will penetrate more and more into software. The success achieved transformations on computers will give computer science new possibilities which nowadays should be taken into account.

Finally I should like to note that the further development of numerical mathematics depends on the standard of research in fundamental branches of mathematics, the importance of the latter essentially increasing at the age of great technological progress. Only a harmonic combination of research in all branches of mathematics will provide the necessary and favourable conditions for self-development of mathematics and its applications.

G. I. Marchuk

Chapter 1

GENERAL INFORMATION FROM THE THEORY OF DIFFERENCE SCHEMES

This chapter presents brief information on the basic problems of the theory of difference schemes, widely applied in the chapters to follow. As our task is to introduce certain modern principles of constructing numerical algorithms, we shall restrict ourselves to the simplest cases that can be easily interpreted. The book is supplied with a list of special literature in which one can find more complicated and specific problems of the theory.

1. 1. - Basic and Adjoint Equations

Let us consider a real Hilbert space Φ of functions φ with an inner product

$$(g, h) = \int_D ghdx, \tag{1.1}$$

where D is the domain of definition of the functions g and h, and x represents a generalized coordinate of an n - dimensional Euclidean space. As usual, the norm of the function φ of Φ is defined from the relation

$$\|\varphi\|^2 = (\varphi, \varphi). \tag{1.2}$$

Let us then consider an operator A acting on the functions of a real Hilbert space Φ.

The operator A is called positive definite on the functions of Φ, if there exists such a constant $\gamma > 0$, that for each function φ the relations hold

$$(A\varphi, \varphi) \geq \gamma(\varphi, \varphi) > 0, \tag{1.3}$$

φ being different from 0 and for the sake of simplicity is defined by $A > 0$. The operator A is called positive semidefinite if there are such non-trivial $\varphi \in \Phi$, elements which turn the inner product $(A\varphi, \varphi)$ into zero. For all the remaining elements there holds the inequality

$$(A\varphi, \varphi) > 0. \qquad (1.4)$$

Below we shall formally denote positive semi-definite operators by $A \geq 0$.

Let us introduce then an adjoint operator A^*, satisfying the Lagrange identity

$$(Ag, h) = (g, A^* h). \qquad (1.5)$$

If is essential to note that $g \in \Phi$ and $h \in \Phi$. The space Φ^*, generally speaking, does not coincide with Φ, though the domain D of definition of basic and adjoint functions is the same. To clarify the fact we shall show that in many problems of mathematical physics g - function belonging to the Hilbert space Φ satisfy some homogeneous boundary conditions. In the application of the Lagrange identity (1.5), as a rule, alongside with the operator A^* those boundary conditions which are satisfied by the adjoint functions h are defined. Further, we shall apply a more convenient notation for adjoint functions. Thus, if the elements of the space of functions Φ are denoted by φ, then those of the adjoint space Φ^* are suitably denoted by φ^*.

In the case of $A = A^*$, the operator A is called self-adjoint. Then $\Phi^* = \Phi$.

Note an important consequence connected with the properties of the adjoint operators. Thus, if $A > 0$, then from the Lagrange identity it follows that $A^* > 0$.

Fourier - series expansions by eigenfunctions of basic and adjoint operators are of great importance for the analysis of algorithms.

Consider the two following spectral problems for $A \geq 0$:

$$Au = \lambda u , \tag{1.6}$$

$$A^* u^* = \lambda u^* \tag{1.7}$$

Assume that each of the homogeneous equations (1.6), (1.7) forms a complete set of orthogonal eigenfunctions $\{u_n\}$ and $\{u_n^*\}$, which can be normalized as follows:

$$(u_n, u_m^*) = \begin{cases} 1, & n = m \\ 0, & n \neq m, \end{cases} \tag{1.8}$$

and eigenvalues λ_n belong to the interval

$$\alpha_A \leq \lambda_n^A \leq \beta_A .$$

We shall call this complete set of eigenfunctions a biorthogonal basis. Then supposing completeness, any function f of Φ and f^* of Φ^* can be represented as a Fourier series

$$f = \sum_n f_n u_n , \tag{1.9}$$

$$f^* = \sum_n f_n^* u_n^* ,$$

where

$$f_n = (f, u_n^*) , \qquad f_n^* = (f^*, u_n) \tag{1.10}$$

Later we shall consider, without stipulating, the spectrum of $A_1 > 0$ and $A_2 \geq 0$ operators to be real. Hence, it is not difficult to establish that in such a case $\lambda_n^{A_1} > 0$ and $\lambda_n^{A_2} \geq 0$.

Of great value for the analysis of numerical algorithms are estimations of norms of operators. The norm of the operator A is defined from :

$$\|A\|^2 = \sup_{\substack{\varphi \in \Phi \\ \varphi \neq 0}} \frac{(A\varphi, A\varphi)}{(\varphi, \varphi)} \tag{1.11}$$

(further for the sake of simplicity the restriction $\varphi \neq 0$ will not be stated). Taking into consideration the relation

$$(A\varphi, A\varphi) = (\varphi, A^*A\varphi),$$

the squared norm of the operator A can be also written as follows;

$$\|A\|^2 = \sup_{\varphi \in \Phi} \frac{(\varphi, A^*A\varphi)}{(\varphi, \varphi)} \tag{1.12}$$

The operator A^*A is symmetric and positive semi-definite. Consider a spectral problem

$$A^*A\Omega = \lambda^{A^*A}\Omega. \tag{1.13}$$

The problem defines a set of eigenfunctions $\{\Omega_n\}$ and eigenvalues $\lambda_n^{A^*A} > 0$. The set $\{\Omega_n\}$ for symmetric operators is complete. Then any function φ of Φ_{A^*A} can be represented as a Fourier series

$$\varphi = \sum_n \varphi_n \Omega_n, \tag{1.14}$$

where

$$\varphi_n = (\varphi, \Omega_n). \tag{1.15}$$

Substitute the series (1.14) into (1.12) and use the condition for function Ω_n orthonormalization. Then we shall have

$$\|A\|^2 = \max_{\{\varphi_n\} \in Q} \frac{\sum_n \lambda_n^{A^*A} \varphi_n^2}{\sum_n \varphi_n^2}, \tag{1.16}$$

where Q is a Hilbert space of Fourier coefficients. It is not difficult to find that

$$\frac{1}{\|A^{-1}\|^2} = \lambda_{\min}^{A^*A} = \alpha_{A^*A}, \tag{1.17}$$

$$\|A\|^2 = \lambda_{max}^{A^*A} = \beta_{A^*A}$$

where $\lambda_{min}^{A^*A}$ and $\lambda_{max}^{A^*A}$ are minimum and maximum eigenvalues respectively from the totality $\lambda_n^{A^*A}$ of the spectral problem (1. 13). The value $\beta_{A^*A} = \lambda_{max}^{A^*A}$ is usually called a spectral radius of the operator A^*A.

In the case of a self-adjoint operator A consider a spectral problem.

$$Au = \lambda^A u . \tag{1. 18}$$

We have

$$\|A\| = \beta_A . \tag{1. 19}$$

If is evident that for the self-adjoint operator

$$\beta_{A^2} = \beta_A^2 \tag{1. 20}$$

Let us consider certain properties of norms of operators the problem of eigenvalues.

1. 1. 1. Energy Norms.

Later we shall always deal with Hilbert spaces of real functions with the norm

$$\|\varphi\|_C^2 = (C\varphi, \varphi), \tag{1. 21}$$

where $C > 0$. It is easy to see that using the Lagrange identity we have the equality

$$(C\varphi, \varphi) = (C^*\varphi, \varphi)$$

and consequently

$$(C\varphi, \varphi) = \left(\frac{C+C^*}{2}\varphi, \varphi\right). \tag{1. 22}$$

The operator $\frac{C+C^*}{2}$ is symmetric and positive. It means that if $C > 0$, then the norm of φ function can always be presented as any inner product with a symmetric operator in the form of the weight function, i.e..

$$\|\varphi\|_C = \|\varphi\|_{\frac{C+C^*}{2}} \tag{1.23}$$

On the basis of Buniakowsky-Schwarz inequality

$$\frac{1}{\|(\frac{C+C^*}{2})^{-1}\|}(\varphi,\varphi) \le (\frac{C+C^*}{2}\varphi,\varphi) \le \|\frac{C+C^*}{2}\|(\varphi,\varphi)$$

one can obtain the following important estimation :

$$\alpha_{\frac{C+C^*}{2}}\|\varphi\|^2 \le \|\varphi\|_C^2 \le \beta_{\frac{C+C^*}{2}}\|\varphi\|^2, \tag{1.24}$$

where $\alpha_{\frac{C+C^*}{2}}$ and $\beta_{\frac{C+C^*}{2}}$ is a maximum and minimum eigenvalue of the symmetric operator $\frac{C+C^*}{2}$,

For simpler and more frequent cases one usually assumes $C = E$. Then we get

$$\|\varphi\|^2 = (\varphi,\varphi).$$

1.1.2. Estimation of the Norm of a single operator

Let us consider a positive semi-definite operator $A \ge 0$. There is the following relation :

$$\|(E+\sigma A)^{-1}\| \le 1 \tag{1.25}$$

for any parameter $\sigma > 0$. This assumption can be proved by the formula

$$\| (E + \sigma A)^{-1} \| = \sup_{\varphi \in \Phi} \frac{((E + \sigma A)^{-1}\varphi, (E + \sigma A)^{-1}\varphi)}{(\varphi, \varphi)} \tag{1.26}$$

Let us take

$$\psi = (E + \sigma A)^{-1} \varphi$$

as a new trial function of (1.26). Then we get

$$\| (E + \sigma A)^{-1} \| = \sup_{\psi \in \Psi} \frac{(\psi, \psi)}{((E + \sigma A)\psi, (E + \sigma A)\psi)} =$$

$$= \frac{1}{\inf\limits_{\psi \in \Psi} \left[1 + 2\sigma \dfrac{(A\psi, \psi)}{(\psi, \psi)} + 2 \dfrac{(A\psi, A\psi)}{(\psi, \psi)} \right.}$$

As $A \geq 0$, the estimation (1.25) follows from the last relation. If $A > 0$, we have

$$\| (E + \sigma A)^{-1} \| < 1. \tag{1.27}$$

1.1.3. Kellogg's Lemma

If $A \geq 0$ and > 0, then

$$\| (E - \sigma A)(E + \sigma A)^{-1} \| \leq 1. \tag{1.28}$$

Let us introduce the notation

$$T = (E - \sigma A)(E + \sigma A)^{-1}$$

Consider the expression for $\| T \|$

$$\| T \|^2 = \sup_{\varphi \in \Phi} \frac{((E - \sigma A)(E + \sigma A)^{-1}\varphi, (E - \sigma A)(E + \sigma A)^{-1}\varphi)}{(\varphi, \varphi)} =$$

$$= \sup_{\psi \in \Psi} \frac{((E-\sigma A)\psi,\ (E-\sigma A)\psi)}{((E+\sigma A)\psi,\ (E+\sigma A)\psi)} =$$

$$= \sup_{\psi \in \Psi} \frac{(\psi,\psi) - 2\ (A\psi,\psi) + \sigma^2 (A\psi, A\psi)}{(\psi,\psi) + 2\ (A\psi,\psi) + \sigma^2 (A\psi, A\psi)} \leq 1$$

The property of positive semi-definiteness of the operator A has been essentially used here. Thus, the lemma is proved.

In the case of $A > 0$ instead of (1.28) we shall get

$$\| (E - \sigma A)(E + \sigma A)^{-1} \| < 1.$$

1.1.4. Estimation of the Norm of Operators

As it was stated above

$$\| A \|^2 = \sup_{\varphi \in \Phi} \frac{(A\varphi, A\varphi)}{(\varphi, \varphi)} = \sup \frac{(A^* A\varphi, \varphi)}{(\varphi, \varphi)} = \beta_{A^* A}$$

Since the squared norm of the operator A coincides with the spectral radius of the self-adjoint operator $A^* A$, to define $\beta_{A^* A}$ one can use the well-known iterative Kellogg's process, if A is a normal operator, that is $AA^* = A^* A$,

$$\| A \|^2 = \lim_{k \to \infty} \frac{(\varphi^{(k+1)}, \varphi^{(k+1)})}{(\varphi^{(k)}, \varphi^{(k)})} \qquad (1.29)$$

where index k denotes the number of the sequential iteration of the following scheme :

$$\varphi^{(k+1)} = A \varphi^{(k)} ,$$

$$\varphi^{(0)} = g .$$

The proof of the convergence of the iterative process (1.29)immediately follows from the Fourier analysis. In fact, let

$$\varphi^{(o)} = \sum_n \varphi_n^{(o)} \Omega_n ,$$

where the Ω_n are eigenvalues of problem (1. 13). Consequently

$$\varphi^{(k+1)} = \sum_n A^{k+1} \varphi_n^{(o)} \Omega_n .$$

Substituting the series into (1. 29) with large k we have

$$\frac{(\varphi^{(k+1)}, \varphi^{(k+1)})}{(\varphi^{(k)}, \varphi^{(k)})} = \frac{\sum_n (\lambda_n^{A^*A})^{k+1} [\varphi_n^0]^2}{\sum_n (\lambda_n^{A^*A})^k [\varphi_n^0]^2} \underset{k>>1}{\Rightarrow} \|A\|^2 \left\{ 1+0\left[\left(\frac{\lambda_{m=1}}{\lambda_m}\right)^{2k}\right]\right\},$$

where $\lambda_m = \beta_{A^*A} = \|A\|^2$ and λ_{m-1} is the eigenvalue preceeding its maximum of the operator A^*A.

1. 1. 5. Calculation of Spectrum Bounds of a Positive Matrix

Consider a problem of finding out maximum and minimum eigenvalues of the operator A, having a positive spectrum

$$Au = \lambda^A u .$$

For this purpose we use Lyusternik's method. We shall introduce an iterative process

$$\varphi^{n+1} = \frac{1}{c_n} A \varphi^{(n)} ;$$

$$\varphi^{(o)} = g ,$$

where c_n is a normalization factor, which is conveniently chosen in the form $c_n = \|\varphi^{(n)}\|_1$. Them

$$\varphi^{n+1} = A \frac{\varphi^{(n)}}{\|\varphi^n\|_1} \quad \text{and} \quad \beta_A = \lim_{n\to\infty} \|\varphi^{(n)}\|_1$$

Here the following norm is used

$$\| \varphi^{(n)} \|_1 = \sum_k |\varphi_k^{(n)}| ,$$

where $\varphi_k^{(n)}$ are vector $\varphi^{(n)}$ components. The constant c_n is usually selected to the order β_A. Then consider the matrix $B = \beta_A E - A$ and the problem

$$Bu = \lambda^B u .$$

It is evident that $B \geq 0$. Then consider again Lyusternik's iterative process

$$\varphi^{(n+1)} = B \frac{\varphi^{(n)}}{\| \varphi^n \|}$$

and get

$$\beta_B = \lim_{n \to \infty} \| \varphi^{(n)} \|_1$$

It is easy to see that operators A and B have a common base and

$$\beta_B = \beta_A - \alpha_A ,$$

Hence

$$\alpha_A = \beta_A - \beta_B .$$

Un this way not only maximum, but also minimum eigenvalues of the matrix A are found. We assumed the matrix A to be of the form in which Lyusternik's method is applicable.

1. 1. 6. Examples

Let us now go over the simplest examples which later will help illustrate methods in numerical mathematics.

1. Let

$$A = -\Delta ,\qquad (1.30)$$

where $\Delta = \frac{\partial^2}{\partial x^2} + \frac{\partial^2}{\partial y^2}$ is the Laplace operator. The operator A is defined on a set of real functions Φ, whose elements satisfy the following requirements. First ,

$$\varphi = 0 \text{ for } \partial D, \qquad (1.31)$$

where ∂D is the boundary of the domain D. For the sake of simplicity it is assumed that the domain D is a unit square $\{0 \le x \le 1, 0 \le y \le 1\}$. Second , the functions (φ) form a Hilbert space with an inner product

$$(a, b) = \int_D ab \; dD \qquad (1.32)$$

where $a \in \Phi$ and $b \in \Phi$, with the norm

$$\|\varphi\| = \sqrt{(\varphi, \varphi)} .$$

Let us show now that on the set of functions Φ the operator A is self-adjoint. Let us indeed consider a functional

$$(A\varphi, \varphi^*) = - \int_D \Delta \varphi \varphi^* \, dD . \qquad (1.33)$$

Using Green's second formula or a double integration by parts we get

$$(A\varphi, \varphi^*) = - \int_{dD} \left(\varphi^* \frac{\partial \varphi}{\partial n} - \varphi \frac{\partial \varphi^*}{\partial n}\right) ds - \int_D \Delta \varphi^* \varphi \, dD. \qquad (1.34)$$

Suppose each function $\varphi^* \in \Phi^*$ satisfies the following boundary condition

$$\varphi^* = 0 \quad \text{on} \quad \partial D \qquad (1.35)$$

The condition (1.35) together with (1.31) makes the integral along ∂D equal to zero. As a result , we have

$$(A\varphi, \varphi^*) = -\int_D \Delta\varphi^* \varphi \, dD = (A\varphi, \varphi) \qquad (1.36)$$

It means that $A = A^*$ and the operator in question is self-adjoint.

Let us then study the problem of definiteness of A. For this purpose we shall consider the functional

$$(A\varphi, \varphi) = -\int_D \Delta\varphi\varphi \, dD. \qquad (1.37)$$

With the help of Green's first formula we get

$$(A\varphi, \varphi) = -\int_{\partial D} \varphi \frac{\partial\varphi}{\partial n} \, ds + \int_D \left[\left(\frac{\partial\varphi}{\partial x}\right)^2 + \left(\frac{\partial\varphi}{\partial y}\right)^2 \right] dD.$$

As φ satisfies condition (1.31), we have

$$(A\varphi, \varphi) = \int_D \left[\left(\frac{\partial\varphi}{\partial x}\right)^2 + \left(\frac{\partial\varphi}{\partial y}\right)^2 \right] dD > 0 \qquad (1.38)$$

for any function φ not identically equal to zero.

Finally, by this example we shall illustrate the problem of eigen-values. As the operator $A = A^*$, the system of eigenfunctions of the problem

$$Au = \lambda u \qquad (1.39)$$

in the case of

$$u = 0 \quad \text{for} \quad \partial D \qquad (1.40)$$

is complete. If D is taken as a square, the system will be of the form

$$u_{mp} = \sin m\pi x \sin p\pi y. \qquad (1.41)$$

The eigen-values of the operator A being :

$$\lambda_{mp} = (m^2 + p^2)\pi^2 > 0. \qquad (1.42)$$

Hence, it follows that

$$2\pi^2 \leq \lambda_n^A \leq \infty .$$

Thus, in this case

$$\alpha_A = \frac{1}{\|A^{-1}\|} = 2\pi^2,$$

$$\beta_A = \|A\| = \infty .$$

Hence if follows in particular that $(A\varphi, \varphi) \geq 2\pi^2 (\varphi, \varphi)$. Consequently, the operator A is positive definite. The system of eigen-function u_n being complete, any function of Φ can be represented as a Fourier series.

$$\varphi(x, y) = \sum_m \sum_p \varphi_{mp} u_{mp}(x, y) = \sum_i \varphi_i u_i(x, y), \tag{1.43}$$

where

$$\varphi_i = (\varphi, u_1) \tag{1.44}$$

when i is a new index of the ordered series.

2. Let us consider now an example of a finite-difference analog of the Laplace operator.

Suppose that φ_{kl} are the function φ values in nodes (x_k, y_l), uniformly covering the domain D with the h step so that

$$x_{k+1} = x_k + h, \quad y_{l+1} = y_l + h .$$

Let us call the points (x_k, y_l) mesh points, their set a grid, and h - mesh size We shall denote, the domain of definition of grid functions (so the functions prescribed in the mesh points are usually called) by D_h, boundary points by dD_h and the set of grid functions φ^h by Φ_h

Such a projection of the operators upon a grid domain leads to finite-difference analogs of the equations, whose methods of construction and some other theoretical problems, such as approximation , numerical stability and convergence of the solutions of an approximate problem to a precise one, will be considered later.

Let $\underline{\varphi}$ be a vector with a polynomial, $\varphi_{k,1}$. - the values of the function in mesh points and

$$A^h = A_k + A_1 , \tag{1.45}$$

where A_k and A_1 are matrices built in the following way.

Let us introduce auxiliary scalar operators

$$\Delta_k \varphi_{kl} = \frac{1}{h} (\varphi_{k+1,1} - \varphi_{k,1}) , \tag{1.46}$$

$$\nabla_k \varphi_{kl} = \frac{1}{h} (\varphi_{k,1} - \varphi_{k-1,1})$$

and similar operators of index 1. Then the components of $(A_k \underline{\varphi}_{kl})$ and $(A_1 \underline{\varphi}_{kl})$ factors will be

$$(A_k \underline{\varphi}_{kl}) = \Delta_k (\nabla_k \varphi_{kl}) ,$$

(1.47)

$$(A_1 \underline{\varphi}_{kl}) = \Delta_1 (\nabla_1 \varphi_{kl})$$

The totality of the mesh points for which $k = 0, n$ with $1 = 0, 1, 2. . . . ,$ n and $1 = 0, n$ with $k = 0, 1, 2. . . ,$ n will be called boundary points of the domain D_h and denoted by ∂D_h . Suppose that

$$\varphi_{kl} = 0 \quad \text{for } \partial D_h . \tag{1.48}$$

As an inner product we take

$$(\underline{a}, \underline{b}) = h^2 \sum_{k=1}^{n-1} \sum_{l=1}^{n-1} a_{kl} b_{kl}. \tag{1.49}$$

Then

$$\|\varphi\| = \sqrt{(\underline{\varphi}, \underline{\varphi})} \tag{1.50}$$

Then consider the functional

$$(A^h \underline{\varphi}, \underline{\varphi}^*) = -h^2 \sum_{k=1}^{n-1} \sum_{l=1}^{n-1} [\Delta_k (\nabla_k \varphi_{kl}) + \Delta_l (\nabla_l \varphi_{kl})] \varphi^*_{kl}. \tag{1.51}$$

Here we have the following identities similar to Green's first and second formulas

$$\sum_{k=1}^{n-1} \Delta_k (\nabla_k \varphi_{kl}) \cdot \varphi^*_{kl} = - \sum_{k=1}^{n} \nabla_k \varphi_{kl} \cdot \nabla_k \varphi^*_{kl},$$

$$\sum_{k=1}^{n-1} \Delta_k (\nabla_k \varphi_{kl}) \cdot \varphi^*_{kl} = \sum_{k=1}^{n-1} \Delta_k (\nabla_k \varphi^*_{kl}) \cdot \varphi_{kl}. \tag{1.52}$$

Formulas (1.52) are valid only for $\varphi \in \Phi_h$ functions satisfying the condition (1.48) and $\varphi \in \Phi^*_h$, satisfying the relation

$$\varphi^*_{kl} = 0 \quad \text{for} \quad \partial D_h. \tag{1.53}$$

Similar equalities take place for the sums of index l. The second relation from (1.52) helps us get

$$(A^h \underline{\varphi}, \underline{\varphi}^*) = (\underline{\varphi}, A^h \underline{\varphi}^*).$$

Hence follows Ah self-adjointness, i. e.

$$A^h = (A^h)^* .. \qquad (1.54)$$

Let us consider the functional

$$(A^h \underline{\varphi} , \underline{\varphi}) = - h^2 \sum_{k=1}^{n-1} \sum_{l=1}^{n-1} [\Delta_k(\nabla_k \varphi_{kl}) + \Delta_l(\nabla_l \varphi_{kl})] \varphi_{kl} .$$

With the help of the first identity (1. 52) for k and l

we get

$$(A^h \underline{\varphi} , \underline{\varphi}) = h^2 \sum_{k=1}^{n-1} \sum_{l=1}^{n-1} [(\nabla_k \varphi_{kl})^2 + (\nabla_l \varphi_{kl})^2] , \qquad (1.55)$$

hence follows

$$(A^h \underline{\varphi} , \underline{\varphi}) > 0 ,$$

At last, consider a spectral problem

$$A^h u = \lambda u \quad \text{in } D ,$$
$$u = 0 \quad \text{for } \partial D_h . \qquad (1.56)$$

The components of the eigenvectors corresponding to (1. 56) are

$$u^{kl}_{mp} = \sin km\pi h \sin lp\pi h. \qquad (1.57)$$

In (1. 57) the indices k, l specify the components of the solution, and m and p are the numbers of eigenvalues which can be ordered as follows :

$$u_{mp}^{kl} = u_l^{k,l}, \quad (i = 1, 2, \ldots).$$

With the obvious relations

$$-\Delta_k (\nabla_k \sin km\pi h) = \frac{4}{h^2} \sin^2 \frac{m\pi h}{2} \sin km\pi h,$$

$$-\Delta_l (\nabla_l \sin lp\pi h) = \frac{4}{h^2} \sin^2 \frac{p\pi h}{2} \sin lp\pi h,$$

the eigen-values will be

$$\lambda_{mp} = \frac{4}{h^2} \left(\sin^2 \frac{m\pi h}{2} + \sin^2 \frac{p\pi h}{2} \right). \tag{1.58}$$

Note that m and p change from 1 to $n-1$. Consequently,

$$\frac{8}{h^2} \sin^2 \frac{\pi h}{2} \leq \lambda_i \leq \frac{8}{h^2}$$

Here the λ_i are ordered λ_{mp}. As, usually, $\frac{\pi h}{2} < 1$ we can write approximately

$$\sin^2 \frac{\pi h}{2} = \frac{\pi^2 h^2}{4} + 0 (h^4)$$

hence, we get

$$2\pi^2 \leq \lambda_i \leq \frac{8}{h^2}. \tag{1.59}$$

Thus, we can write

$$\alpha_A = \frac{1}{\|A^{-1}\|} = 2\pi^2, \quad \beta_A = \|A\| = \frac{8}{h^2}. \tag{1.60}$$

The basis of eigen-vectors (1.57) can be used to present the vector

φ_{kl} as a series . Then we get

$$\varphi_{kl} = \sum_i \varphi_i^{kl} \quad u_i^{kl} \qquad , \tag{1.61}$$

where

$$\varphi_i^{kl} = (\varphi_{kl} , u_i^{kl}) . \tag{1.62}$$

The examples considered above give the necessary understanding of some operators and their properties.

1. 2. Approximation

Let us consider a certain problem of mathematical physics in the operator form

$$\begin{aligned} A \varphi &= f, \\ a \varphi &= g, \end{aligned} \tag{2.1}$$

where A is linear operator, $\varphi \in \Phi$ and $f \in F$.
Here Φ and F are Hilbert spaces with the domains of definition of elements in $D + \partial D$ and D respectively; is a linear operator of the bountary condition, $g \in G$, G is a Hilbert space of functions with the definition domain ∂D.

Along with the equation (2. 1) , let us consider a similar equation in a finite-dimensional euclidean space

$$\begin{cases} A^h \varphi^h = f^h , \text{ in } D_h \\ a^h \varphi^h = g^h , \text{ for } \partial D_h \end{cases} \tag{2.2}$$

where A^h is a linear operator depending on the mesh size h, $\varphi^h \in \Phi_h$, $f^h \in F_h$, and Φ_h and F_h are euclidean spaces with the domain of definition of elements $D_h + \partial D_h$ and D_h, respectively. Here D_h is a set of inner mesh-points of the D domain, and ∂D_h is a set of the meshpoints, on which the boundary condition of the problem is approximated, a^h is a linear operator on the grid $g^h \in G_h$, G_h is an euclidean space of the vectors with the domain of definition ∂D_h.

Let us introduce the Hilbert norm of the vector in grid spaces F_h, G_h, Φ_h. Then we shall denote by $(\xi)_h$ the totality of values of any function of problem (2.1) after projection on the grid domain D_h, ∂D_h or $D_h + \partial D_h$. Then the following definition is usually used: problem (2.2) <u>approximates</u> problem (2.1) with the <u>order h^n</u> for the solution φ, if

$$
\begin{aligned}
&\| (A\varphi)_h - A^h (\varphi)_h \|_{F_h} \leq M_1 h^n, \\
&\| (f)_h - f^h \|_{F_h} \leq M_2 h^n, \\
&\| (a\varphi)_h - a^h (\varphi)_h \|_{G_h} \leq M_3 h^n, \\
&\| (g)_h - g^h \|_{G_h} \leq M_4 h^n.
\end{aligned}
\qquad (2.3)
$$

where M_i are some constants different from ∞.

For the cases where the solution of problem (2.1) is smooth enough, errors in approximation are conveniently measured by the maximum norm peculiar to the space of continuous and differentiable functions. To this end the Taylor-series functions participating in the formulation of the problem is used.

Later we shall assume that reduction of problem (2. 1) to (2. 2) is made and moreover, the boundary condition of (2. 2) is used to eliminate the solution in boundary points of the $D_h + \partial D_h$ domain. As a result we have an equivalent problem

$$A\varphi^h = f^h , \qquad (2.4)$$

where φ^h is now the domain of definition of the solution D_h. The solution φ^h in boundary points is to be found after solving equation (2. 4) as a result of a solution of Eq. (2. 2) with respect to the unknowns.

Thus, in some cases it is convenient to use form (2. 4) in writing an approximation problem, and in others form (2. 2) is more proper. So, as a result of the reduction applied with certain approximation, a problem with a continuous argument (2. 1) is reduced to a problem in linear algebra (2. 4). Further task is to solve a system of algebraic equations.

E x a m p l e. Consider problems

$$\begin{aligned} -\Delta\varphi &= f \quad \text{in } D, \\ \varphi &= 0 \quad \text{for } \partial D. \end{aligned} \qquad (2.5)$$

The domain of definition D is assumed to be a square $\{0 \leq x \leq 1 , 0 \leq y \leq 1\}$, and f - a smooth function. Let us cover square D with a uniform grid along x and y with mesh size h. The mesh-points of the domain will be denoted by two indices (k, 1), where the former $0 \leq k \leq n$ corresponds to the points of the x coordinate , and $0 \leq 1 \leq n$ to those of y. Let us consider the following approximations:

$$\varphi_{xx} \Rightarrow \Delta_k(\nabla_k \varphi_{kl}) , \qquad \varphi_{yy} \Rightarrow \Delta_l (\nabla_l \varphi_{kl}) ,$$

where difference operators Δ_k, Δ_l, ∇_k and ∇_l are defined by (1. 46). Then (2. 5) can be approximated by

$$-[\Delta_k(\nabla_k \varphi_{kl}) + \Delta_l(\nabla_l \varphi_{kl})] = f_{kl} \quad \text{in } D_h,$$

$$\varphi_{kl} = 0 \quad \text{for} \quad \partial D_h, \tag{2. 6}$$

where D_h is a set of mesh-points coinciding with the boundary D . With the help of (1. 46) problem (2. 6) can be reduced to

$$-\Delta^h \varphi^h_{kl} = f^h_{kl} \quad \text{in } D_h,$$

$$\varphi^h_{kl} = 0 \quad \text{for } \partial D_h, \tag{2. 7}$$

where

$$\Delta^h \varphi_{kl} = \frac{1}{h^2}(\varphi_{k+1,l} + \varphi_{k-1,l} + \varphi_{k,l+1} + \varphi_{k,l-1} - 4\varphi_{kl}),$$

$$f^h_{kl} = \frac{1}{h^2}\int_{x_{k-1/2}}^{x_{k+1/2}} dx \int_{y_{l-1/2}}^{y_{l+1/2}} f dy .$$

The operator Δ^h is usually called a difference analog of Laplace operator.

Let us introduce a space of solutions Φ_h. Assume $D_h + \partial D_h = \{0 \le k \le n, 0 \le l \le n\}$ to be domain of definition of the elements from Φ_h. The values f^h_{kl} belong to F_h, all $D_h = \{1 \le k \le n-1, 1 \le l \le n-1\}$ being their domain of definition. Applying the Taylor-series expansion in the vicinity of (x_k , y_l) points , as well as assuming the continuity of the solution and derivatives with respect to (x , y) up to the second order, we get :

$$\varphi(x,y) = \varphi_{kl} + \left(\frac{\partial\varphi}{\partial x}\right)_{kl}(x-x_k) + \left(\frac{\partial\varphi}{\partial y}\right)_{kl}(y-y_l) + \left(\frac{\partial^2\varphi}{\partial x^2}\right)_{kl}\frac{(x-x_k)^2}{2} +$$

$$+ \left(\frac{\partial^2}{\partial x\partial y}\right)_{kl}(x-x_k)(y-y_l) + \left(\frac{\partial^2\varphi}{\partial y^2}\right)_{kl}\frac{(y-y_l)^2}{2} + 0\,(h^3)\,.$$

There is a similar expansion for the function f(x, y) expansions as well. These expansions are considered in the domain $\{x_{k-1} \leq x \leq x_{k+1}\,,$ $y_{l-1} \leq y \leq y_{l+1}\}$. Let us substitute the expansions obtained for φ and f into (2. 7) , and estimate the result in the energy norm. Then we have

$$\|(\Delta\varphi)_h - \Delta^h(\varphi)^h\|_{F_h} \leq M_1 h^2 \,,$$
$$\|(f)_h - f^h\|_{F_h} \leq M_2 h^2 \,, \tag{2.8}$$

where M_1 and M_2 are constants. If is of interest to note if f^h_{kl} is chosen equal to f (x_k , y_l) , then $M_2 = 0$ in the second relation (2. 8) and we get an exact approximation of the right part of equation (2. 5) in the given metric.

As a result of the analysis (2. 8), we come to the conclusion that problem (2. 7) approximates the basic problem (2. 5) accurate to the second order relative to h.

Until now have considered the approximation of a problem in space variables. Similarly, one can consider an approximation problem of an evolutionary equation

$$\frac{\partial \varphi}{\partial t} + A\varphi = f \quad \text{in } D,$$

$$a\varphi = g \quad \text{for} \quad \partial D, \tag{2.9}$$

$$\varphi = \varphi^0 \quad \text{with} \quad t = 0.$$

For the sake of simplicity, using the above considerations, we shall carry out a two-stage approximation of (2.9). We shall first approximate this problem in the $D_h + \partial D_h$ domain in space variables. As a result, we get a differential equation in time, and a difference equation in space variables.

Consider a new evolutionary equation

$$\frac{d\varphi^h}{dt} + A\varphi^h = f^h, \tag{2.10}$$

where Λ, f^h and φ^h are the functions of the time t. Later the index h in (2.10) will be omitted as non-essential, supposing we deal with a difference analog in space variables of the basic problem in mathematical physics.

Generally speaking (2.10) is a system of ordinary differential equations for the vector φ^h components, which are approximate solutions in the mesh-points of the D_h domain.

Thus, consider the following Cauchy problem :

$$\frac{d\varphi}{dt} + \Lambda\varphi = f, \tag{2.11}$$

$$\varphi = g \quad \text{with} \quad t = 0.$$

Suppose that the operator Λ does not depent on time. For the sake of explicitness consider the simplest methode of approximation of the problem (2. 11) in respect to time.
Schemes of the first and second accuracy in t are currently most applied difference schemes.

First, consider the simplest explicit scheme of the first order accuracy in a grid space T_τ

$$\frac{\varphi^{j+1} - \varphi^{j}}{\tau} + \Lambda \varphi^{j} = f^{j}, \quad \varphi^{0} = g, \qquad (2.12)$$

where $\tau = \Delta t$, f^j is a certain approximation of the function f. For simplicity one can take $f^j = f(t_j)$. If the simplest implicit approximation scheme is considered, we have

$$\frac{\varphi^{j+1} - \varphi^{j}}{\tau} + \Lambda \varphi^{j+1} = f^{j}, \quad \varphi^{0} = g \qquad (2.13)$$

and f^j is chosen as $f(t_{j+1})$. Both (2. 12) and (2. 13) are of the first order accuracy in time. It becomes quite obvious if one applies the Taylor - series expansion in time, allowing the existence of time derivatives from the solution to the second order.

Solving (2. 12) and (2. 13) in respect to the unknown, we get a recurrent relation

$$\varphi^{j+1} = T\varphi^{j} + \tau S f^{j}, \qquad (2.14)$$

where T is a step operator , defined in the following way :

$$T = \begin{cases} E - \tau\Lambda & \text{for scheme } (2.12) \\ (E + \tau\Lambda)^{-1} & \text{for scheme } (2.13) \end{cases}$$

and a source operator

$$S = \begin{cases} E & \text{- for scheme (2.12)} \\ T & \text{- for scheme (2.13)} \end{cases}$$

One should note the fact that all single-layer difference schemes for evolutionary equations are reduced to (2. 14) .

The Crank - Nicolson difference scheme of the second order of accuracy is of certain interest in applications

$$\frac{\varphi^{j+1} - \varphi^j}{\tau} + \Lambda \frac{\varphi^{j+1} + \varphi^j}{2} = f^j , \quad \varphi^0 = g, \tag{2.15}$$

where $f^j = f(t_{j+1/2})$. Scheme (2. 14) is canonical for (2. 15), where

$$T = (E + \frac{\tau}{2}\Lambda)^{-1} (E - \frac{\tau}{2}\Lambda),$$

$$S = (E + \frac{\tau}{2}\Lambda)^{-1}.$$

In some cases difference equations (2. 12), (2. 13) and (2. 15) are conveniently written in the form of a system of two equations, where one approximates the equation itself in D , and the other - the boundary condition on ∂D.

In this case we have

$$\begin{aligned} L^{h\tau} \varphi^{h\tau} &= f^{h\tau} \quad \text{in } D, \\ l^{h\tau} \varphi^{h\tau} &= g^{h\tau} \quad \text{on } \partial D. \end{aligned} \tag{2.16}$$

It is supposed here that $L^{h\tau}$ approximates the operator

$$L = \frac{\partial}{\partial t} + A ,$$

and the operator l approximates the boundary condition within the interval $0 \leq t \leq T$. Similarly, $f^{h\tau}$ and $g^{h\tau}$ approximate in the respective (generally speaking, different) norms f and g, i. e.

$$\|(L\varphi)_{h\tau} - L^{h\tau}\varphi_{h\tau}\|_{F_{h\tau}} \leq M_1 h^n + N_1 \tau^p ,$$

$$\|(l\varphi)_{h\tau} - l^{h\tau}\varphi_{h\tau}\|_{G_{h\tau}} \leq M_2 h^n + N_2 \tau^p , \qquad (2.17)$$

$$\|f_{h\tau} - f^{h\tau}\|_{F_{h\tau}} \leq M_3 h^n + N_3 \tau^p ,$$

$$\|g_{h\tau} - g^{h\tau}\|_{G_{h\tau}} \leq M_4 h^n + N_4 \tau^p .$$

The difference equation in the canonical form (2. 14), by introducing vector- functions and new operators, acting in the $D_h \times T_\tau$ space, can be written as

$$L \underline{\varphi} = \underline{f} . \qquad (2.18)$$

Essentially speaking, an evolutionary equation with regard to boundary conditions and initial data can be reduced to the problem in linear algebra (2. 18). Note that for the analysis of approximation in different cases one can apply either a metric space D_h or $D_h \times T_\tau$. In a particular case, a boundary value elliptic problem, an integral equation etc., can be reduced to equation (2. 18). The condition of approximation can be again written as (2. 17), where only h - the maximum value from the totality of (Δx_i) of steps in geometrical variables, is the approximation index.

Esample . Consider a problem

$$A\varphi = \frac{\partial\varphi}{\partial t} - \Delta\varphi = f \quad \text{in } D,$$

$$\varphi = 0 \quad \text{for } \partial D, \tag{2.19}$$

$$\varphi = g \quad \text{with} \quad t = 0.$$

D x T is the domain of definition of the solution, where D , as above, is a square , and $T = \{0 \leq t \leq T_o\}$. Let us change from D to D_n, from ∂D to ∂D_h and from T to T_τ. Let T_τ be the set of t_j points, so that $t_{j+1} - t_j = \tau$.

Then the following will be an approximation to the problem (2. 19):

$$A^{h\tau}\varphi^j_{kl} = f^j_{kl} \quad \text{in } D_h, \tag{2.20}$$

$$\varphi^j = 0 \quad \text{on } \partial D_h ,$$

$$\varphi^0 = g_{kl}$$

1. Consider the simplest explicit approximation

$$A^{h\tau}\varphi^j_{kl} = \frac{\varphi^{j+1}_{kl} - \varphi^j_{kl}}{\tau} - \Delta^h \varphi^j_{kl} \tag{2.21}$$

$$f^j_{kl} = \frac{1}{h^2}\int_{x_{k-1/2}}^{x_{k+1/2}} dx \int_{y_{l-1/2}}^{y_{l+1/2}} f(x,y,t)\, dy, \tag{2.22}$$

$$g_{kl} = \frac{1}{h^2} \int_{x_{k-1/2}}^{x_{k+1/2}} dx \int_{y_{l-1/2}}^{y_{l+1/2}} g(x,y)\,dy. \qquad (2.23)$$

Problem (2. 21) is obviously solvable in respect to the solution. We have

$$\varphi_{kl}^{j+1} = \varphi_{kl}^{j} + \tau\Delta^h \varphi_{kl}^{j} + \tau f_{kl}^{j} \quad \text{in } D_h, \qquad (2.24)$$

Under the condition

$$\begin{aligned} \varphi_{kl}^{j} &= 0 \quad \text{on } \partial D_h, \\ \varphi_{kl}^{j} &= g_{kl} \quad \text{at} \quad = 0. \end{aligned} \qquad (2.25)$$

The recurrent relation (2. 24) is represented as

$$\varphi_{kl}^{j+1} = T\varphi_{kl}^{j} + \tau f_{kl}^{j}, \qquad (2.26)$$

where

$$T = E + \tau\Delta^h = E + \tau(A_1 + A_2)$$

is a step operator, and the operators A_i are defined by (1. 47). Let us estimate its norm. To that end, consider the maximum eigenvalue for the problem

$$\begin{cases} Tu = \lambda^T u & \text{in } D, \\ u = 0 & \text{on } \partial D_h. \end{cases} \qquad (2.27)$$

The following relation is obvious

$$\lambda_n^T = 1 + \tau \lambda_n^{\Delta^h} \tag{2.28}$$

$$\max_n |\lambda_n^T| = \max\left\{ \left|1 - \frac{8\tau}{h^2}\right|, \ \left|1 - 2\tau\pi^2\right| \right\}$$

is

$$\|T\| = \max\left\{ \left|1 - \frac{8\tau}{h^2}\right|, \ \left|1 - 2\tau\pi^2\right| \right\}$$

2. Along with the explicit approximation of first order accuracy in τ, one can consider an implicit approximation of the first order in τ and second order in h, then instead of (2. 21) we take

$$A^{h\tau} \underline{\varphi}^h = \frac{\varphi_{kl}^{j+1} - \varphi_{kl}^j}{} - \Delta^h \varphi_{kl}^{j+1}, \tag{2.29}$$

and the values f_{kl}^j and g_{kl} are again defined by (2. 22). In this case equation (2. 20) is unsolvable, and we get the operator equation

$$(E - \tau\Delta^h)\, \varphi_{kl}^{j+1} = \varphi_{kl}^j + \tau f_{kl}^j \tag{2.30}$$

This equation must be solved if

$$\begin{aligned} \varphi_{kl}^j &= 0 \quad \text{on } \partial D_h, \\ \varphi_{kl}^j &= g_{kl} \quad \text{at } t = 0. \end{aligned} \tag{2.31}$$

Write equation (2. 30) in the form

$$\varphi_{kl}^{j+1} = T\,(\varphi_{kl}^{j} + \tau f_{kl}^{j}), \tag{2.32}$$

where

$$T = (E - \tau\Delta^{h})^{-1}.$$

In this case the norm of the operator T is

$$\| T \| = \frac{1}{1 + \frac{8\tau}{h^2}} \tag{2.33}$$

3. At last, consider the approximation by the Crank-Nicolson difference scheme. In this case operators and functions in (2. 20) are defined the following way :

$$(A^{h\tau}\varphi^{h\tau})_{kl} = \frac{\varphi_{kl}^{j+1} - \varphi_{kl}^{j}}{\tau} - \Delta^{h}\,\frac{\varphi_{kl}^{j} + \varphi_{kl}^{j+1}}{2} \tag{2.34}$$

and

$$f_{kl}^{j} = \frac{1}{h^2}\int_{x_{k-1/2}}^{x_{k+1/2}} dx \int_{x_{l-1/2}}^{y_{l+1/2}} f(x, y, t_{j+1/2})\, dy,$$

(2. 35)

$$g_{kl}^{j} = \frac{1}{h^2}\int_{x_{k-1/2}}^{x_{k+1/2}} dx \int_{y_{l-1/2}}^{y_{l+1/2}} g(x, y)\, dy.$$

Then we have the problem

$$(E - \frac{\tau}{2}\Delta^h)\varphi_{kl}^{j+1} = (E + \frac{\tau}{2}\Delta^h)\varphi_{kl}^{j} + \tau f_{kl}^{j},$$

$$\varphi_{kl}^{j} = 0 \quad \text{on } \partial D_h, \tag{2.36}$$

$$\varphi_{kl}^{j} = g_{kl} \quad \text{at} \quad t = 0. \tag{2.37}$$

In this case equation (2.36) is formally solved in respect to the unknown φ_{kl}^{j+1} in the form :

$$\varphi_{kl}^{j+1} = T\varphi_{kl}^{j} + Sf_{kl}^{j}, \tag{2.38}$$

where

$$T = (E - \frac{\tau}{2}\Delta^h)^{-1}(E + \frac{\tau}{2}\Delta^h),$$

$$S = (E - \frac{\tau}{2}\Delta^h)^{-1}.$$

The norm of the step operator is

$$\| T \| = \left| \frac{1 - \frac{8\tau}{h^2}}{1 + \frac{8\tau}{h^2}} \right| \tag{2.39}$$

1. 3. Numerical Stability

Now let us turn to the notion of numerical stability of difference schemes. We shall not aim at a possible generalization of the definition, as we are mainly interested in the simplest algorithmic approaches to the analysis of the quality of difference schemes, approximating the problems of mathematical physics. Different aspects of stability theory and a number of important generalizing results can be found in some monographs.

To find out the main definitions and concepts of the stability theory let us first consider an explicit difference scheme (2.12)

$$\varphi^{j+1} = (E - \tau\Lambda)\varphi^j + \tau f^j \qquad (3.1)$$

Suppose the operator $\Lambda > 0$ generates a complete set of eigen-functions $\{u_n\}$ and a set of eigen-values $\{\lambda_n > 0\}$ corresponding to the spectral problem

$$\Lambda u = \lambda u.$$

Introduce the following Fourier series :

$$\varphi^j = \sum_n \varphi_n^j u_n , \qquad (3.2)$$

$$f^j = \sum_n f_n^j u_n ,$$

where

$$\varphi_n^j = (\varphi^j, u_n^*), \quad f_n^j = (f^j, u_n^*),$$

u_n^* are eigenfunctions of the adjoint spectral problem. Substitute (3.2) into (3.1) and multiply the result scalarly by u_n^*. Then, if $\tau > 0$ we get expressions for Fourier coefficients

$$\varphi_n^{j+1} = (1 - \tau\lambda_n)\varphi_n^j + \tau f_n^j . \qquad (3.3)$$

Supposing that

$$\varphi^0 = \sum_n g_n u_n ,$$

we obtain the initial condition

$$\varphi_n^0 = g_n \tag{3.4}$$

The solution of (3. 3), (3. 4) is obtained from recurrent elimination of the unknowns. As a result, we have

$$\varphi_n^j = r_n^j g_n + \tau \sum_{i=1}^{j} r_n^{j-i} f_n^{i-1} , \tag{3.5}$$

where

$$r_n = 1 - \tau \lambda_n . \tag{3.6}$$

Equality (3. 5) is estimated in modulus

$$|\varphi_n^j| \leq |r_n|^j |g_n| + \tau \sum_{i=1}^{j} |r_n|^{j-i} |f_n^{i-1}| .$$

We reinforce the latter substituting $\max_j |f_n^j|$ for $|f_n^{i-1}|$. Then we have

$$|\varphi_n^j| \leq |r_n|^j |g_n| + \frac{1 - |r_n|^j}{1 - |r_n|} \tau |f_n| , \tag{3.7}$$

where

$$|f_n| \quad \max_j \quad |f_n^j| .$$

John Neumann introduced a so-called <u>spectral critetion of stability.</u>

It means that if for each harmonic φ_n^j of the Fourier series of (3. 2), there holds the relation

$$|\varphi_n^j| \leq C_1 \, |g_n| + C_2 \, |f_n| \quad (n = 1, 2, \ldots), \tag{3. 8}$$

where C_1, C_2 are constants independent of j, then the difference scheme (3. 1) is announced to be numerically stable. Let us see what conditions should be applied to the parameters of the difference scheme (2. 12) to satisfy relation (3. 8). Relation (3. 7) analysis shows that stability criterion (3. 8) is fulfilled if the condition

$$|r_n| < 1 \quad (n = 1, 2, \ldots). \tag{3. 9}$$

is imposed upon the parameter r_n.

Suppose that the spectrum of the operator Λ is situated in the interval

$$0 < \alpha_\Lambda \leq \lambda_n \leq \beta_\Lambda . \tag{3. 9'}$$

Then, in accordance with (3. 6), relation (3. 9) will hold if

$$\tau < \frac{2}{\beta_\Lambda} \tag{3. 10}$$

Relation (3. 10) will be the constructive condition for the stability of the difference scheme (3. 1).

Note that condition (3. 10) is sufficient for stability. In fact, suppose that

$$\tau = \frac{2}{\beta_\Lambda} .$$

In this case, relation (3. 7) is very likely convert into the following :

$$|\varphi_n^j| \leq |g_n| + j\tau |f_n| . \tag{3.11}$$

Let $j\tau \leq T$, where T is fixed. It means that with a small τ a large number of j steps are considered and with $\tau \to 0$, $j \to \infty$, but so that T remains fixed. Then, with such an approach we again have stable schemes according to Neumann.

In conclusion, two important facts should be stressed. First, it should be noted that stability according to Neumann is based on the spectrum analysis of the operator of the problem. It means that with such an approach estimation of the maximum eigenvalue of the problem or its upper bound is the necessary element of the algorithm. Second, the spectral stability criterion establishes stability and correctness of the solution in respect to every harmonic of the Fourier series, but it does not say anything about the correctness of the solution as a whole. Meanwhile, it is the solution φ^j, which is the main object of our consideration. All this has stimulated the investigators to find other stability definitions, connected with the norms of the operators of the problem. We should stress as well that up to now stability analysis according to Neumann plays a significant role in applications.

Now consider other difference schemes, based on implicit difference approximations. In the case of an implicit scheme of the first order approximation (2.13) we get an expression similar to (3.7),

$$|\varphi_n^j| \leq |r_n|^j |g_n| + \frac{1-|r_n|^j}{1-|r_n|} \tau |r_n| |f_n| , \tag{3.12}$$

where

$$r_n = \frac{1}{1+\lambda_n} .$$

It is obvious that for the given difference scheme with $\lambda_n > 0$ the absolute stability holds, as

$$|r_n| < 1 \quad (n = 1, 2, \ldots).$$

Similarly, the absolute stability of the Crank - Nicolson scheme (2. 15) according to Neumann can be established. In this case the estimation for Fourier coefficients of the solution is

$$|\varphi_n^j| \leq |r_n|^j \, |g_n| + \frac{1 - |r_n|^j}{1 - |r_n|} \tau \mu_n \quad |f_n|, \tag{3. 13}$$

where

$$r_n = \frac{1 - \frac{\tau}{2}\lambda_n}{1 + \frac{\tau}{2}\lambda_n}, \qquad \mu_n = \frac{1}{1 + \frac{\tau}{2}\lambda_n}$$

Hence

$$|r_n| < 1$$

with any $\lambda_n > 0$.

Now let us come to a more general definition of the notion of numerical stability. For this purpose let us consider the problem

$$\begin{aligned} \frac{\partial \varphi}{\partial t} + A\varphi &= f \quad \text{in } D, \\ a\varphi &= g \quad \text{on } \partial D, \end{aligned} \tag{3. 14}$$

which is approximated by the difference problem

$$\begin{aligned} \varphi^{j+1} &= T\varphi^j + \tau S f^j, \\ \varphi &= g \quad \text{with } j = 0. \end{aligned} \tag{3. 15}$$

We say that the difference scheme (3. 15) is <u>stable</u>, if with the fixed parameter h, characterizing the difference approximation, there holds the following relation for any j :

$$\|\varphi^j\|_{\Phi_h} \leq C_1^h \|g\|_{G_h} + C_2^h \|f\|_{F_h}, \tag{3.16}$$

where the C_1^h, and C_2^h constants are independent of j.

The definition of numerical stability involves notion of the correctness of problems with continuous argument. One can say that numerical stability establishes a continuous dependence of the solution on the input data for the problems of discrete argument.

In fact, we choose, as the input data of (3. 15)

$$f = f_*, \quad \text{and} \quad g = g_* .$$

We get some solution of (3. 15) and denote it by φ_*. Then we take as the input data

$$f = f_* + \xi \quad , \quad g = g_* + \delta .$$

Then for the difference of the solutions

$$\varepsilon = \varphi_* - \varphi$$

we have the following problem :

$$\varepsilon^{j+1} = T\varepsilon^j + \tau S \xi ,$$

$$\varepsilon^0 = \delta .$$

Then the stability condition becomes

$$\|\varepsilon^{j+1}\|_{\Phi_h} \leq C_1^h \|\delta\|_{G_h} + C_2^h \|\xi\|_{F_h} .$$

Hence it follows that small variations of the solution φ correspond to those of the input data f and g.

It is easy to see that the definition of stability as it is given in (3.16) already relates the solution itself with a priori information of the input of the problem. Such a definition is more convenient for stability analysis of many problems than the one due to Neumann. Let us consider stability of (2.12) from this point of view. To this end we rewrite the recurrent relation (3.1) as

$$\varphi^{j+1} = T\varphi^j + \tau f^j, \quad \varphi^0 = g, \tag{3.17}$$

where

$$T = E - \tau\Lambda, \tag{3.18}$$

The formal solution of (3.17) is

$$\varphi^{j+1} = T^j g + \tau \sum_{i=1}^{j} T^{j-1} f^{i-1}. \tag{3.19}$$

We shall estimate solution (2.28) by the norm using the Cauchy-Buniakowski inequality and the triangle inequality. Then we get

$$\|\varphi^j\| \leq \|T\|^j \|g\| + \tau \sum_{i=1}^{j} \|T\|^{j-i} \|f^{i-1}\|. \tag{3.20}$$

We replace $\|f^{j-1}\|$ by the maximal value in all j.

Let

$$\max_j \|f^j\| = \|f\|,$$

Then

$$\|\varphi^j\| \leq \|T\|^j \|g\| + \frac{1 - \|T\|^j}{1 - \|T\|} \tau \|f\|. \tag{3.21}$$

If we put

$$\| T \| < 1 , \tag{3.22}$$

then scheme (2.12) will be stable in the sense of the definition (3.16). It is natural that condition (3.22) is a sufficient condition of stability. One could get finer and weaker criteria through the norms of the powers of the step operators $\| T^i \|$ $(i = 1, 2, \ldots, j)$. However, such a weakining of the condition makes difficult a constructive procedure for establishing the stability criterion.

As a rule, in practical calculations the sufficient condition of the form (3.22) is usually applied.

Let us consider a case with the operator $\Lambda = \Lambda^* > 0$

and denote

$$J = \frac{(T\varphi, T\varphi)}{(\varphi, \varphi)} \tag{3.23}$$

Then

$$J = 1 - 2\tau \frac{(\Lambda\varphi, \varphi)}{(\varphi, \varphi)} \quad \tau^2 \frac{(\Lambda\varphi, \Lambda\varphi)}{(\varphi, \varphi)} .$$

Let

$$\varphi = \sum_n \varphi_n u_n ,$$

where $\{u_n\}$ is the base of the operator Λ.

Then

$$J = 1 - 2\tau\bar{\lambda} + \tau^2 \overline{\lambda^2} , \tag{3.24}$$

where

$$\bar{\lambda} = \frac{\sum_n \lambda_n \varphi_n^2}{\sum_n \varphi_n^2} , \qquad \overline{\lambda^2} = \frac{\sum_n \lambda_n^2 \varphi_n^2}{\sum \varphi_n^2}$$

Let us find conditions to be satisfied by τ, so that $J \leq 1$, i.e.

$$1 - 2\tau\bar{\lambda} + \tau^2 \overline{\lambda^2} \leq 1 ,$$

then
$$\tau \leq 2 \frac{\bar{\lambda}}{\overline{\lambda^2}} = 2 \frac{\sum \lambda_n \varphi_n^2}{\sum \lambda_n^2 \varphi_n^2}$$

Hence , if $\beta_\Lambda = \|\Lambda\| = \max \lambda_n = \lambda_1$, then

$$\tau \leq \frac{2}{\lambda_1} \frac{\varphi_1^2 + \sum\limits_{n \neq 1} \frac{\lambda_n}{\lambda_1} \varphi_n^2}{\varphi_1^2 + \sum\limits_{n \neq 1} \frac{\lambda_n^2}{\lambda_1^2} \varphi_n^2} . \tag{3.25}$$

As
$$\frac{\varphi_1^2 + \sum\limits_{n \neq 1} \frac{\lambda_n}{\lambda_1} \varphi_n^2}{\varphi_1^2 + \sum\limits_{n \neq 1} \frac{\lambda_n^2}{\lambda_1^2} \varphi_n^2} \geq 1 ,$$

We get following sufficient condition of stability :

$$\tau \leq \frac{2}{\beta_\Lambda}$$

In this case
$$\|T\| \leq 1 ,$$

and the calculation will be stable in the sense of (3. 16). Note that in the case of a self-adjoint operator, sufficient conditions of numerical stability according to Neumann (3. 10) and those in the sense of definition (3. 16) coincide. Generally speaking , in the case of not self-adjoint operator A , there is not a similar correspondence in stability criteria.

Really , if Neumann's method involves the maximum eigenvalue of the operator Λ , the spectral norm stability involves the maximum eigenvalue of the operator $(\Lambda^*\Lambda)^{1/2}$.

Similarly, we can consider the stability of implicit difference equations (2. 13) and (2. 15). In these cases we get

$$\|\varphi^j\| \leq \|T\|^j \|g\| + \frac{1 - \|T\|^j}{1 - \|T\|} \tau \|S\| \|f\| , \tag{3.26}$$

where

$$T = \begin{cases} (E + \tau\Lambda)^{-1} & \text{for } (2.13) \\ (E + \frac{\tau}{2}\Lambda)^{-1} (E - \frac{\tau}{2}\Lambda) & \text{for } (2.15) \end{cases}$$

and

$$S = \begin{cases} (E + \tau\Lambda)^{-1} & \text{for } (2.13) \\ (E + \frac{\tau}{2}\Lambda)^{-1} & \text{for } (2.15) \end{cases}$$

As stated above, for the self-adjoint operator $\Lambda > 0$ the sufficient condition of stability for these schemes coincides with Neumann's criterion considered above. It means that in this case difference schemes will be absolutely stable in the sense of the definition (3. 16).

We have considered the principal scheme for the research of numerical stability of the difference scheme, assuming the operator A to be time independent. Such an assumption is quite natural for a number of problems in mathematical physics. If also allows us to consider a number of further constructive approaches widely used in numerical mathematics. Indeed the research of stability leads eventually to the estimation of the norm of the operator with the step t . As it was stated above in § 1. 1. 4 , the squared norm of the operator T coincides with the spectral radius of the self-adjoint positive operator $T^* T$. To define the spectral radius one can apply Kellogg's interative process.

$$\|T\|^2 = \lim_{k \to \infty} \frac{(T^* T\varphi^{(k)}, \varphi^{(k)})}{(\varphi^{(k)}, \varphi^{(k)})},$$

where $\varphi^{(k)}$ are the elements of the following :

$$\varphi^{(k+1)} = T\varphi^{(k)} . \qquad (3.27)$$

Thus, the problem of defining the norm of the operator T , is reduced to a successive implementation of the recurrent relation (3. 27). It is this way that is constructively most developed for the use in computers. In the case of a self-adjoint operator T.

$$\|T\| = \beta_T .$$

The following observation is noteworthy. When one studies stability of a difference scheme one sometimes applies the method for the definition of the spectral radius of the infinite periodic - in space - variables problem. For problems with non-periodic boundary conditions, it is quite obligatory to estimate the spectral radius with the help of Kellogg's method for the operators T , in whose construction real boundary conditions have already been considered.

If the operator Λ changes with time, the problem of stability research is hampered to a great extent, as in such a case the norm of the operator T changes with time as well and it is necessary to extablish a spectral radius for each step, the latter depending in the number of the time step. In this case it is appropriate to construct absolutely stable difference analogs of problems. In this respect some methods will be discussed below.

In conclusion it ought to be noted that if the evolutionary equation approximation is investigated in terms of $D_h \times T_\tau$ space, the stability definition would be rather given in terms of the same space. Really, let equation (2. 16) approximate the basic evolutionary problem and let the difference problem be of the form :

$$L^{h\tau} \varphi^{h\tau} = f^{h\tau} \quad \text{in } D_h , \tag{3.28}$$

$$l^{h\tau} \varphi^{h\tau} = g^{h\tau} \quad \text{on} \quad \partial D_h . \tag{3.29}$$

Then the stability criterion is

$$\| \varphi^{h\tau} \|_{\Phi_{h\tau}} \leq C_1^h \, \| f^{h\tau} \|_{F_{h\tau}} + C_2^h \, \| g^{h\tau} \|_{G_{h\tau}} , \tag{3.30}$$

where C_1^h and C_2^h are constants with h fixed, independent of τ .

Now let the basic problem of mathematical physics be approximated by the difference equation (2. 18) so that the boundary conditions are already taken into account in its construction. Then the stability criterion is conveniently introduced in the form :

$$\| \varphi^{h\tau} \|_{\Phi_{h\tau}} \leq C^h \, \| f^{h\tau} \|_{F_{h\tau}} \tag{3.31}$$

where C^h with h fixed is independent of τ .

Of course, if numerical stability is studied with any h , $h \to 0$ including , then it is supposed that with small h's there are $C_1 > 0$ and $C_2 > 0$, already independent of h, i. e.

$$C_1 = \max_h C_1^h \quad \text{and} \quad C_2 = \max_h C_2^h .$$

Thus, with the constructive approach to the establishment of stability of some or other difference scheme it is always preferrable to define C_1^h and C_2^h constants for h fixed. However, in the cases when passages to the limit are studied with $h \to 0$, $\tau \to 0$, judgement of stability should be connected with C_1 and C_2 constants independent either of h or of τ . It is this view of stability that we shall use in the investigation of the convergence of approximate problem solutions to accurate ones, with simultaneous tendency of h and τ towards zero.

A similar consideration refers to the definition of numerical stability in the form (3. 31).

Example 1. Let us investigate the numerical stability of difference schemes approximating the equation of heat conductivity.

In the case of an explicit scheme (2. 21) we had the norm of the step operator in the form (2. 28)

$$\| T \| = \left| 1 - \frac{8\tau}{h^2} \right| .$$

Now let $\| T \| < 1$. It is in this case that the difference scheme is numerically stable. To realize the assumption a time step τ should be properly chosen. Obviously, if we choose

$$\tau \leq \frac{h^2}{4} \qquad (3.32)$$

then $\|T\| \leq 1$, and the calculation will be stable. Thus, the explicit difference scheme (2.21) is stable with the condition (3.32). Such schemes are usually called conditionally stable.

Let us consider now an implicit scheme (2.30). In this case

$$\|T\| = \frac{1}{1 + \frac{8\tau}{h^2}} < 1$$

and, consequently, the difference scheme will be stable with any $\tau > 0$. Such schemes are usually called unconditionally or absolutely stable.

Similarly, for the Crank-Nicolson scheme (2.36) we have :

$$\|T\| = \left| \frac{1 - \frac{8\tau}{h^2}}{1 + \frac{8\tau}{h^2}} \right| < 1$$

Thus this scheme is also absolutely stable.

1.4. The Convergence Theorem

The present section will deal with one of the most important theorems of numerical mathematics, known as the equivalence theorem, finally formulated by P. Lax. The point is that from approximation and stability of the difference scheme follows the convergence of the approximate problem to the solution of the exact one.

Let us consider an evolutionary problem

$$\frac{\partial \varphi}{\partial t} + A\varphi = f \text{ in } D \times T , \tag{4.1}$$

with the boundary conditions :

$$a\varphi = g \quad \text{for } \partial D \times T \tag{4.2}$$

and the initial data

$$\varphi = \varphi^{0} \quad \text{with } t = 0 \tag{4.3}$$

Let us write the problem (4. 1) - (4. 3) as follows ;

$$\begin{aligned} L\varphi &= f, \\ l\varphi &= g. \end{aligned} \tag{4.4}$$

Let us consider a space of functions with the definition domain Dx T and cover this domain with a grid. Then we project the solution of the problem (4. 4) upon the grid domain $\bar{D}_h \times T_\tau$, where $\bar{D}_h = D_h + \partial D_h$ and consider an approximate problem approximating (4. 4) in the form :

$$\begin{aligned} L^{h\tau}\varphi^{h\tau} &= f^{h\tau} \quad \text{in } \bar{D}_h \times T_\tau, \\ l^{h\tau}\varphi^{h\tau} &= g^{h\tau} \quad \text{in } dD_h \times T_\tau. \end{aligned} \tag{4.5}$$

Suppose now that the following approximation is valid :

$$\left\| (L\varphi)_{h\tau} - L^{h\tau}\varphi_{h\tau} \right\|_{F_{h\tau}} \leq M_1 h^k + N_1 \tau^p ,$$

$$\left\| (l\varphi)_{h\tau} - l^{h\tau}\varphi_{h\tau} \right\|_{G_{h\tau}} \leq M_2 h^k + N_2 \tau^p , \tag{4.6}$$

$$\| f_{h\tau} - f^{h\tau} \|_{F_{h\tau}} \leq M_3 h^k + N_3 \tau^p ,$$

$$\| g_{h\tau} - g^{h\tau} \|_{G_{h\tau}} \leq M_4 h^k + N_4 \tau^p .$$

Then assume a certain numerical stability of a difference problem

$$\| \varphi^{h\tau} \|_{\Phi_{h\tau}} \leq C_1 \| f^{h\tau} \|_{F_{h\tau}} + C_2 \| g^{h\tau} \|_{G_{h\tau}} . \qquad (4.7)$$

Let C_1 and C_2 be independent of h and τ. Then provided that we have approximation (4. 6), stability (4. 7) and linearity L, l, $L^{h\tau}$ and l^h there is a convergence

$$\| \varphi_{h\tau} - \varphi^{h\tau} \|_{\Phi_{h\tau}} \leq M h^k + N \tau^p . \qquad (4.8)$$

Let us prove the theorem as follows. Consider the identities

$$\| L^{h\tau} \varphi^{h\tau} - L^{h\tau} \varphi_{h\tau} \|_{F_{h\tau}} = \| L^{h\tau} \varphi^{h\tau} - (L\varphi)_{h\tau} + (L\varphi)_{h\tau} - L^{h\tau} \varphi_{h\tau} \|_{F_{h\tau}} ,$$

$$\| l^{h\tau} \varphi^{h\tau} - l^{h\tau} \varphi_{h\tau} \|_{h\tau} = \| l^{h\tau} \varphi^{h\tau} - (l\varphi)_{h\tau} + (l\varphi)_{h\tau} - l^{h\tau} \varphi_{h\tau} \|_{G_{h\tau}} ,$$

and taking (4. 4) and (4. 5) into account we write :

$$\| L^{h\tau} \varphi^{h\tau} - L^{h\tau} \varphi_{h\tau} \|_{F_{h\tau}} = \| f^{h\tau} - f_{h\tau} + (L\varphi)_{h\tau} - L^{h\tau} \varphi_{h\tau} \|_{F_{h\tau}} ,$$

$$\| l^{h\tau} \varphi^{h\tau} - l^{h\tau} \varphi_{h\tau} \|_{G_{h\tau}} = \| g^{h\tau} - g_{n\tau} + (l\varphi)_{h\tau} - l^{h\tau} \varphi_{h\tau} \|_{G_{h\tau}} \qquad (4.9)$$

If triangle inequalities are taken for the norms :

$$\| L^{h\tau}\varphi^{h\tau} - L^{h\tau}\varphi_{h\tau} \|_{F_{h\tau}} \leq \| f^{h\tau} - f_{h\tau} \|_{F_{h\tau}} + \| (L\varphi)_{h\tau} - L^{h\tau}\varphi_{h\tau} \|_{F_{h\tau}} , \tag{4.10}$$

$$\| l^{h\tau}\varphi^{h\tau} - l^{h\tau}\varphi_{h\tau} \|_{G_{h\tau}} \leq \| g^{h\tau} - g_{h\tau} \|_{G_{h\tau}} + \| (l\varphi)_{h\tau} - l^{h\tau}\varphi_{h\tau} \|_{G_{h\tau}}$$

Applying the conditions of approximation (4. 6) we get :

$$\| L^{h\tau} (\varphi^{h\tau} - \varphi_{h\tau}) \|_{F_{h\tau}} \leq (M_1 + M_3) h^k + (N_1 + N_3) \tau^p , \tag{4.11}$$

$$\| l^{h\tau} (\varphi^{h\tau} - \varphi_{h\tau}) \|_{G_{h\tau}} \leq (M_2 + M_4) h^k + (N_2 + N_4) \tau^p .$$

Now consider the equalities

$$L^{h\tau} \varepsilon^{h\tau} = \Theta^{h\tau} , \quad l^{h\tau} \varepsilon^{h\tau} = \eta^{h\tau} , \tag{4.12}$$

where

$$\varepsilon^{h\tau} = \varphi^{h\tau} - \varphi_{h\tau} , \quad \Theta^{h\tau} = L^{h\tau} (\varphi^{h\tau} - \varphi_{h\tau}) . \tag{4.13}$$

As the difference scheme (4. 12) is numerically stable according to the theorem condition, we get

$$\| \varepsilon^{h\tau} \|_{\Phi_{h\tau}} \leq C_1 \| \Theta^{h\tau} \|_{F_{h\tau}} + C_2 \| \eta^{h\tau} \|_{G_{h\tau}} , \tag{4.14}$$

and, considering (4. 13) and (4. 11), we have

$$\|\varphi^{h\tau} - \varphi_{h\tau}\|_{\Phi_{h\tau}} \leq \left[C_1 (M_1 + M_3) + C_2 (M_2 + M_4)\right] h^k +$$

$$+ \left[C_1 (N_1 + N_3) + C_2 (N_2 + N_\psi)\right] \tau^p$$

or, finally,

$$\|\varphi^{h\tau} - \varphi_{h\tau}\|_{\Phi_{h\tau}} \leq Mh^k + N\tau^p . \qquad (4.15)$$

Thus the convergence theorem is proved. The assumption of the theorem included a rather rigid condition that C_1, and C_2 are independent of h and τ.

Particularly unpleasant is the requirement of independence of these constants of h . As it was already mentioned in the previous section, the values C_1^h and C_2^h are defined with h fixed. Moreover, with $h \to 0$ these constants, in many cases, tend to infinity as follows:

$$C_1^h = \frac{C_1}{h^m} \quad \text{and} \quad C_2^h = \frac{C_2}{h^m} ,$$

where $m \geq 0$. If this fact is taken into account, then the approximate solution convergence to the exact one will be evaluated in the following way :

$$\|\varphi^{h\tau} - \varphi_{h\tau}\|_{\Phi_{h\tau}} \leq Mh^{k-m} + N\tau^p h^{-m}.$$

If $k>m$ and $\tau^p < h^m$, then there is convergence. Naturally, the convergence theorem can also be formulated for the cases when C_1 and C_2 depend both on h and on τ.

Let us turn to the case of convergence in stationary problems of mathematical physics. Let the problem be

$$\begin{aligned} A\varphi &= f \quad \text{in } D \\ a\varphi &= g \quad \text{for } \partial D. \end{aligned} \tag{4.16}$$

Let problem (4. 16) be approximated by the following difference scheme:

$$\begin{aligned} A^h \varphi^h &= f^h \quad \text{in } \overline{D}_h , \\ a^h \varphi^h &= g^h \quad \text{on } \partial D_h . \end{aligned} \tag{4.17}$$

Suppose, there is the following approximation

$$\begin{aligned} \| (A\varphi)_h - A^h \varphi_h \|_{F_h} &\leq M_1 h^k , \\ \| (a\varphi)_h - a^h \varphi_h \|_{G_h} &\leq M_2 h^k , \\ \| f_h - f^h \|_{F_h} &\leq M_3 h^k , \\ \| g_h - g^h \|_{G_h} &\leq M_4 h^k . \end{aligned} \tag{4.18}$$

Besides there is an estimate a priori of the problem solution (4. 17)

$$\| \varphi^h \|_{\Phi_h} \leq C_1 \| f^h \|_{F_h} + C_2 \| g^h \|_{G_h} , \tag{4.19}$$

where C_1 and C_2 are constants independent of h. Then , similar to the above , we get a convergence

$$\| \varphi_h - \varphi^h \| \leq Mh^k . \qquad (4.20)$$

Thus, in the study of difference stationary problems of mathematical physics the role of stability condition is performend by close to it correctness condition using a priori estimations. Here lies a profound inner relationship of difference equations for stationary and evolutionary problems. After establishing approximation and a priori estimates (of stability in the case of evolutionary problems), both of the problems become principally equivalent in their formulations and are investigated by same methods.

G. I. Marchuk

Chapter 2

METHODS OF SOLUTION OF NONSTATIONARY PROBLEMS

In this chapter we shall deal with methods of solution of nonstationary problems in mathematical physics. We shall concentrate on a solution of complex problems and their reduction to simple ones using the method of finite differences. Our main object will be an evolution problem in mathematical physics

$$\frac{\partial \Phi}{\partial t} + A\Phi = f,$$
$$\Phi = g \quad \text{at} \quad t = 0, \qquad (*)$$

where $A \geq 0$, and the solution of the problem $(*)$, the functions f and g possess a necessary smoothness in the domain of definition of the solution D x T. It will be assumed that at the boundary of dD the solution of the problem satisfies some boundary conditions.

2.1 Approximation - Stability Relation

The problem of approximating difference equations by finite-difference ones and stability of difference equations are closely related. Indeed, let us assume that it is required to find an approximate solution to a problem in mathematical physics given imput data of the problem.

The approach to approximation of stability formulated in the previous sections appears in most instances too general judge of individual properties or the algorithm being developed. One of the reasons is generality of assumptions made when one investigates properties of the solution of an approximate problem. Thus, when one judges of approximation one usually uses estimates which are valid for a whole class of problems but not for an individual problem, a theoretical estimate of the operator's norm being given from the worst function of the class, i. e. the function which results in a maximum error. In practical calculations, however , we have to do with specific functions defined by the input data of the problem.

Therefore it is expected that investigation of an approximate solution of a problem studied may allow us to construct effective algorithms of an approximate calculation in a different aspect. This thesis is easily illustrated if we consider an evolution problem

$$\frac{\partial \Phi}{\partial t} + A \Phi = 0 ,$$
$$\Phi = g \quad \text{at} \quad t = 0, \tag{1. 1}$$

where $A = A \geq 0$, and the operator A does not depend on t, Φ and g belong to the Hilbert space L_2 .

Let us approximate (1. 1) by

$$\frac{\varphi^{j+1} - \varphi^{j}}{\tau} + A\varphi^{j} = 0, \tag{1. 2}$$
$$\varphi^{0} = g .$$

Naturally, a necessary condition for approximation to (1. 1) by the difference problem (1. 2) is , in a sense, smallness of the expression $\tau A \varphi^{j}$. This follows immediately from Tailor's expansion in a series of the initial problem (1. 1) and from substitution of the series into (1. 2). Here , surely, it is natural , when one uses this method, to make an assumption that the solution is sufficiently smooth. Thus, let

$$\Phi^{j+1} = \Phi^j + \tau \left(\frac{\partial \Phi}{\partial t} \right)^j + \frac{\tau^2}{2} \left(\frac{\partial^2 \Phi}{\partial t^2} \right)^j + \dots \qquad (1.3)$$

Substituting (1.3) into (1.2) we get

$$\left(\frac{\partial \Phi}{\partial t} \right)^j + A\Phi^j = -\tau \left(\frac{1}{2} \frac{\partial^2 \Phi}{\partial t^2} \right)^j \dots \qquad (1.4)$$

Using (1.1) the last equality is transformed and written as

$$\frac{\partial \Phi}{\partial t} + A\Phi = -\frac{\tau}{2} A^2 \Phi + \dots \quad \text{at} \quad t = t_j . \qquad (1.5)$$

From here it follows that the approximation condition may be chosen, so as to emphasize smallness of the remainder due to the mutual compensation of the terms in the left-hand side of (1.4) and, considering that for symmetric operators

$$(A^2 \Phi^j, \Phi^j) = (A \Phi^j, A \Phi^j),$$

we have

$$\frac{\tau}{2} (A\Phi^j, A\Phi^j) \ll (A\Phi^j, \Phi^j). \qquad (1.6)$$

Naturally, the condition (1.6) will be only necessary. From here follows necessity of the following criterion of approximation to the problem in question :

$$\tau \ll \frac{2(A\Phi^j, \Phi^j)}{(A\Phi^j, A\Phi^j)} \qquad (1.7)$$

On the other hand the condition of numerical stability also yields a restriction upon τ, which will be obtained in a constructive form. To this end let us consider the difference equation (1.2) and solve it with respect to φ^{j+1}. We get

$$\varphi^{j+1} = \varphi^j - \tau A \varphi^j. \qquad (1.8)$$

Make up the functional

$$J^{j+1} = (\varphi^{j+1}, \varphi^{j+1}) = \|\varphi^j\|^2 .$$

Then we have

$$J^{j+1} = q^j J^j , \tag{1.9}$$

where

$$q^j = 1-2\tau \frac{(A\varphi^j, \varphi^j)}{(\varphi^j, \varphi^j)} + \tau^2 \frac{(A\varphi^j, A\varphi^j)}{(\varphi^j, \varphi^j)} . \tag{1.10}$$

Obviously, following will be the stability condition

$$q^j \leq 1 . \tag{1.11}$$

From here immediately follows the condition for

$$\tau \leq \frac{2(A\varphi^j, \varphi^j)}{(A\varphi^j, A\varphi^j)} . \tag{1.12}$$

If we assume that the approximate solutions φ^j and the exact one Φ differ but little and join (1. 7) and (1. 12) we arrive at the following algorithme of choice of τ in the step j. It is required that the following conditions be fulfilled simultaneously :

$$\tau_j \ll \frac{2(A\varphi^j, \varphi^j)}{(A\varphi^j, A\varphi^j)} , \qquad \tau_j \leq \frac{2(A\varphi^j, \varphi^j)}{(A\varphi^j, A\varphi^j)}$$

From here it follows that the algorithm's stability imposes a less restriction upon τ_j than approximation. Hence the approximation condition

$$\tau_j \ll \frac{2(A\varphi^j, \varphi^j)}{(A\varphi^j, A\varphi^j)} \qquad (1.13)$$

automatically ensures stability of the algorithm as well. We would like to draw attention to this fact.

When one considers implicit schemes of the type

$$\frac{\varphi^{j+1} - \varphi^j}{\tau} + A\varphi^{j+1} = 0, \quad \varphi^o = g, \qquad (1.14)$$

which are unconditionally stable at $A \geq 0$, the choice of the parameter τ should be made from the approximation condition. In this case we have the necessary approximation condition of the form

$$\frac{\tau}{2}(A^2\Phi^j, \Phi^j) \ll (A\Phi^j, \Phi^j)$$

or, according to the definition introduced above,

$$\tau \ll \frac{2(A\Phi^j, \varphi^j)}{(A\varphi^j, A\varphi^j)}. \qquad (1.15)$$

Here, as in the previous case, Φ^j can be replaced by φ^j. Then we are led to a constructive criterion of choice of the steps

$$\tau_j \ll \frac{2(A\varphi^j, \varphi^j)}{(A\varphi^j, A\varphi^j)} \qquad (1.16)$$

Here, again, we get estimate of the step τ_j from the approximation condition only.

From the above analysis it follows that the choice of the parameter τ for schemes of the first - order approximation is dictated by the accuracy of approximation. This is a decisive point when schemes of higher order accuracy are used in numerical calculations. For such schemes, as is easily found, a limiting condition for the step τ will be a condition of numerical stability.

Apparently, in future, numerical schemes will be constructed with an automatic adjustment to optimal conditions. This means that the choice of τ_j will not be made from a priori consideration of errors on classes of solutions but from a posteriori estimates of the type (1. 13) and (1. 16).

2. 2. Difference schemes of second-order accuracy with time-dependent operators.

Schemes of second-order approximation with respect to τ should be specially mentioned in numerical methods. Most common at present is the Crank-Nicolson difference scheme. Let us consider the evolution equation

$$\frac{\partial \Phi}{\partial t} + A\Phi = 0. \qquad \Phi = g \quad \text{at} \quad t = 0, \tag{2. 1}$$

The difference equation corresponding to (2. 1) is written in the form

$$\frac{\varphi^{j+1} - \varphi^{j}}{\tau} + A\,\frac{\varphi^{j+1} + \varphi^{j}}{2} = 0, \quad \varphi^{o} = g. \tag{2. 2}$$

It is easy to find that if the solution is sufficiently smooth the problem (2. 1) is approximated by the approximate problem (2. 2) with second-order accuracy with respect to τ. (2. 2) is generally called a sheme with central differences in time or the Crank-Nicolson difference scheme. It is curious to note that the scheme (2. 2) is the result of an alternate application of explicit and implicit schemes of first- order accuracy written for the intervals $t_j \leq t \leq t_{j+1/2}$ and $t_{j+1/2} \leq t \leq t_{j+1}$, respectively, (if A is a linear operator):

$$\frac{\varphi^{j+1/2} - \varphi^{j}}{\tau/2} + A\varphi^{j} = 0, \tag{2. 3}$$

$$\frac{\varphi^{j+1}-\varphi^{j+1/2}}{\tau/2} + A\varphi^{j+1} = 0$$

Eliminating the unknowns $\varphi^{j+1/2}$ from the system of difference equations we get the Crank-Nicolson scheme.

Let us assume now the operator A is dependent on time and in the problem (2.1) it is approximated by a difference operator. In this case by the solution of the problem φ we shall understand the vector-function whose components are approximate solutions at mesh points of space and Λ is the matrix approximating the operator A. Thus we have a problem in linear algebra

$$\frac{\varphi^{j+1}-\varphi^{j}}{\tau} + \Lambda^{j}\frac{\varphi^{j+1}+\varphi^{j}}{2} = 0, \quad \varphi^{0} = g, \tag{2.4}$$

where

$$(\Lambda^{j}\varphi, \varphi) \geq 0 \text{ or } \Lambda^{j} \geq 0 \tag{2.5}$$

for any functions of a Hilbert space.

Eq. (2.4) is formally solved with respect to φ^{j+1}. Then we have

$$\varphi^{j+1} = (E + \frac{\tau}{2}\Lambda^{j})^{-1}(E - \frac{\tau}{2}\Lambda^{j})\varphi^{j} \tag{2.6}$$

or

$$\varphi^{j+1} = T^{j}\varphi^{j}, \tag{2.7}$$

where
$$T^{j} = (E + \frac{\tau}{2}\Lambda^{j})^{-1}(E - \frac{\tau}{2}\Lambda^{j}) \tag{2.8}$$

is a step operator.

In order to prove numerical stability it is not necessary to evaluate the norm of the step operator - T^{j}. We shall do differently. Let us form inner product of Eq. (2.4) and $1/2\,(\varphi^{j+1}+\varphi^{j})$.

We get

$$\frac{(\varphi^{j+1}, \varphi^{j+1})-(\varphi^{j},\varphi^{j})}{2\tau} + (\Lambda^{j} \frac{\varphi^{j+1}+\varphi^{j}}{2}, \frac{\varphi^{j+1}+\varphi^{j}}{2}) = 0 \tag{2.9}$$

Since, by assumption, the operator A^j is positive-semidefinite (see (2. 5)), we have

$$\| \varphi^{j+1} \| \leq \| \varphi^{j} \| \tag{2.10}$$

i. e. stability is ensured.

Estimate of the norm of the step operator is of importance in the analysis of difference schemes. For this purpose we consider Eq. (2. 7).

$$\varphi^{j+1} = T^j \varphi^j .$$

Let us evaluate this equation by the norm using the Bunfakowski-Schwarz inequality. As a result we get

$$\| \varphi^{j+1} \| \leq \| T^j \| \, \| \varphi^j \| . \tag{2.11}$$

Comparing (2. 10) to (2. 11) we arrive at the important conclusion that

$$\| T^j \| \leq 1.$$

When the operator Λ^j is skew - symmetric, i. e. the equality is valid

$$(\Lambda^j \varphi, \varphi) = 0 ,$$

then, in place of (2. 10), we have a strict equality

$$\| \varphi^{j+1} \| = \| \varphi^j \| . \tag{2.12}$$

We shall use of this fact later when we shall consider different applications.

Analogously to the foregoing it can be shown that in this case

$$\| T^j \| = 1 .$$

Let us discuss now approximation of the Crank-Nicolson difference scheme when the operator A is time-dependent. To this end we expand the operator T^j in powers of the parameter τ.

Then we have

$$T^j = E - \tau\Lambda^j + \frac{\tau^2}{2} (\Lambda^j)^2 - \dots . \tag{2.13}$$

Let us introduce the operator H accroding to the equality

$$H\Phi = \frac{\partial \Phi}{\partial t} + A\Phi \tag{2.14}$$

and the approximating operator

$$H_\tau \Phi^j = \frac{\Phi^{j+1} - \Phi^j}{\tau} + \Lambda^j \frac{\Phi^{j+1} + \Phi^j}{2} = \frac{(E + \frac{\tau}{2}\Lambda^j)}{\tau} \left\{ \Phi^{j+1} - \left[E - \tau\Lambda^2 + \frac{\tau^2}{2} (\Lambda^j)^2 - \dots \right] \Phi^j \right\}. \tag{2.15}$$

Then we introduce a norm convenient for estimating approximation to the operator H.

$$\| (H\Phi)^j - H_\tau^j \Phi^j \|_{C_\tau} = \| H_\tau^j \Phi^j \|_{C_\tau} = \max_{t_\tau} | H_\tau^j \Phi^j | . \tag{2.16}$$

In order to find the norm (2.16) we expand the solution of the initial equation (2.1) into a Tailor's series. We have

$$\Phi^{j+1} = \Phi^j + \tau \Phi_t^j + \frac{\tau^2}{2} \Phi_{tt}^j + \dots . \tag{2.17}$$

Taking into consideration

$$\Phi_t = - A\Phi , \quad \Phi_{tt} = A^2\Phi - A_t \Phi , \tag{2.18}$$

where $A_t = \frac{\partial A}{\partial t}$, the Tailor's series (2.17) will be reduced

$$\Phi^{j+1} = \Phi^j - \tau A^j \Phi^j + \frac{\tau^2}{2} \left[(A^j)^2 \Phi^j - A_t^j \Phi^j \right] - \dots . \tag{2.19}$$

Substitute (2.19) into (2.16). Then , considering (2.15); we get

$$\| (H\Phi)^j - H^j_\tau \Phi^j \|_{c_\tau} = \max_{t_j} \Big| \Lambda^j \Phi^j - A^j \Phi^j + \tag{2.20}$$

$$+ \frac{\tau}{2}\Big\{ (A^j)^2 - A^j_t - {}^j A^j \Big\} \Phi^j + 0\,(\tau^2) \Big| .$$

If we choose

$$\Lambda^j = A^j = A(t_j) , \tag{2.21}$$

as the approximating operator A^j , then from (2. 20) it follows that

$$\| (H\Phi)^j - H^j_\tau \Phi^j \|_{c_\tau} = \frac{\tau}{2} \max_{t_j \in c_\tau} | A^j_t \Phi^j | + 0\,(\tau^2) ,$$

and we have the first - order approximation. It will be noted that in a special case, when Λ is independent of t, approximation of the form (2. 21) ensures the second-order approximation in τ .

Let the approximating operator Λ^j be chosen in the form

$$\Lambda^j = A^j + \frac{\tau}{2} A^j_t , \tag{2.22}$$

In this case we have

$$\| (H\Phi)^j - H^j_\tau \Phi^j \|_{c_\tau} = 0\,(\tau^2) .$$

In particular, approximation by the Crank - Nicolson scheme will be of the second order in τ, if in addition to (2. 22) we can assume one of the following forms :

$$A^j = A^{j=1/2} , \tag{2.23}$$

$$A^j = \frac{1}{2} (A^{j+1} + A^j). \tag{2.24}$$

In different applications, especially in a numerical solution of quasilinear equations, one can use one of the three above-mentioned forms of approximation of the operator A : (2. 22) and (2. 23) or (2. 24) ensuring the second order of accuracy.

Finally, it should be noted that when one chooses the approximating operator Λ^j in the form (2. 22), (2. 23) and (2. 24), the step operator T^j of the form (2. 8) ensures the second-order approximation. Later we shall make use of this fact.

2. 3 Inhomogeneous evolution equations

In the previous section we dealt with homogeneous equations. Let us consider now inhomogeneous equations

$$\frac{\partial \varphi}{\partial t} + A\varphi = f, \tag{3. 1}$$

$$\varphi = g \quad \text{at} \quad t = 0.$$

Difference approximation to Eq. (3. 1) on the basis of the Crank-Nicolson difference scheme under the assumptions made in § 2. 2 has the form

$$\frac{\varphi^{j+1} - \varphi^j}{\tau} + \Lambda^j \cdot \frac{\varphi^{j+1} + \varphi^j}{2} = f^j, \tag{3. 2}$$

$$\varphi^o = g,$$

where $F^j = f(t_j + 1/2)$.

It is easy to find that the difference problem (3. 2) approximates (3. 1) to within τ^2. Let us write the formal solution of (3. 2) at each interval

$$\varphi^{j+1} = T^j\,{}^j + \tau(E + \frac{\tau}{2}\Lambda^j)^{-1} f^j. \tag{3. 3}$$

In the previous section in the case of a homogeneous equation it is shown that if $\Lambda^j \geq 0$ the following estimate holds

$$\| T^j \| \leq 1 . \tag{3. 4}$$

Naturally, this estimate of the operator's norm does not depend on the right-hand side f. Hance it holds in this case as well. From Eq. (3. 3), it follows that

$$\|\varphi^{j+1}\| \leq \|T^j\| \|\varphi^j\| + \tau \| (E + \frac{\tau}{2} \Lambda^j)^{-1} \| \| f^j \| . \tag{3. 5}$$

In order to establish stability we make use of the estimate (2. 5) from Chapter I. As $\tau > 0$ and

$$(\Lambda^j \varphi^j , \varphi^j) \geq 0, \quad (\Lambda^j \varphi^j , \Lambda^j \varphi^j) \geq 0. \tag{3. 6}$$

we have

$$\| (E + \frac{\tau}{2} \Lambda^j)^{-1} \| \leq 1, \tag{3. 7}$$

Thus, considering (3. 4) and (3. 7), we transform the inequality (3. 5) as follows :

$$\| \varphi^{j+1} \| \leq \| \varphi^j \| + \tau \| f^j \| \tag{3. 8}$$

setting $\| \varphi^o \| = \| g \|$ and $\| f \| = \max\limits_j \| f^j \|$ and using the recurrence relation (3. 8-, we get

$$\| \varphi^j \| \leq \| g \| + C \| f \| , \tag{3. 9}$$

where

$$C = J\tau = T < \infty \tag{3. 10}$$

is a time interval of definition of a solution.

In this way, the relation (3. 9) establishes stability of the difference scheme. In addition, this relation is an a priori estimate of the norm of a solution if we take (3. 10) into account.

2. 4 Splitting-up methods for nonstationary problems

Very often when one has to solve a complicated problem of mathematical physics it is possible to reduce it to a sequence of simpler problems which can be effectively solved on a computer. Such reduction is possible when the initial positive semi-definite operator of the problem can be presented as a sum of simplest positive semi-definite operators. This procedure will be called a splitting-up method. Theory of the splitting-up methods has been developed in extense for the case when the initial operator is representable as a sum of two simpler operators. Therefore we begin presentation of splitting-up methods just with this case.

Let us consider the evolution equation

$$\frac{\partial \varphi}{\partial t} + A\varphi = 0 , \qquad (4.1)$$

$$\varphi = g \quad \text{if} \quad t = 0,$$

where the operated $A \geq 0$ is independent of time can be presented as

$$A = A_1 + A_2 \qquad (4.2)$$

if

$$A_1 \geq 0 , \quad A_2 \geq 0. \qquad (4.3)$$

Let us assume then that the solution to (4. 1) possesses the necessary smoothness. Now we consider three most efficient splitting-up methods.

2. 4. 1 The method of universal algorithm

Let us write the approximate formulation of the problem (4. 1)-(4. 3) as follows :

$$(E + \frac{\tau}{2} A_1)(E + \frac{\tau}{2} A_2) \frac{\varphi^{j+1} - \varphi^j}{\tau} + A\varphi^j = 0 \qquad (4.4)$$

$$\varphi^o = g.$$

If the solution is sufficiently smooth it is not difficult to show that (4. 4) approximates the initial-value differential problem (4. 1) - (4. 3) to an accuracy of second- order smallness in τ. Indeed, through algebraic transformations, equation (4. 4) is reduced to

$$(E + \frac{\tau^2}{4} A_1 A_2) \frac{\varphi^{j+1} - \varphi^j}{\tau} + \frac{\varphi^{j+1} + \varphi^j}{2} = 0 \qquad (4.5)$$

$$\varphi^o = g.$$

It is seen that the difference equation (4. 5) is equivalent in accuracy to the Crank-Nicolson difference scheme if the solution is sufficiently smooth

$$\frac{\varphi^{j+1} - \varphi^j}{\tau} + A\frac{\varphi^{j+1} + \varphi^j}{2} = 0 \qquad (4.6)$$

$$\varphi^o = g.$$

This follows from the fact that scheme (4. 6) itself is of second order approximation in τ and therefore in this sense (4. 5) and (4. 6) are equivalent.

Let us analyse now stability of the difference equation (4. 4). For this purpose (4. 4). will be written as

$$(E + \frac{\tau}{2} A_1)(E + \frac{\tau}{2} A_2)\varphi^{j+1} = (E - \frac{\tau}{2} A_1)(E - \frac{\tau}{2} A_2)\varphi^j$$

$$\varphi^o = g; \qquad (4.7)$$

and the difference equation from (4. 7) will be solved with respect to φ^{j+1} Then we have

$$\varphi^{j+1} - (E + \frac{\tau}{2} A_2)^{-1}(E+ \frac{\tau}{2} A_1)^{-1}(E-\frac{\tau}{2} A_1)(E-\frac{\tau}{2} A_2)\varphi^j. \qquad (4.8)$$

Then from the unknown φ^j we get over to ψ^j according to the formula

$$\psi^j = (E + \frac{\tau}{2} A_2)\varphi^j . \qquad (4.9)$$

In this case for the new unknown ψ^j we come to the relation

$$\psi^{j+1} = T\psi^j , \qquad (4.10)$$

where

$$T = (E + \frac{\tau}{2} A_1)^{-1}(E- \frac{\tau}{2} A_1)(E-\frac{\tau}{2} A_2) (E + \frac{\tau}{2} A_2)^{-1} \qquad (4.11),$$

is a step operator.

Using (4.10) we get the estimate in the energy norm

$$\|\psi^{j+1}\| \leq \|T\| \|\psi^j\| . \qquad (4.12)$$

Let us find the norm of the operator T

$$\|T\| \leq \|T_1\| \|T_2\| , \qquad (4.13)$$

where

$$T_\alpha = (E - \frac{\tau}{2} A_\alpha) (E + \frac{\tau}{2} A_\alpha)^{-1}$$

Here we use commutativity property

$$(E + \frac{\tau}{2} A_\alpha)^{-1}(E - \frac{\tau}{2} A_\alpha) = (E - \frac{\tau}{2} A_\alpha) (E + \frac{\tau}{2} A_\alpha)^{-1} ,$$

which follows from the obvious identity.

$$(E - \frac{\tau}{2} A_\alpha)^{-1}(E - \frac{\tau}{2} A_\alpha) = (E - \frac{\tau}{2} A_\alpha) (E + \frac{\tau}{2} A_\alpha)^{-1} \qquad (4.14)$$

Indeed, by multiplying the left-hand side and the right-hand side of

(4. 14) by $(E - \frac{\tau}{2} A_\alpha)$ and using commutativity of the operators $(E - \frac{\tau}{2} A_\alpha)$ and $(E + \frac{\tau}{2} A_\alpha)$ (which can be verified by a direct multiplication) we arrive at the property being proved.

Thus, the problem of determining stability is reduced to that of finding norms the operators T_α.

Applying Kellogg's lemma to evaluating norms of the operators T_1 and T_2 in the relation (4. 13) we come to the conclusion

$$\| T \| \leq 1. \qquad (4.15)$$

hence,

$$\| \varphi^{j+1} \| \leq \| \varphi^j \| \qquad (4.16)$$

Thus, stability of (4. 10) is proved. However, our ultimate aim is to determine stability of the initial difference problem (4. 4). For this purpose we use relation (4. 9) and rewrite (4. 16).

$$\| (E + \frac{\tau}{2} A_2) \varphi^{j+1} \| \leq \| (E + \frac{\tau}{2} A_2) \varphi^j \|. \qquad (4.17)$$

We introduce the notation

$$\| (E + \frac{\tau}{2} A_2) \varphi \| = (C_2 \varphi, \varphi)^{1/2} = \| \varphi \|_{C_2}, \qquad (4.18)$$

where

$$C_\alpha = (E + \frac{\tau}{2} A_\alpha^*) (E + \frac{\tau}{2} A_\alpha).$$

It is not difficult to see that $C_\alpha > 0$, hence, $\| \cdot \|_{C_2}$ is , in fact, the norm.

So, in this metric takes place the condition of absolute stability

$$\| \varphi^{j+1} \|_{C_2} \leq \| \varphi^j \|_{C_2} \qquad (4.19)$$

It will be noted that the difference scheme of the universal splitting-up algorithm allows a convenient implementation on a computer. Indeed, let us write the difference equation (4. 4) as follows :

$$F^j = A\varphi^j ,$$

$$(E + \frac{\tau}{2} A_1) \xi^{j+1/2} = - F^j, \qquad (4.20)$$

$$(E + \frac{\tau}{2} A_2) \xi^{j+1} = \xi^{j+1/2} ,$$

$$\varphi^{j+1} = \varphi^j - \tau \xi^{j+1}$$

Here $\xi^{j+1/2}$ and ξ^{j+1} are some auxiliary values allowing reduction of the problem (4. 4) to a sequence of simplest problems (4. 20) which are solved successively. It should be noted that the first and the last equations of (4. 20) are explicit relations. It means that we must invert operators only when we solve the second and third equations of (4. 20) in which the simplest operators A_1 and A_2 are present.

Let us consider the inhomogeneous problem

$$\frac{\partial \varphi}{\partial t} + A\varphi = f, \qquad (4.21)$$

$$\varphi = g \text{ if } t = 0 ,$$

where

$$A = A_1 + A_2 , \quad A_1 \geq 0, \qquad A_2 \geq 0 .$$

Then the scheme of the universal algorithm is written as

$$(E + \frac{\tau}{2} A_1) (E + \frac{\tau}{2} A_2) \frac{\varphi^{j+1} - \varphi^j}{\tau} + A\varphi^j = f^j, \qquad (4.22)$$

$$\varphi^{o} = g ,$$

where

$$f^{j} = f(t_{j+1/2}). \tag{4.23}$$

It can be shown that under condition (4. 23) the difference problem (4. 22) approximates the initial value problem (4. 21) with second order approximations in τ. Let us discuss now stability of the difference scheme. For this purpose we transform equation (4. 22) as follows :

$$\psi^{j+1} = T\psi^{j} + \tau (E + \frac{\tau}{2} A_1)^{-1} f^{j} , \tag{4.24}$$

where

$$\psi^{j} = (E + \frac{\tau}{2} A_2) \varphi^{j}. \tag{4.25}$$

Equation (4. 24) will be evaluated by the energy norm

$$\| \psi^{j+1} \| \leq \| T \| \|\psi^{j}\| + \tau \|(E+ \frac{\tau}{2} A_1)^{-1} \| \, \| f^{j} \|. \tag{4.26}$$

Since for the homogeneous equation it was found that

$$\| T \| \leq 1 ,$$

we have

$$\| \psi^{j+1} \| \leq \|\psi^{j}\| + \tau \|(E + \frac{\tau}{2} A_1)^{-1} \| \, \| f^{j} \| . \tag{4.27}$$

Let as make the following obvious transformation

$$\| f^{j} \| = \|(E + \frac{\tau}{2} A_2)^{-1} (E + \frac{\tau}{2} A_2) f^{j} \| \leq \|(E+\frac{\tau}{2} A_2)^{-1} \| \, \|(E+\frac{\tau}{2} A_2) f^{j} \| \tag{4.28}$$

Taking (4. 18), (4. 25) and (4. 28) into account we get the desired condition

$$\|\varphi^{j+1}\|_{C_2} \leq \|\varphi^j\|_{C_2} + \tau \|(E+\frac{\tau}{2}A_1)^{-1}\| \, \|(E+\frac{\tau}{2}A_2)^{-1}\| \, \|f^j\|_{C_2} .$$

Let us use then the estimation of the norm

$$\|(E+\frac{\tau}{2}A_\alpha)^{-1}\| \leq 1,$$

at $A_\alpha \geq 0$. As a result we have

$$\|\varphi^{j+1}\|_{C_2} \leq \|\varphi^j\|_{C_2} + \tau\|f^j\|_{C_2} . \tag{4. 29}$$

From here, using recurrence relations, we obtain

$$\|\varphi^j\|_{C_2} \leq \|g\|_{C_2} + j\tau\|f\|_{C_2} , \tag{4. 30}$$

where

$$\|f\|_C = \max_j \|f^j\|_C .$$

It means that (4. 22) is a stable scheme on the interval $0 \leq t_j \leq T = J\tau$.

2. 4. 2 The predictor - corrector method

We shall formulate now another splitting-up technique, a so-called predictor-corrector method. The point of this approximate method is that the whole interval $0 \leq t \leq T$ is split into a number of intervals and within each elementary interval $t_j \leq t \leq t_{j+1}$ the problem (4. 1) is solved in two steps. First one finds approximate solution to the problem at the moment $t_{j+1/2} = t_j + \tau/2$ using a scheme of first order accuracy which has a sufficient "reserve" of stability. Then on the whole interval (t_j, t_{j+1}) one writes the basic equation of second-order approximation which is a corrector. It is essential that in constructing the corrector one uses the "rough" solution at $t_{j+1/2}$ found with the aid of the predictor. Thus, the predictor - corrector scheme can be written as follows :

$$\frac{\varphi^{j+1/4} - \varphi^{j}}{/2} + A_1 \varphi^{j+1/4} = 0$$

$$\frac{\varphi^{j+1/2} - \varphi^{j+1/4}}{/2} + A_2 \varphi^{j+1/2} = 0 , \qquad (4.34)$$

$$\frac{\varphi^{j+1} - \varphi^{j}}{\tau} + A \varphi^{j+1/2} = 0$$

previded that $\varphi^{o} = g$.

Let us study the predictor-corrector scheme more carefully. First of all we eliminate the auxiliary function $\varphi^{j+1/4}$ from the first two equations of (4. 34). Then (4. 34) is reduced to the two equations

$$(E + \frac{\tau}{2} A_1)(E + \frac{\tau}{2} A_2) \varphi^{j+1/2} = \varphi^{j} , \qquad (4.35)$$

$$\frac{\varphi^{j+1} - \varphi^{j}}{\tau} + A \varphi^{j+1/2} = 0.$$

$$\frac{\varphi^{j+1} - \varphi^{j}}{\tau} + A \varphi^{j+1/2} = 0.$$

Eliminating $\varphi^{j+1/2}$, from (4. 35) , we get

$$\frac{\varphi^{j+1} - \varphi^{j}}{\tau} + A(E + \frac{\tau}{2} A_2)^{-1} (E + \frac{\tau}{2} A_1)^{-1} \varphi^{j} = 0, \qquad (4.36)$$

$$\varphi^{j} = g .$$

To investigate the question of approximation we rewrite equation (4. 36) as follows :

$$(E + \frac{\tau}{2} A_1)\ (E + \frac{\tau}{2} A_2) \frac{\varphi^{j+1} - \varphi^j}{\tau} + \Lambda \varphi^j = 0 \ ,$$

where

$$\Lambda = (E + \frac{\tau}{2} A_1)\ (E + \frac{\tau}{2} A_2)\ A\ (E + \frac{\tau}{2} A_2)^{-1} (E + \frac{\tau}{2} A_2)^{-1}.$$

By means of a series expansion in powers of τ it is easy to find

$$\Lambda = A + 0\ (\tau^2)$$

and by the method of estimates used in the case of the universal algorithm we conclude that the predictor - corrector method has second order approximation in τ.

Let us study stability of this method. For this purpose (4. 36) will be written as

$$(E + \frac{\tau}{2} A_1)\ (E + \frac{\tau}{2} A_2)\ \frac{\varphi^{j+1} - \varphi^j}{\tau} + A\,{}^{j} = 0, \tag{4. 37}$$

where

$$\Phi^j = (E + \frac{\tau}{2} A_2)^{-1} (E + \frac{\tau}{2} A_1)^{-1} \varphi^j. \tag{4. 38}$$

As for the difference equation (4. 27), above it has been shown to be stable in the following metric :

$$\| \Phi^{j+1} \|_C \leq \| \Phi^j \|_C \ . \tag{4. 39}$$

Let us substitute relation (4. 38) into (4. 39). Then, taking account of (4. 21) , we get

$$\| (E + \frac{\tau}{2} A_1)^{-1} \varphi^{j+1} \| \leq \| (E + \frac{\tau}{2} A_1)^{-1} \varphi^j \| \tag{4. 40}$$

or

$$\| \varphi^{j+1} \|_{C_1^{-1}} \leq \| \varphi^j \|_{C_1^{-1}} \tag{4. 41}$$

where

$$C_1^{-1} = (E + \frac{\tau}{2} A_1^*)^{-1} (E + \frac{\tau}{2} A_1)^{-1}.$$

Thus we have proved stability of the system (4. 41).

Let us consider now an inhomogeneous problem. In this case we formulate the predictor-corrector method as follows :

$$\frac{\varphi^{j+1/4} - \varphi^j}{\tau/2} + A_1 \varphi^{j+1/4} = 0 ,$$

$$\frac{\varphi^{j+1/2} - \varphi^{j+1/4}}{\tau/2} + A_2 \varphi^{j+1/2} = 0 , \tag{4. 42}$$

where

$$\frac{\varphi^{j+1} - \varphi^j}{\tau} + A \varphi^{j+1/2} = f^j ,$$

$$f^j = f(t_{j+1/2}) . \tag{4. 43}$$

If f^j chosen in the form (4. 43), one can show that (4. 42) approximates the initial problem with second order accuracy in τ . Stability of (4. 42) is established in the same way. As a result we get

$$\|\varphi^j\|_{C_1^{-1}} \leq \|g\|_{C_1^{-1}} + \tau j \, \|(E + \frac{\tau}{2} A)^{-1}\| \, \|f\|_{C_1^{-1}} , \tag{4. 44}$$

where

$$\|f\|_{C_1^{-1}} = \max_j \|f^j\|_{C_1^{-1}} .$$

Since

$$\|(E + \frac{\tau}{2} A)^{-1}\| \sim 1 + 0(\tau)$$

and $\tau^2 j \sim 0(\tau)$, the asymptotic estimate of (4. 44) for $j \gg 1$ will be

$$\| \varphi^j \|_{C_1^{-1}} \leq \| g \|_{C_1^{-1}} + \tau j \; \| f \|_{C^{-1}} \tag{4.45}$$

hence, if $0 \leq t_j \leq T$, we again get stability of the difference scheme.

2.4.3 Component - wise splitting up method

Let us consider now a method of complete component-wise splitting. Let us find an approximation solution to the problem (4. 1) - (4. 3). A scheme consisting of a successive solution of the simplest Crank-Nicolson difference schemes will be regarded as an algorithm. Schemes of this kind were suggested by N. N. Yanenko and studied at length by the author in this paper at SYNSPADE - 1970.

$$\begin{aligned} &\frac{\varphi^{j+1/2} - \varphi^j}{\tau} + A_1 \frac{\varphi^{j+1/2} + \varphi^j}{2} = 0, \\ &\frac{\varphi^{j+1} - \varphi^{j+1/2}}{\tau} + A_2 \frac{\varphi^{j+1} + \varphi^{j+1/2}}{2} = 0. \end{aligned} \tag{4.46}$$

The system of difference equations (4. 46) after eliminating the auxiliary function $\varphi^{j+1/2}$ will reduce to one equation

$$\varphi^{j+1} = T\varphi^j, \tag{4.47}$$

where

$$T = (E + \frac{\tau}{2} A_2)^{-1} (E - \frac{\tau}{2} A_2)(E + \frac{\tau}{2} A_1)^{-1} (E - \frac{\tau}{2} A_1). \tag{4.48}$$

Let us study first the problem of approximation. For this purpose we expand operator T in powers of τ. After obvious transformations we get

$$T = E - \tau A + \frac{\tau^2}{2}(A_1^2 + 2A_2A_1 + A_2^2) - \ldots . \tag{4.49}$$

If the operators A_1 and A_2 are commutative, i. e. $A_1A_2 = A_2A_1$, the above formula (4. 49) can be written as

$$T = E - \tau A + \frac{\tau^2}{2} A^2 - \dots. \tag{4.50}$$

If the operators A_α are noncommutative, one can achieve second order approximation by a special organization of computation. This is a two-cycle computation. Thus, for instance, one solves first the problem.

$$\frac{\varphi^{j-1/2} - \varphi^{j-1}}{\tau} + A_1 \frac{\varphi^{j-1/2} + \varphi^{j-1}}{2} = 0, \tag{4.51}$$

$$\frac{\varphi^{j} - \varphi^{j-1/2}}{\tau} + A_2 \frac{\varphi^{j} + \varphi^{j-1/2}}{2} = 0$$

and then

$$\frac{\varphi^{j+1/2} - \varphi^{j}}{\tau} + A_2 \frac{\varphi^{j+1/2} + \varphi^{j}}{2} = 0, \tag{4.52}$$

$$\frac{\varphi^{j+1} - \varphi^{j+1/2}}{\tau} + A_1 \frac{\varphi^{j+1} + \varphi^{j+1/2}}{2} = 0 .$$

Within the cycle the difference schemes (4. 51), (4. 52) are alternately used. In a similar way (see above) it can be shown that for the whole computation, by use of (4. 51) and (4. 52), we have.

$$\varphi^{j+1} = T_c \varphi^{j-1} , \tag{4.53}$$

where

$$T_c = E - 2\tau A + \frac{(2\tau)^2}{2} A^2 - \dots.$$

If we compare the operator T_c with the step operator of the total Crank-Nicolson difference scheme

$$\frac{\varphi^{j+1} - \varphi^{j-1}}{2\tau} + A \frac{\varphi^{j+1} + \varphi^{j-1}}{2} = 0$$

we find that operator T_c for the two-cycle splitting-up scheme coincides, to within τ^2, with the step operator of the initial Crank-Nicolson difference scheme applied to a double time interval. This result is true for both commutative and noncommutative operators A_α. Hence this method does not require commutativity of the operators.

Let us discuss now numerical stability of the method. For this purpose we shall consider the relation (4. 47) and evaluate it in the energy norm

$$\| \varphi^{j+1} \| \leq \| T \| \| \varphi^j \| .$$

Since, as it is shown above,

$$\| T \| \leq 1$$

If $A_\alpha \geq 0$, we get

$$\| \varphi^{j+1} \| \leq \| \varphi^j \| \qquad (4.54)$$

From here it immediately follows that

$$\| \varphi^j \| \leq \| g \| \qquad (4.55)$$

If we use the two-cycle method we get estimates of the form (4. 55) at each step of the cycle. It means that the two-cycle method is absolutely stable.

Let us consider now an inhomogeneous problem and approximation of its solution with a two-cycle complete splitting. To this end we write the system of difference equations of the form (4. 51), (4. 52) in a more convenient form

$$\begin{aligned} &(E + \frac{\tau}{2} A_1) \varphi^{j-1/2} = (E - \frac{\tau}{2} A_1)(\varphi^{j-1} + 2\tau f^j) \\ &(E + \frac{\tau}{2} A_2) \varphi^j = (E - \frac{\tau}{2} A_2) \varphi^{j-1/2}, \\ &(E + \frac{\tau}{2} A_2) \varphi^{j+1/2} = (E \frac{\tau}{2} A_2) \varphi^j , \end{aligned} \qquad (4.56)$$

$$(E + \frac{\tau}{2} A_1)\varphi^{j+1} = (E - \frac{\tau}{2} A_1)\varphi^{j+1/2} ,$$

where $f^j = f(t_j)$.

Solving these equations with respect to φ^{j+1}, we get

$$\varphi^{j+1} = T_c \varphi^{j-1} + 2\tau T_c \quad f^j \tag{4.57}$$

where

$$T_c = T_1 T_2 T_2 T_1 \tag{4.58}$$

and

$$T_\alpha = (E + \frac{\tau}{2} A_\alpha)^{-1} (E - \frac{\tau}{2} A_\alpha) \tag{4.59}$$

By expanding in powers the small parameter τ we reduce the expression (4.57) to

$$\varphi^{j+1} = (E - 2\tau A + \frac{(2\tau)^2}{2} A^2)\varphi^{j-1} + 2\tau(E - \tau A) f^j + 0(\tau^3) \tag{4.60}$$

and transform it as follows

$$\frac{\varphi^{j+1} - \varphi^{j-1}}{2\tau} + A(E - \tau A)\varphi^{j-1} = (E - \tau A) f^j + 0(\tau^2) \tag{4.61}$$

Let us eliminate φ^{j-1} from the last relation. To this end we make use of the expansion of the solution in a Tailor's series in the vicinity of the point t_{j-1} . To within τ^2 have

$$\varphi^j = \varphi^{j-1} + \frac{\partial \varphi^{j-1}}{\partial t} \tau + 0(\tau^2) . \tag{4.62}$$

Let us eliminate the derivative $\frac{\partial \varphi}{\partial t}$ using the equation

$$\frac{\partial \varphi^{j-1}}{\partial t} = - A\varphi^{j-1} + f^j + 0(\tau) . \tag{4.63}$$

Let us substitute (4. 63) into (4. 62). Then we get

$$\varphi^{j} = (E - \tau A)\varphi^{j-1} + \tau f^{j} + 0\,(\tau^{2}).$$

Hence

$$(E - \tau A)\varphi^{j-1} = \varphi^{j} - \tau f^{j} + 0\,(\tau^{2}) . \tag{4. 64}$$

After substituting (4. 64) into (4. 61) we get

$$\frac{\varphi^{j+1} - \varphi^{j-1}}{2\tau} + A\varphi^{j} = f^{j} + 0\,(\tau^{2}) . \tag{4. 65}$$

Evidently, Eq. (4. 65) approximates the intial equation (4. 1) on the interval ($t_{j-1} \le t \le t_{j+1}$) to within second order in τ. Thus we have found difference approximation of the inhomogeneous evolution equation of second order accuracy by applying the two-cycle method.

Stability of the method is elementally proved in the energy norm. Indeed , let us evaluate (4. 57) by the norm

$$\|\varphi^{j+1}\| \le \|T_c\|\,\|\varphi^{j-1}\| + 2\tau\,\|(E + \tau A_1)^{-1}\|\,\|(E + \tau A_2)^{-1}\|\,\|f^{j}\| . \tag{4. 66}$$

Above it was found that

$$\|T_c\| \le 1$$

$$\|(E + \tau A_{\alpha})^{-1}\| \le 1.$$

Hence , we have

$$\|\varphi^{j+1}\| \le \|\varphi^{j-1}\| + 2\tau\,\| f^{j} \| . \tag{4. 67}$$

Using the recurrence relation (4. 63) we get

$$\|\varphi^{j}\| \le \| g \| + \tau j\,\| f \| , \tag{4. 68}$$

where

$$\| f \| = \max_{t_j} \| f^j_. \| .$$

From (4. 68) follows numerical stability of the scheme on any finite time interval.

Let us consider now the splitting-up method for implicit difference approximations. To the end we analyse the problem

$$\frac{\partial \varphi}{\partial t} + A\varphi = 0 \text{ in } D \times T ,$$

$$\varphi = g \text{ if } t = 0 \text{ in } D.$$

Let $A = \sum_{\alpha=1}^{n} A_\alpha$ and all $A_\alpha \geq 0$ and consider the splitting-up algorithm in the form

$$\frac{\varphi^{j+\frac{1}{n}} - \varphi^j}{\tau} + A_1 \varphi^{j+\frac{1}{n}} = 0 ,$$

$$\cdots\cdots\cdots\cdots\cdots\cdots\cdots\cdots\cdots\cdots$$

$$\frac{\varphi^{j+1} - \varphi^{j+\frac{n-1}{n}}}{\tau} + A_n \varphi^{j+1} = 0 . \qquad (*)$$

We show now that such an algorithm is absolutely stable. Indeed, let us look at the equation

$$\frac{\varphi^{j+\frac{\alpha}{n}} - \varphi^{j+\frac{\alpha-1}{n}}}{\tau} + A_\alpha \varphi^{j+\frac{\alpha}{n}} = 0$$

and multiply it scalarly by $\varphi^{j+\frac{\alpha}{n}}$.

As a result we obtain

$$(\varphi^{j+\frac{\alpha}{n}} - \varphi^{j+\frac{\alpha-1}{n}}, \varphi^{j+\frac{\alpha}{n}}) + \tau (A_\alpha \varphi^{j+\frac{\alpha}{n}}, \varphi^{j+\frac{\alpha}{n}}) = 0 .$$

Considering positive semidefiniteness of the operators A , we have

$$\left(\varphi^{j+\frac{\alpha}{n}} - \varphi^{j+\frac{\alpha-1}{n}}, \varphi^{j+\frac{\alpha}{n}}\right) \leq 0 .$$

or

$$\left(\varphi^{j+\frac{\alpha}{n}}, \varphi^{j+\frac{\alpha}{n}}\right) \leq \left(\varphi^{j+\frac{\alpha}{n}}, \varphi^{j+\frac{\alpha-1}{n}}\right).$$

But, since

$$\left(\varphi^{j+\frac{\alpha}{n}}, \varphi^{j+\frac{\alpha-1}{2}}\right) \leq \frac{1}{2}\left[\left(\varphi^{j+\frac{\alpha}{n}}, \varphi^{j+\frac{\alpha-1}{n}}\right) + \left(\varphi^{j+\frac{\alpha-1}{n}}, \varphi^{j+\frac{\alpha-1}{n}}\right)\right],$$

then

$$\left\|\varphi^{j+\frac{\alpha}{n}}\right\|^2 \leq \left\|\varphi^{j+\frac{\alpha-1}{n}}\right\|^2, \qquad \alpha = 1, 2, \ldots, n.$$

Using the recurrence inequality we have

$$\|\varphi^{j+1}\| \leq \|\varphi^{j}\| .$$

This means that under the assumptions made the computation by the splitting-up scheme (✱) will be absolutely stable.

It is not difficult to see that the system (✱) approximates the initial - value problem to within first order in τ .

Let us see the inhomogeneous problem

$$\frac{\partial \varphi}{\partial t} + A\varphi = f \quad \text{in DxT},$$

$$\varphi = g \quad \text{if} \quad t = 0 \quad \text{in D}.$$

The splitting-up scheme for this problem will be considered in the form

$$\frac{\varphi^{j+\frac{1}{n}} - \varphi^j}{\tau} + A_1 \varphi^{j+\frac{1}{n}} = 0 ,$$
$$\cdots\cdots\cdots\cdots\cdots\cdots \qquad (**)$$
$$\frac{\varphi^{j+1} - \varphi^{j+\frac{n-1}{n}}}{\tau} + A_n \varphi^{j+1} = f^j .$$

Such a splitting-up scheme approximates the basic inhomogeneous equation to within first order in τ .

Stability of the scheme will be proved as follows. Let us multiply scalarly each of the equation, respectively, by $\varphi^{j+1/n}, \ldots, \varphi^{j+1}$ Then, analogously to the previous, we have

$$\| \varphi^{j+\frac{\alpha}{n}} \| \leq \| \varphi^{j+\frac{\alpha-1}{n}} \| , \alpha = 1, 2, \ldots, n-1.$$

Let us consider the final equation of (**) in more detail. After the above procedure we have

$$(\varphi^{j+1}, \varphi^{j+1}) = (\varphi^{j+\frac{n-1}{n}}, \varphi^j) - \tau (A_n \varphi^{j+1}, \varphi^{j+1}) + \tau (f^j, \varphi^{j+1}).$$

Taking into consideration that $A_n \geq 0$, we get

$$(\varphi^{j+1}, \varphi^{j+1}) \leq (\varphi^{j+\frac{n-1}{n}}, \varphi^{j+1}) + \tau (f^j, \varphi^{j+1}).$$

and , using the Buniakowski-Schwarz inequality,

$$(\varphi^{j+\frac{n-1}{n}}, \varphi^{j+1}) \leq \| \varphi^{j+\frac{n-1}{n}} \| \, \| \varphi^{j+1} \| ,$$

$$(f^j + \varphi^{j+1}) \leq \| f^j \| \, \| \varphi^{j+1} \|$$

Hence,

$$\| \varphi^{j+1} \|^2 \leq \| \varphi^{j+\frac{n-1}{n}} \| \, \| \varphi^{j+1} \| + \tau \| f^j \| \, \| \varphi^{j+1} \| .$$

Cancelling $\|\varphi^{j+1}\|$, we arrive at the following inequality

$$\|\varphi^{j+1}\| \leq \|\varphi^{j+\frac{n-1}{n}}\| + \tau \|f^j\| .$$

Eliminating the solution with fractional indices, we have

$$\|\varphi^{j+1}\| \leq \|\varphi^j\| + \tau \|f^j\|.$$

Considering that

$$\|\varphi^o\| = \|g\|,$$

by eliminating intermediate values of the solution, we get

$$\|\varphi^{j+1}\| \leq \|g\| + \tau j \|f\| ,$$

where

$$\|f\| = \max_j \|f^j\|$$

From here follows absolute stability of the difference scheme for any instant of time of the interval

$$0 \leq t_j \leq T .$$

This splitting-up algorithm is generalized for the case of time dependence of the operator A . In such a case at each cycle of computation by the splitting-up scheme instead of A we should take any difference approximation of this operator at each interval $t_j \leq t \leq t_{j+1}$.

2.4.4 A solution of problems with time-dependent operators.

Thus, we have considered the three splitting-up techniques : the method of universal algorithm, the predictor-corrector method and the method of successive splitting based on two-cycle procedure. It is remarkable that all the three methods are equivalent in accuracy and absolutely stable if $A_\alpha \geq 0$.

However we should keep in mind one restriction which was imposed on the operators A_α, at the very beginning, that is their independence of time. Due to this restriction we could make a complete analysis of stability assuming only that the A_α are positive semi-definite operators. Unfortunately, if the operators are time-dependent it is generally impossible to make such an analysis of stability. It is pleasant to note that the method of successive splitting is an exception.

If the operators A_α are time-dependent, then for the other two methods, one can choose approximation of the operators A_α in time, for example,

$$\Lambda_\alpha^j = A_\alpha(t_{j+1/2}),$$

which retain second order approximation, but stability in other norms is to be established. Most often it is energy norm. In both cases we have the relations

$$\|\varphi^{j+1}\| \leq \|S^j\| \|\varphi^j\| + \tau \|f^j\|,$$

where

$$S^j = E - \tau(E + \frac{\tau}{2}\Lambda_2^j)^{-1}(E + \frac{\tau}{2}\Lambda_1^j)^{-1}\Lambda^j$$

for the scheme of universal algorithm, and

$$S^j = E - \tau\Lambda^j(E + \frac{\tau}{2}\Lambda_2^j)^{-1}(E + \frac{\tau}{2}\Lambda_1^j)^{-1}$$

for the predictor-corrector scheme.

Unfortunately, $\Lambda_\alpha^j \geq 0$ does not follow in the energy metric if we let $\| S^j \| < 1$. Therefore investigation of the schemes' stability in this case consists of an estimation of the norm of a very involved operator S^j at each computation step. Owing to this fact the two methods in question become less valuable if the operators A_α are time dependent.

For the time dependent A_α operators the method of successive splitting has an advantage over the other methods because it can be effectively used to a wide range of problems.

2. 4. 5 An example

Let us illustrate splitting-up methods for nonstationary problems taking as an example the following problem :

$$\frac{d\varphi}{dt} - \Delta^h \varphi = 0 \quad \text{in} \quad D_L \ ,$$

$$\varphi = g_1 \qquad \text{for} \quad dD_h \ , \qquad (4.69)$$

$$\varphi = g_2 \qquad \text{if} \quad t = 0 \ .$$

Here it is assumed that approximation of the problem in space variables has been made. Hence, we deal with a system of ordinary differential equations for the components of the solution φ_{kl} and the function g_{kl}. For the sake of simplicity the indices of the functions φ_k and g will be omitted. The operator $A = -\Delta^h$ will be presented as as a sum of two operators

$$A = A_1 + A_2 \ , \qquad (4.70)$$

where

$$A_1 = -\Delta_k \nabla_k, \quad A_2 = -\Delta_1 \nabla_1 . \tag{4.71}$$

It is not difficult to find norms of the operators A_1 and A_2 by the same procedure which we used in § 1. 1 when we looked for the norm of the operator A.

It is easy to obtain

$$\pi^2 \leq \lambda_n^{A_1} \leq \frac{4}{h^2}$$

and, analogously

$$\pi^2 \leq \lambda_n^{A_2} \leq \frac{4}{h^2}$$

Thus

$$\| A_1 \| = \| A_2 \| = \frac{4}{h^2}$$

and

$$\frac{1}{\| A_1^{-1} \|} = \frac{1}{\| A_2^{-1} \|} = \pi^2 .$$

Let us examine now schemes of realization and stability of the basic splitting-up algorithms.

I. If we use the method of universal algorithm (4. 4), the scheme of realization is of the form (4. 20) :

$$\begin{gathered} f_{kl}^j = - A_{kl}^h \varphi_{kl}^j , \\ -\frac{\tau}{2} A_k \xi_{kl}^{j+1/2} + \xi_{kl}^{j+1/2} = F_{kl}^j , \quad (l = 1, 2, \dots, n-1), \\ -\frac{\tau}{2} A_1 \xi_{kl}^{j+1} + \xi_{kl}^{j+1} = \xi_{kl}^{j+1/2} , \quad (k = 1, 2, \dots, n-1), \\ \varphi_{kl}^{j+1} = \varphi_{kl}^j - \tau \xi_{kl}^{j+1} . \end{gathered} \tag{4.72}$$

Since the step operator has the form (4. 11) and the operators A_α in our case are commutative, we have

$$\| T \| = \left(\frac{1 - \frac{2\tau}{h^2}}{1 + \frac{2\tau}{h^2}} \right)^2 < 1.$$

Thus stability of the difference scheme (4. 68) is ensured.

2. Let us make use of the predictor-corrector (4. 34), we have

$$-\frac{\tau}{2} A_k \varphi_{kl}^{j+1/4} + \varphi_{kl}^{j+1/4} = \varphi_{kl}^{j}, \tag{4. 73}$$

$$(1=1,2,\ldots,n-1)$$

$$-\frac{\tau}{2} A_1 \varphi_{kl}^{+1/2} + \varphi_{kl}^{j+1/2} = \varphi_{kl}^{j+1/4},$$

$$(k = 1,2\ldots,n-1),$$

$$\varphi_{kl}^{j+1} = \varphi_{kl}^{j} + \tau A^h \varphi_{kl}^{j+1/2}$$

The norm of the step operator for the scheme in question, due to the proved general statement (3. 39) is strictly less than unity. Therefore stability of (4. 73) is valid.

3. In the case of complete componentwise splitting of (4. 46) we have

$$-\frac{\tau}{2} A_1 \varphi_{kl}^{j+1/2} + \varphi_{kl}^{j+1/2} = \frac{\tau}{2} A_k \varphi_{kl}^{j} + \varphi_{kl}^{j},$$

$$(1 = 1,2,\ldots,n-1), \tag{4. 74}$$

$$-\frac{\tau}{2} A_1 \varphi_{kl}^{j+1} + \varphi_{kl}^{j+1} = \frac{\tau}{2} A_1 \varphi_{kl}^{j+1/2} + \varphi_{kl}^{j+1/2}, \tag{4.74}$$

$$(k = 1, 2, \ldots, n-1).$$

Here $A = A_k + A_1$ and A_k is obtained from $\nabla_k \Delta_k$, $A_{\bar{1}}$ from $\nabla_1 \Delta_1$, with using of boundary conditions.

It was shown in (4.64) that the norm of the step operator in the case of the scheme under consideration and more general schemes is strictly less than unity. Therefore the scheme (4.74) is stable.

In all the above examples stability is proved in the energy norm. Hence each of the schemes considered is a set of equations with difference operators which depend either on the index k or on the index 1, only. Eventually the problem is reduced to a solution of the simplest three-point equations of the form

$$a_n \varphi_{k+1} - b_k \varphi_k + c_k \varphi_{k+1} = -f_k, \tag{4.75}$$

where a_n, b_n and c_k and f_n are prescribed values, where

$$a_K + c_K < b_K.$$

For solving difference equations (4.75) one usually uses a factorization method (See § 5.4).

2.5. Multi- component splitting of problems

Until now it has been assumed that the initial operator A is a sum of two simpler operators. When we solve complicated problems in mathematical physics we often have to do with the splitting of operators into a number of components. In a general case we have

$$A = \sum_{\alpha=1}^{n} A \quad , \tag{5.1}$$

where $A_\alpha \geq 0$. Since in the previous paragraph we have dealt with $n = 2$, here we shall only deal with $n > 2$.

First of all we can find out that a trivial extension of the above splitting-up methods to the case $n = 2$ is generally impossible. Therefore our object will be to extend splitting-up algorithms to this case making assumptions which allow such an extension.

2.5.1 The method of universal algorithm

Under the assumption (5.1) it can be presented as

$$\prod_{\alpha=1}^{n} (E + \frac{\tau}{2} A_\alpha) \frac{\varphi^{j+1} - \varphi^j}{\tau} + A\varphi^j = f^j , \tag{5 2}$$

$$\varphi^o = g,$$

where

$$f^j = f\,(t_{j+1/2}).$$

The scheme of the operational algorithm is as follows :

$$\begin{aligned} F^j &= - A\varphi^j + f^j , \\ (E + \frac{\tau}{2} A_1)\, \xi^{j+1/n} &= F^j , \\ (E + \frac{\tau}{2} A_2)\, \xi^{j+1/n} &= \xi^{j+1/n} \end{aligned} \tag{5.3}$$

....

$$(E+\frac{\tau}{2}A_n)\xi^{j+1}=\xi^{j+\frac{n-1}{n}}$$

$$\varphi^{j+1}=\varphi^j-\tau\xi^{j+1} \qquad (5.3)$$

It is not difficult to check that the universal splitting-up algorithm has second order accuracy in τ if a solution is sufficiently smooth. Numerical stability will be achieved if the condition is fulfilled :

$$\|T\| < 1, \qquad (5.4)$$

where T is the step operator defined by

$$T = E-\tau\prod_{\alpha=n}^{1}(E+\frac{\tau}{2}A_\alpha)^{-1}A. \qquad (5.5)$$

Unfortunately from the condition $A_\alpha \geq 0$ stability in some norm does not follow as was the case for $n = 2$. To establish stability one usually uses the following simple algorithmic technique. If $f^j = 0$, the homogeneous equation (5.2) solved with respect to φ^{j+1} becomes

$$\varphi^{j+1} = T\varphi^j. \qquad (5.6)$$

T is assumed to be time-independent operator (of the index j). Therefore solving (5.6) with the initial condition

$$\varphi^0 = g \qquad (5.7)$$

and the fixed τ, allowing the necessary approximation, we shall look after the norm $\|\varphi^j\|$. If this norm does not increase, it follows that

$$\|T\| < 1 \qquad (5.8)$$

hence, it will be reckoned that the condition for numerical stability can be fulfilled. Then one can go over to a solution of the inhomogeneous problem. Indeed, if the condition (5. 8) can be fulfilled, Eq. (5. 2) will be rewritten as

$$\varphi^{j+1} = T\varphi^{j} + \tau \prod_{\alpha=n}^{1} (E + \frac{\tau}{2} A_{\alpha})^{-1} f^{j}. \qquad (5.9)$$

From here

$$\| \varphi^{j+1} \| \leq \| T \| \| \varphi^{j} \| + \tau \prod_{\alpha=1}^{1} \| (E + \frac{\tau}{2} A_{\alpha})^{-1} \| \| f^{j} \|$$

or from the condition of inequality

$$\| \varphi^{j+1} \| \leq \| \varphi^{j} \| + \tau \| f^{j} \| .$$

By the recurrence relation we come to the stability condition in the energy metric

$$\| \varphi^{j} \| \leq \| g \| + \tau j \| f \|, \qquad (5.10)$$

where

$$\| f \| = \max_{tj} \| f^{j} \| .$$

It will be noted that when solving the homogeneous equation (5. 6) we used the initial condition (5. 7), though it is not necessary. We can choose as the initial condition any function and look after the computation. If the computation is stable the norm of the operator T will be less than or equal to unity. Otherwise round-off errors will cause an increase ot the norm of the solution beginning with some j.

2. 5. 2 The predictor-corrector method

In this case the splitting-up scheme is as follows :

$$\begin{aligned}
&(E + \frac{\tau}{2} A_1) \varphi^{j+1/2n} = \varphi^j, \\
&(E + \frac{\tau}{2} A_2) \varphi^{j+2/2n} = \varphi^{j+1/2n}, \\
&\cdots\cdots\cdots\cdots\cdots\cdots \\
&(E + \frac{\tau}{2} A_n) \varphi^{j+1/2} = \varphi^{j+\frac{n-1}{2n}}
\end{aligned} \tag{5.11}$$

$$\frac{\varphi^{j+1} - \varphi^j}{\tau} + A \varphi^{j+1/2} = f^j$$

where we let $A_\alpha \geq 0$ and $f^j = f(t_{j+1/2})$.

The system of equations (5. 11) reduces to one equation

$$\frac{\varphi^{j+1} - \varphi^j}{\tau} + A \prod_{\alpha=n}^{1} (E + \frac{\tau}{2} A_\alpha)^{-1} \varphi^j = f^j \tag{5.12}$$

provided that

$$\varphi^0 = g \tag{5.13}$$

The predictor-corrector method in this case has second-order accuracy in τ if the solution is sufficiently smooth. Eq. (5. 12) will be written as

$$\varphi^{j+1} = T \varphi^j + \tau f^j , \tag{5.14}$$

where

$$T = E - A \prod_{\alpha=n}^{1} (E + \frac{\tau}{2} A_\alpha)^{-1} \tag{5.15}$$

is a step operator.

In a similar way the condition of numerical stability reduces eventually to evaluation of the norm of the operator T. For this purpose we can use the above method by which we estimated the norm of the solution to the problems (5. 6), (5. 7). Unfortunately in this case like in the previous one we did not manage to prove stability of the scheme at $A_\alpha \geq 0$.

In order to complete the analysis of the two above-mentioned splitting-up schemes we shall examine the simplest case where the operators A_α are commutative and have a common basis. This requirement in addition to the condition $A_\alpha \geq 0$ seems sufficient to prove stability of the schemes. Indeed, in the case of commutativity, the step operators T for both schemes coincide with each other. For simplicity we shall consider the homogeneous problem (5. 6), (5. 7) and look for the solution in the spectral form

$$\varphi^j = \sum_k \varphi_k^j u_k , \tag{5.16}$$

where the u_n are eigenfunctions of the problem (1. 6) (chapter 1) and $\varphi_k^j = (\varphi^j, u_k^*)$, where the u_k^* are eigenfunctions of the adjoint problem (1. 7) (Chapter 1).

Since u_n is a common basis, then

$$Au_k = \lambda_k u_k ; \quad A_\alpha u_k = \lambda_k^\alpha u_k , \tag{5.17}$$

$$\lambda = \sum_{\alpha=1}^{n} \lambda_\alpha .$$

Substituting (5. 16) and the respective expansions for the function g into (5. 6), (5. 7) we get for the Fourier coefficients φ_k^j the following expressions :

$$\varphi_k^{j+1} = T_k \varphi_k^j , \tag{5.18}$$

$$\varphi_o^o = g_k ,$$

where

$$T_k = 1 - \frac{\tau\lambda_k}{\prod_{\alpha=1}^{n}(1+\frac{\tau}{2}\lambda_k^{\alpha})} \tag{5.19}$$

The expression for T_k from (5. 19) will be rewritten as

$$T_k = \frac{\mu_k - \frac{\tau}{2}\lambda_k}{\mu_k + \frac{\tau}{2}\lambda_k}, \tag{5.20}$$

where the μ_k are positive constants $\lambda_k^{\alpha} \geq 0$. From (5. 20) it follows that

$$|T_k| \leq 1, \tag{5.21}$$

which proves the statement.

The method of universal algorithm as well as the prodictor-corrector method for the n-component splitting of the operator can also be applied when the operator A is time dependent. However, in this situation the a priori formulation of the stability condition appears to be a more complicated problem. Therefore it is difficult to say to what extent in this case the application of the two above schemes is in general worth while. This fact encouraged the author to formulate a universal approach to the solution of different complicated and rather general problems using splitting-up techniques. In what follows this new approach, called a two cycle method of successive splitting, will be discussed in detail.

2. 5. 3. The method of successive splitting based on elementary Crank-Nicolson difference schemes

We shall attempt to build a difference analog of the problem accurate to. second order in τ and absolutely stable in time. According to multicomponent splitting it will be assumed that

$$A^j = \sum_{\alpha=1}^{n} A_\alpha^j , \qquad (5.22)$$

where all the Λ_α^j are positive semi-definite operators so that $\Lambda_\alpha^j \geq 0$. Let us consider a system of equations :

$$(E + \frac{\tau}{2} \Lambda_\alpha^j) \Phi^{j+\frac{\alpha}{n}} = (E - \frac{\tau}{2} \Lambda_\alpha^j) \Phi^{j+\frac{\alpha-1}{n}} , \qquad (5.23)$$

$$(\alpha = 1, 2, \ldots, n) .$$

When $\Lambda_\alpha^j \geq 0$ and are commutative, (5. 23) is unconditionally stable and has second order approximation. This can be easily established by the Fourier method. However, for noncommutative operators Λ_α^j, as is easily seen, (5. 23) will be , generally speaking, of first order accuracy in τ and therefore it will be of less interest for applications than the following scheme of second order accuracy :

$$\Phi^{j+\frac{\alpha}{2n}} = (E - \frac{\tau}{2} \Lambda_\alpha^j) \Phi^{j+\frac{\alpha-1}{2n}} , \quad (a = 1, 2, \ldots, n),$$

$$(E + \frac{\tau}{2} \Lambda_\alpha^j) \Phi^{j+\frac{\alpha}{2n}} , \quad \Phi^{j+\frac{\alpha-1}{2n}} , \quad (\alpha = n+1, n+2, \ldots, 2n). \qquad (5.24)$$

Later we shall try to find a special construction of the method of complete splitting using (5. 23). This construction will give the solution to the Cauchy problem for the positive semi-definite and noncommutative operators Λ_α^j which have second order approximation. In fact this is in a sense the final solution of the splitting-up problem.

We note that the system of equations (5. 23) reduces to one equation

$$\Phi^{j+1} = \prod_{\alpha=1}^{n} (E + \frac{\tau}{2}\Lambda_\alpha^j)^{-1} (E - \frac{\tau}{2}\Lambda_\alpha^j)\Phi^j \tag{5. 25}$$

Using (5. 25) we find the estimation by the norm

$$\|\Phi^{j+1}\| \leq \prod_{\alpha=1}^{n} \|(E + \frac{\tau}{2}\Lambda_\alpha^j)^{-1} (E - \frac{\tau}{2}\Lambda_\alpha^j)\| \|\Phi^j\|. \tag{5. 26}$$

By Kellogg's lemma we have

$$\|\Phi^{j+1}\| \leq \|\Phi^j\| \leq \dots \leq \|g\| . \tag{5. 27}$$

It the operator is skew-symetric, we have

$$\|\Phi^{j+1}\| = \|\Phi^j\| = \dots \|g\|. \tag{5. 28}$$

Thus we have proved absolute stability of this scheme.

To determine the order of approximation we shall expand the expression

$$T^j = \prod_{\alpha=1}^{n} (E + \frac{\tau}{2}\Lambda_\alpha^j)^{-1} (E - \frac{\tau}{2}\Lambda_\alpha^j)$$

in powers of the smaller parameter τ . Since

$$T^j = \prod_{\alpha=1}^{n} T_\alpha^j ,$$

we shall expand first the operator T_α^j a series. Then, as in (2. 13), we have

$$T_\alpha^j = E - \tau\Lambda_\alpha^j + \frac{\tau^2}{2} (\Lambda_\alpha^j)^2 \dots \tag{5. 29}$$

As a result we get

$$T^j = E-\tau\Lambda^j + \frac{\tau^2}{2}\Big[(\Lambda^j)^2 + \sum_{\alpha=1}^{n}\sum_{\beta=\alpha+1}^{n}(\Lambda_\alpha^j\Lambda_\beta^j - \Lambda_\beta^j\Lambda_\alpha^j) + 0(\tau^3) \tag{5.30}$$

When the operators Λ_α^j are commutative, the expression under the sign of the double sum disappears and we have

$$T^j = E - \tau\Lambda^j + \frac{\tau^2}{2}(\Lambda^j)^2 + 0(\tau^3). \tag{5.31}$$

Comparing (5. 31) to (2. 13) we find that in this particular case the scheme (5. 23) is accurate to the second order in τ. If the operators Λ_α^j are noncommutative, the splitting -up scheme is of first order accuracy in τ. In order to build a scheme accurate to the second order in τ for the noncommutative case the scheme (5. 23) should be modified and substituted by

$$\Phi^j = \prod_{\alpha=1}^{n} T_\alpha^j \Phi^{j-1}, \quad \Phi^{j+1} = \prod_{\alpha=n}^{1} T_\alpha^j \Phi^j. \tag{5.32}$$

Algorithmically it means that first the system (5. 23) is solved on the interval $t_{j-1} \le t \le$ for $\alpha = 1,2,\dots,n$, then a similar system is solved on the interval $t_j \le t \le t_{j+1}$ in the inverse sequence $\alpha = n, n-1, \dots, 1$.

$$(E + \frac{\tau}{2}\Lambda_\alpha^j)\Phi^{j+\frac{\alpha}{n}-1} = (E - \frac{\tau}{2}\Lambda_\alpha^j)\Phi^{j+\frac{\alpha-1}{n}1},$$

$$(\alpha = 1,2,\dots,n) \tag{5.33}$$

$$(E + \frac{\tau}{2}\Lambda_\alpha^j)\Phi^{j+\frac{\alpha}{n}} = (E - \frac{\tau}{2}\Lambda_\alpha^j)\Phi^{j+\frac{\alpha-1}{n}},$$

$$(\alpha = n, n-1, \ldots, 1).$$

It is obvious that for whole cycle of (5. 33) we have

$$\Phi^{j+1} = T^j \Phi^{j-1}, \tag{5. 34}$$

where

$$T^j = \sum_{\alpha=1}^{n} T_\alpha^j \sum_{\alpha=n}^{1} T_\alpha^j = E - 2\tau\Lambda^j + \frac{(2\tau)^2}{2}(\Lambda^j)^2 + 0(\tau^3). \tag{5. 35}$$

Thus within the interval $t_{j-1} \le t \le t_{j+1}$ the scheme (5, 33) is accurate to the second order in τ as is the case with (2. 13) for a double interval.

In conclusion it will be remarked that the difference scheme (5. 33) is absolutely stable for $\Lambda_\alpha^j \ge 0$. Hence we have come, in a sense, to an optimal algorithm of multicomponent splitting.

2. 6. A General Approach to Component-wise Splitting

To solve many problems in mathematical physics one has to split initial differential , integral and integro-differential equations into simpler ones, subsequently reducing the latter to a difference form using the algorithms described in this chapter. In doing so one has each time to consider a question of aprroximation to initial equations by difference equations, and this is the object of our discussion. Let us take some problem in mathematical physics.

$$\frac{\partial \varphi}{\partial t} + A\varphi = 0 \text{ in } D \times T, \tag{6. 1}$$

$$\varphi = g \text{ if } t = 0 \text{ in } D.$$

Suppose that

$$A = \sum_{\alpha=1}^{n} A_\alpha, \tag{6. 2}$$

where $A_\alpha \geq 0$. The solution φ and the function g are assumed to be sufficiently smooth. Then, on each interval $\theta_j \{t_j \leq t \leq t_{j+1}\}$, (6.1) is written as

$$\frac{\partial \varphi_\alpha}{\partial t} + A_c \varphi_\alpha = 0 \quad \text{in} \quad D \times \theta,$$

$$\varphi_\alpha^j = \alpha_{\alpha-1}^{j+1} \quad \text{in} \quad D, \tag{6.3}$$

$$(\alpha = 1, 2, \ldots, n).$$

where

$$\varphi_o^{j+1} = \varphi^j, \quad \varphi_n^{j+1} = \varphi^{j+1} \tag{6.4}$$

Earlier it has been shown that if one applies the Crank-Nicolson difference scheme to each equation, one comes to the system of difference equations of the second order approximation

$$\frac{\varphi^{j+\frac{\alpha}{n}} - \varphi^{j+\frac{\alpha-1}{n}}}{\tau} + A_\alpha \frac{\varphi^{j+\frac{\alpha}{n}} + \varphi^{j-\frac{\alpha-1}{n}}}{2} = 0, \tag{6.5}$$

$$(\alpha = 1, 2, \ldots, n),$$

where we use the notation

$$\varphi^{j+\frac{\alpha}{n}} = \varphi_\alpha^{j+1}, \quad \varphi^{j+1} = \varphi_n^{j+1} \tag{6.6}$$

Let us suppose then that each of the operators A_α is in turn representable in the form

$$A_\alpha = \sum_{\beta=1}^{m_\alpha} A_{\alpha\beta}, \tag{6.7}$$

where $A_{\alpha\beta} \geq 0$. There is a question if it is reasonable to "split", first, the operator A into A_α and then, in turn, the operators A into $A_{\alpha\beta}$?. Is it not easier to represent the operator A as a set of operators $A_{\alpha\beta}$ right away ? In this connection it should be remarked that though these two approaches seem equivalent, in many cases it is more convenient to convert, first, a complex problem in mathematical physics into simpler ones which further can be independently reduced to difference problems (see supplement). Let us analyse any problem of (6. 3) and , considering (6. 7), split it into even simpler ones

$$\frac{\varphi_\alpha^{j+\frac{\beta}{m_\alpha}} - \varphi_\alpha^{j+\frac{\beta-1}{m_\alpha}}}{\tau} + A_{\alpha\beta}\frac{\varphi_\alpha^{j+\frac{\beta}{m_\alpha}} + \varphi_\alpha^{j+\frac{\beta-1}{m_\alpha}}}{2} = 0 \text{ in } D, \tag{6.8}$$

$$\begin{pmatrix} \alpha = 1,2,\dots,n \\ \beta = 1,2,\dots,m \end{pmatrix},$$

where

$$\varphi_1^j = \varphi^j, \quad \varphi_\alpha^j = \varphi_{\alpha-1}^{j+1}, \quad (\alpha \neq 1).$$

It is not difficult to see that the system of the split equations (6. 8) approximates the initial-value problem (6. 1) to within the second order in τ. The proof of such a statement is based on the fact that, using (6. 2) and (6. 6), one can change the ordering of the components of splitting-up , by writting

$$A = \sum_{\alpha=1}^{n} \sum_{\beta=1}^{m_\alpha} A_{\alpha\beta} = \sum_{\gamma=1}^{p} A_\gamma$$

and in this event we gave the problem

$$\frac{\varphi^{j+\frac{\gamma}{p}} - \varphi^j}{\tau} + \frac{\varphi^{j+\frac{\gamma}{p}} + \varphi^{j+\frac{\gamma-1}{p}}}{2} = 0, \tag{6.9}$$

$$(\gamma = 1, 2, \dots, p),$$

which, as was shown in §2. 5, approximates the problem (6. 1) accurate to the second order in τ . The result is also true of the case when $A_{\alpha\beta}$ is dependent on time. Then one should make approximation to the operators $A_{\alpha\beta} = \Lambda^j_{\alpha\beta}$ to within the second order in τ on each interval $t_j \leq t \leq t_{j+1}$. If the $\Lambda_{\alpha\beta}$ are non-commutative, then, using the two-cicle procedure described in § 2. 5, we obtain a difference scheme accurate to the second order for each interval $t_{j-1} \leq t \leq t_{j+1}$. To summarize, we can assert the following. When an evolution problem of the form (6. 1) , under the condition $A_\alpha \geq 0$, is reduced to particular evolution problems (6. 3) and these are regarded as a set of new evolution problems, the approximation to the initial - value problem will be accurate to the first order in τ provided that at least one of the elementary problems is reduced to difference schemes accurate to the first order. If every such problem has approximation of second-order accuracy, then, using the two-cycle procedure with respect to α and β, we again come to approximation of second order accuracy in τ. It should be noted that if the operators A are non-commutative, then without using the two-cycle procedure we derive approximation of (6. 1) accurate to the first order.

Indeed, let us consider the case of non-commutative operators. Then the following is an initial-value problem :

$$\frac{\partial \varphi}{\partial t} + \sum_{\alpha=1}^{n} A_\alpha \varphi = 0 \quad \text{in } Dx\theta_j , \qquad (6.10)$$

$$\varphi = \varphi^j \text{ if } t = t_j .$$

(6. 10) is reduced to the system

$$\frac{\partial \varphi}{\partial t} + A_\alpha \varphi_\alpha = 0 , \quad \alpha^j_\alpha = \varphi^{j+1}_{\alpha-1} . \qquad (6.11)$$

Let $A_\alpha = \sum_{\beta=1}^{m_1} A_{\alpha\beta}$, where $A_{\alpha\beta_i} A_{\alpha\beta_j} \neq A_{\alpha\beta_j} A_{\alpha\beta_i}$.

Then every problem of (6. 11) is solved by the two-cycle method.

$$\frac{\varphi_\alpha^{j+\frac{\beta}{2m_\alpha}} - \varphi_\alpha^{j+\frac{\beta-1}{2m_\alpha}}}{\tau/2} + A_{\alpha\beta} \frac{\varphi^{j+\frac{\beta}{2m_\alpha}} + \varphi^{j+\frac{\beta-1}{2m_\alpha}}}{2} = 0, \tag{6. 12'}$$

$$(\beta = 1, 2, \ldots, m\ ),$$

$$\frac{\varphi_\alpha^{j+\frac{\beta}{2m_\alpha}} - \varphi_\alpha^{j+\frac{\beta-1}{2m_\alpha}}}{\tau/2} + A_{\alpha, 2m_\alpha+1-\beta} \frac{\varphi^{j+\frac{\beta}{2m_\alpha}} + \varphi^{j+\frac{\beta-1}{2m_\alpha}}}{2} = 0, \tag{6. 12''}$$

$$(\beta = m_\alpha + 1,\ m_\alpha + 2, \ldots, 2m_\alpha).$$

The initial conditions for each of the systems (6. 12) are taken in the form

$$\varphi_\alpha^j = \varphi_\alpha^j,\quad \varphi_\alpha^j = \varphi_\alpha^{j+1/2} \tag{6. 13}$$

It is easy to find that (6. 12) approximates, on the interval $t_j \leq t \leq t_{j+1}$, any of the problems (6. 11) to within τ^2.

In order for the whole algorithm to lead to a solution of (6. 1) to within τ^2 it is also necessary alternate the basic cycles. Thus, instead of (6. 11) on the interval $t_{j-1} \leq t \leq t_j$ we should have

$$\frac{\partial \varphi_\alpha}{\partial t} + A_\alpha \varphi_\alpha = 0, \quad (\alpha = 1, 2, \ldots, n), \tag{6. 14}$$

$$\varphi_1^{j-1} = \varphi^{j-1},\quad \varphi_\alpha^{j-1} = \varphi_{\alpha-1}^j,\ (\alpha > 1)$$

and, on the next interval $t_j \leq t \leq t_{j+1}$:

$$\frac{\partial \varphi_\alpha}{\partial t} + A_{n-\alpha+1}\varphi_\alpha = 0, \qquad (\alpha = 1, 2, \ldots, n) , \tag{6. 15}$$

$$\varphi_1^j = \varphi^j , \ \varphi_\alpha^j = \varphi_{\alpha-1}^{j+1} , \quad \alpha > 0 .$$

It is assumed that each problem of (6. 13) and (6. 14) is solved by the two-cycle method of the form (6. 12). Note that for the condition $A_{\alpha\beta} \geq C$ the component-wise splitting-up method is absolutely stable.

2. 7 Hyperbolic Equations

Hyperbolic equations hold a prominent place in applications. Numerical methods for such equations are studied to full advantage. Hyperbolic equations have the following characteristic features. First, the domain of dependence of a solution for such equations is bounded by a characteristic cone so that the region outside D x T space does not affect the solution in the point under consideration. Second, among the solutions of the initial-value problem there may be non- smooth solutions as well and this should be kept in mind when one develops numerical schemes. Since a great amount of excellent investigations are devoted to constructing difference schemes for hyperbolic problems we shall discuss only some of the methods most widely used in recent years.

Let us look at the problem

$$\frac{\partial^2 \varphi}{\partial t^2} + A\varphi = 0 \quad \text{in} \quad D \times T , \tag{7. 1}$$

$$\varphi = g$$

$$\frac{\partial \varphi}{\partial t} = p \quad \text{if} \quad t = 0 \quad \text{in } D. \tag{7.1}$$

It will be assumed that the operator A is independent of time and the functions g and p allow sufficient smoothness of the solution of the periodic problem. Let the operator A be positive, i. e. there exist $\gamma > 0$ for all $\varphi \neq 0$, so that

$$(A\varphi, \varphi) \geq \gamma^2 (\varphi, \varphi) \tag{7.2}$$

By the way, we must note that for symmetric positive definite operators $\gamma = \alpha_A$, where α_A is a minimum eigenvalue of the operator A spectrum.

Let us consider difference approximation to the equation of (7. 1) in the form

$$\frac{\varphi^{j+1} - 2\varphi^j + \varphi^{j-1}}{\tau^2} + A\varphi^j = 0\,, \tag{7.3}$$

It is easy to show that the difference scheme (7. 3) approximates the initial equation of (7. 1) to within quantities of the second-order of smallness with respect to τ. We apply initial data to (7. 3). In order not to distort the second order of approximation we take, along with the condition,

$$\varphi^o = g, \tag{7.4}$$

the following relation :

$$\varphi^1 = (1 - \frac{\tau^2}{2} A) g + \tau p. \tag{7.5}$$

(7. 5) is derived by expanding the problem (7. 1) into Taylor's series in the vicinity of $t = 0$ with a subsequent elimination of derivatives

using the equation and the known initial conditions in (7. 1).

The problem of (7. 3), (7. 4), (7. 5) is fully formulated. Our object now is to analyse numerical stability of (7. 3), using the spectral method. Let u_n and u_n^* be eigenfunctions and $\lambda_n \geq 0$ eigenvalues of spectral problems

$$\begin{aligned} Au &= \lambda u, \\ A' u' &= \lambda u' \end{aligned} \qquad (7.6)$$

Next, it will assumed that $\{u_n\}$ forms a basis. Then we seek a solution to the equation in the form

$$\varphi^j = \sum_n \varphi_n^j u_n , \qquad (7.7)$$

where

$$\varphi_n^j = (\varphi^j, u_n^*)$$

Substituting the Fourier series (7. 7) into (7. 3) and multiplying the result by φ_n^j we get for the Fourier coefficients the following expression

$$\frac{\varphi_n^{j+1} - 2\varphi_n^j + \varphi_n^{j-1}}{\tau^2} + \lambda_n \varphi_n^j = 0. \qquad (7.8)$$

A solution to (7. 8) is sought in a form of the power function

$$\varphi_n^j = \eta_n^j \qquad (7.9)$$

Note that in the left- hand side of (7. 9) j is an index while in the right - hand side it is a power.

Substituting (7. 9) into (7. 8) we get the characteristic equation for η_n:

$$\eta_n^2 - 2\left(1 - \frac{\tau^2 \lambda_n}{2}\right)\eta + 1 = 0 . \tag{7.10}$$

It is easily seen that if

$$\left| 1 - \frac{\tau^2 \lambda_n}{2} \right| \leq 1 \tag{7.11}$$

the roots of (7. 10) are complex conjugate and equal to unity in modulus i. e.

$$|\eta_n| = 1 . \tag{7.12}$$

From the condition (7. 11)

$$\tau^2 \leq \frac{4}{\lambda_n} , \qquad (n = 1, 2, \ldots ,) , \tag{7.13}$$

Evidently, (7. 13) will be fulfilled for all λ_n if τ are taken such that

$$\tau \leq \frac{2}{\sqrt{\beta_A}} , \tag{7.14}$$

where β_A is the upper bound of the operator A spectrum. For symmetric operators $\beta_A = \| A \|$ hence,

$$\tau \leq \frac{2}{\sqrt{\| A \|}} \tag{7.15}$$

Let us proceed to implicit difference schemes

$$\frac{\varphi^{j+1} - 2\varphi^{j} + \varphi^{j-1}}{\tau^2} + A\,\frac{\varphi^{j+1} + \varphi^{j-1}}{2} = 0 \qquad (7.16)$$

The scheme (7. 16) is accurate to the second order with respect to τ and in combination with (7. 4), (7. 5) it approximates (7. 1) to within the second order. For (7. 16) the characteristic equation is of the form.

$$\eta_n^2 - \frac{2}{1+\frac{\tau^2\lambda_n}{2}}\,\eta_n + 1 = 0 \qquad (7.17)$$

hence,

$$\eta_n = \frac{1}{1+\frac{\tau^2\lambda_n}{2}} \pm \sqrt{\left(\frac{1}{1+\frac{\tau^2\lambda_n}{2}}\right)^2 - 1} \qquad (7.18)$$

From here it follows that with any

$$\|\eta_n\| = 1 \qquad (7.19)$$

Thus, (7. 16) is an unconditionally stable scheme.

Let

$$A = \sum_{\alpha=1}^{n} A_\alpha, \qquad (7.20)$$

where $A_\alpha \geq 0$. For an approximate solution of (4. 1) we make use of difference approximation of the form

$$B\,\frac{\varphi^{j+1} - 2\varphi^{j} + \varphi^{j-1}}{\tau^2} + A\varphi^{j} = 0\,, \qquad (7.21)$$

where

$$B = \prod_{\alpha=1}^{n} \left(E + \frac{\tau^2}{2} A_\alpha \right) \qquad (7.22)$$

From (7. 21) and (7. 22) it follows that (7. 21) approximates the initial equation of (7. 1) to within quantities of the second order with respect to τ . Since the equation (7. 21) can be reduced to

$$\frac{\varphi^{j+1} - 2\varphi^j + \varphi^{j-1}}{\tau^2} + B^{-1} A \varphi^j = 0 , \qquad (7.23)$$

from the analysis made above there follows stability of (7. 21), (7. 22), provided that

$$\tau \leq \frac{2}{\sqrt{\beta_{B^{-1}A}}} , \qquad (7.24)$$

In this way the problem of choice of the parameter τ satisfying the stability condition reduces to calculating a maximum eigenvalue of the problem (under the assumption that all eigenvalues $\beta_{B^{-1}A}$ are positive) :

$$Au = \lambda Bu. \qquad (7.25)$$

This problem is solved by the iterative process

$$B\varphi^k = A \frac{\varphi^{k-1}}{\|\varphi^{k-1}\|} . \qquad (7.26)$$

In this connection

$$\beta_{B^{-1}A} = \lim_{k \to \infty} \|\varphi^k\| . \qquad (7.27)$$

The operational scheme of the difference system corresponding to (7. 21) is written as

$$(E + \frac{\tau^2}{2} A_1)\, \xi^{j+\frac{1}{n}} = A \varphi^j ,$$

$$(E + \frac{\tau^2}{2} A_2)\, \xi^{j+\frac{2}{n}} = \xi^{j+\frac{1}{n}} , \qquad (7.28)$$

. . . .

$$(E + \frac{\tau^2}{2} A_n)\, \xi^{j+1} = \xi^{j+\frac{n-1}{n}} ,$$

$$\varphi^{j+1} = 2\varphi^j - \varphi^{j-1} + \tau^2 \xi^{j+1} .$$

This problem is solved successively with $j = 2, 3 \ldots .$, and using the initial data (7. 4) and (7. 5).

(7. 28) is a splitting-up scheme.

To conclude, we consider a wave equation

$$\frac{\partial^2 \varphi}{\partial t^2} - a^2 \Delta \varphi = 0 \quad \text{in} \quad D \times T , \qquad (7.29)$$

$$\varphi = f$$

$$\text{if} \quad t = 0 \quad \text{in } D ,$$

$$\frac{\partial \varphi}{\partial t} = F$$

where a^2 is squared velocity of propagation of wave perturbation . The problem (7. 29) will be called periodic in geometric variables. Using our notation

$$A = -a^2 \Delta .$$

Let us assume that instead of the differential operator we consider its second-order difference approximation in all variables x_α. Then

$$A\varphi = -a^2 \sum_{\alpha=1}^{n} \frac{\varphi_{k_\alpha+1} - 2\varphi_{k_\alpha} + \varphi_{k_\alpha-1}}{\Delta x_\alpha^2}. \qquad (7.30)$$

If $\Delta x_\alpha = h$, the spectral problem

$$Au = \lambda u$$

defines the upper bound of the difference operator A spectrum in the form

$$\beta_A = \frac{4na^2}{h^2}$$

Thus, in this case the explicit scheme (7. 3) requires fulfilment of the condition (7. 14) or

$$\tau \le \sqrt{n}\,\frac{h}{a}\,. \qquad (7.31)$$

If we consider the scheme (7. 21), (7. 22) , where

$$A_\alpha \varphi = -a^2 \frac{\varphi_{k_\alpha+1} - 2\varphi_{k_\alpha} + \varphi_{k_\alpha-1}}{\Delta x_\alpha^2}$$

then, applying the spectred analysis , we get

$$|\eta_n| = 1\,. \qquad (7.32)$$

This means that the difference scheme of (7. 29) on the basis of the splitting-up algorithm (7.21) will be unconditionally stable.

The algorithm considered extends fairly simply to inhomogeneous hyperbolic equations.

G. I. Marchuk

Contents

G. I. Marchuk

References

The symbol (*) indicates the papers which are very close with a theory of the spletting-up method.

1 - Monographs and text books.

Babuska I. Prager M., Vitasek E.
Numerical processes in differential equations-Interscience ,1966

Bahvalov N. S.
Foundations of numerical analysis - (1970) Moscow (Russian).

Berezin I. S., Zhidkov N. P.
Computing methods - Pergamon Press - Oxford 1965, 2 vols.

Wasow W., Forsythe G.
Finite difference methods for partial differential equations - J. Wiley and Sons (1959).

Voevodin V. V.
Numerical methods of algebra. Theory and algorithms. "Nauka" Moscow 1966 (Russian)

Godunov S. K.
Lectures on difference methods for the solution of the equations of gas-dynamics - Novosibirsk 1962 (Russian).

Godunov S. K., Ryabenki V. S.
The theory of difference schemes. An Introduction. North Holland - Amsterdam 1964.

D'jakonov E. G.
(*) Iterative methods for the solution of discrete analogues of boundary value problems for ellyptic equations (Intern. Spring School on Numerical Math., Kiev 1966 (Russian) I. K. A. N. SSSR V. Z. Acad. Nauk. SSSR/Kiev 1970.

Il'in V. P.
Difference schemes for the solution of ellyptic equations - Izd. NGU. Novosibirsk (1970) (Russian)

Kantorovich L. V.
Functional analysis and applied mathematics - Uspehi Matem. Nauk 3, 6, 89-185 (1948) (Russian).

Kantorovich L. V., Krylov R. I.
Approximate methods of higher analysis, FM., Moscow-Leningrad (1962).

Collatz L.
The numerical treatment of differential equations, 3rd. ed. Springer Verlag, Berlin (1960).

- Funktional analysis und Numerische Mathematik , Springer-Verlag, Berlin (1964).

Krasnosel'skii M. A. , Vainikko G. M., Zabreiko P. P., Rutickii Ja. B., Stecenko V. Ja.
Approximate solutions of operator equations Izdat "Nauka" Moscow (1969) (Russian)

Courant R.
Partial differential equations - "Courant-Hilbert" vol. 11, New York 1962.

Lions J.
(*) Résolution itérative d'inéquations variationnelles per décomposition et éclatement - Collège de France (1967)

Marchuk G. I.
Computational methods for nuclear reactors - Atomizdat, Moscow 1961 (Russian)

Marchuk G. I.
(*) Numerical methods for weather forecasting-Hidrometizdat 1967 (Russian)

- (*) Methods and Problems of numerical analysis, (Russian) Int. Congress of Math. Nice (1970)

Marchuk G. I., Lebedev V. I.
(*) Numerical methods in transport theory - Atomizdat (1971) Moscow (Russian)

Mikhlin S. G.
Variationsmethoden der Mathematischen Physik-Akademie Verlag-Berlin 1962

Richtmyer R.
Difference methods for Initial-Value problems Intersciance Publ. Inc. New York (1957)

Richtmyer R. , Morton K.
Difference methods for initial-value problems, New York 1967

Rozdestvenski B. L., Ianenko N. N.
(*) Systems of quali-linear equations - "Nauka" Moscow 1968 (Russian)

Ryabenki V. S., Philippov A. F.
On stability of difference equations - Gozudarst. Izdat. Tehn-Teor. Lit. Moscow 1956 (Russian)

Samarskii A. A.
(*) Lectures on difference schemes - Moscow 1969 (Russian)

Saul'yev V. K.
Integration of equations of parabolic type by the method of nets - Pergamon Press - London 1964

Smirnov V. K.
Lehrgang der höherer Mathematik - Deutscher Verlag - Berlin 1956

Sobolev S. L.
Lectures on the theory of cubature formulas part I (1964) part II (1965) Novosibirsk (Russian), Izd NGU

Tihonov A. N., Samarskii A. A.
Equations of mathematical physics - "Nauka" Moscow 1966 (Russian)

Wilkinson J. H.
The algebraic eigenvalue problem - Oxford, Claredon Press (1965)

Faddeev V. K., Faddeeva V. N.
Numerische Methoden der linearen Algebra - Veb. Deutscher Verlag-Berlin (1964)

Forsythe G and Möller F. B.
Computer solution of linear algebraic systems, Prentice-Hall, Inc. Engle wood Cliffs, N. Y. 1967

Ianenko N. N.
(*) The method of fractional steps for solving multidimensional problems of mathematical physics. - "Nauka" Novosibirsk 1967 (Russian)

- (*) Introduction to difference methods of mathematical physics - part I and 2 Novosibirsk 1968 (Russian), Izd. NGU.

G. I. Marchuk

2. Additional literature

Babuska I.
The finite element method foe elliptic differential equations. Numerical solution of partial differential equations - II SYN-SPADE - 1970 - Acc. Press New York, London , 69-106 (1967)

Belotserkovskii O. M. , Chuskin P. I.
A numerical method of integral relations
Zh. vych. mat. i mat. fiz. 2, 5 (1962) 731-759 (Russian)

Birkoff G. , Varga R. , Young D.
Alternating direction implicit methods. Advantages in Comp. , Vol. 3, Academic Press, New York-London 189-273 (1962)

Birkoff G. , Schultz M. H. , Varga R. S.
Piecewise Hermite interpolation in one and two variables with applications to partial differential equations, Num. Math. 11 (1968) 232-256

Bryan K.
A scheme for numerical integration of the equations of motion on an irregular grid free of non linear instability - Mon. Wea. Review, v. 94, 1, 39 - 40 (1966)

Byleev N. N.
Numerical methods for the solution of two-and three - dimensional diffusion equation
Mat. Sb. T 51, 2, 227-238 (1960) (Russian)

Varga R. S.
Matrix iterative analysis - Prentice-Hall-New Jersey (1962)

Wachspress E. L.
(*) Extended application of alternating direction implicit iteration model problem theory
SIAM. J. v. 11, 3, 994 - 1016 (1963)

Gunn J. E.
(*) The solution of elliptic difference equations by semi-explicit iterative techniques . SIAM. J. Numer. Anal. v. 2, 1, 24-25 (1965)

Godunov S. K. , Prokopov G. P.
Variational approach to the solution of large systems of linear equations appearing in strongly elliptical problems. Preprint - Inst. of appl. Math, Acad Nauk SSSR, Moscow 1968

Dorodnizin A. A.
A computing method for solving some non-linear problems of aerohydrodynamics.
Trudy 3-rd All Union Math. Symposium 3, 447-453 (1958)
- A contribution to the problem of computing eigenvalues and eigenvectors of matrics.
Dokl. Acad. Nauk. SSSR 126 (1959) 1170-1171 (Russian)

D'jakonov E. G.
(*) Difference schemes with a " disintegrating "operator for multidimensional stationary equations.
Zh. vych. mat. Mat. fiz. 2, 4 (1962) 549-568 (Russian)
- (*) The construction of iterative methods based on the use of spectrally equivalent operators.
Zh. vych. mat. mat. fiz. 6, 1 (1966) 12-34 (Russian)

Douglas J. , Rachford H.
(*) On the numerical solution of heat conduction problems in two and three space variables.
Trans. mev. Math. Soc. v. 82, 2, 421 (439) (1956)

Kellogg
(*) Another alternating - direction - implicit method
SIAM J. v. 11, 4, 976-979 (1963)

Konovalov A. N
(*) Numerical methods for problems of the theory of elasticity. - Izd. NGU Novosibirsk 1968 (Russian)

Krasnosel'skii M. A. , Krein S. G.
An iteration process with minimal residuals - Mat. Sb. (N. S) 31 (73) 315-334 (1952) (Russian)

Kreiss H. O.
Initial boundary value problem for partial differential and difference equations in one space dimension? Numerical solution of partial differential equations. II SYNSPADE - 1970 - Academic Press New York-London , 401-410 (1971)

Courant R. , Friedrichs K. , Lewy H.
Uber di partiellen Differenzengleichungen der mathematischen Physik - Math Ann. T. 100, 32 (1928)

Kurihara Y., Holloway J. L.
Numerical integration of a nine-level global primitive equations model formulated by the box method. - Mon. Wea. Rev. v. 95, 8, 509-530 (1967)

Ladyzenskaya O. A.
The mothod of finite differences in the theory of partial differential equations - Uspehi Mat. Nauk (N. S.) 12 (1957) 123-145 (Russian).

Lax P. D., Wendroff W.
On the stability of difference schemes with variable coefficients - Comm. Pure Appl. Math. v. 15, 4, 363-371 (1962)

Lax P. D., Richtmyer R. D.
Survay of the stability of linear finite difference equations- Comm. Pure Appl. Math. v. 9, 2, 267-293 (1956)

Lebedev V. I.
On the mesh method for a certain system of partial differential equations - Izv. Acad. Nauk. SSSR Ser. Mat. 22 (1958) 717 - 734 (Russian)
Dirichlet and Neumann problems on triangular and hexagonal grids. - Dokl. Acd. Nauk. SSSR 138 (1961) 33-36 (Russian)

Lyusternik L. A.
On difference approximations of the Laplace operator - Uspehi Mat. Nauk (N. S.) 9, 2 (1954) 3 - 66 (Russian)

Marchuk G. I.
(*) Numerical solution of Poincarè problem for the oceanic circulation - Dokl. Acad. Nauk. SSSR T 185, θ (1969) 1041-1044 (Russian)
\- (*) On the theory of the splitting-up method. Numerical solution of partial differential equations - SYNSPADE - 1970 - Academic Press, New York-London 469-500 (1971)

Marchuk G. I., Kuznezov Ju. A.
On the question of optimal iteration processes - Dokl. Acad. Nauk. SSSR 181 (1968) 1331-1334 (Russian)

Marchuk G. I., Sultangazin U. M.
(*) Convergence of a decoupling method for the radiation transfer equation - Dokl. Acad. Nauk. SSSR 161 (1965) 66-69 (Russian)
\- (*) Solving the kinetic transfer equation by the separation method - Dokl. Acad. Nauk. SSSR 163 (1965) 857-860 (Russian)

Marchuk G. I.
(✱) On the foundation of the separation method for the equations of radiation transfer - Zh. vych. mat. fiz. 5, 5 (1965) 852-863 (Russian)

Marchuk G. I., Ianenko N. N.
(✱) Application of the fractional steps method to the solution of problems of mathematical physics - Proc. All Union Conference of Num. Math., Moscow February 1965 - Proc. Congress IFIP, New York May 1965 Proc. "Some questions of applied and numerical math. "Nauka" Novosibirsk 5-22 (1966)

Aubin J. P., Burchard H. G.
Some aspects of the method of hypercircle applied to elliptic variational problems. - Numerical solution of partial differential equations. II SYNSPADE 1970. Academic Press New York-London 1-67 (1971)

Oganesyan L. A., Rukhovets. L. A.
Variational-difference schemes for linear second-order elliptic equations in a two-dimensional region with piecewise smooth boundary. - Zh. vych mat. fiz. 8. 1 (1968) 97-114 (Russian)
Analysis of the rate of convergence of variational-difference schemes for second-order elliptic equations on a two-dimensional domain with smooth boundary. Zh vych. mat. fiz. 9, 5 (1969) 1102-1120 (Russian)

Rivkind V. Ja.
An approximate method of solving the Dirichlet problem and estimates of the rate of convergence of solutions of the difference equations to solutions of elliptic equations with dicontinuous coefficients - Vestik Leningrad Univ., Sez. Mat. Meh 13 (1964) 3, 37-52 (Russian)

Peaceman D. W., Rachford H. H.
(✱) The numerical solution of parabolic and elliptic differential equations - SIAM J. v. 3, 1 (1955) 28-42

Raviart P. A.
Sur l'approximation de certains equations d'évolution linéaires et non linéaires - J. de Mathem. Pures et Appl., v. 46, 1(1967) 11-107

Richtmyer R. D.
Nonlinear stability of difference schemes - Proc. "Some questions of applied and numerical mathematics" "Nauka" Novosibirsk 1966, (Russian) 54-59

Samarskii A. A.
(*) An economical algorithm for the numerical solution of systems of differential and algebraic equations - Zh. vych. mat. mat. fiz. 4, 3 (1964) 580-585
(*) Necessary and sufficient conditions for the stability of double layer difference schemes.
Doll. Acad. Nauk SSSR 181 (1968) 808-811 (Russian)

Strang W.
Difference methods for mixed boundary problem - Duke Math. J., v. 27, 2 (1960) 221-232

Tihonov A. N., Samarskii A. A.
Homogeneous difference schemes - Zh. vych. mat. mat fiz. 1, 1 (1961) 5-63 (Russian)

Thomée V.
On maximum-norm stable difference operators. Numerical solution of partial differential equations. Proc. Intern. Symposium 1965, Academic Press 125-151 New York.

Fedorenko R. F.
The rate of convergence of an itarative process - Zh. vych. mat. mat fiz. 4, 3 (1964) 559-564 (Russian)

Phillips N. A.
An example of non-linear computational instability. The atmosphere and the sea in motion- Scientific Contribution to the Rossby Memorial Volume - The Rockfeller Inst.

Frankel S. P.
Convergence rates of iterative treatments of partial differential equations - Math. Tables and Other Aids Comput. v. 4, 30 (1950) 65-75

Hubbard B. E.
(*) Alternating direction schemes for the heat equation in a general domain. - SIAM J. Num. Anal. 2, 3 (1966) 448-463

Young D. M.
Iterative methods for solving partial difference equations of elliptic type- Trans. Math. Soc. 76, 1 (1954) 93-111.

CENTRO INTERNAZIONALE MATEMATICO ESTIVO

(C. I. M. E.)

U. MOSCO

AN INTRODUCTION TO THE APPROXIMATE SOLUTION OF VARIATIONAL INEQUALITIES

Corso tenuto ad Erice dal 27 giugno al 7 luglio 1971

AN INTRODUCTION TO THE APPROXIMATE SOLUTION OF VARIATIONAL INEQUALITIES *

by

Umberto Mosco

(University of Rome)

These lectures are aimed to offer an elementary introduction to the theory of variational inequalities of elliptic type, with special emphasis to the finite-dimensional approximation of the solutions.

More specific introductory remarkes will be given at the beginning of each chapter.

CHAPTER 1

<u>Minimum problems and variational inequalities : convexity, monotonicity and fixed-points</u>

1 : The direct formulation

2 : The weak formulation

3 : The linearized formulation

4 : The fixed-point formulation

5 : The epigraph formulation

6 : Minimum problems in normed spaces

7 : Monotone operators and variational inequalities : the linearization lemma

8 : Variational inequalities and fixed-points

9 : Minimization of non-differentiable functionals and mixed variational inequalities

* Partially supported by CNR, Com. Mat., GNAFA

U. Mosco

CHAPTER 1

Minimum problems and variational inequalities: convexity, monotonicity and fixed-points

U. Mosco

CHAPTER I

Minimum problems and variational inequalities : convexity, monotonicity and fixed-points

The aim of this introductory chapter is to bring to light some simple geometric features underlying the theory of the so called variational inequalities which are also inherent in the problem of minimizing a convex functional on a convex set.

Most of the characterizations of the solutions of these problems, indeed, give rise to inequalities involving the differential of the given functional, which is a monotone map from the space where the functional is defined to its dual. For minimum problems in function spaces, these inequalities should be seen as the analogue of the Euler condition of the calculus of variations, in the presence of unilateral constraints on the solution.

The variational inequality approach consists in dealing directly with such inequalities, without assuming "a priori" that the monotone operator involved is the differential of a convex functional. Moreover, even in this special case, they can often be used, as the Euler equation is, to investigate the properties of the solution of the original minimum problem - for example, the regularity, i.e., how smooth that solution is - and they also occur in various approximate methods of solution, alternative to the direct Ritz approach.

Therefore, it is of some interest that some basic general features of convex optimization are shared by the monotone inequalities we are going to discuss in our lectures.

Let us recall some of the specific aspects of convexity.

First, any local minimum is actually a global minimum. The relevance of this property is apparent: only a "local" investigation of the functional and the constraints is required, which in terms,for instance, of computer performances means less information to be memorized.

Another motivation for setting (if possible) an infinite dimensional minimum problem in a convex framework is that "good" topologies are then allowed. It is well known, in fact, that convex functionals keep their semicontinuity in topologies weak enough to make the set of constraints compact in a suitable function space, under reasonable boundedness assumptions most of the times intrinsic in the problem at hand : the existence of the minimum is then an immediate consequence of Weierstrass theorem.

Another specific feature of convexity is that some linearization is always possible. This is, on the other hand, the underlying reason for the existence of good topologies. As we shall see later, a linearization of the problem is the basic tool in the existence theory for infinite dimensional variational inequalities.

The equivalent formulations of a convex minimum problem that we shall discuss in the present chapter, can be tentatively named as follows:

I the <u>direct</u> formulation

II the <u>gradient</u> or <u>weak</u> formulation

III the <u>linearized</u> formulation

IV the <u>fixed-point</u> and the <u>iterative fixed-point</u> formulation.

V the <u>epigraph</u> formulation.

Further methods that should be mentioned are the

VI <u>duality</u> and <u>minimax</u> methods and the methods based on

VII <u>penalization</u> and <u>regularization</u>

We should perhaps also say that many of these approaches are often adopted at the same time and can be variously combined with the finite-dimensional approximate methods of solution we shall talk about in the following chapters.

We shall first intuitively sketck I ... V in Sections 1 ... 5 below, postponing the rigorous proof of their equivalence to the later sections, where we shall also discuss further properties of monotone variational inequalities. A partial account of the duality theory will be given in the Section 7 of Chapter 3, whereas the other subjects mentioned in VI and VII above will not be treated at all in our lectures, though important they are.

Let us mention that a detailed discussion of the methods based on penalization and regularization devices with regard to variational inequalities, can be found in J. L. Lions [1] , while an account of the duality and minimax methods can be found in J. L. Lions, R. Glowinski and R. Tremolieres [1] .

1 . Direct formulation

Let us consider the problem of minimizing a real valued convex function F on a convex subset K of the n-dimensional euclidean space E^n , i. e., the problem

$$\text{I} \qquad u \in K : \quad F(u) \leqslant F(v) \qquad \underline{\text{for all}}\ v \in K \ .$$

If we introduce for any fixed vector v of the space the (possibly empty) level set L(v) of F on K,

$$L(v) = \{ w \in K : \qquad F(w) \leqslant F(v) \}$$

then I can be equivalently written as

$$u \in \bigcap_{v \in K} L(v)$$

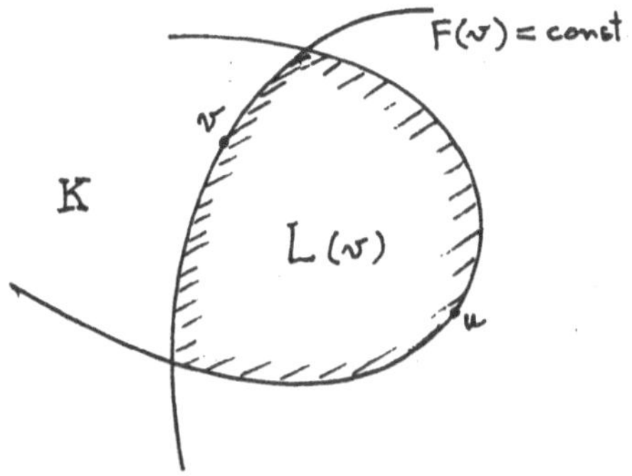

Remark 1 : The set of all solutions u of problem I is convex, for L(v) is convex for every v . It is also closed provided F is lower semicontinuous and K is closed, for then L(v) is also closed. ■

Let us notice that throughout this chapter we disregard any question about the existence of solutions. In other words, we are assuming that a solution does exist and we are only concerned with the characterization of it in terms of F and K, together with the properties of the set of all solutions.

2. Weak formulation

Let us assume now that F is differentiable on E^n and let

$$\nabla F(v) = \left(\frac{\partial F}{\partial v_1}, \dots, \frac{\partial F}{\partial v_n} \right)$$

be the gradient of F at the point $v \equiv (v_1, \dots, v_n)$ of E^n :

$$F : E^n \longmapsto E^n$$

and

$$(\nabla F(v) \mid w) = \sum_{i=1}^{n} \frac{\partial F}{\partial v_i} \cdot w_i \qquad , \quad w \equiv (w_1, \dots, w_n) .$$

Remark 2 : $\nabla F(z)$ is a vector at the point z pointing toward the direction of maximum increase of F : due to the convexity of F, hence of the level sets of F, at any point z all directions w - z pointing to the same half space where $\nabla F(z)$ is oriented, i. e., all vectors w - z such that

$$(\nabla F(z) \mid w - z) \geqslant 0 \qquad ,$$

are directions of increasing (≡ non-decreasing) F : we shall call them forward directions at z

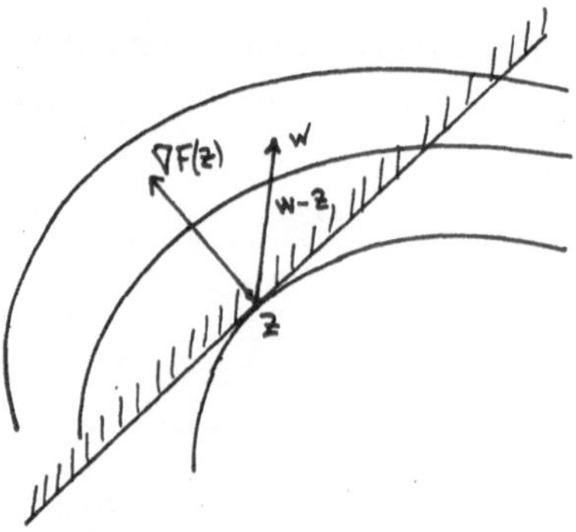

Let us now introduce for any fixed v of E^n the set

$$M(v) = \{ z \in K : \quad (\nabla F(z) \mid v - z) \geq 0$$

of all z of K from which the given v is seen in a forward direction.

Clearly, a vector u minimizes F on K if and only if <u>all</u> vectors v of K are seen forward from u, that is, I is equivalent to the inclusion

$$u \in \bigcap_{v \in K} M(v)$$

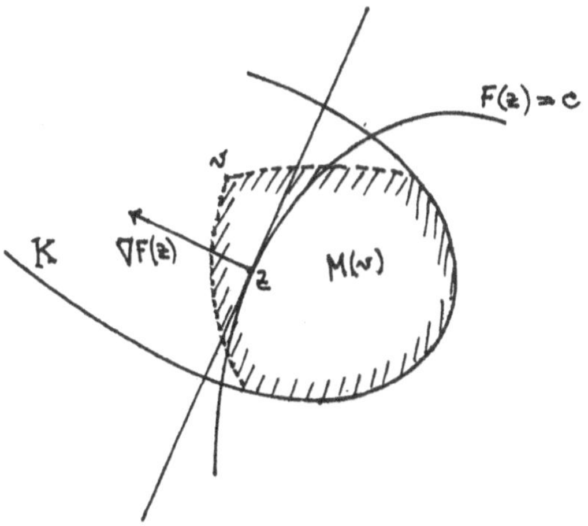

which is to say that problem I above is equivalent to the problem

II $\quad u \in K \; : \; (\nabla F(u) \mid v - u) \geq 0 \qquad$ <u>for all</u> $v \in K$.

Remark 3 : If u belongs to the interior of K, then II is equivalent to

$$u \in K : \qquad (\nabla F(u) \mid w) = 0 \qquad \text{for all } w ,$$

which is the "weak" form of the equation

$$\nabla F(u) = 0 \quad .$$

[In fact, if $u \in \text{int}\, K$, then the vector $v = u \pm \rho w$ belongs to K for all $\rho > 0$ sufficiently small, whatever is the vector w of the space; therefore, we can put any such v into inequality II and we get $\pm \rho (\nabla F(u) \mid w) \geq 0$, with $\rho \geq 0$, hence $(\nabla F(u) \mid w) = 0$.] ■

3. The linearized formulation

Both I and II can be viewed as a system of infinite non-linear inequalities in u. As we shall see below, however, it is also possible to characterize a solution u of I and II by means of a system of linear inequalities.

It is now convenient to consider, for any given point v of the space, the set

$$N(v) = \{ w \in K : \quad (\nabla F(v) \mid v - w) \geq 0 \}$$

of all w of K which are seen backward from v,

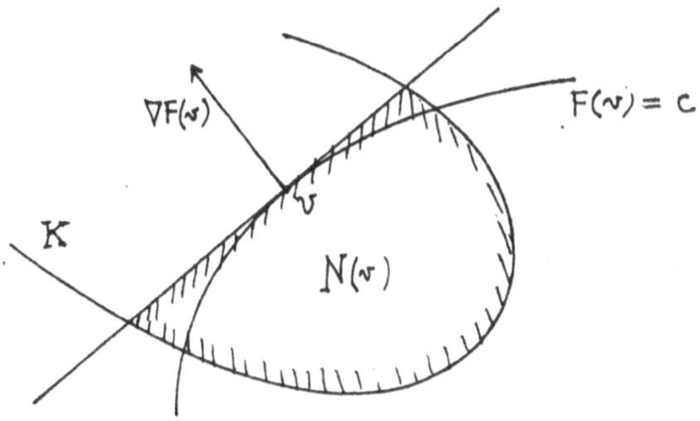

Note that "backward" here does'nt necessarily mean "direction of non-increasing F" : it simply denotes a direction $w - v$ pointing to the opposite half space with respect to $\nabla F(v)$.

Clearly, a vector u minimizes F on K if and only if u is seen backward from <u>all</u> points v of K, i.e.,

$$u \in \bigcap_{v \in K} N(v) \quad .$$

Therefore, problem I is equivalent to

III. $u \in K$: $\quad (\nabla F(v) \mid v - u) \geq 0 \quad$ <u>for</u> <u>all</u> $v \in K$,

which is indeed a system of linear inequalities in u.

<u>Remark</u> 4 : While it is not immediately recognizible that the set of all solutions of problem II is convex, this property is again apparent for problem III above, as it was for the direct minimum problem I, even

if ∇F is replaced in III by <u>any</u> map A of E^n into itself. ■

4 The fixed-point formulation

Let u be a vector of K that minimizes F on K and let us assume that $\nabla F(u) \neq 0$. If we move backward from u to some vector

$$u - \rho \nabla F(u) \quad , \quad \rho > 0 ,$$

then we leave the convex K normally to a supporting hyperplane

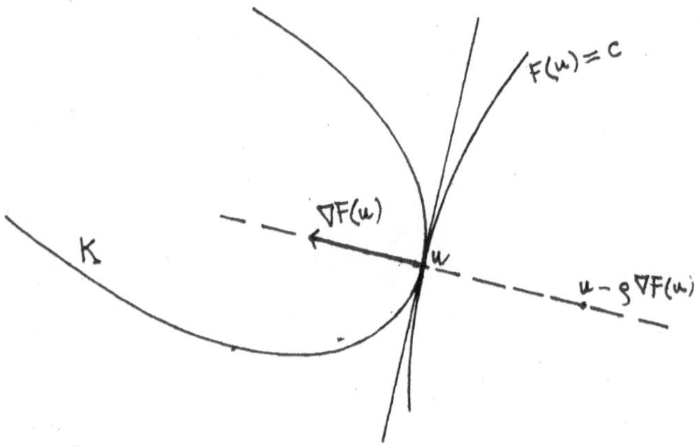

[The vector $u - \rho \nabla F(u)$ cannot stay in K, because that would contradict the fact that F attains its minimum at u on K.]

Therefore, if we now project the vector $u - \rho \nabla F(u)$ on K we fall again in the point u we started with, that is, we have

IV $\quad u = P_K(u - \rho \nabla F(u)) , \qquad \rho > 0 ,$

where P_K denotes the minimum distance projection onto K. This amounts to say that u is a fixed-point of the map

$$P_K (I - \rho \nabla F) \qquad , \quad \rho > 0 ,$$

$I \equiv$ identity map in E^n. Note that if $\nabla F(u) = 0$, then IV reduces to $u = P_K u$, a trivial consequence of $u \in K$.

Let us suppose now that u is a fixed-point of the map $P_K(I - \rho \nabla F)$ and let us show that then u minimizes F on K.

Let us distinguish two cases :

1st : $\qquad u - \rho \nabla F(u) \in K$:

then ,

$$u = P_K(u - \rho \nabla F(u)) = u - \rho \nabla F(u)$$

hence

$$\nabla F(u) = 0 \quad ,$$

which implies that u minimizes F ;

2d $\qquad u - \rho \nabla F(u) \notin K$, in particular, $\nabla F(u) \neq 0$:

then, the projection u of $u - \rho \nabla F(u)$ on K will certainly be

on the boundary of K

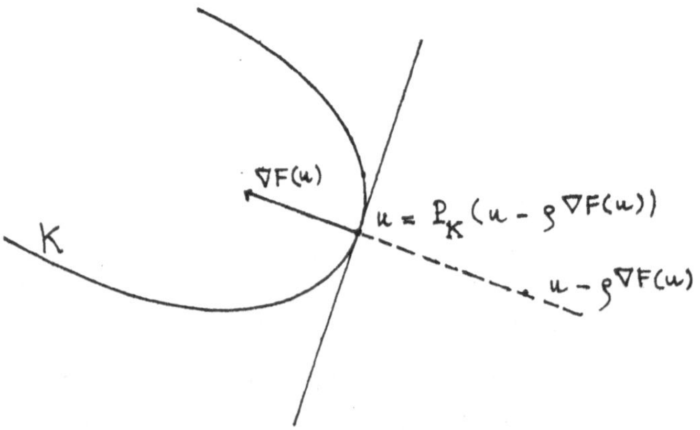

the hyperplane at u normal to ∇F(u) will be a support hyperplane of the convex set K and K will be entirely contained in the forward half space at u : again, the conclusion is that u minimizes F on K.

Remark 5 : The fixed-point formulation IV suggests an iterative algorithm for the search of a minimizing u, namely

$$\mathrm{IV}_n \qquad u_{n+1} = P_K(u_n - \varrho \nabla F(u_n)) \qquad , \quad \varrho > 0 ,$$

which can be expected to yield a convergent sequence of approximate u_n, provided the map

$$P_K(I - \varrho \nabla F)$$

is a contraction on K. Algorithm IV_n is well known in optimization theory as a "projected gradient" method. We shall come back to this point in Chapter 2. ■

5 . The epigraph formulation

All the characterizations of a minimizing vector u discussed so far require the function F be differentiable. Even if F is not differentiable, however, it is possible to give a characterization of u by means of a system of inequalities, which involve the epigraph of F, that is, the subset

$$\text{epi } F = \{ [v, \beta] \quad : \quad F(v) \leq \beta \}$$

of the product space $E^n \times \mathbb{R}$.

In fact, let $\tilde{K}$ be the intersection of epi F with the "cylinder" $K \times \mathbb{R}$ of the space $E^n \times \mathbb{R}$, i. e.,

$$\tilde{K} = \{ [v, \beta] \quad : \quad v \in K \quad , \quad F(v) \leq \beta \}$$

Clearly, u minimizes F on K if and only if $[u, F(u)]$ minimizes the function

$$\Phi([v, \beta]) = \beta$$

on the convex set $\tilde{K}$. Note that the level sets of Φ are the half spaces of $E^n \times \mathbb{R}$: $\{ [w, \gamma] \quad : \quad \gamma \leq \beta \}$, $\beta \in \mathbb{R}$.

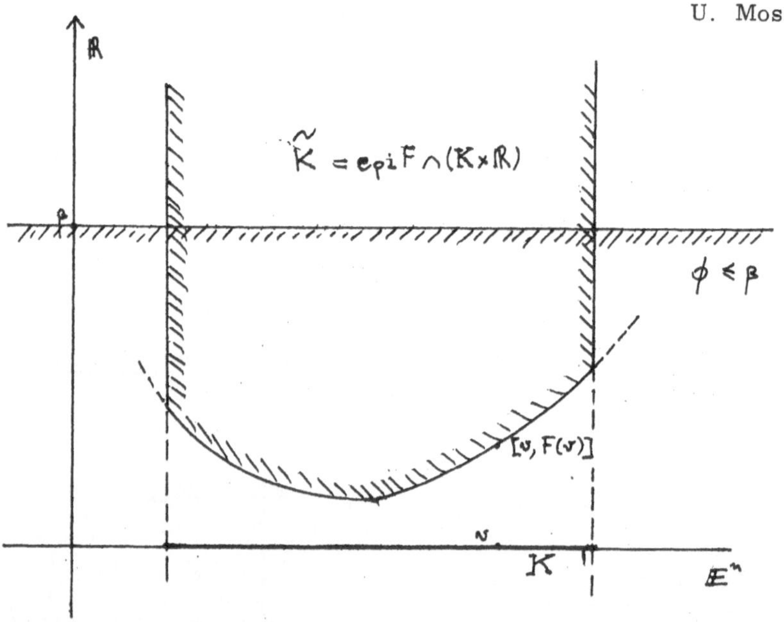

The convex function Φ in clearly differentiable on $E^n \times \mathbb{R} \simeq E^{n+1}$, with a constant gradient given by

$$\nabla \Phi([v, \gamma]) = [0, 1]$$

for all $[v, \gamma] \in E^n \times \mathbb{R}$, that is,

$$\nabla \Phi = 0 \times 1 .$$

Therefore, $\tilde{u} = [u, F(u)]$ minimizes the function Φ on the convex set $\tilde{K}$ if and only if $\tilde{u}$ is a solution of the system of inequalities

$$V \qquad \tilde{u} \in \tilde{K} : \quad (\nabla \Phi(\tilde{u}) \mid \tilde{v} - \tilde{u}) \geq 0 \qquad \forall \tilde{v} \in \tilde{K} .$$

The equivalence of our original minimum problem I with problem V above can be also checked by a direct computation (see Section 9).

6 Minimum problems in normed spaces

In this section we shall prove the equivalence of problems I-III of sections 3-5 in an infinite dimensional normed space framework.

Let X be a real normed space, X^* the dual of X, (v^*, v) the pairing between $v \in X$ and $v^* \in X^*$. Let

$$F : X \longmapsto \mathbb{R}$$

be a convex function that we shall assume to be (Gateaux) differentiable, with a differential

$$DF : X \longmapsto X^*$$

given by

$$(DF(v), w) = \frac{d}{dt} F(v + tw) \Big|_{t=0} \qquad , \quad v, w \in X$$

We have the following

PROPOSITION 1. *Let F be a differentiable convex function on X K a convex subset of X. Then, I II and III below are equivalent*

I	$u \in K$	:	$F(u) \leqslant F(v)$	$v \in K$
II	$u \in K$	:	$(DF(u),\ v - u) \geqslant 0$	$v \in K$
III	$u \in K$	:	$(DF(v),\ v - u) \geqslant 0$	$v \in K$

Furthermore, the set of all solutions u is convex (possibly empty) and, if K is closed, also closed.

Proof :

I ⇒ II : we use the definition of DF : in fact, if u minimizes F on K, then for each $v \in K$, $t = 0$ minimizes the real function

$$t \mapsto F(u + t(v - u))$$

on an interval $[0, \delta)$, $\delta > 0$, therefore we must have

$$(1) \qquad 0 \leqslant \frac{d}{dt} F(u + t(v - u)) \Big|_{t = 0^+} = (DF(u),\ v - u)$$

[Note that we have not used the convexity of F, nor even that F attaines a global minimum at u : the proof above shows indeed the necessity of II for any differentiable F having a local minimum at $u \in K$].

II ⇒ I : it is an immediate consequence of the following property of a differentiable convex F

(2) $F(v) \geqslant F(u) + (DF(u), v - u)$, $u, v \in X$

[This can be proved as follows : by the convexity of $t \mapsto F(u + tv)$, the derivative appearing in the definition of DF is the non-increasing limit of the differential quotient of F as t decreases to 0 : hence, for all $t > 0$ and all $u, w \in X$ we have

$$F(u + tw) - F(u) \geqslant t(DF(u), w),$$

which for $t = 1$ and $w = v - u$ gives (2) above .]

II $\Rightarrow$ III : it is a consequence of the following basic property of DF :

(3) $(DF(u) - DF(v), u - v) \geqslant 0$ $u, v \in X$

i.e., $DF : X \mapsto X^*$ is a <u>monotone</u> operator.

[(3) can be easily proved by writing inequality (2) above for the given vectors u and v and also for the interchanged pair v, u and then adding up the two inequalities obtained].

III $\Rightarrow$ II : it is a consequence of the continuity of

$$t \mapsto \frac{d}{dt} F(u + t w) \quad , u, w \in X ,$$

hence also of

$$t \mapsto (DF(u + t w), w) \quad , u, w \in X ,$$

which is a well known elementary property of differentiable convex functions on the real line.

In fact, we can replace the vector v in III with the vector

$$u + t(v - u)$$

that belongs to K for all $0 \leq t \leq 1$, and we get

$$(DF(u + t(v - u)), v - u) \geq 0,$$

which to the limit $t \downarrow 0$ yields II, by the continuity of $t \mapsto (DF(u + t w), w)$ at 0^+. ■

Remark 6 : It should be remark that only the right side differentiability of F at the point $u \in K$ along any vector $v - u$, $v \in K$, has been used in the proof above. The differentiability of F at u is required, instead, in the following corollary of Proposition 1. ■

Corollary of Proposition 1 : Under the assumptions of Proposition 1, a vecotr u which belongs to the interior of K minimizes the functional F on K if and only if u is a solution of the problem

$$(DF(u), w) = 0 \qquad \forall w \in X$$

Proof : See Remark 3 of Section 2. ■

Both properties (2) and (3) of DF, on which the proof of Proposition 1 has been based, characterize the differential of a convex function.

We have in fact the following lemma (see also Remark 11 of Chapter 3) :

LEMMA 1 : Let F be a differentiable function on X, with $DF: X \longmapsto X^*$. Then (i), (ii) and (iii), below are equivalent

(i) F is convex

(ii) $F(v) \geq F(u) + (DF(u), v - u) \quad , \quad u, v \in X$

(iii) DF is monotone, i. e.,

$$(DF(u) - DF(v), u - v) \geq 0 \quad , \quad u, v \in X .$$

Proof : The implications (i) $\Longrightarrow$ (ii) $\Longrightarrow$ (iii) have been shown in the proof of Proposition 1.

The proof of (iii) $\Longrightarrow$ (ii) is based on the formula

$$F(v) - F(u) = (DF(u + \bar{t}(v - u), v - u) \quad , \quad 0 < \bar{t} < 1,$$

that is obtained by integrating the function of t

$$\frac{d}{dt} F(u + t(v - u) = (DF(u + t(v - u)), v - u)$$

from 0 to 1 and then applying the mean value theorem. We find indeed

$$F(v) - F(u) = (DF(u),\ v - u) + (DF(u + \bar{t}(v - u)) - DF(u),\ v - u) \geq (DF(u),\ v - u),$$

since

$$(DF(u + \bar{t}(v - u)) - DF(u),\ \bar{t}\,(v - u)) \geq 0$$

by the monotonicity of DF.

(ii) $\Rightarrow$ (i) : Let

$$u = \lambda v_1 + (1 - \lambda)\, v_2 \qquad , \qquad v_1,\ v_2 \in X,\ 0 < \lambda < 1.$$

By (ii) we have

$$F(v_1) \geq F(u) + (DF(u),\ v_1 - u)$$

$$F(v_2) \geq F(u) + (D(u),\ v_2 - u)$$

By multiplying the first of these inequalities by λ and the latter one by $(1 - \lambda)$ and adding up ,we find

$$\lambda F(v_1) + (1 - \lambda)\, F(v_2) \geq F(u)$$

which is the convexity inequality . ∎

7 . <u>Monotone operators and variational inequalities : the linearization lemma</u>.

Any problem such as II of Proposition 1, that is

$$\text{II} \qquad u \in K : \qquad (Au, v - u) \geq 0 \qquad \forall v \in K ,$$

with K a convex subset of the normed space X and A a map of K into X^*, is called a <u>variational inequality.</u>

As we said in the introduction and the discussion up to now should have shown, variational inequalities involving monotone mappings arise naturally in connection with the minimization of a convex functional subject to convex constraints and share indeed many important properties with these problems, even if the map A is not the differential of a convex functional.

For instance, whatever the map A is, any <u>local</u> solution of II i.e., a vector u of K such that for some $\delta > 0$

$$(Au, v - u) \geq 0 \qquad \text{for all} \quad v \in K_\delta$$

where

$$K_\delta = K \cap \{ v : \| v - u \| < \delta \}$$

is actually a global solution of II . [This can be seen by replacing the vector v in the inequality above with the vector $u + \varepsilon (v - u)$ that belongs to K for $\varepsilon > 0$ small enough .]

Moreover, the solutions of variational inequalities involving any monotone mapping having some mild continuity property can still be characterized, like the solutions of convex minimum problems, by means of a system of infinite linear inequalities.

In fact, if we inspect the proof of the equivalence II $\Longleftrightarrow$ III of Proposition 1, we note that only the following properties of the map $A = DF$ have been used (see also Remark 6) :

$$A : X \longmapsto X$$

is <u>monotone</u>, i. e.,

$$(Au - Av, u - v) \geq 0 \qquad u, v \in X$$

and <u>hemicontinuous</u>, i. e.,

$$t \longmapsto (A(u + tw), w) \text{ is continuous at } 0^+. \quad u, w \in X$$

Therefore, we can state the following basic lemma, which is essentially due to G. J. Minty [1] (see also F. E. Browder [8]) :

<u>Linearization Lemma</u> : <u>Let A be a monotone and hemicontinuous map of the normed space X to its dual X^*. Then, for any convex subset K of X, problems II and III below are equivalent</u>

$$\text{II} \qquad u \in K \qquad (Au, v - u) \geq 0 \qquad v \in K$$

$$\text{III} \qquad u \in K \ : \qquad (Av, v - u) \geq 0 \qquad v \in K .$$

Corollary : The common set of all solutions u of problems II and III above is convex : it is also closed, provided K is closed.

Perhaps the most important consequence of this "linearization" of problem II is that the inequalities in III are stable under limits in u in the weak topology of X, contrary to what occurs for inequalities II, where we can take weak limits in u only if A has some compacteness property. An essential use of this property of the equivalent linearized problem III will be made in Chapter 3, when we shall deal with the existence and stability of the solutions of variational inequalities such as II .

Remark 7 : If the solution u is an interior point of K , for example if $K = X$, then II reduces to

$$u \in K : \qquad (Au, w) = 0 \qquad \forall w \in X ,$$

which is the weak form of the equation

$$Au = 0$$

(cfr. the Corollary of Proposition 1) .

Even in this case, however, the linearization lemma is meaningful, for it states the equivalence of this (non-linear) equation with the system of linear inequalities satisfied by the solution u :

$$(Av, v - u) \geq 0 \qquad \forall v \in X \qquad ■$$

Remark 8 : An interesting result due to T. Kato [1] affirms that a monotone hemicontinuous map on a convex open domain of a Banach space X is always demicontinuous , i. e., continuous from the strong topology of X to the weak* topology of X^* (this property is perhaps less surprising if we think of the case A = DF, F being a differentiable convex functional). Let us remark, however, that what is really required to have the equivalence II $\Leftrightarrow$ III is the monotonicity and hemicontinuity of A only on the set K and not on the whole of X, and hemicontinuity on a convex subset K of X is in general a weaker property than demicontinuity. For the boundedness properties of monotone operators see also F. E. Browder [3] and R. T. Rockafellar [4] . ▨

Remark 9 : If A = DF, F a differentiable convex functional on X, then for any finite set

$$v_0, v_1, \ldots, v_n$$

of vectors of X, we have

$$F(v_1) \geq F(v_0) + (DF(v_0), v_1 - v_0)$$

$$F(v_2) \geq F(v_1) + (DF(v_1), v_2 - v_1)$$

$$F(v_n) \geq F(v_{n-1}) + (DF(v_{n-1}), v_n - v_{n-1})$$

$$F(v_0) \geq F(v_n) + (DF(v_n), v_0 - v_n)$$

(see (2) of Section 6) and adding up all these inequalities we find

$$0 \geq (DF(v_0), v_1 - v_0) + (DF(v_1), v_2 - v_1) + \dots$$

$$\dots + (DF(v_{n-1}), v_n - v_0) + (DF(v_n), v_0 - v_n).$$

Therefore, not only the monotonicity condition is satisfied by the map $A = DF$, indeed a whole family of "cyclic" inequalities are also satisfied, each one corresponding to a finite subset of X : the map DF is <u>cyclically</u> <u>monotone</u> . As Rockafellar has shown, this property is the basic one occurring in the characterization of the monotone mappings which are the differential of convex functionals, see R. T. Rockafellar [1] . ■

8 . Variational inequalities and fixed-points

Let us come back now to the relation between variational inequalities and fixed-points. Even if the reduction of a general variational inequality to a fixed-point statement can be realized in any smoothly normed Banach space (see Remark 13 below), we shall assume for the sake of simplicity that our problem takes place in a Hilbert space framework.

We shall make use of the following tools, both of which depend on the specific inner product of the given Hilbert space V and not merely on the topology induced by it :

(i) the duality Riesz isomorphism J of V onto V^*

(ii) the weak characterization of the Riesz projection P_K on the convex set K.

The Riesz isomorphism J :

If V is a (real) Hilbert space, V^* its dual (v^*, v) the pairing between $v \in V$ and $v^* \in V^*$ and $(u \mid v)$ the inner product in V, then

$$J : V \longmapsto V$$

is the map defined by the identity

$$(Ju, v) = (u \mid v) \qquad , \; u, v \in V :$$

the map J is an (isometric) isomorphism of V onto V^*.

We can use the inverse of J

$$J^{-1} : V^* \longmapsto V ,$$

$$(v^*, v) = (J^{-1} v^* \mid v) \qquad v \in V, v^* \in V^* ,$$

to represent any given map

$$A : V \longmapsto V$$

by the map

$$\mathcal{A} = J^{-1}A : V \longmapsto V \qquad :$$

$$(Au, v) = (\mathcal{A}u \mid v) \qquad u, v \in V \quad .$$

Remark 10 : If $A = DF$, F being a differentiable real valued functional on V, then $\mathcal{A} = J^{-1} A$ is the gradient ∇F of F :

$$(DF(u), v) = (\nabla F(u) \mid v) \qquad u, v \in V . \blacksquare$$

The Riesz projection P_K :
If K is a convex subset of V and $z \in V$, the vector

$$u = P_K z$$

is defined to be, if it exists, the unique solution of the minimum problem

$$u \in K : \quad \| u - z \| \leq \| v - z \| \qquad \forall v \in K ,$$

where $\| w \| = (w \mid w)^{1/2}$.

It is well known and it can be elementary proved by using the parallelogram identity in V, that any vector z of V has a projection $u = P_K z$ on K, provided K is closed.

It is also clear that a vector u is the solution of the minimum problem above if and only if u is the sol of the problem

$$u \in K : \frac{1}{2} \| u - z \|^2 \leq \frac{1}{2} \| v - z \|^2 \qquad \forall v \in K .$$

Lemma 2: Let K be a convex subset of a Hilbert space V. Then, given $z \in V$, we have

$$u = P_K z$$

<u>if and only if</u>

$$u \in K \; : \; (u - z \mid v - u) > 0 \qquad \forall\, v \in K .$$

Proof: The functional

$$F(v) = \frac{1}{2} \| v - z \|^2 = \frac{1}{2} (v - z \mid v - z)$$

is differentiable on V, with

$$DF(u) = J(u - z)$$

(that is, $\nabla F = I - z$, $I \equiv$ identity of V) .

Therefore ,

$$(DF(u), v - u) = (J(u - z), v - u) = (u - z \mid v - u)$$

and the lemma follows as a special case of the equivalence I $\Longleftrightarrow$ II of Proposition 1 . ∎

The weak characterization of P_K turns out to be useful to prove that P_K does not increase distances, a property we shall make use of later on :

<u>Corollary of Lemma</u> 2 : P_K <u>is non-expansive</u>, i. e.,

$$\| P_K z_1 - P_K z_2 \| \leqslant \| z_1 - z_2 \| \qquad , z_1, z_2 \in V$$

Proof : Let us write the weak characterizations of

$$u_1 = P_K z_1 \quad \text{and} \quad u_2 = P_K z_2,$$

i. e:

$$(u_1 - z_1 \mid v - u_1) \geqslant 0 \qquad \forall \ v \in K$$

$$(u_2 - z_2 \mid v - u_2) \geqslant 0 \qquad \forall \ v \in K$$

and replace $v = u_2$ in the first inequality, $v = u_1$ in the latter one. Adding up, we then find

$$(u_1 - z_1 - u_2 + z_2 \mid u_2 - u_1) \geqslant 0 \ ,$$

that is

$$(u_1 - u_2 \mid u_1 - u_2) \leqslant (z_1 - z_2 \quad u_1 - u_2)$$

hence, by Schwarz inequality,

$$\| u_1 - u_2 \| \leqslant \| z_1 - z_2 \| \ . \qquad \blacksquare$$

Now we are ready to prove the fixed-point characterization of a variational inequality sketched in Section 4 .

PROPOSITION 2 : *Let* K *be a convex set of a Hilbert space* V *and* A *a map of* K *into* V^* . *Then,* II *and* IV *below are equiva-*

lent :

II $\quad u \in K : \quad (Au, v - u) \geq 0 \qquad \forall v \in K$

IV $\quad u = P_K [I - \rho J^{-1} A] u \quad , \quad \rho > 0$

where I is the identity map of V , J the canonical isomorphism of V^* onto V .

Remark 11 : In terms of the map $\mathcal{A} = J^{-1} A$, II and IV above can be written respectively as

$$u \in K : \qquad (\mathcal{A} u \mid v - u) \geq 0 \quad \forall v \in K$$

and

$$u = P_K (I - \rho \mathcal{A}) u \quad , \qquad \rho > 0 . \quad \blacksquare$$

Proof of Proposition 2 : It suffices to write the weak characterization of

$$u = P_K z \quad , \qquad \text{with} \qquad z = u - \rho J^{-1} Au .$$

In fact, we find

$$u \in K : \quad (u - (u - \rho J^{-1} A u) \mid v - u) \geq 0 \quad \forall v \in K$$

which is to say, since $\rho > 0$,

$$u \in K : \quad (J^{-1}Au \mid v - u) \geq 0 \qquad \forall v \in K ,$$

where

$$(J^{-1}Au \mid v - u) = \langle Au, v - u) \quad . \quad ■$$

Remark 12 : The fixed-point characterization IV of problem II is not intrinsic : if we change the inner product in V to an equivalent one, then the dual V^* of V, hence also A and problem II, does not change, whereas P_K and $\mathcal{A} = J^{-1}A$ will change. The choice of the inner product will also effect the range of values of ρ which make the map $P_K(I - \rho \mathcal{A})$ a contraction in V (see Remark 5).

We shall come back to this point in the following chapter, when we shall discuss the iterative methods of solution of problem II . ■

Remark 13 : The weak characterization of the Riesz projection P_K , hence also Proposition 2 above, holds in any Banach space, X whose norm is Gateaux differentiable (outside the origin) . In that case we can take J to be any duality mapping of X , that is ,

$$J = D \varphi(\| \cdot \|)$$

where φ is a suitable continuous increasing function of the norm (we have chosen above $\varphi(r) = \frac{1}{2} r^2$) . For more details on duality mappings see A. Beurling - A . E. Livingstone [1] , F. E. Browder [4] , [7] , E. Asplund [1] . ■

Remark 14 : Proposition 2 shows that a variational inequality can always be reduced to a fixed-point problem. The converse reduction, however, is also possible : indeed, a vector u is a fixed point of a map

$$U : K \longmapsto K$$

if and only if u is a solution of the variational inequality

$$u \in K : \quad (\mathcal{R} u \mid v - u) \geqslant 0 \qquad \forall v \in K$$

where

$$\mathcal{R} = I - U .$$

In fact, by replacing v = Uu in the inequality aboven, we obtain

$$- \| u - Uu \|^2 \geqslant 0 , \quad \text{that is} , \qquad Uu = u .$$

Let us also remark that the same conclusion can be drawn if U is a map of K into V , such that for any u ∈ K, there exists a vector v ∈ K with

$$Uu - u = \lambda (v - u) \qquad , \qquad \lambda > 0$$

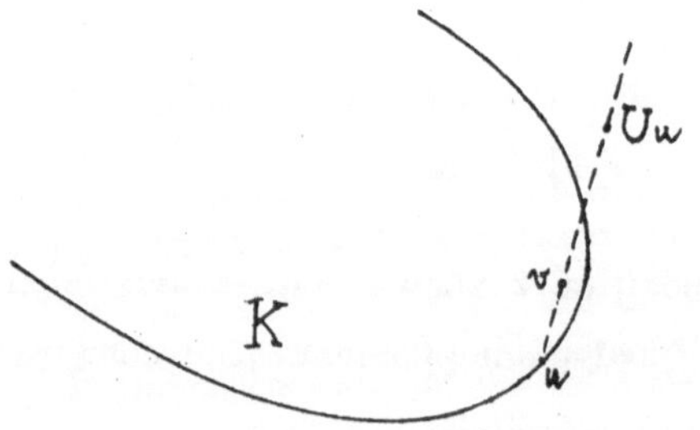

The fixed points of these so-called <u>inward</u> mappings, as well as those of <u>outward</u> mappings ($\lambda < 0$ in the condition above) have been investigated by using the variational inequality approach by F. E. Browder [11] [14] ■

<u>Remark</u> 15 Proposition 2 is essentially due to H. Brezis [2] .

It should be remarked, however, that the connection between variational inequalities and fixed points was already present in the theory as it had been previously developed by G. Stampacchia [1] and J. L. Lions - G. Stampacchia [1] for bilinear forms in Hilbert spaces. The existence results obtained by these authors were based, in fact, on iterative methods and reduction to a fixed point statement for a suitable contractive mapping. We shall come back on this case with more details in the following Chapter 2 . ■

9. Minimization of non-differentiable functionals and mixed variational inequalities

Let us go back to the problem of minimizing a convex functional F on a convex subset K of a normed space X , dropping now the assumption, made in Section 6, that F is differentiable. Indeed, many variational problems arising from the applications, as we shall see later on, lead to the minimization of functionals which appear to be the <u>sum</u> of a differentiable F and a non-differentiable G .

By allowing the functional G to take the value $+\infty$, we can also assume that the <u>indicator function</u> ρ_K of the constraint set K has been preliminarily incorporated into the nondifferentiable component of the functional at hand.

Let us recall that ρ_K is the functional on X given by

$$\rho_K(v) = 0 \quad \text{if} \quad v \in K, \quad = +\infty \text{ if } \quad v \notin K.$$

Clearly, minimizing $F + G$ on X is the same thing as minimizing $F + G$ on the subset of X,

$$K \equiv \text{dom } G \equiv \{ v : G(v) < +\infty \}$$

(dom G is also called the effective domain of $G : X \mapsto (-\infty, +\infty)$).

Proposition 1 of Section 6 can then be generalized as follows

POPOSITION 3 : Let $F : X \mapsto R$ be a differentiable convex functional and $G : X \mapsto (-\infty, +\infty]$ be convex, $G \not\equiv +\infty$. Then, I' II' below are equivalent

I' $\quad u \in X : \quad F(u) + G(u) \leqslant F(v) + G(v) \qquad \forall v \in X$

II' $\quad u \in X : \quad (DF(u), v - u) \geqslant G(u) - G(v) \qquad \forall v \in X$

Proof: (see also M. Sibony [1]) If I' holds, for every v of X and all t, $0 < t < 1$, we have

$$F(u) + G(u) \leqslant F(u + t(v - u) + G(u + t(v - u) \leqslant F(u + t(v - u) + (1 - t)G(u) + tG(v),$$

where the convexity of G has been used. Since $F(u) < +\infty$, we deduce from the inequality above

$$t^{-1} [F(u + t(v - u)) - F(u)] \geqslant G(u) - G(v)$$

that gives II' as $t \downarrow 0$, by the differentiability of F

To prove that II' in turn implies I', it suffices to use the inequality

$$F(v) \geq F(u) + (DF(u), v - u) \qquad v \in X$$

which is a consequence of the convexity of F . ∎

Remark 16 : As the proof above shows, a vector u minimizing F + G always satisfies the inequality II' even if the differentiable functional G is not convex . ∎

Another way of looking at the inequality II' above is to regard it as a unification of the direct and weak formulations of minimum problems.

In fact, while II' obviously reduces to the problem

$$u \in X : \qquad G(u) \leq G(v) \qquad \forall v \in X$$

when $F \equiv 0$, on the other hand, it is easy to verify that II' is equivalent to the variational inequality

$$u \in K : \quad (DF(u), v - u) \geq 0 \qquad \forall v \in K$$

when G is taken to be the indicator function of the set K .

Similarly, we can generalize both the direct minimum problems and the variational inequalities discussed so far, by introducing inequalities of the form

$$\text{II}'' \qquad u \in X : \quad (Au, v - u) \geq F(u) - F(v) \qquad \forall\, v \in X ,$$

with $A : X \mapsto X^*$ and $F : X \mapsto (-\infty, +\infty]$

It turns out, however, that such "mixed" variational inequalities only apparently are more general then the original ones.

In fact, by making use of the epigraph formulation discussed in Section 5, we can equivalently write problem II'' above as a variational inequality as those considered so far. In fact, let us consider now the product space

$$\tilde{X} = X \times \mathbb{R} ,$$

and the inequality :

$$\tilde{\text{II}} \qquad \tilde{u} \in \tilde{K} : \quad (\tilde{A}\tilde{u}, \tilde{v} - \tilde{u}) \geq 0 \qquad \forall\, \tilde{v} \in \tilde{K}$$

where $\tilde{A}$ is the map $A \times 1$ of $\tilde{X}$ into its dual $\tilde{X}^* \simeq X^* \times \mathbb{R}$, i.e.,

$$\tilde{A}([v, \beta]) = [Av, 1] , \qquad v \in X, \quad \beta \in \mathbb{R},$$

and $\tilde{K}$ is the epigraph of F :

$$\tilde{K} = \text{epi } F = \{ [v, \beta] \in X \times \mathbb{R} : F(v) \leq \beta \} .$$

The following lemma holds :

<u>Lemma</u> 3 : <u>Let</u> $A : X \mapsto X^*$, $F : X \mapsto (-\infty, +\infty]$,

$F \not\equiv +\infty$. Then, a vector u of X is a solution of problem II'' above and $F(u) = \alpha$, $\alpha \in \mathbb{R}$, if and only if the vector $\tilde{u} = [u, \alpha]$ of $\tilde{X}$ is a solution of problem $\tilde{II}$.

Proof : If u is a solution of II'', then for every $\beta \geq F(v)$ we have

$$0 \leq (Au, v-u) + F(v) - F(u) =$$

$$= (Au, v-u) + 1 \cdot (\beta - F(u)) =$$

$$= (\tilde{A}\tilde{u}, \tilde{v} - \tilde{u})$$

where $\tilde{u} = [u, F(u)]$, $\tilde{v} = [v, \beta]$, therefore $\tilde{II}$ holds.

Conversely, if $\tilde{u} = [u, \alpha]$ is a solution of $\tilde{II}$, then $\alpha \geq F(u)$ and for all $\tilde{v} = [v, \beta]$ with $\beta \geq F(v)$ [note that when $F(v) = +\infty$ the inequality in II'' is trivially satisfied], we have

$$0 \leq (\tilde{A}\tilde{u}, \tilde{v} - \tilde{u}) = (Au, v-u) + \beta - \alpha .$$

Therefore, by taking $v = u$ and $\beta = F(u)$, we find, in particular, that $\alpha \leq \beta = F(u)$, hence

$$\alpha = F(u)$$

and, moreover, for all $\beta \geq F(v)$

$$0 \leqslant (Au, v - u) + \beta - F(u)$$

which yields II' when $\beta = F(v)$. ■

Remark 17 : If we now assume that the vector $\tilde{u} = [u, \alpha]$ is a solution of the analog of problem $\tilde{II}$ with $\tilde{K}$ replaced by the intersection $\tilde{K}'$ od $\tilde{K}$ with a "strip" X x I, were I = [a, b] with $a \leqslant \inf F$,

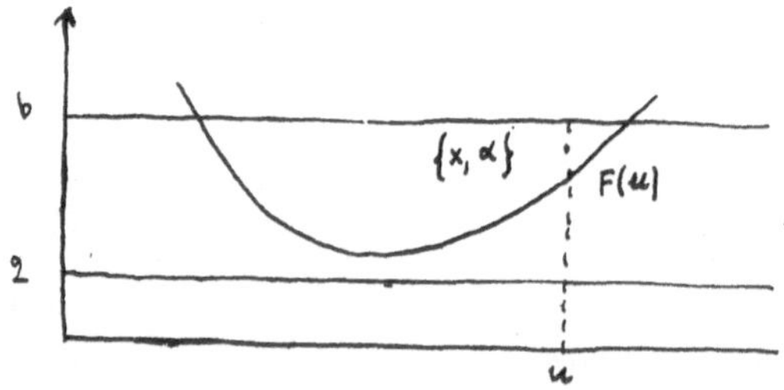

then we can still conclude that u is a solution of the problem

$$u \in X : (Au, v - u) \geqslant F(u) - F(v) \quad \forall v \in X, \qquad F(v) \leqslant b$$

and $\alpha = F(u)$.

We shall make use of this fact later. ■

Remark 18 : A different approach to the minimization of non differentiable convex F consists in making the subdifferential ∂F of F take the role of the differential DF . Let us recall that ∂F is the, in general multivalued, mapping of X into 2^{X^*}, which associates each u of X with the set (possibly empty) of all subgradients u^* of F at u, i.e., of all $u^* \in X^*$ such that $v \mapsto F(u) + (u^*, v - u)$ defines a support hyperplane of F at the point u (not necessarily a tangent one). In other words, for each u of X ,

$$\partial F(u) = \{ u^* \in X^* : \; F(v) \geqslant F(u) + (u^*, v - u) \quad \forall v \in X \}$$

It should be noted, indeed, that all statements about the vector $u = DF(u)$ we have made so far, involve u only as a subgradient of F. This approach yields naturally to variational inequalities involving <u>multivalued</u> <u>monotone</u> mappings. We refer to F. E. Browder [14] R. T. Rockafellar [6] .

* * *

As general reference on the theory of convex functions, let us only mention here R. T. Rockafellar [7] , J. Stoer and C. Witzgall [1] J. J. Moreau [1] , A. Ioffe - V. Tikhomirov [1] .

Monotonicity properties of operators in Hilbert or Banach spaces have been investigated M. M. Vainberg and R. I. Kachurovskii 1 R. I. Kachurovskii [1] [2] , E. H. Zarantonello [1] , G. J. Minty [1] , [2] , [3] , F. E. Browder see ref. quoted in F. E. B. [15], T. Kato [1] R. T. Rockafellar [3] [4] and others. More references and a survey of the theory and its applications can be found in R. I. Kachurovskii [3] and F. E. Browder loc. cit.

Specific references to variational inequalities will be given in the following chapters.

CHAPTER 2

Some typical problems and existence theorems

U. Mosco

CHAPTER 2

The existence results and the methods of approximation of the solutions of variational inequalities such as

$$\text{II} \qquad u \in K \; : \quad (Au, v - u) \geqslant 0 \qquad \forall\, v \in K \;,$$

where K is a convex subset of a normed space X and A a monotone map of K into X^*, are essentially based on the characterizations of problem II we discussed in the preceding chapter.

In fact, the three main ideas underlying this existence and approximation theory can be summarized as follows :

(i) Reduction of II to a fixed-point problem and application of a fixed-point theorem such as Brouwer's or Schauder's theorem or, more constructively, the contraction principle which also yields an iterative algorithm for the approximation of the solution ;

(ii) Reduction of II to a direct minimum problem (when the map A is the differential of some convex functional F), what makes it possible to apply the classical existence theorems of the calculus of variations. The solution can be evaluated by combining a finite-dimensional approximation of Ritz type together with some method of finite-dimensional convex optimization ;

(iii) Preliminary restriction of II to finite-dimensional subspaces of X, where a solution can be found by applying any one of the foregoing methods, and then linearization of the problem to get a solution in the whole space. The constructive aspect of this approach comes out again from a combination of a finite-dimensional discretization of Ritz-Galerkin type and iterative algorithms or methods of convex optimization for the evaluation of the solution of the finite-dimensional approximate problem.

In the present chapter we shall present some of the existence and approximation results that can be found along the lines mentioned in (i) and (ii) above, while a more general existence theorem, based on the Linearization Lemma of Chapter 1, and a description of the discretization procedures, will be postponed to the following Chapter 3 .

Moreover, as a motivation to all problems discussed so far in an abstract framework, we shall describe below some typical examples of minimum problems and variational inequalities involving integral functionals and partial differential operators. Most of these problems arise in the mathematical description of the equilibrium of a physical system subject to unilateral constraints.

As everywhere else in these lectures, we shall confine ourselves to problems of elliptic stationary type and we refer to the lectures of J. L. Lions and R. Glowinski at this Course for all that concerns variational inequalities of evolutive type.

1 . Some variational inequalities

Everywhere in this section we shall denote by Ω a bounded open subset of the n-dimensional euclidean space E^n, by Γ its boundary.

We shall begin our list of examples with a quite classical probelm, that does'nt involve unilateral constraints.

Example 1 : The Dirichlet problem :

$$(1) \qquad \begin{cases} -\Delta u = f & \text{in } \Omega \\ u = 0 & \text{on } \Gamma \end{cases}$$

where $-\Delta$ is the Laplace operator and f is a given function on Ω, say, $f \in L^2(\Omega)$.

The variational (or weak) solution of (1) is the function u that minimizes the energy integral

$$\frac{1}{2} \sum_{i=1}^{n} \int_{\Omega} \left| \frac{\partial v}{\partial x_i} \right|^2 dx - \int_{\Omega} f \, v \, dx$$

over the appropriate Sobolev space $H_0^1(\Omega)$: this is, indeed, the classical Dirichlet principle .

[The space $H_0^1(\Omega)$ is made of all functions $f \in L^2(\Omega)$ whose distribution derivatives $\frac{\partial v}{\partial x_i}$, $i = 1, \dots, n$ still belongs to $L^2(\Omega)$. and whose trace on the boundary Γ of Ω vanishes . With the norm ·

$$\| v \|_{H_0^1(\Omega)} = \left(\| v \|^2_{L^2(\Omega)} + \sum_{i=1}^{n} \left\| \frac{\partial v}{\partial x_i} \right\|^2_{L^2(\Omega)} \right)^{1/2} ,$$

$H_0^1(\Omega)$ is a reflexive Banach space, whose dual,

$$H^{-1}(\Omega) ,$$

can be identified with all distributions T on Ω that can be (non uniquely) written as

$$T = g_0 - \sum_{i=1}^{n} \frac{\partial g_i}{\partial x_i} , \qquad g_0, g_i \in L^2(\Omega) ,$$

the duality pairing between $v \in H_0^1(\Omega)$ and $T \in H^{-1}(\Omega)$ being given by

$$(T, v) = \int_\Omega g_0 v \, dx + \sum_{i=1}^{n} \int_\Omega \frac{\partial g_i}{\partial x_i} \cdot \frac{\partial v}{\partial x_i} \, dx . \quad]$$

The functional

$$F_0(v) = \frac{1}{2} \sum_{i=1}^{n} \int_\Omega \left| \frac{\partial v}{\partial x_i} \right|^2 dx$$

is differentiable on $H_0^1(\Omega)$, with differential

$$DF_0 \equiv - \triangle : \quad H_0'(\Omega) \longmapsto H^{-1}(\Omega)$$

given by

$$(- \triangle u, v) = \frac{d}{dt} F_0 (u + tv) \Big|_{t=0} = \sum_{i=1}^{n} \int_{\Omega} \frac{\partial u}{\partial x_i} \frac{\partial v}{\partial x_i} dx ,$$

$$, \quad u, \ v \in H_0'(\Omega)$$

The identity above, indeed, provides the variational definition of the Laplace operator

$$- \triangle = - \sum_{i=1}^{n} \frac{\partial^2}{\partial x_i^2}$$

and the bilinear form at the right hand

$$(2) \qquad a(u, v) = \sum_{i=1}^{n} \int_{\Omega} \frac{\partial u}{\partial x_i} \frac{\partial v}{\partial x_i} dx$$

is called the Dirichlet form .

The variational solution of the Dirichlet problem can then be characterized as the solution of the problem

$$u \in H_0^1(\Omega) : \qquad a(u, v) = (f, v) \qquad \forall \ v \in H_0^1(\Omega)$$

as it follows, for instance, from the Corollary of Proposition 1 of Chapter 1 .

As for the existence of u, it suffices to show that the energy integral above attains its minimum in the space $H_0^1(\Omega)$, what can be done, for instance, by applying the general theorem on the existence of minima we shall give in Section 4 of the present Chapter.

Remark 1 : To replace the Laplace operator $-\Delta$, in the problem above, by any 2^d order elliptic partial differential operator of the form

$$L(v) = \sum_{i=1}^{n} \frac{\partial}{\partial x_i} \left(a_{ij}(x) \frac{\partial v}{\partial x_j} \right) ,$$

where the $a_{ij}(x)$ are bounded measurable functions on Ω satisfying the ellipticity condition

$$\sum_{i,j=1}^{n} a_{ij}(x) \xi_i \xi_j \geq c \, |\xi|^2 , \quad c > 0 ,$$

$$\forall \xi = (\xi_1, \dots, \xi_n) \in \mathbb{R}^n$$

amounts only to replace the Dirichlet form a(u, v) in problem (2) with the form

$$a(u, v) = \sum_{i,j=1}^{n} \int_\Omega a_{ij}(x) \frac{\partial u}{\partial x_i} \frac{\partial v}{\partial x_j} \, dx$$

However, this generalized form is not the differential of the functional

$$\frac{1}{2} a(v, v) = \frac{1}{2} \sum_{i,j} \int_{\Omega} a_{ij}(x) \frac{\partial v}{\partial x_i} \frac{\partial v}{\partial x_j} dx ,$$

unless it is symmetric, that is, the coefficients a_{ij} satisfies

$$a_{ij}(x) = a_{ji}(x) \qquad \text{a.e.} \qquad \forall \; i, j = 1, \ldots, n .$$

The proof of the existence of the solution of this generalized Dirichlet problem cannot be obtained, as before, by using the direct variational formulation. We must appeal, indeed, to a well known theorem of Lax and Milgram that will be recalled later. ∎

The existence theorems for variational inequalities we are going to discuss in our lectures can be seen, and in fact so they were first obtained, as a generalization of the Lax-Milgram theorem to problems involving unilateral constraints .

The simplest example of problems of this type is the following

Example 2 : The capacity problem :

Given a compact subset E of Ω, we look for the function u that minimizes the Dirichlet integral $F_0(v)$ over the cone of all $v \in H_0^1(\Omega)$, such that

$$v \geq 1 \text{ on } E \qquad \text{in the sense of } H_0^1(\Omega) ,$$

[We say that $v \geq 1$ on E in the sense of $H_0^1(\Omega)$ if v is the limit in the norm of $H_0^1(\Omega)$ of a sequence of smooth functions which are $\geqslant 1$ on E]

Therefore, our problem now is

$$(3) \quad \begin{cases} u \in H_0^1(\Omega), \quad u \geqslant 1 \quad \text{on } E : \\ F_0(u) \leqslant F_0(v) \quad \forall v \in H_0^1(\Omega), \quad v \geqslant 1 \quad \text{on } E \end{cases}$$

The solution u is called the equilibrium potential and the value of the minimum is the capacity of the set E in Ω.

By applying Proposition 1 of Chapter 1, now we find that the solution u of (3) can be characterized by means of the variational inequality

$$(4) \quad \begin{cases} u \in H_0^1(\Omega), \quad u \geqslant 1 \quad \text{on } E : \\ a(u, v - u) \geqslant 0 \quad \forall v \in H_0^1(\Omega), \quad v \geqslant 1 \quad \text{on } E \end{cases}$$

where a(u, v) is the Dirichlet form (2).

Remark 2. The variational inequality (4) is the only possible formulation of this capacity problem whenever a(u, v) is the generalized Dirichlet form associated with a non-symmetric elliptic operator L as in Example 1. In this case, (4) has to be taken as the definition of the equilibrium potential u on E relative to the operator L in Ω.

The capacity theory for non-symmetric second order elliptic p.d.o. was started by G. Stampacchia [1] and his variational inequality approach to this problem was extended to other variational problems with unilateral constraints, both of stationary and evolutive type, by J. L. Lions and G. Stampacchia [1]. Many other problems of this type arising in physics and engineering have been (investigated since then) by many authors in the light of the theory of variational inequalities. The main reference in this regard is the recent book of C. Duvaut and J. L. Lions [1].

Let us also mention that many unilateral problems of mechanics and hydrodynamics have been also investigated by J. J. Moreau [3], by using the methods of convex analysis. ∎

Example 3 : The "obstacle" problem.

We now want to minimize the Dirichlet integral $F_0(v)$ over the cone of all functions v of $H_0^1(\Omega)$ which are $\geq$ a. e. of a preassigned function ψ in Ω. The function ψ will be called the obstacle and we assume that ψ is such that the cone just defined is not empty. For instance, we may have $\psi \in H^1(\Omega)$ [that is, $\psi \in L^2(\Omega)$ and all distribution derivatives ψ_{x_i} of ψ also belonging to $L^2(\Omega)$], the trace of ψ on Γ being ≤ 0 a. e.

The minimizing u is the solution of the variational inequality

$$(5) \quad \begin{cases} u \in H_0^1(\Omega) , \quad u \geq \psi \quad \text{a. e. in } \Omega : \\ a(u, v - u) \geq 0 \quad \forall v \in H_0^1(\Omega), \quad v \geq \psi \quad \text{a. e. in } \Omega \end{cases}$$

It can be shown that if I is the closed subset of Ω where, formally, $u = \psi$, then the solution u satisfies the conditions

$$\begin{array}{llll} u \geq \psi & \text{a. e.} & \text{and} & -\Delta u \geq 0 \text{ in } \Omega , \\ u = \psi & \text{on I} & \text{and} & -\Delta u = 0 \text{ in } \Omega - I \end{array}$$

Thus, u, is super-harmonic all over Ω and harmonic outside the set I where it "touches" the obstacle.

A similar problem which includes the capacity problem, consists in minimizing F_0 over all functions v which are $\geqslant \psi$ only on a given compact subset E of Ω (and now $v \geqslant \psi$ on E has an analogue meaning than the condition $v \geqslant 1$ on E in Example 2). If E is a (n-1)-dimensional manifold in Ω, the problem at hand may be called a "thin obstacle" problem.

Problems of this type have been first considered by J. L. Lions and G. Stampacchia [1]. The regularity of the solution has been also investigated by many authors. See H. Lewy and G. Stampacchia [1], [2], [3], H. Brezis and G. Stampacchia [1], H. Lewy [1], [2], H. Brezis [5], G. Stampacchia [2], [3], [4]. An application to hydraulics has been recently given by C. Baiocchi [1].

Example 4 : The "boundary obstacle" problem.

The problem now consists in minimizing the functional

$$F(v) = \frac{1}{2} \sum_{i=1}^{n} \int_{\Omega} \left| \frac{\partial v}{\partial x_i} \right|^2 dx - \int_{\Omega} f v \, dx$$

where F is a given function on Ω, say, $f \in L^2(\Omega)$, on the cone of all $v \in \overset{\cdot}{H}{}^1(\Omega)$ such that

$$v \geqslant h \quad \text{a.e.} \quad \text{on } \Gamma ,$$

where h is a preassigned function on Γ.

It can be seen that the associated variational inequality satisfied by the solution u corresponds to a boundary value problem for the Laplacian $-\Delta$, with unilateral constraints on the boundary Γ. Formally, this problem can be stated as follows

$$(6)\quad \begin{cases} -\Delta u = f & \text{in } \Omega \\ u - h \geq 0\,, \quad \frac{\partial u}{\partial n} \geq 0\,, \quad (u-h)\cdot\frac{\partial u}{\partial n} = 0 & \text{a.e. on } \Gamma\,. \end{cases}$$

A detailed discussion of this example can be found in C. Duvaut and J. L. Lions [1] . Let us only remark here, following J. L. Lions , R. Glowinski and R. Tremoliers [1] , that the conditions (6) can be interpreted as describing the stationary equilibrium of a fluid in a region Ω surrounded by a membrane Γ that allows the fluid to come in and prevents it to leave Ω .

If u(x) is the pressure of the fluid inside Ω and h(x) is the external pressure applied on the boundary Γ, when the fluid comes in, then $\frac{\partial u}{\partial n} \geq 0$, while $u = h$ on Γ ; on the other hand, if $u > 0$, then the fluid is pushed out, however Γ forbids the outcome, hence $\frac{\partial u}{\partial n} = 0$. The regularity of the solution u has been studied, in particular, by H. Brezis-G. Stampacchia [1] , H. Brezis [5], H. de Veiga [1] , [2] .

Example 5 : <u>A problem in nonlinear elasticity</u>

The energy integral

$$F(v) = \frac{1}{2} \sum_{i=1}^{n} \int_{\Omega} \left| \frac{\partial v}{\partial x_i} \right|^2 dx - \int_{\Omega} f\, v\, dx$$

now has to be minimized on the convex set of all $v \in H_0^1(\Omega)$ such that

$$(17)\qquad |\operatorname{grad} v| \leq 1 \qquad \text{a.e. in } \Omega\,.$$

This is a problem occurring in the elastic-plastic torsion of a bar and it has been studied by many authors, see in particular B. D. Annin [1] , H. Lanchon and C. Duvaut [1] , H. Lanchon [1] , T. W. Ting [1] and C. Duvaut - J. L. Lions, loc. cit., where more information can be found.

The variational inequality characterizing the solution u can be formally interpreted as follows : There is a "plasticity region" Ω_0 in Ω where

$$|\operatorname{grad} u| = 1 \quad ;$$

outside Ω_0, that is, in the region $\Omega - \Omega_0$ where

$$|\operatorname{grad} u| < 1 \quad ,$$

the function u satisfies the equation

$$- \Delta u = f \quad ;$$

moreover, u and its derivatives $\frac{\partial u}{\partial x_i}$ satisfy certain matching conditions at the interface between Ω_0 and $\Omega - \Omega_0$. Like the obstacle problem, this too is a free boundary problem.

Let us also mention that the present problem can be equivalently stated in the form of a "two-obstacles" problem, that is, condition (17) above can be replaced by a condition of the type

$$\psi_1 \leqslant v \leqslant \psi_2 \qquad \text{a.e. in } \Omega \quad ,$$

where ψ_1 and ψ_2 are two suitable functions. We refer to T. W.

Ting, loc. cit. and H. Brezis-M. Sibony [2] for more details on this point

The regularity of the solution has been studied by H. Brezis-G. Stampacchia, loc. cit.; see also H. Brezis [5]. For the numerical solution of this problem see R. Glowinski 3 , J.F. Bourgat [1] , M. Nedelec [1] , M. Goursat [1] , M. Sibony [2] .

Example 6 : A Bingham's fluid

In its direct variational formulation, the problem consists in minimizing the non differentiable functional

$$\frac{1}{2} \sum_{i=1}^{n} \int_{\Omega} \left| \frac{\partial v}{\partial x_i} \right|^2 dx - \int_{\Omega} fv \, dx + g \int_{\Omega} | \operatorname{grad} v | \, dx , \quad g > 0 ,$$

over the space $H_0^1(\Omega)$

The functional above is the sum of the same energy integral $F(v)$ occurring in Example 5, which is obviously differentiable on $H_0^1(\Omega)$, and the non differentiable term

$$G(v) = g \int_{\Omega} | \operatorname{grad} v | \, dx .$$

The minimizing u is thus characterized by the mixed variational inequality

$$u \in H_0^1(\Omega) : a(u, v - u) \geq G(u) - G(v) \qquad \forall \, v \in H_0^1(\Omega)$$

where

For the numerical solution see R. Glowinski [1] , M. Goursat [1] .

$$a(u, v) = \sum_{i=1}^{n} \int_{\Omega} \frac{\partial u}{\partial x_i} \frac{\partial v}{\partial x_i} dx - \int_{\Omega} f v dx ,$$

cfr. Proposition 3 of Chapter 1 .

The physical motivation of this problem as well a discussion of the properties of the solution u can be found in C. Duvaut - J. L. Lions, loc. cit..

Example 7 : A problem of elasticity with friction on the boundary.

Another problem involving a non differentiable functional is the minimization of

$$\frac{1}{2} \sum_{i=1}^{n} \int_{\Omega} \left| \frac{\partial v}{\partial x_i} \right|^2 dx + \int_{\Omega} |v|^2 dx - \int_{\Omega} f v dx + c \int_{\Gamma} |v| \, d\Gamma$$

over the whole space $H^1(\Omega)$. The non differentiable term is the boundary integral

$$G(v) = c \int_{\Gamma} |v| \, d\Gamma$$

and the solution u is characterized, as in Example 6, by the mixed variational inequality

$$u \in H^1(\Omega) : \quad a(u, v - u) \geq G(u) - G(v) \quad \forall \, v \in H^1(\Omega) ,$$

where now we have

$$a(u, v) = \sum_{i=1}^{n} \int_{\Omega} \frac{\partial u}{\partial x_i}\frac{\partial v}{\partial x_i}\, dx + \int_{\Omega} uv\,dx - \int_{\Omega} fv\,dx \quad .$$

Problems of this type occur in the theory of elastic bodies subjected to unilateral boundary constraints. Once again we refer to C. Duvaut - J. L. Lions, loc. cit. or J. L. Lions, R. Glowinski and R. Tremoliers, loc. cit.

Example 8 : Inequalities involving non linear generalization of the Laplace operator .

A generalization of all problems considered so far, which is quite natural from a mathematical point of view though of no direct physical interest, consists in replacing the Dirichlet integral with the functional

$$F(v) = \frac{1}{p} \sum_{i=1}^{n} \int_{\Omega} \left| \frac{\partial v}{\partial x_i} \right|^{p} dx \quad .$$

For every $p \geqslant 2$ this is indeed a differentiable convex functional on the Sobolev space

$$H^{1,p}(\Omega) = \left\{ v \in L^p(\Omega) \; , \; v_{x_i} \in L^p(\Omega) \quad , \; i = 1, \dots, n \right\}$$

and its differential is the monotone operator

$$(18) \qquad Au = - \sum_{i=1}^{n} \int_{\Omega} \frac{\partial}{\partial x_i} \left(\left| \frac{\partial u}{\partial x_i} \right|^{p-2} \frac{\partial u}{\partial x_i} \right)$$

from $H^{1,p}(\Omega)$ to its dual, associated with the form

$$a(u, v) = \sum_{i=1}^{n} \int_\Omega \left| \frac{\partial u}{\partial x_i} \right|^{p-2} \frac{\partial u}{\partial x_i} \frac{\partial v}{\partial x_i} \, dx \quad .$$

[We have in fact

$$(DF(u), v) = \frac{d}{dt} \left[\frac{1}{p} \sum_{i=1}^{n} \int_\Omega | u_{x_i} + t v_{x_i} |^p \, dx \right] \Big|_{t=0} =$$

$$= \sum_{i=1}^{n} \int_\Omega | u_{x_i} + t v_{x_i} |^{p-2} (u_{x_i} + t v_{x_i}) v_{x_i} \, dx \Big|_{t=0} =$$

$$= \sum_{i=1}^{n} \int_\Omega | u_{x_i} |^{p-2} u_{x_i} v_{x_i} \, dx \quad .]$$

Let us remark that another natural family of convex functionals that generalize the Dirichlet integral is given by

$$F(v) = \frac{1}{p} \int_\Omega | \operatorname{grad} v |^p \, dx \quad ,$$

whose differential is now the operator

$$Av = - \operatorname{div} (| \operatorname{grad} v |^{p-2} \operatorname{grad} v)$$

Like the opetarot (18), this operator obviously reduces to $-\Delta$ when $p = 2$. Therefore, they can both be considered as natural non linear generalization of the Laplace operator.

It should be also noted that these operators are all duality mappings of the spaces $H^{1,p}(\Omega)$ suitable normed, cfr. Section 8 of Chapter 1, Remark 12 .

Example 9 : Inequalities involving more general non linear second order elliptic, partial differential operators.

Let us consider the form

$$a(u, v) = \int_\Omega a_0(x; u, u_x)\, v\, dx + \sum_{i=1}^{n} \int_\Omega a_i(x; u, u_x) v_{x_i}\, dx$$

$u_x \equiv (u_{x_1}, \ldots, u_{x_n})$, where the functions

$$a_i(x; \xi) = a_i(x; \xi_0, \xi_1, \ldots, \xi_n), \quad x \in \Omega, \quad \xi \in \mathbb{R}^{n+1}, \; i = 0, 1, \ldots, n$$

are measurable in x for fixed ξ and continuous in ξ for x fixed a.e. in Ω.

If the functions $a_i(x; \xi)$ are of polynomial growth at ∞ in ξ, that is

$$(19) \qquad |a_i(x; \xi)| \leqslant c(1 + |\xi|^{p-1}), \quad x \in \Omega, \quad \xi \in \mathbb{R}^{n+1}, i = 1, \ldots n$$

for some p with $1 < p < +\infty$, $c > 0$, then it is easy to verify that for every function u of the Sobolev space $H^{1,p}(\Omega)$ (see the Example 8) we have

$$a_i(x; u, u_x) \in L^{p'}(\Omega) \quad , \quad p' = p/(p-1)$$

and an estimate such as

$$|a(u, v)| \leq \gamma(\|u\|_{1,p}) \|v\|_{1,p} , \quad u, v \in H^{1,p}(\Omega)$$

holds, with $\gamma(r)$ a continuous function of $r > 0$, $\|\cdot\|_{1,p}$ being the norm in $H^{1,p}(\Omega)$.

[By taking the Sobolev imbedding theorem into account, the growth condition (19) could be obviously weakened, see the ref. quoted below .]

Therefore, the identity

$$(Au, v) = a(u, v) \quad , \quad u, v \in H^{1,p}(\Omega)$$

defines a map A from $H^{1,p}(\Omega)$ to its dual, formally

$$Au = a_0(x; u, u_x) - \sum_{i=1}^{n} \frac{\partial}{\partial x_i} a_i(x; u, u_x) .$$

Clearly, this map A is monotone, that is

$$(Au - Av, u - v) = a(u, u - v) - a(v, u - v) \geq 0 ,$$

provided the functions $a_i(x; \xi)$, satisfy a weak ellipticity conditions of type

$$(20) \quad \sum_{i=0}^{n} [a_i(x; \xi) - a_i(x; \xi')](\xi_i - \xi_i') \geq 0 , \quad \xi, \xi' \in \mathbb{R}^{n+1}.$$

Let us notice that if there exists a function $\Phi(x; \xi)$ such that

$$a_i(x; \xi) = \frac{\partial}{\partial_i} \Phi(x; \xi) \quad , \qquad i = 0, 1, \ldots, n,$$

then A is the differential of the multiple integral

(21) $$F(u) = \int_\Omega \Phi(x; u, u_x)\, dx \qquad ,$$

since then we have

$$a(u, v) = \frac{d}{dt} F(u + tv) \Big|_{t=0} = (DF(u), v) \qquad .$$

The monotonicity condition (20) is then equivalent to the convexity of $\Phi(x; \xi)$ in ξ (see also the Remark below) .

Thus, variational inequalities involving differential operators as the operator A above arise, for instance, whenever we minimize an integral functional like (21) subject to a convex set of constraints.

There is an extensive literature on the partial differential operators of type described above, and of higher order too, and on the related boundary value problems. We only refer here to the papers of F. E. Browder [1] , [2] , to J. Leray and J. L. Lions [1] and to P. H. Hartman and G. Stampacchia [1] , where variational problems with unilateral constraints are studied in detail. Surveys of the theory and its applications can be found in F. E. Browder [6] [7] J. L. Lions [1], R. I. Kachurovski [3] .

Remark 3 : The application of the theory of monotone operators to the boundary value problems for partial differential operators of elliptic type has brought to a natural generalization of the theory. In fact, it is more convenient in many applications concerning an operator such as the A considered above, to require the functions $a_i(x; u, u_{x_1}, \dots, u_{x_n})$ be monotone in the arguments corresponding to the highest derivatives only - $u_{x_1}, \dots, u_{x_n}$ in our example above - and handle the lower order terms - u in the example - by using a compactness argument. When $A = DF$, F being the integral functional (21), this corresponds to restrict the convexity assumption on the highest order derivatives appearing in the integrand of F, as it is indeed natural in many problems of calculus of variations.

The operators that arise in this way can be described in the simplest case by adding a compact operator to a monotone one and in general by allowing a more sophisticated intertwining between the monotone and compact components. These so-called <u>semimonotone</u> operators are also discussed in the references quoted above.

In this regard let us also mention a more general class of operators, the <u>pseudo-monotone</u> operators which has been introduced by H. Brezis [1], [2], [3], see also J. L. Lions [1].

Example 10 : <u>Minimal surfaces with obstacles</u>

Let us consider the functional

$$F(v) = \int_\Omega \sqrt{1 + |\operatorname{grad} v|^2}\, dx$$

that gives the area of the surface $v = v(x)$, $x \in \Omega$. We can minimize F over the cone of all v which are $\geqslant$ then a given function ψ on a compact subset E of Ω,

The variational inequality that characterizes the minimizing surface $u = u(x)$ involves the Euler operator

$$DF(u) = - \sum_{i=1}^{n} \frac{\partial}{\partial x_i} \left(\frac{u_{x_i}}{\sqrt{1 + |\text{grad } u|^2}} \right)$$

The natural Sobolev space here is $H^{1,1}(\Omega)$, that is, the space of all $v \in H^1(\Omega)$ with all first derivatives $v_{x_i} \in L^1(\Omega)$. However, this space is not reflexive. Therefore, the problem mentioned above cannot be handled with the standard methods discussed in our lectures and ad hoc techniques have been indeed developed.

From the extensive literature concerning minimal surfaces, let us only quote the papers by J. C. C. Nitsche [1], M. Miranda [1], H. Lewy and G. Stampacchia [3], E. Giusti [1], R. Temam [1], that specifically concern the obstacle problem.

2 Finite-dimensional and iterative existence theorems

By taking into account the relation between variational inequalities and fixed-point problems (see Proposition 2 of Chapter 1) and making use of the classical Brouwer's fixed-point theorem, we can easily prove the following finite-dimensional existence theorem, due to P. H. Hartman and G. Stampacchia [1],

Theorem 1 : Let K be a non-empty closed convex subset of the euclidean space E^n, $\mathcal{A}$ a continuous map of K into E^n. Let us suppose, furthermore, that either K is bounded or the following coerciveness condition holds :

(c) There exists a bounded open convex subset B of E^n and a vector $v_0 \in K \cap B$, such that

$$(\mathcal{A} v \mid v - v_0) > 0 \quad \underline{\text{for all}} \quad v \in K \cap \partial B ,$$

∂B <u>being the boundary of</u> B . <u>Then, there exists at least one solution</u> u <u>of the problem</u>

$$\text{II} \qquad u \in K : (\mathcal{A} u \mid v - u) \geq 0 \qquad \forall \; v \in K$$

Proof : Let us first assume that K is bounded, hence compact.

By Proposition 2 of Chapter 1, u is a solution of problem II above if and only if u is a fixed-point of the map

$$P_K(I - \mathcal{A})$$

of K into itself, where P_K is the Riesz projection on K. Since $P_K : E^n \longmapsto K$ is continuous, as it follows from the Corollary 2 of Lemma 2 of Chapter 1, the map $P_K(I - \mathcal{A})$ too is continuous, provided $\mathcal{A}$ is such . The existence of a fixed-point u of this map is now a consequence of Brouwer's theorem .

Now let us replace the compacteness assumption with the coerciveness condition (c). By what we have just proved, there exists a solution $\bar{u}$ of the problem

$$\bar{u} \in K \cap \bar{B} : \quad (\mathcal{A}\bar{u}, v - \bar{u}) \geq 0 \qquad \forall \; v \in K \cap \bar{B},$$

where $\bar{B} = B \cup \partial B$ Since we are assuming that condition (c) holds, $\bar{u}$ cannot belong to the boundary ∂B of B , for we should have $(\mathcal{A}\bar{u} \mid v_0 - \bar{u}) < 0$ in contradiction with the inequality above. This means that $\bar{u}$ is a <u>local</u> solution of problem II, hence it is also a global solution of II (cfr. Section 1 of Chapter 1) . ■

<u>Remark</u> 4 : For a map $\mathfrak{A}$ which is the gradient of a functional F , the coerciveness condition (c) is related to the growth at ∞ of F, see Lemma 2 of the following Section 4 . ■

<u>Remark</u> 5 : Theorem 1 could be extended to infinite dimensional spaces by making use of the Schauder or Tychonoff fixed point theorems, see F. E. Browder [8] [11] [14]. However, the continuity assumption needed by the map $\mathfrak{A}$ would then be too strong for direct application to problems of the type mentioned in the preceding section. ■

Theorem 1 cannot be considered a "constructive" existence theorem, because it relies on the deep though non constructive, Brower's theorem.

However, we can easily convert Theorem 1 into a "constructive" theorem, whenever we can replace Brouwer's theorem with a constructive fixed point theorem. The main example is obviously given by the well known contraction principle.

We haven in fact, the following <u>iterative existence theorem</u> :

THEOREM 2 : <u>Let K be a closed convex subset of a Hilbert space V , $\mathfrak{A}$ a map of K into V , such that</u>

(∗) $\qquad I - \rho\,\mathfrak{A}$ <u>is a contraction for some</u> $\rho > 0$.

<u>Then, there exists a unique solution u of the problem</u>

II $\quad u \in K : (\mathcal{A}u \mid v - u) \geq 0 \qquad \forall \; v \in K$

and $u = \lim u_n$ in V, where the sequence (u_n) is given by the iterative scheme

$$(IV_n) \qquad u_{n+1} = P_K(u_n - \rho \mathcal{A} u_n), \qquad u_0 \in K$$

Proof : Since $P_K : V \mapsto K$ is non-expansive (see Coroll. 2 of Lemma 2, Chapt. 1), the map $P_K(I - \rho \mathcal{A})$ is a contraction provided $I - \rho \mathcal{A}$ is a contraction, which is, for a suitable $\rho > 0$, our assumption ($*$) . Therefore, there exists a unique fixed point u yielded by the iterative scheme (IV_n), which is also a solution of problem II, again by Proposition 2 of Chapter 1 . ∎

Theorem 2 above can be integrated by the following lemma that gives a simple sufficient condition for $I - \rho \mathcal{A}$ be a contraction.

LEMMA 1 : Let $\mathcal{A}$ be a map of a subset K of a Hilbert space V into V, such that

(i) $\mathcal{A}$ is lipschiztian, i. e., there exists $L > 0$ such that

$$\| \mathcal{A}u - \mathcal{A}v \| \leq L \; \| u - v \| , \qquad u, v \in K$$

(ii) $\mathcal{A}$ is strongly monotone, i. e., there exists $c > 0$ such that

$$c \, \| u - v \|^2 \leq (\mathcal{A}u - \mathcal{A}v \mid u - v), \quad u, v \in K.$$

Then, the map $I - \rho \mathcal{A}$ is a contraction in V for all ρ satisfying the bound

$$0 < \rho < \frac{2c}{L^2} \; .$$

Proof. An elementary computation shows that

$$\|(I-\rho\mathcal{A})u-(I-\rho\mathcal{A})v\|^2 =$$

$$= \|u-v\|^2 - 2\rho(\mathcal{A}u-\mathcal{A}v \mid u-v) + \rho^2\|\mathcal{A}u-\mathcal{A}v\|^2 .$$

Therefore, by (i) and (ii)

$$\|(I-\rho\mathcal{A})u-(I-\rho\mathcal{A})v\|^2 \leqslant (1-2\rho c + L^2\rho^2)\|u-v\|^2 ,$$

hence $I-\rho\mathcal{A}$ is a contraction, that is, $1-2\rho c+L^2\rho^2<1$ (note that $c \leqslant L$), provided $-2\rho c + L^2\rho^2 < 0$, which is to say, $0<\rho<2c/L^2$. ▨

Remark 6 : The minimum of the function $1-2c\rho+L^2\rho^2$ of ρ is attained at the value

$$\rho_{opt} = \frac{c}{L^2} \quad ,$$

that gives the contraction constant

$$(1-2c\,\rho_{opt}+L^2\rho_{opt}^2)^{1/2} = (1-\frac{c^2}{L^2})^{1/2} . \quad ▨$$

Iterative methods for solving variational inequalities have been considered by J. L. Lions and G. Stampacchia [1] , H. Brezis and M. Sibony [1] , M. Sibony [1] , [2] , J. P. Dias-M. Sibony [1] , G. Stampacchia [5], See also J. L. Lions, R. Glowinski and R. Tremoliers, loc. cit.

The special case of variational inequalities involving bilinear forms in Hilbert spaces will be treated with some more details in the following section.

For further references on the iterative methods for the solution of equations involving monotone operators see F. E. Browder and W. V. Petryshyn [1] , R. I. Kachurovskii [3] .

3 Variational inequalities for bilinear forms in Hilbert spaces

An important class of problems to which the existence and approximation results of the previous section apply are the variational inequalities involving a coercive continuous bilinear form in a Hilbert space V.

Let us recall that a bilinear form a(u, v) on V is continuous (on V x V) if and only if there exists a constant $L > 0$, such that

(22) $$|a(u, v)| \leq L \|u\| \|v\| , \quad u, v \in V$$

Moreover, a(u, v) is said to be coercive on V if there exists a constant $c > 0$, such that

(23) $$c \|v\|^2 \leq a(v, v) \qquad v \in V .$$

[As in Section 8 of Chapter 1, we shall denote by $(\cdot | \cdot)$ the inner product in V and we put $\|\cdot\| = (\cdot | \cdot)^{1/2}$].

Let us also recall that any continuous bilinear form a(u,v) on V can be represented in the given inner product of V by means of a bounded linear operator

$$\mathcal{A} : V \longmapsto V$$

given by the identity

$$a(u, v) = (\mathcal{A} u, v) \qquad a u, \quad v \in V$$

Such operator $\mathcal{A}$ is obviously lipschiztian in V, while the coerciveness of $a(u, v)$ is nothing else than the strong monotonicity of $\mathcal{A}$ stated in Lemma 1. Let us also notice that $\mathcal{A}$ satisfies the conditions (i) and (ii) of that lemma with the same constants L and c that appear in (22) and (23) respectively.

The following theorem is then an immediate consequence of Theorem 2 above

THEOREM 3 : _Let_ $a(u, v)$ _be a coercive continuous bilinear form on a Hilbert space_ V, K _a closed convex subset of_ V. _Then, for any given_ f _in the dual_ V^* _of_ V, _there exists a unique solution_ u _of the problem._

$$(24) \qquad u \in K : \qquad a(u, v-u) \geq (f, v-u) \qquad \forall \; v \in K$$

Proof : In terms of the inner product of V the problem above can be written as

$$(25) \qquad u \in K : \quad (\mathcal{A}u - \tilde{f} \mid v-u) \geq 0 \qquad \forall \; v \in K$$

where $\mathcal{A}$ is the operator that represents $a(u, v)$ in V and $\tilde{f}$ is the vector of V, such that

$$(f, w) = (\tilde{f} \mid w) \qquad \forall \; w \in V ,$$

i.e., $\tilde{f} = J^{-1}f$, J being the Riesz isomorphism of V onto V^*

As we already remarked, the operator $\mathcal{A}$, hence also the operator $\mathcal{A} - \tilde{f}$, is lipschiztian and strongly monotone in V. Therefore, by Lemma 1, the map $I - \rho(\mathcal{A} - \tilde{f})$ is a contraction for $\rho > 0$ small enough and then the existence and uniqueness of the solution u of problem (25) follows from Theorem 2'. ∎

Remark 7 : A further consequence of Theorem 2 is the following iterative algorithm yielding the solution u of problem (24)

$$(27) \qquad u_{n+1} = P_K \left[u_n - \rho(\mathcal{A} u_n - \tilde{f})\right], \qquad u_0 \in K,$$

where

$$0 < \rho < \frac{2c}{L^2} \qquad \left(\rho_{opt} = \sqrt{1 - c^2/L^2}\right)$$

The constants c and L are those appearing in the inequalities (22) and (23) satisfied by the form $a(u, v)$.

We can also write this scheme in two successive steps,

$$(28) \qquad \begin{cases} u_{n+1/2} = \mathcal{A} u_n - \tilde{f} \\ u_{n+1} = P_K(u_n - \rho\, u_{n+1/2}) \end{cases}$$

By using the weak characterization of P_K (see Lemma 2 of Chapter 1), the iterative scheme (28) can be also written in the following weak form :

$$\begin{cases} u_{n+1/2} = \mathcal{A} u_n - \tilde{f} \\ u_{n+1} \in K : (u_{n+1} - [u_n - \rho\, u_{n+1/2}] \mid v - u_{n+1}) \geq 0 \quad \forall v \in K, \end{cases}$$

which is to say, in terms of the original form a(u, v) :

$$
(29) \quad \begin{cases} u_{n+1} \in K : \ (u_{n+1} \mid v - u_{n+1}) \geq \\ \geq (u_n \mid v - u_{n+1}) - \rho \left[a(u_n, v - u_{n+1}) - (f, v - u_{n+1}) \right] \ \forall v \in K \end{cases}
$$

Let us remark that this is a variational inequality such as the one we started with, with the bilinear form a(u, v) replaced by the inner product of V and the given functional f of V^* replaced by the functional g_n, given by

$$
g_n(w) = (u_n \mid w) - \rho \left[a(u_n, w) - (f, w) \right] . \blacksquare
$$

<u>Remark</u> 8 : The iterative algorithm that gives the solution u of problem (24), both in the strong form (27) or the weak form (29), depends on the specific inner product that has been used. In fact, as we already noted in Chapter 1, Remark 5, a change of the inner product of V to an equivalent one does not affect the form a(u, v), nor even the given vector f of V^*, hence it leaves our problem invariant.

On the other hand, such a change modifies the projection P_K and the best constants c and L in (22) and (23), hence the convergence range $0 < \rho < 2c/L^2$ of the ρ' allowed in the algorithms

Moreover, in the strong form (27) of the algorithm, the change of $\mathfrak{R}$ and $\tilde{f}$ must be also taken into account.

To make the role of the inner product appear explicitely, let us introduce a continuous bilinear form b(u, v) on V and let us suppose that b(u, v) is symmetric, i. e.

$$b(u, v) = b(v, u) \quad , \quad u, v \in V ,$$

and coercive on V . Under these assumptions,

$$(30) \qquad (u \mid v)_b = b(u, v)$$

will define an inner product in V, which is equivalent to the original one. Clearly, any equivalent inner product in V arises in such a way.

If we denote by P_K^b and J_b the Riesz projection on K and the Riesz isomorphism of V onto V^* relative to the new inner product (30), then the strong form (27) of our iterative algorithm takes the form

$$(31) \qquad u_{n+1} = P_K^b \left[u_n - \rho J_b^{-1} (Au_n - f) \right]$$

where $A : V \mapsto V^*$ is the bounded linear operator defined by the identity

$$(Au, v) = a(u, v) = (\mathfrak{A} u \mid v), \quad u, v \in V$$

The weak form of (31) now is

$$(32) \quad \begin{cases} u_{n+1} \in K : \quad b(u_{n+1}, v - u_{n+1}) \geq \\ \geq b(u_n, v - u_{n+1}) - \rho \left[a(u_n, v - u_{n+1}) - (f, v - u_{n+1}) \right] , \forall v \in K \end{cases}$$

The term in square brackets above is not affected by the change of the inner product, in fact it is the same in (32) than it was in (29).

However, its representation as a vector $u_{n+1/2}$ of V in the two-steps scheme (28) depends on the inner product, and we should now write $u_{n+1/2}$ as

$$u_{n+1/2} = J_b^{-1} (Au_n - f)$$

Thus, the explicit evaluation of $u_{n+1/2}$ requires an inversion of the Riesz isomorphism J_b induced by the form b(u, v). By carefully choosing the form b(u, v) we may hope to make this inversion "simple" and consequently simplify the whole algorithm. Some regularity of the solution could possibly facilitate this choice. For more details on this point we refer to J. L. Lions, R. Glowinski and R. Tremoliers, loc. cit., where a more detailed account of the iterative methods discussed in this section can be also found. ▨

Similar results for variational inequalities involving lipschiztian, strongly monotone operators can be also deduced from Theorem 2 and we refer to the papers of H. Brezis and M. Sibony mentioned in the preceding section.

4. Direct existence theorem

As we said in the introductory remarks of the present chapter, when a variational inequality involves a mapping A which is the differential of a convex functional F , then the existence of solutions can be proved by applying the direct existence theorems of the calculus of variations to the equivalent minimum problem for the functional F.

This approach is also fruitful from a general point of view because the theorem we shall obtain along these lines will be proved to hold, in Chapter 3, even if we drop the assumption that the map A is the differential of a functional.

Let us now recall the following basic existence theorem of minima.

THEOREM 4 : Let K be a non-empty closed subset of a reflexive Banach space X, F a lower semicontinuous convex function on K. Let us suppose either that K is bounded or the following coerciveness condition (c_0) holds : (c_0) : There exists a vector $v_0 \in K$ and a constant $R > 0$, with $F(v_0) < +\infty$ and $\| v_0 \| < R$, such that

$$F(v) > F(v_0) \qquad \text{for all } v \in K, \text{ with } \| v \| = R$$

Then, there exists at least one solution u of the problem

$$\text{I} \qquad u \in K \ : \qquad F(u) \leqslant F(v) \qquad \text{for all} \qquad v \in K$$

Proof : The proof is based on the following two well known results of functional analysis : closed convex subsets of a normed space are also closed in the weak topology of the space; bounded subsets of a reflexive Banach space are relatively compact in that topology . Therefore bounded closed convex subsets of X are weakly compact.

Such are then, if K is bounded, all level sets

$$L(v) = \{ z \in K \ : \ F(z) \leqslant F(v) \}$$

of F on K . Thus, we have

$$(33) \qquad \bigcap_{v \in K} L(v) \neq \emptyset \qquad ,$$

which is to say, problem I above has a solution u. If the boundedness assumption of K is replaced by the coerciveness condition (c_0), we can still draw the same conclusion as before, provided we show that some non-empty level set of F on K is bounded.

Now, if v_0 and R are the vector and the constant, respectively, appearing in (c_0), it is easy to verify that all vectors $z \in L(v_0)$ are bounded in norm by R.

[In fact, if there exists $z_1 \in L(v_0)$ with $\|v_1\| > R$, then it would also exist $z_2 \in L(v_0)$ such that $\|z_2\| = R$, and this would contradict the condition $F(z_2) > F(v_0)$.] ▨

Addendum 1 to Theorem 4 : The set of all solutions u of problem I above, under the assumptions of Theorem 4, is a bounded closed convex subset of K.

In fact, this set is nothing else than the set (33), which is bounded closed and convex for some (or all) sets $L(v)$ are such. ▨

Addendum 2 to Theorem 4 : The solution u of problem I is unique, provided F is strictly convex on K, i.e.,

$$2F\left(\frac{u_1 + u_2}{2}\right) < F(u_1) + F(u_2) \quad , \qquad u_1 \neq u_2, \quad u_1, u_2 \in K .$$

In fact, if u_1 and u_2 were two distinct solutions of I, since $(u_1 + u_2)/2 \in K$ we would have both

$$F(u_1) \leqslant F((u_1 + u_2)/2)$$

$$F(u_2) \leqslant F((u_1 + u_2)/2) ,$$

hence

$$F(u_1) + F(u_2) \leqslant 2\ F(\ (u_1 + u_2)/2\ )\ ,$$

that contradicts the strict convexity of F . ∎

Let us now suppose that the functional F is differentiable on X and let us ask how the properties of F involved in the assumption of Theorem 4 may be expressed in terms of the differential DF of F :

We already know, from Section 6 of Chapter 1, that the convexity of F is equivalent to the monotonicity of the map DF . Moreover, the differentiability of F clearly assures its lower semicontinuity, as it follows trivially from the inequality

$$F(v) \geqslant F(u) + (DF(u),\ v - u) \quad , \qquad u, v \in X\ ,$$

that implies $\liminf F(v_j) \geqslant F(u)$ whenever v_j converges (weakly or strongly) to a given u .

More interesting to investigate is the connection between the coercivity of F and the coercivity of DF and we shall do that in the following four lemmas.

We are now assuming that F is a differentiable convex functional on a normed space X , $DF : X \mapsto X^*$ the differential of F and K an unbounded convex subset of X .

Lemma 2 : Let DF satisfy the condition : (d_0) There exists $v_0 \in K$ and $R > 0$, with $\| v_0 \| < R$, such that

$$(DF(v),\ v - v_0) > 0 \qquad \forall\ v \in K,\ \ \| v \| = R\ .$$

<u>Then,</u> F <u>satisfies the condition</u> :

(c_0) <u>There exists</u> $v_0 \in K$ <u>and</u> $R > 0$, <u>with</u> $F(v_0) < +\infty$ <u>and</u> $\| v_0 \| < R$, <u>such that</u>

$$F(v) > F(v_0) \qquad \forall\, v \in K, \quad \| v \| = R .$$

Proof : Let v_0 and R be such that (d_0) holds and let us fix $R' > R$. Let z be a vector of K with $\| z \| = R'$. Since $v_0 \in K$ and $\| v_0 \| < R < \| z \|$, the vector

$$v = \frac{\bar{\varepsilon}}{1+\bar{\varepsilon}}\, v_0 + \frac{1}{1+\bar{\varepsilon}}\, z$$

belongs to K and $\| v \| = R$ for a suitable $\bar{\varepsilon} > 0$. Moreover, since

$$z - v = \bar{\varepsilon}\,(v - v_0) \qquad ,$$

we have

$$\begin{aligned} F(z) &\geq F(v) + (DF(v),\ z - v) = \\ &= F(v) + \bar{\varepsilon}\,(DF(v),\ v - v_0) \end{aligned}$$

and thus, in consequence of (d_0), we find

$$F(z) > F(v) \quad \cdot$$

Therefore, condition (c_0) is satisfied by taking $v_0 = z_0$ to be any vector that minimizes F on $K \cap \{z : \|z\| \leqslant R\}$. ■

If the dimension of X is finite, and only then, the coerciveness condition (d_0) satisfied by DF implies that the stronger condition (c_1) below is satisfied by the functional F

Lemma 3 : <u>If X is a finite-dimensional normed space, then condition (d_0) of Lemma 2 implies that (c_1) below holds</u> :

$$(c_1) \qquad F(v) \longrightarrow +\infty \quad \text{as} \quad \|v\| \longrightarrow +\infty, \qquad v \in K .$$

Proof . The map $v \mapsto (DF(v), v - v_0)$ is continuous on X, hence it attains its minimum m on the compact subset

$$K \cap \{v : \|v\| = R\}$$

of X, and $m > 0$ because we are assuming that (d_0) holds . Thus,

$$(34) \qquad (DF(v), v - v_0) \geqslant m > 0 \quad \forall v \in K, \quad \|v\| = R .$$

Now, for every $z \in K$ with $\|z\| > R$ we can find, as in the proof of Lemma 2, a vector

$v \in K$, with $\|v\| = R$, such that

$$z - v = \frac{\|z - v\|}{\|v - v_0\|} (v - v_0)$$

and

$$F(z) \geqslant F(v) + \frac{\|z - v\|}{\|v - v_0\|} \cdot (DF(v), v - v_0) .$$

Therefore, if z_0 is any vector that minimizes F on $K \cap \{ v : \| v \| = R \}$, by taking (34) into account we find

$$F(z) \geq F(z_0) + \frac{\| z \| - R}{R + \| v_0 \|} \cdot m \quad ,$$

hence $\| z \| \longrightarrow \infty$ implies $F(z) \longrightarrow + \infty$. ∎

Remark 9 : The lemma above is false if X has infinite dimension, as the following simple example shows : $X = l_2$, space of all sequence $v \equiv (v_k)_k$ such that

$$\| v \| = (\sum_{n=1}^{\infty} v_k^2)^{1/2} < + \infty ,$$

$$F(v) = \frac{1}{2} \sum_{k=1}^{\infty} \frac{v_k^2}{k^2} \quad , \qquad v \equiv (v_k)_k \in l_2 \quad ,$$

hence

$$DF(v) \simeq (\frac{v_k}{k^2})_k \ .$$

Then, we have

$$(DF(v), v) = \sum_{n=1}^{\infty} \frac{v_k^2}{k^2} > 0 \qquad \text{if} \qquad v \neq 0,$$

and $F(v) > 0 \quad \forall v \neq 0$, according to Lemma 2 above,

whereas

$$F(v^{(n)}) = \frac{1}{2n} \longrightarrow 0 \quad \text{on the sequence}$$

$$v^{(n)} = \underbrace{(0, \ldots, 0, \sqrt{n}, 0, \ldots}_{n}) \quad ,$$

though

$$\| v^{(n)} \| = \sqrt{n} \longrightarrow \infty \quad . \blacksquare$$

Thus, in order that F satisfy the coerciveness condition (c_1) in an arbitrary normed space X, DF must be coercive in a stronger sense than (d_0). We have indeed the following

Lemma 4 : <u>Let</u> DF <u>satisfy the condition</u>

(d_1) <u>There exists</u> $v_0 \in K$, <u>such that</u>

$$(DF(v), v - v_0) \longrightarrow +\infty \quad \text{as} \quad \| v \| \longrightarrow \infty, \; v \in K .$$

<u>Then,</u> F <u>satisfies the coerciveness condition</u> (c_1) <u>of the preceding lemma</u> .

Proof . Let v_0 be the vector that appears in condition (d_1) and let $R > 0$ be such that $\| v_0 \| < R$ and (34) holds for some $m > 0$. Such a constant R now exists in consequence of our assumption (d_1). Thereafter, the proof is the same as that of Lemma 2 . $\blacksquare$

The coerciveness condition (c_1) is not stable under a correction of $F(v)$ by an affine function $(v_0^*, v) + c$, $v_0^* \in X^*$, $c \in \mathbb{R}$, as condition (d_1) is not stable under the addition of a constant term v_0^* to $DF(v)$. In other words, the coerciveness (c_1) of the map $v \longmapsto DF(v) - v_0^*$ would depend on the given vector v_0^*.

Coerciveness conditions which are stable under the above mentioned corrections are those given in the following

Lemma 5 : Let DF satisfy the condition

$$(d_2) \qquad \frac{(DF\ (v),\ v)}{\| v \|} \longrightarrow +\infty \qquad \text{as} \qquad \| v \| \longrightarrow \infty, \qquad v \in K.$$

Then, F satisfies the condition

$$(c_2) \qquad \frac{F(v)}{\| v \|} \longrightarrow +\infty \qquad \text{as} \qquad \| v \| \longrightarrow \infty, \qquad v \in K$$

Proof : It is easy to verify that the lemma at hand is invariant under addition of affine functions to F . Therefore, it suffices to prove the lemma with regard to the functional

$$\tilde{F}(v) = F(v) - (DF(0),\ v) - F(0)\ .$$

$\tilde{F}$ still is a differentiable convex function on X, with $D\tilde{F} = DF - DF(0)$ and we have

$$\tilde{F}(0) = 0, \qquad D\tilde{F}(0) = 0 \ .$$

Therefore, we have

$$\tilde{F}(v) = \int_0^1 \frac{d}{dt}\,\tilde{F}(tv)\,dt = \int_0^1 (D\tilde{F}(tv),\ v)\ dt \qquad ,$$

where

$$(D\tilde{F}(tv),\ v) = (D\tilde{F}(tv) - D\tilde{F}(0),\ v) \geqslant 0 \quad , \qquad t > 0\ ,$$

by the monotonicity of DF . Thus,

$$\tilde{F}(v) \geq \int_{1/2}^{1} (D\tilde{F}(tv), v)\, dt = \frac{1}{2} (D\tilde{F}(\bar{t}v), v)$$

for a suitable $\bar{t}$, with $\frac{1}{2} < \bar{t} < 1$, hence also

$$\frac{\tilde{F}(v)}{\|v\|} \geq \frac{1}{2} \frac{(D\tilde{F}(\bar{t}v), \bar{t}v)}{\|\bar{t}v\|}, \qquad v \neq 0 .$$

As $\|v\| \longrightarrow \infty$, we also have $\|\bar{t}v\| \longrightarrow \infty$, since $\bar{t}$ is positively bounded from below. Therefore, (c_2) follows from (d_2) . ■

At this point we know how to formulate the properties of F , required in the direct existence theorem given at the beginning of this section, in terms of the differential DF of F .

On the other hand, we also know that a solution u of the minimum problem considered in that theorem can be characterized by means of the variational inequality.

$$u \in K : (DF(u), v - u) \geq 0 \qquad \forall v \in K .$$

Therefore, the existence theorem below is nothing else than Theorem 4 mentioned above, whenever the map A is the (Gateaux) differential of a function F on X :

THEOREM 5 <u>Let A be a monotone hemicontinuous map from a reflexive Banach space X to its dual X^*, K a non-empty closed convex subset of X . Let us suppose either that K is bounded or</u>

that A satisfies the following coerciveness condition on K :

(d_0) There exists $v_0 \in K$ and $R > 0$, with $\| v_0 \| < R$, such that

$$(Av, v - v_0) > 0 \qquad \forall v \in K, \quad \| v \| = R .$$

Then, there exists at least one solution u of the variational inequality

$$\text{II} \qquad u \in K : \quad (Au, v - u) \geq 0 \qquad \forall v \in K$$

Proof of Theorem 5 under the additional assumption $A = DF$.

If A is the Gateaux differential of a function F on X , then such an F , as we know from the discussion above, is convex, lower semicontinuous and, in case K is unbounded, it satisfies the coerciveness condition (c_0) of Theorem 4 . Therefore, there exists a vector u of K that minimizes F on K . By Proposition 1 of Chapter 1, any such minimizing u is a solution of problem II above . ▨

Remark 10 If we want a solution u of the inequality

$$u \in K : \quad (Au, v - u) \geq (f, v - u) \qquad \forall v \in K$$

to exist whatever f is given in X^*, then we must assume in place of the coerciveness condition (d_0) above that the stronger condition (d_2) holds. The latter condition is indeed stable under the addition of a constant vector to A , as we already remarked. ▨

Theorem 5 in its general form is due to P. H. Hartman and G. Stampacchia [1] and F. E. Browder [5] . The proof of Theorem 5, under the additional assumption that A is bounded, will be given in Chapter 3 and it will be based on the Linearization Lemma of Chapter 1 and a "stability theorem" for solutions of II under perturbation of the convex set K that will be proved in that Chapter.

Since now, however, we can easily prove that the set of all solutions u of problem II above has the same properties than the set of all solutions of the minimum problem considered in Theorem 4 .

We have in fact the following

Addendum 1 to Theorem 5 : Under the assumptions of the theorem, the set of all solutions of the inequality II is a bounded closed convex subset of K .

Proof : We already proved in Chapter 1, as a corollary of the Linearization Lemma of Section 7 (where the hemicontinuity of f was used) that this set is closed and convex . Moreover, if K is unbounded, any solution u of II is bounded in norm by the constant R appearing in condition (d_0) of the theorem, as it can be shown easily by taking the convexity of the set of all solutions into account . ∎

Addendum 2 to Theorem 5 . The solution u of II is unique, provided A is strictly monotone on K, i. e.,

$$(Au - Av, u - v) > 0 \qquad , u \neq v , u, v \in K.$$

Proof : If u_1 and u_2 are solutions of II , then we have both

$$(Au_1, u_2 - u_1) \geqslant 0$$

$$(Au_2, u_1 - u_2) \geqslant 0 \quad ,$$

hence

$$(Au_1 - Au_2, u_1 - u_2) \leq 0 ,$$

that, by the monotonicity of A, implies

$$(Au_1 - Au_2 , u_1 - u_2) = 0$$

Therefore, if $u_1 \neq u_2$ this would contradict the strict monotonicity of A . ■

Remark 11 . If A = DF, then A is strictly monotone if and only if F is strictly convex (cfr. Lemma 1 of Section 6 of Chapter 1) . In fact, if A is strictly monotone and the convex function F were not strictly convex, then the directional derivative of F would be constant on a line segemnt joining two points u_1, u_2, and this would contradict the strict monotonicity of A [the proof could be also achieved by remarking that in the implications (iii) $\Longrightarrow$ (ii) $\Longrightarrow$ (i) of the lemma quoted above one now has strict inequalities everywhere, when $u \neq v$].

On the other hand, if F is strictly convex, then

$$F(u + t(v - u)) < F(u) + t \left[F(v) - F(u)\right] \qquad u \neq v,$$

for every $0 < t < 1$. Therefore, as in the proof of (2) of Chapter 1, Sec. 6, we have

$$F(v) - F(u) > \frac{1}{t} \left[F(u + t(v - u)) - F(u)\right] \geq$$

$$\geq (DF(u), v - u)$$

for every $u \neq v$.

By interchanging the role of u and v and adding up the two inequalities obtained, we find

$$0 > (DF(u) - DF(v), v - u) \qquad u \neq v \ . \ \blacksquare$$

U. Mosco

CHAPTER 3

Convergence of convex sets and of solutions of variational inequalities

Solutions of variational inequalities enjoy a remarkable "stability" property under perturbations of the convex set involved in the inequality. We shall describe this stability in Section 3 below, by introducing a suitable topology in the space of all closed convex subsets of a normed space. The topology which turns out to be the appropriate one to describe the perturbation we mentioned above will be discussed in the following Section 2 .

This stability is also the property underlying the infinite-dimensional existence theorem we shall prove in Section 4. This theorem, indeed, can be obtained from the existence theorem for variational inequalities in an euclidean space, by approximating the initial problem with a family of finite-dimensional problems.

Under suitable assumption on the map A this procedure can be easily converted into a "constructive" one, yielding a method of Ritz-Galerkin type for the numerical solution of variational inequalities . The solution $u(x)$ of the initial problem can be found as the limit in an appropriate norm of a family of approximate solutions u_h , which can be computed numerically by solving a variational inequality in euclidean spaces of increasing dimensions.

This approximation will be first sketched in Section 1 below and then discussed with more details in Section 5 and Section 6 .

Finally in Section 7 we shall give a short account of a duality theory for variational inequalities . We shall restrict ourselves to variational inequalities on cones, by relating this subject to the so called complementarity systems .

U. Mosco

1 . The Ritz-Galerkin approximation

Let us consider a variational inequality such as

$$(1) \qquad u \in K \ : \quad (Au, v - u) \geq 0 \quad \forall v \in K$$

in a normed space X . We shall speak of the elements of X as of functions.

Now let X_h be some finite-dimensional subspace of X, depending on a parameter h to be specified, and let

$$(2) \qquad \varphi_1^h(x), \ldots\ldots\ldots, \quad \varphi_n^h(x) \quad , \qquad n = n_h \quad ,$$

be a basis in X_h , $n = n_h$ being the dimension of X_h. Thus, the functions $v_h(x)$ in X_h are those of the form

$$(3) \qquad v_h(x) = \sum_q v_q^h \varphi_q^h(x) \qquad ,$$

where $(v_q^h)_q$ are the components of $v_h(x)$ in the chosen basis (2), q ranging from 1 to $n = n_h$.

Let us now choose a convex subset K_h of X_h , which may be conveniently assumed to realize an approximation of the given K . The choice of such a finite-dimensional approximation of K is crucial and we shall come back on this point in a moment.

We then replace the initial problem (1) with the following approximate problem

$$(4) \qquad u_h \in K_h \ : \ (Au_h, v_h - u_h) \geq 0 \qquad \forall v_h \in K_h \ ,$$

where, for sake of simplicity, the operator A has been left unchanged.

This means that the approximate solution $u_h(x)$ in K_h is required to satisfy exactly the given inequality for all $v_h(x)$ of K_h.

If we now take the expressions (3) of $u_h(x)$ and $v_h(x)$ in the basis $(\varphi_q^h(x))_q$ that has been chosen to span the subspace X_h and replace them into (4), we then find that the approximate problem above takes the form of the following discrete problem in the space R^n :

$$(5) \quad u^h = (u_q^h) \in C^h : \quad \sum_s A_s^h(u_1^h, \ldots, u_n^h)\,(v_s^h - u_s^h) \geq 0$$

$$\forall\, v^h = (v_q^h) \in C^h ,$$

where the vector field A_s^h is given by

$$(6) \quad A_s^h(u_1^h, \ldots, u_n^h) = (A(\sum_q u_q^h \varphi_q^h), \varphi_s^h) , \qquad s = 1, \ldots, n ,$$

the pairing at right hand being that between X and its dual X , and C^h the convex subset of R^n

$$C^h = \{ v^h = (v_q^h) \in R^n : v_h(x) = \sum_q v_q^h \varphi_q^h(x) \in K_h \} .$$

When A is a linear map, then the system of inequalities (5) reduces to the system of <u>linear</u> inequalities

$$(7) \quad u^h = (u_q^h) \in C^h . \quad \sum_{qs} A_{qs}^h\, u_q^h(v_s^h - u_s^h) \geq 0 \quad \forall v^h = (v_s^h) \in C^h,$$

with the n x n matrix (A^h_{qs}) given by

(8) $\quad A^h_{qs} = (A\ \varphi^h_q,\ \varphi^h_s) \qquad , \qquad q,s = 1,\dots,n$.

Let us point out that we have been considering so far three distinct problems, first, the initial problem (1) in the space X ; second, the approximate problem (4) in the finite - dimensional subspace X_h of X; third, the discrete problem (5) in the n-dimensional vector space R^n, $n = n_h$ being the dimension of X_h .

To write down the approximate probem (4) we only need to specify the subspace X_h of X and the convex subset K_h of X_h . To write the discrete probem (5) we must additionally choose a basis in the subspace X_h.

The choice of an approximate convex set K_h is thus a basic additional element of the Galerkin scheme of approximation of the inequalities we are dealing with. When these reduce to the equation $Au = 0$, and this is the case, e.g., when $K = X$, then only the choice of the subspace X_h has to be made.

The choice of a family of "good" approximants K_h should be submitted to the following two general requirements: first, the corresponding approximate problem can be converted into a discrete probelm as "easy to solve" as possible; second, the dependence of K_h on the parameter h is such that the approximate solution $u_h(x)$ converges to the initial solution u(x) as K_h approaches K .

It is trivial to realize that once the subspace X_h has been chosen, a naif choice of K_h may be not always successeful. For example, if X is a separable Hilbert space, X_h the subspace spanned by the first n vectors of a given orthonormal system of X, K the one-dimensional a subspace spanned by a vector v_0 of X which has infinitely many non-zero components with respect to the given basis, then what in general may be the most natural candidate for K_h, namely $K_h = K \cap X_h$, is obviously a bad choice, for this K_h consists of the single vector 0 .

2 . A convergence for convex sets and convex functions .

As we pointed out in the preceding section, the finite-dimensional approximation of a given variational inequality, as far as the approximation of the convex set involved in the inequality and the convergence of the corresponding approximate solutions are concerned, can be dealt with by suitably, defining a topology in the family of all closed convex subsets of a normed space . The topology we are looking for, however, must be weak enough as to allow a family of finite dimensional K_h to converge to a possibly infinite-dimensional K .

To investigate what a "good" converge

$$K = \lim K_h \tag{9}$$

would be like, let us consider the basic variational problem consisting in the orthogonal projection of a given vector z of a Hilbert space V onto a closed linear subspace M of V .

A "good" convergence can then be reasonably requested to fulfill the following requirement

"For any closed linear subspace M of V if (M_h) is any sequence of closed linear subspaces of V such that

$$(10) \qquad M = \lim M_h$$

then for every z of V the orthogonal projection $P_{M_h} z$ of z on M_h converges strongly to the orthogonal projection $P_M z$ of z on M.

In other words, the map

$$M \longmapsto P_M \quad ,$$

where P_M is the orthogonal projection operator on M, must be continuous at any M for the strong topology of operators.

Moreover, the requirement above can be assumed to be invariant under the orthogonal complementation $M \longmapsto M^{\perp}$ of subspaces of V.

Therefore, the convergence (10) too must be stable under $\perp$, which is to say, we must have

$$M = \lim M_h \quad \text{if and only if} \quad M^{\perp} = \lim M_h^{\perp} .$$

Let us now assume that the convergence (9) has been indeed so defined as to satisfy requirement "   " above and let us derive some necessary condition satisfied by any sequence (M_h) that converges to a given M according to (10).

A first condition is the following inclusion

$$M \subset s\text{-}\lim\inf M_h$$

where $s\text{-}\lim\inf M_h$ denotes the lim inf of the sequence (M_h) in the strong topology of V .

(11) $\quad$ $\text{s-lim inf } M_h = \{ v \in V : v = \text{strong lim } v_h, \quad v_h \in M_h \quad \forall h \}$.

In fact, by property " .... " above, for any z of M the vector $z_h = P_{M_h} z$ belongs to M_h and converges strongly to $P_M z = z$.

The following lemma shows at one time that the condition above is not stable under orthogonal complementation and how it must be strenghtened to become such.

We shall denote below by w-limsup M_h the sequential limsup of the equence (M_h) in the weak topology of V, i.e.,

(12) $\quad$ $\text{w-limsup } M_h = \{ v \in V : v = \text{weak lim } v_{h_j}, \quad v_{h_j} \in M_{h_j} \quad \forall j ,$

$$(M_{h_j})_j \text{ subsequence of } (M_h)_h \quad \}$$

LEMMA 1 <u>Let (M_h) be a sequence of closed linear subspaces of V, $M_h \neq V$, and M a closed linear subspace of V. Then we have</u>

(13)
$$M \subset \text{s-liminf } M_h$$

<u>if and only if</u>

(14)
$$\text{w-limsup } M_h^{\perp} \subset M^{\perp}$$

Proof : We shall base the proof on the following well known fornula

(15) $$\operatorname{dist}(v, H) = \max_{\substack{w \in H^{\perp} \\ \|w\| \leq 1}} (w \mid v) .$$

which gives the distance of a vector v of V from a closed linear subspace H of V .

[dist (v, H) can be seen as the norm of v in the quotient space V/H , whose normed dual is isometric to $H^{\perp}$; hence (15) is a special case of the well known dual formula

$$\| v \| = \max_{\substack{v^* \in X^* \\ \|v^*\| \leq 1}} (v^*, v)$$

satisfied by the norm of any normed space X .]

Let us prove that (13) implies (14) . Let $v_0 \in$ w-limsup M_h , i.e.,

$$v_0 = \text{weak} \lim v_{h_j} , \qquad v_{h_j} \in M_{h_j}^{\perp} \quad \forall j ,$$

$(M_{h_j}^{\perp})_j$ being a subsequence of $(M_h^{\perp})_h$.

Let $r > 0$ be such that

$$\| v_{h_j} \| \leq r \qquad \text{for all } j .$$

Then, for any v we have, in consequence of (15),

$$(r^{-1} v_{h_j} \mid v) \leq \operatorname{dist}(v, M_{h_j}) ,$$

and, by our assumption (13), dist $(v, M_h) \longrightarrow 0$.

Therefore,

$$(v_0 \mid v) = \lim (v_{h_j} \mid v) = 0 \qquad \text{for all} \quad v \in M,$$

that is, $v_0 \in M^{\perp}$.

Conversely, let us suppose that (14) holds and let us prove that for every v of M we have dist $(v, M_h) \longrightarrow 0$, what clearly proves (13).

Again as a consequence of (15), we can find for every h a vector $v_h \in M_h^{\perp}$, with

$$\| v_h \| \leq 1 \qquad ,$$

such that

$$\text{dist}(v, M_h) = (v_h \mid v) .$$

Let $(M_{h_j})_j$ be an arbitrary subsequence of $(M_h)_h$. By the boundedness of the sequence $(v_h)_h$, there exists a subsequence of $(v_{h_j})_j$, which we still call $(v_{h_j})_j$, that converges weakly to some vector v_0 of V and this v_0, by our assumption (14), belongs to $M^{\perp}$. Therefore we have, since $v \in M$,

$$\text{dist}(v, M_{h_j}) = (v_{h_j} \mid v) \longrightarrow (v_0 \mid v) = 0 .$$

It follows, by the arbitrariness of $(M_{h_j})_j$, that

$$\text{dist}(v, M_h) \longrightarrow 0 . \blacksquare$$

<u>Corollary</u> : <u>Let</u> $M = \lim M_h$ <u>mean that both the inclusions</u> (13) <u>and</u> (14) <u>above hold. Then, we have</u> $M = \lim M_h$ <u>if and only if</u> $M^{\perp} = \lim M_h^{\perp}$.

We shall also see in Section 6 that if we actually define the convergence of a sequence of subspaces as in the corollary above, then the requirement " ... " we stipulated at the beginning of our discussion is satisfied : that is, $M = \lim M_h$ does imply indeed that

$$P_M z = \text{strong} \lim P_{M_h} z$$

for every z of V.

The discussion made up to now leads us to the following general definition

<u>Definition</u> 1 : A sequence (K_h) of convex subsets of a normed space X <u>converges</u> to a (closed convex) subset K of X, and then we write

$$K = \lim K_h \qquad \text{in } X \quad ,$$

if both the following inclusions holds

$$\text{w-limsup } K_h \subset K \subset \text{s-liminf } K_h \quad ,$$

where the limits wre defined as in (11) and (12) above.

<u>Remark</u> 1 : If $K_h \subset K$ for every h, than $K = \lim K_h$ is equivalent to the single inclusion $K \subset \text{s-lim inf } K_h$. On the other hand, if $K \subset K_h$ for every h, than $K = \lim K_h$ reduces to the condition $\text{w-limsup } K_h \subset K$.

It could be shown that if X is a reflexive Banach space, then a Hausdorff topology can be introduced in the space of all closed convex subsets of X which reduces on sequences of sets to the convergence defined above. However, we shall not discuss this problem here, and we refer to J. L. Joly [1] for a general investigation of the topologies for convex sets and convex functions and the action of polarity on them.

The basic property of the convergence we have just defined (and of the topology mentioned above) is indeed its stability under polarity, as we already showed in the special case of Lemma 1 .

To make this point more specific, let us consider the Young-Fenchel transform

$$f \mapsto f^*$$

which associates every l. s. c. convex function

$$f : X \mapsto (-\infty, +\infty] \qquad (f \not\equiv +\infty)$$

with its polar function

$$f^* : X^* \mapsto (-\infty, +\infty] \qquad ,$$

given by

$$(16) \qquad f^*(v^*) = \sup_{v \in X} [(v^*, v) - f(v)] \quad , \qquad v^* \in X^* .$$

It can be shown that f^* is again a l. s. c. convex function and $f \mapsto f^*$ is bijective and involutory, that is, $f^{**} = f$

Let us note, in particular, that when $f = \delta_K$ is the indicator function of a closed convex subset of X (see Section 9 of Chapter 1), then, $f^* = \sigma_K$, where

(17) $$\sigma_K(v^*) = \sup_{v \in K} (v^*, v)$$

is the support function of K. As a special case of this, if K = M is a closed subspace of X and $f = \delta_M$, then $f^* = \delta_{M^\perp}$, where $M^\perp$ is the annihilator of M in X^* (hence, the orthogonal subspace of M, if X is an Hilbert space and X^* is identified with X). More generally, if H is a closed convex cone with vertex at zero, then $(\delta_H)^* = \delta_{H^*}$, where

(18) $$H^* = \{v^* \in X^* : (v^*, v) \leq 0 \quad \forall\ v \in H\}$$

is the polar cone of H.

If (f_h) is a sequence of l.s.c. convex functions on X we now define

(19) $$f = \lim f_h \quad \text{in } X,$$

where f is also a l.s.c. convex function on X, to be equivalent to the limit

$$\text{epi } f = \lim \text{epi } f_h \quad \text{in } X \times R$$

in the product space X x R, according to Definition 1 above. Let us recall that for any l. s. c. convex function $g : X \mapsto (-\infty, +\infty]$, epi g is the *epigraph* of g, that is the closed convex subset

$$\text{epi } g = \{ (v, \beta) \in X \times R : g(v) \leqslant \beta \}$$

of the product space $X \times R$.

It is easy to verify that if $f = \delta_K, f_h = \delta_{K_h}$, with K, K_h closed convex subsets of X, then $f = \lim f_h$ if and only if $K = \lim K_h$.

Again, if X is a reflexive Banach space a Hausdorff topology can be define on the space of all l. s. c. convex functions on X, induced from the analog topology for closed convex sets we mentioned above, which reduces on sequences to the convergence (19).

The stability of this topology under polarity can now be stated as follows

THEOREM 1 *The Young-Feuched bijection* $f \mapsto f^*$ *is bicontinuous. In particular, for any sequence* (f_h) *of l. s. c. convex functions on the reflexive Banach space* X, *we have* $f = \lim f_h$ *in* X *if and only if* $f^* = \lim f_h^*$ *in* X^*.

This theorem generalizes Lemma 1 to which it reduces when $f_h = \delta_{M_h}$. Another special case of the theorem above, which we shall need later, is obtained by taking f to be the indicator function of a closed convex cone H, with vertex at 0 :

COROLLARY : *Let* (H_h) *be a sequence of closed convex cones, with vertex at 0, in a reflexive Banach space* X, (H_h^*) *the sequence of the polar cones in* X^*. *Then we have*

$$H = \lim H_h \qquad \text{in } X$$

if and only if

$$H_h = \lim H_h \qquad \text{in } X \quad ,$$

H *being the polar cone of* H .

This result could be also proved directly, along the lines of the proof of Lemma 1 .

For the proof of Theorem 1 see J. L. Joly, loc. cit. and U. Mosco [6] .

Let us now give a few examples of sequences of convex sets that converge according to Definition 1. By X we shall always denote a reflexive Banach space, even if for some of the results stated below the reflexivity of the space is not needed.

(a) *Let* $M_1 \subset M \subset \ldots \subset M_h \subset \ldots .$

be subspaces of X . *Then ,*

$$\lim M_h = M$$

where M = closure of $\bigcup_h M_h$ *in* X .

(b) *Let* $M_1 \supset M \supset \ldots \supset M_h \supset \ldots .$

be subspaces of X . *Then,*

$$\lim M_h = M ,$$

where $M = \bigcap_h M_h$.

The Ritz approximation suggests the following example

(c) *Let* K *be a closed convex subset of* X , *whose interior* $\overset{\circ}{K}$ *in* X *is not empty, and* (X_h) *an increasing sequence of subspaces of* X, *such that* X = closure of $\bigcup_h X_h$. *Then* ,

$$K = \lim K \cap X_h$$

More generally we have

(c') *Let* K *be as in* (c) *above and* (S_h) *a sequence of closed convex subsets of* X *converging to* S *in* X . *Let us assume, furthermore, that* $\overset{\circ}{K} \cap S \neq \emptyset$. *Then,*

$$K \cap S = \lim K \cap S_h \quad .$$

For the proof we refer to U. Mosco [4], Lemma 1.4. (+)

The last example leads us to the general problem of the continuity of the intersection operation : under which assumptions does $K = \lim K_h$, $S = \lim S_h$ imply $K \cap S = \lim K_h \cap S_h$? At the end of Section 1 we saw that if we drop the assumption $\overset{o}{K} \neq \emptyset$ the conclusion of (c) above may be false. Therefore, some condition on the sequences involved must be imposed. We refer to the papaer of J. L. Joly already quoted, where the problem raised above is investigated also in the form it takes for convex functions. Then, it is the problem of the continuity of the so-called inf-convolution $f \nabla g$, which is, roughly speaking, the polar operation of the sum $f + g$, see J. J. Moreau [1]. Joly introduces some notion of angle, called codistance $\theta(K_1, K_2)$ between two convex sets K_1 and K_2 to find an alternative condition to the one required in (c') above, which is of the type

$$\liminf \; \theta(K, S_h) > 0 \quad .$$

That a condition of this kind may be required to draw the conclusion of (c') can be also inferred from trivial examples such as that sketched below :

(+) As J. L. Joly kindly pointed out to the author, the additional hypotesis $\overset{o}{K} \cap S \neq \emptyset$ was erroneously omitted in that lemma : the vector $u_0 \in \overset{o}{K}$ that appears in the proof must be replaced, indeed, by a vector $u_0 \in \overset{o}{K} \cap S$.

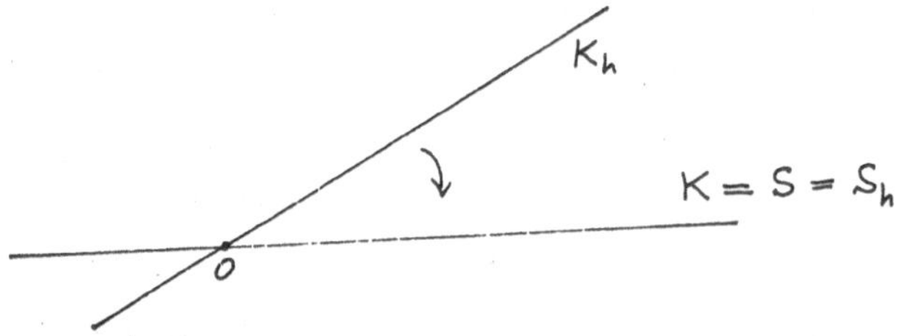

(d) Let X be a Hilbert space, (p_h) a sequence of simmetric linear operator on X such that

$$v = \text{strong} \lim p_h v \qquad \text{for every} \quad v \in X \ .$$

If K is a bounded closed convex subset of X, then

$$p_h K = \{ v \in X \ : \ v = p_h w \ , \ w \in K \}$$

is a bounded closed convex subset of X and

$$K = \lim p_h K \ .$$

For the proof see the author's paper [4], Lemma 1.5. [That $p_h K$ is closed for each given h can be seen as follows : let $v = \lim_j p_h w_j$, $w_j \in K$; since K is bounded and weakly closed there exists a subsequence of (w_j') of (w_j) that converges weakly to a vector $w_0 \in K$; thus, for every $z \in X$ we have

$$0 = \lim_j (w_j' - w_0 \mid p_h z) = \lim_j (p_h w_j' - p_h w_0 \mid z) = (v - p_h w_0 \mid z)$$

therefore, $v = p_h w_0$.]

An important special case of (d) is the following

(d') Let X be a Hilbert space, (X_h) an increasing sequence of finite-dimensional subspaces of X with $\bigcup_h X_h$ dense in X, and, for each h, p_h the orthogonal projection of X onto X_h. Then, the conclusion of (d) above holds.

It can be shown by simple examples that if K is unbounded then the projected sets $p_h K$ may not converge to K, even if K is a closed linear subspace of X (see Remark 1.3 of Ref. quoted above).

A sufficient condition is then the inclusion $p_h K \subset K$ for every h. Let us also remark that if K is unbounded then the projected set $p_h K$ need not even be closed : think of the orthogonal projection of the epigraph of the real function $y = 1/x$, $x > 0$, on the x-axis of the euclidean plane x, y.

Further examples of converging sequences of convex sets in Sobolev spaces have been also considered in the author's papers [4] and [5], in connection with some perturbed boundary value problems for partial differential operators. In this regard see also L. Boccardo [1].

In the following Section 6 we shall consider some examples arising in the theory of internal or external approximation of Sobolev spaces.

Finally, let us also mention the applications to approximation theory that have been given by J. L. Joly in his paper quoted above.

3. The "stability" theorem

Let us now study how the solution u of a variational inequality, involving a map A and a convex set K, depends on the set K : this is the problem of the continuity of the map $K \mapsto u$ and we shall study it with respect to the topology for convex sets described in the preceding section and the weak or strong topology in the space X .

We shall first assume that the map A is kept fixed while K is allowed to vary and we shall mention later how joint perturbations of A and K can be taken into account.

Let us consider the "initial" problem

$$(21) \qquad u \in K : \qquad (Au, v - u) \geqslant 0 \qquad \forall v \in K$$

and a family of "perturbed" or "approximate" problems of the form

$$(22) \qquad u_h \in K_h : \qquad (Au_h, w - u_h) \geqslant 0 \qquad \forall w \in K_h .$$

We shall assume $(K_h)_h$ be a sequence of sets. However, any directed family could be also allowed, with only minor changes in what follows.

The main result concerning the problem considered above can be stated as follows

THEOREM 2 : Let X be a reflexive Banach space and

(i) A a bounded monotone and hemicontinuous map of a domain $D(A)$ of X in X ,

(ii) $(K_h)_h$ a sequence of subset of $D(A)$, which converges to a convex subset K of $D(A)$, according to Definition 1 of Section 2 .

Let us assume, furthermore, that there exists for every h a solution u_h of the perturbed problem (22) and that the sequence $(u_h)_h$ is bounded in X .

Then the initial problem (21) has at least one solution ; moreover, if this solution, u , is unique, then

$$u_h \text{ converges weakly to } u \text{ in } X$$

and

$$(Au_h) - Au,\ u_h - u) \longrightarrow 0 \quad .$$

Proof . Let us prove first :

(a) Any weak limit u in X of a subsequence $(u_{h_j})_j$ of approximate solutions is a solution of the initial problem.

To simplify our notation, let us call (u_h) the subsequence at hand. Thus, we have for every h

$$u_h \in K_h : \qquad (Au_h,\ w - u_h) \geqslant 0 \qquad \forall\ w \in K_h \quad .$$

By our assumption (ii), we have $w\text{-}\limsup K_h \subset K$, hence

$$u \in K$$

On the other hand, we also have $K \subset s\text{-}\liminf K_h$, therefore every $v \in K$ is the strong limit of a sequence (v_h), with $v_h \in K_h$ for all h . We can put $w = v_h$ in the inequality above and make v appear in that inequality, by writing it as

$$(Au_h, v - u_h) \geqslant (Au_h, v - v_h) .$$

Now we use the monotonicity of A at the left hand to get

$$(Av, v - u_h) \geqslant (Au_h, v - v_h)$$

and then we go to the limit in h : since A is bounded, the sequence $(Au_h)_h$ is bounded, for $(u_h)_h$ is such; thus we find

$$(Av, v - u) \geqslant 0 .$$

Therefore we can conclude that u is a solution of the linearized problem

$$u \in K : \quad (Av, v - u) \geqslant 0 \qquad \forall v \in K$$

and since we are under the assumption of the Linearization Lemma of Chapater 1, this also means that u is a solution of problem (21) .

The second step in our proof is

(b) <u>There exists a solution</u> u <u>of problem</u> (21) <u>and if the solution of</u> (21) <u>is unique then the whole sequence</u> $(u_h)_h$ <u>converges weakly to</u> u <u>in</u> X .

Since X is reflexive, the existence of a solution u follows from (a) and our assumption that there exists a bounded sequence (u_h) of approximate solutions. The weak convergence of u_h to u is then an obvious consequence of the uniqueness if u, once we take (a) again into account.

The final step is the proof of

$$\text{(c)} \qquad (Au_h) - Au,\ u_h - u) \longrightarrow 0 \ .$$

Still in consequence of the hypotesis $K \subset s\text{-Lim inf } K_h$, we can find for every h a vector $z_h \in K_h$ that converges strongly to u.

On the other hand, since u_h is a solution of (22), we have for all h

$$(Au_h,\ z_h - u_h) \geqslant 0 \ .$$

We now write this inequality by making u to appear in it,

$$(Au_h,\ u - u_h) \geqslant (Au_h,\ u - z_h)$$

and then we go to the limit in h. Since A is bounded and u_h converges weakly to u, we find that

$$\operatorname{Lim}_h \sup \ (Au_h,\ u_h - u) \leqslant 0$$

hence also

$$\operatorname{Lim}_h \sup \ (Au_h - Au,\ u_h - u) < 0,$$

which, by the monotonicity of A, is clearly equivalent to (c) above.

<u>Remark</u> 2 . Many operators A arising in the applications have the property that the weak convergence of a sequence u_h to a vec-

tor u of X together with the convergence of the form

$$(Au_h - Au, u_h - u)$$

to zero imply the strong convergence of u_h to u in X . Operators of this kind are often said to be of type (S) see F. E. Browder [15] . This is the case, for instance, if A is strongly monotone :

(23) $$c \|v - u\|^2 \leqslant (Av - Au, v - u) \qquad c > 0 ,$$

or, more generally, if A satisfies a condition such as

(24) $$\gamma(\| u - v\|) \leqslant (Au - Av, u - v)$$

with $\gamma : R_+ \mapsto R_+$ any continuous and strictly increasing function at 0^+, with $\gamma(0) = 0$. [If the space X is uniformly convex, then this condition can be somewhat weakened, see H. Brezis - M. Sibony [1].] Let us also remark that the analogue of condition (c) for minimum problems involving a functional F is that the joint weak convergence of u_h to u and of $F(u_h)$ to F(u) should imply the strong convergence of u_h to u and it is well known that this is the case, for instance, if F is the norm of a uniformly convex Banach space .

In this regard, see also the Corollary of Theorem 2 below .

Remark 3 . If a simultaneous approximation of the map A must be also taken into account, then problem (22) should be replaced by the problem

(25) $$u_h \in K_h : (A_h u_h, v - u_h) \geqslant 0 \qquad \forall v \in K_h \quad ,$$

where A_h is a suitable perturbation of the given A , with domain $D(A_h)$ in X containing K_h and range in X^*

Indeed, if the maps A_h's are uniformly bounded, monotone and hemicontinuous and satisfy the convergence sondition

$$(26) \qquad \text{graph } A \subset \text{s-Lim inf graph } A_h \qquad \text{in } X \times X^*$$

in the strong topology of this product space, then the analogue of Theorem 1, with (22) replaced by (25) above, can be proved along the same lines, see U. Mosco [4] .

[The maps (A_h) an uniformly bounded in X if for any bounded subset B of X , there exists a bounded subset B' of X^* , such that

$$A_h(B \cap D(A_h)) \subset B' \qquad \text{for all } h \quad .]$$

The following result is included in Theorem 2, as it can be shown by using the epigraph formulation of minimum problems we discussed in Chapater 1 :

Corollary of Theorem 2 : Let $f : X \mapsto (-\infty, +\infty]$ be convex and $f_h : X \mapsto (-\infty, +\infty]$ be such that $f = \lim f_h$ in X according to the definition of Section 2 . Let us suppose that there exists for each h a vector u_h that minimizes f_h and that the sequence $(u_h)_h$ is bounded . Then there exists a vector u minimizing f ; moreover, if u is the unique minimizing vector, then u_h converges weakly to u and $f_h(u_h)$ converges to $f(u)$.

Proof Apply Theorem 2 to the map 0×1 of the product space $X \times R$ into $X^* \times R$ and to the convex subsets $K = \text{epi } f$

and K_h = epi f of X x R and use Lemma 3 of Section 9 of Chapter 1 . ∎

Similarly, we can consider a mixed variational inequality such as

$$u \in X \ : \ (Au, v - u) \geq f(u) - f(v) \qquad \forall \ v \in X \ ,$$

and the perturbed inequality

$$u_h \in X : (Au_h, v - u_h) \geq f_h(u_h) - f_h(v) \qquad \forall \ v \in X \ ,$$

or even take $A = A_h$ to depend on h too, and use the epigraph formulation together with Theorem 1, or its generalization with $A = A_h$, to obtain a result on the convergence of the perturbed solution u_h . We refer to. U. Mosco [4] for more details on this point. See also Theorem 4 below .

In the following section we shall apply Theorem 2 in order to obtain further existence results for variational inequalities and related problems.

In Section 6 we shall apply Theorem 2 to prove the convergence of certain schemes of finite-dimensional approximation of the solutions of variational inequalities or minimum problems.

Let us also mention that Theorem 2 could be also useful to investigate the continuous dependence of the solution of boundary value problems on the, possibly unilateral, constraints imposed on the solution.

Some applications of this type can be found in the author's papers [4] , [5] and in L. Boccardo [1] .

4 : Further existence theorems

We shall assume throughout this section that X is a separable reflexive Banach space .

The reason for the separability assumption, which is indeed unnecessary for the general validity of the existence. results we shall prove below, must. be found in the fact that we shall deduce our results from the "stability" theorem of Section 3 and in that theorem only sequences of perturbed sets K_h were allowed . To remove this assumption we should rely, instead then on Theorem 2, on the analogue stability result for directed families (K_h) .

Let us now prove the general existence theorem, namely Theorem 5, we stated in the last section of Chapter 2, In addition to the separability of the space, we shall now prove that theorem by assuming the map A to be bounded .

THEOREM 3 Let A be a bounded monotone hemicontinuous map of X into X^* , K a closed convex subset of X . Let us suppose either that K is bounded or that A satisfies the following coerciveness condition on K

(d_0) There exists $v_0 \in K$ and $R > 0$, with $\| v_0 \| < R$, such that

$$(Av, v - v_0) > 0 \quad \text{for all} \quad v \in K, \quad \| v \| = R$$

Then, there exists a solution u of the problem

$$(27) \qquad u \in K \quad : \quad (Au, v - u) \geqslant 0 \qquad \forall \; v \in K$$

Proof : By the separability of X , we can find an increasing sequence of finite-dimensional closed convex subsets K_h of K, with $v_0 \in K_1$, such that

$$K = \text{closure of } \bigcup_h K_h \quad \text{in } X .$$

We shall now apply the finite-dimensional existence theorem of Chapter 2 to prove the existence, for every h, of a solution u_h of the problem

$$(28) \qquad u_h \in K_h \; : \; (Au_h, v - u_h) \geq 0 \qquad \forall v \in K_h ,$$

such that

$$\| u_h \| \leq R \qquad \text{for all } h ,$$

R being the constant appearing in condition (d_0) if K is unbounded or any positive constant large enough, if K is bounded .

Let h be fixed and let X_h be an n-dimensional subspace of X, $n = n_h$, that containes K_h.

Let

$$\pi_h : E^n \longmapsto X$$

be an injective map with $\pi_h(E^n) = X_h$, and let

$$\pi_h^* : X^* \longmapsto E^n$$

be the transpose of π_h. Then

$$(w^*, v) = (y \mid x), \qquad v = \pi_h x, \quad y = \pi_h^* w^* .$$

The set

$$C_h = \pi_h^{-1} K_h$$

is a closed convex subset of E^n, bounded if K_h is such.

Moreover, the map

$$A_h = \pi_h^* A \pi_h : E^n \longmapsto E^n$$

is continuous. [In fact, π_h is obviously continuous from E^n to X, A is continuous from X to X^* endowed with the weak topology (see Remark 8 of Chapter 1) and π_h^*, in turn, is continuous from X^* with that topology to E^n.] Finally if K_h, hence C_h too, is not bounded, then by the assumption (d_0) we have

$$(A_h x \mid x - x_0) = (A\pi_h x, \pi_h x - \pi_h x_0) = (Av, v - v_0) > 0 ,$$

where $x_0 = \pi_h^{-1} v_0 \in C_h$, for all x such that $v = \pi_h x \in K$ and $\| v \| = R$, in particular, for all $x \in C_h$ such that $\| \pi_h x \| = R$.

We now are in position to apply Theorem 1 of Chapter 2, with $K = C_h$ and $B = \{ x \in E^n : \| \pi_h x \| \leqslant R \}$ and we find that there exists a solution $x \in C_h \cap B$ of

$$x \in C_h : (A_h x \mid y - x) \geqslant 0 \qquad \forall\, y \in C_h ,$$

hence a solution $u_h = \pi_h x$ of problem (28), with

$$\| u_h \| = \| \pi_h x \| \leqslant R .$$

Thus, we have shown that there exists a bounded sequence (u_h) of approximate solutions. The conclusion of the theorem is now a consequence of the stability theorem. ▨

Remark 4 : If the map A is only defined on K , which is not an open domain, then the hemicontinuity assumption must be strenghtened . In fact, in this case the demicontinuity of A does'nt necessarily follows from the hemicontinuity (cfr. Remark 8 . of Chapter 1), nor even we can affirm that A is continuous from the finite-dimensional sections $K \cap X_h$ of K (X_h being any finite-dimensional subspace of X) to the weak topology of X^* . This latter property, however, was needed in the proof above to show the continuity of A_h . Therefore, it must be assumed "a priori" , in place of the hemicontinuity, when A has not an open domain . ▨

Remark 5 : We know from Chapter 1 that the set of all solutions of problem (27) is closed and convex. Under the assumption of Theorem 3 above, we also have that this set is bounded . In fact this is obviously the case if K is bounded. If K is unbounded, then any solution u is bounded in norm by the constant R that appears in condition (d_0) of the theorem . In fact, if $\bar{u}$ is a solution of (27) with $\| \bar{u} \| > R$, since there exists at least one solution u with $\| u \| \leqslant R$ and the set of all solutions is convex, there would also exist a solution $\tilde{u}$ with $\| \tilde{u} \| = R$; hence

$$(A\tilde{u}, \tilde{u} - v_0) \leqslant 0$$

and this contradicts (d_0) . ▨ -

Remark 6 . The coerciveness condition (d_0) can be obviously replaced by the stronger condition

(d_1) There exists $v_0 \in K$, such that

$$(Av, v - v_0) \longrightarrow +\infty \quad \text{as } \|v\| \longrightarrow \infty , \qquad v \in K$$

(cfr of Section 4 of Chapter 2). ■

We shall now deduce from Theorem 3 an existence theorem for inequalities of type

$$(29) \qquad u \in X \ : \ (Au, v - u) \geq F(u) - F(v) \qquad \forall v \in X ,$$

by using the epigraph formulation of this problem we already discussed in Sections 5 and 9 of Chapter 1 .

THEOREM 4 Let A be a bounded monotone hemicontinuous map of X into X^* , F a l.s.c. convex functional on X with values in $(-\infty, +\infty]$. Let us suppose that A and F satisfy the following coerciveness condition :

There exists $R > 0$ and $v_0 \in X$, with $\|v_0\| < R$ and $F(v_0) < +\infty$, such that

$$(d_0') \qquad (Av, v - v_0) + F(v) - F(v_0) > 0$$

$$\text{for all } v \in X \text{ with } \|v\| = R$$

Then, problem (29) above has a solution u .

Remark 7 : Condition (d_0') can be obviously replaced by the stronger condition :

(d_1') <u>There</u> <u>exists</u> $v_0 \in X$, <u>with</u> $F(v_0) < +\infty$, <u>such</u> <u>that</u>

$$(Av, v - v_0) + F(v) \longrightarrow +\infty \quad \text{as} \quad \|v\| \longrightarrow +\infty ,$$

(cfr Remark 6 above).

Let us also notice that if the effective domain of F is bounded, then (d_0') is automatically satisfied, for it suffices then to choose any v_0 with $F(v_0) < +\infty$ and R large enough, so that $F(v) = +\infty$ whenever $\|v\| = R$.

Proof of Theorem 4 . In terms of the epigraph formulation of Sections 5 and 9 of Chapter 1, problem (29) can be equivalently written as follows

$$(30) \qquad \tilde{u} \in \tilde{K} \; : \; (\tilde{A}\tilde{u}, \tilde{v} - \tilde{u}) \geq 0 \quad . \quad \forall \tilde{v} \in \tilde{K}$$

where

$$\tilde{A} = A \times 1$$

and

$$\tilde{K} = \text{epi } F \; .$$

As we know from Section 7 of Chapter 1, to find a solution u of the inequality above, it suffices to find a vector $\tilde{u}$ of $\tilde{K}$ which is a <u>local</u> solution of (30) .

To find such a local solution $\tilde{u} = [u, \alpha]$ let us consider the auxiliary problem

(31) . $\tilde{u} \in \tilde{K}_R \quad : \quad (\tilde{A}\tilde{u}, \ \tilde{v} - \tilde{u}) \geqslant 0 \qquad \forall \ \tilde{v} \in \tilde{K}_R$

where

$$\tilde{K}_R = \tilde{K} \cap (B_R \times I) = \{ [v, \beta] \in X \times \mathbb{R} : F(v) \leqslant \beta ,$$

$$\| v \| \leqslant R , \quad a \leqslant \beta \leqslant b \}$$

is the intersection of the epigraph $\hat{K}$ of F with the "cylinder" $B_R \times I$. Here

$$B_R = \{ v \in X \ : \ \| v \| \leqslant R \}$$

is the closed ball of X whose radius is the constant R that appears in condition (d_0'), while

$$I = [a, b]$$

is a closed bounded interval of the real line, which we assume to have been closen large enough, as we shall specify later .

Since the map $\tilde{A}$ is obviously bounded, monotone and hemicontinuous in $X \times \mathbb{R}$ and $\tilde{K}$ is a bounded closed convex subset of $X \times \mathbb{R}$, the existence of a solution $\tilde{u} = [u, \alpha]$ of problem (31) is now a consequence of Theorem 3.

To prove that $\tilde{u} = [u, \alpha]$ is also a <u>local</u> solution of our <u>initial</u> problem (30), it suffices now to show that u does not belong to the boundary of the "cylinder" $B_R \times I$, which is to say, that we have

(32) $\|u\| < R$ and $a < \alpha < b$.

Let us now suppose that a and b were so chosen as to satisfy

(33) $$a < - c_0 R - c_1 ,$$

where $c_0 > 0$ and $c_1 > 0$ are such that

$$F(v) \geqslant - c_0 \|v\| - c_1 \qquad \text{for all} \quad v \in X$$

[any l.s.c. convex F can be bounded from below in this way], and

(34) $$b > \|A v_0\| (R + \|v_0\|) + F(v_0) ,$$

v_0 being the vector appearing in condition (d'_0)

Let us now interpret $\tilde{K}_R$ as the intersection

$$\tilde{K}_R = (\text{epi } F_R) \cap (X \times I)$$

with the "strip" $X \times I$ of the epigraph of the functional F_R obtained by putting $F \equiv +\infty$ outside the ball B_R . Note that $F_R \not\equiv +\infty$, for $F_R(v_0) = F(v_0) < +\infty$. Moreover, by our previous choice (33) of a we now have,

$$a \leqslant \inf F_R, \quad \text{actually} \quad a < \inf F_R .$$

[In fact, by (33), we have

$$a < - c_0 R - c_1 \leqslant - c_0 \; \| v \| \; - c_1 \leqslant F(v) \quad \text{for all} \quad \| v \| < R$$

hence $a < \inf F_R$]

Thus, by Lemma 3 and Remark 17 of Section 9 of Chapter 1, since $\tilde{u} = [u, \alpha]$ is a solution of inequality (31), then u is a solution of the mixed inequality

(35) $$u \in X \; : \; (Au, \; v - u) \geqslant F_R(u) - F_R(v) \quad \forall v \in X, \; F_R(v) \leqslant b$$

and

$$\alpha = F_R(u) \quad .$$

By putting now $v = v_0$ in (35), [note that $F_R(v_0) = F(v_0) < b$ by (34)] we find that $F_R(u) < + \infty$, hence, $F_R(u) = F(u)$ and then

(36) $$(Au, \; v_0 - u) \geqslant F(u) - F(v_0) \quad .$$

By condition (d_0''), this implies $\| u \| < R$, which is the "first strict bound in (32) we had to prove. It remains to prove the bounds on α. Again by inequality (36) above and the monotonicity of A, we have

$$\alpha = F_R(u) = F(u) \leqslant (Av_0, \; v_0 - u) + F(v_0) \leqslant$$

$$\leqslant \; \| Av_0 \| \; (\| v_0 \| \; + R) + F(v_0)$$

hence $\alpha < b$, by our choice (34) of b . On the other hand, as we have already seen, we also have

$$\alpha = F_R(u) \geqslant \inf F_R > a . \qquad ■$$

Reamrk 8 . Theorem 4 was deduced above from Theorem 3 . However, it clearly implies in turn both Theorem 3 (which is obtained by taking $F \equiv 0$) and the direct existence theorem, Theorem 4 , of Chapter 2 (which is obtained by taking $A \quad 0$) . ■

Existence theorems for mixed variational inequalities such as (29) have been given by C. Lescarret [1] , J. Lions and G. Stampacchia, [1] , F. E. Browder [8] , [9] , R. T. Rockafellar [6]. Theorem 4 is essentially due to F. E. Browder, loc. cit. See also U. Mosco [2] , where the proof given above is taken from .

In the remaining of the present section we shall apply Theorem 3 to prove the existence of fixed-points of so-called inward non-expansive mappings

$$U : K \longmapsto V$$

of a closed convex subset K of a Hilbert space V into V .

Let us recall that U is said to be an inward mapping if for every $v \in K$ the vector Uv belongs to some ray

$$v + \lambda (z - v) , \qquad z \in K , \qquad \lambda \geqslant 0 .$$

Clearly, any mapping of K into K is an inward mapping, for then we can take $z = Uv$ and $\lambda = 1$.

The following theorem, that generalizes Schauder's theorem, is due to F. E. Browder [11] :

THEOREM 5 . Let U be an inward non-expansive mapping of a bounded closed convex subset K of a Hilbert space V into V , $K \neq \emptyset$. Then, U has a fixed point in K . Moreover, the set of all fixed points of U is a closed convex subset of K .

Proof. We know from Section 4 of Chapter 1 that u is a fixed point of U, i. e.,

$$Uu = u \quad ,$$

if and only if u is a solution of the variational inequality

$$(37) \qquad u \in K \ : \ (\mathcal{A} u \mid v - u) \geqslant 0 \qquad \forall \ v \in K \ ,$$

where

$$\mathcal{A} = I - U \ , \qquad I = \text{identity map of } V \ .$$

By taking Lemma 2 below into account, the existence of a solution u of (37) and the properties of the set of all solutions of (37) come as a consequence of Theorem 3 above and the Addendum 1 to Theorem 5 of Chapter 1 . ∎

LEMMA 2 . Let $U : K \mapsto V$ be non-expansive. Then, the map $\mathcal{A} = I - U$ is monotone and lipschiztian, in particular, bounded and continuous .

Proof. We have for every v, w of K

$$(\mathcal{A}v - \mathcal{A}w \mid v - w) = \|v - w\|^2 - (Uv - Uw \mid v - w)$$

$$\geqslant \|v - w\|^2 - \|Uv - Uw\| \, \|v - w\| \geqslant (1 - c) \|v - w\|^2 ,$$

provided

$$\|Uv - Uw\| \leqslant c \, \|v - w\| .$$

Therefore, if U is non-expansive $(c = 1)$, then $\mathcal{A} = I - U$ is monotone. Moreover,

$$\|\mathcal{A}v - \mathcal{A}w\|^2 = \|v - w\|^2 - 2(Uv - Uw \mid v - w) + \|Uv - Uw\|^2$$

$$\leqslant 4 \, \|v - w\|^2$$

hence $\mathcal{A}$ is lipschiztian . ∎

Remark 9 . If U is a contraction ($c < 1$ in the inequality above), then $\mathcal{A} = I - U$ is strongly monotone. More refined relations between "monotonicity" and "non-expansiveness" can be found in F. E. Browder [13] and Z. Opial [1] . For the construction of fixed-point, see also F. E. Browder-W. V. Petryshyn [1] [2] , S. Kaniel [1] .

An expository reference for fixed-points of non-expansive mappings and iteration methods of solution is D. G. de Figuerido [1] .

5. Finite-dimensional approximation I : the discrete problem .

Most of the methods that can be used for the numerical solution of variational inequalities arising in the applications can be put, essentially into the Ritz-Galerkin framework we discussed at the beginning of this chapter.

Let us briefly summarize the situation, trying again to point out which are the important choices that must be done in order to write down : first, an approximate problem yielding an approximate solution $u_h(x)$; second, a discrete problem allowing in practice the numerical evaluation of $u_h(x)$. We shall postpone to the following section the discussion of the convergence of the approximate solution $u_h(x)$ to the solution $u(x)$ of the given problem. Thus, the discretization parameter h must be always considered in the present section as fixed.

We shall assume for convenience, as in Section 1, that the elements of the space X are functions.

We start with an initial problem

$$(38) \qquad u \in K \; : \; (Au, v - u) \geq 0 \qquad \forall v \in K \; ,$$

that involves a normed space X, a map A of X into X^* and a convex subset K of X

[These should be all considered as our data. However, the same problem can be given a different formalization leading to different X , A and K . A new formalization can arise intrinsically from the problem at hand, or else it can be suggested by a technical motivation.

We shall see an example of that in connection with the usual finite-difference method in problems involving partial differential operators .]

To write down an approximate problem

$$(39) \qquad u_h \in K_h \; : \; (Au_h, v_h - u_h) \geq 0 \qquad \forall \; v_h \in K_h$$

we must conveniently choose a finite-dimensional subspace X_h of X

and a convex subset K_h of X_h.

[We could also take into account an approximation A_h of the map A, A_h being a map from X to X^* with a domain $D(A_h)$ that containes K_h. Then the approximate problem becomes

$$u_h \in K_h \quad : \quad (A_h u_h, v_h - u_h) \geq 0 \qquad \forall v_h \in K_h$$

Here too the pairing (.,.) is the duality pairing between X and X^* .]

The further choice of a basis

$$(40) \qquad \{\varphi_1^h (x), \ldots, \quad \varphi_n^h (x)\}$$

in X_h; $n = n_h$ being the dimension of X_h, allows us to associate with the approximate problem (39) a discrete problem

$$(41) \quad u^h \in C^h \quad : \quad (A^h u^h \mid v^h - u^h) \geq 0 \quad \forall v^h \in C^h$$

in the euclidean space $E^{n h}$

Here A^h is a map of E^n into E^n and C^h a convex subset of E^n, which are obtained from A and K_h as we now show.

Let

$$\pi_h \quad : \quad E^n \longmapsto X$$

be the injective map that brings the vector

$$\dot{v}^h = (v_q^h)_q$$

of E^{nh} into the function

(42) $$v_h(x) = \sum_q v_q^h \varphi_q^h(x)$$

of X .

Clearly

$$\pi_h(E^n) = X_h ,$$

hence π_h is a (norm) isomorphism of E^n onto X_h .

There is then a unique convex subset C^h of E^n , such that

$$\pi_h C^h = K_h ,$$

obviously, $C^h = \pi_h^{-1} K_h$.

[The actual determination of C^h can present in practice some difficulties, depending on the choice of X_h and the basis $\{\varphi_q^h\}$, hence of π_h , that has been done. In this respect see also the following Remark 10 .]

Moreover, the map π_h has a transpose

$$\pi_h^* : X^* \longmapsto E^n ,$$

whose kernel is the annihilator of X_h in X^* , and

$$A^h = \pi_h^* A \pi_h$$

is a map of E^n into E^n.

For all vectors v^h and w^h of E^n

we have

$$(A^h v^h \mid w^h) = (\pi_h^* A \pi_h v^h \mid w^h) =$$

$$= (A \pi_h v^h, \pi_h w^h) = (Av_h(x), w_h(x))\cdot$$

were

$$v_h(x) = \pi_h v^h , \qquad w_h(x) = \pi_h w^h .$$

Thus, <u>a vector</u> $u^h \equiv (u_q^h) \in E^n$ <u>is a solution of the discrete problem</u> (41) <u>if and only if the function</u> $u_h(x) = \pi_h u^h$, <u>i.e.</u>

(43) $$u_h(x) = \sum_q u_q^h \varphi_q^h(x)$$

<u>is a solution of the approximate problem</u> (39)

[We have in fact

$$(A^h u^h \mid v^h - u^h) = (Au_h, v_h - u_h) ,$$

were

$v_h = \pi_h v^h$, $u_h = \pi_h u^h$, and π_h is a 1-1 map of C^h onto K_h .]

To write down the discrete problem (41) explicitely, let us compute the components

$$(A^h u^h)_s = (A^h u^h \mid e_s^h) \quad , \quad s = 1, \dots, n$$

of the vector $A^h u^n$ of E^n in the canonical basis

$$e_s^h \equiv (\delta_{qs}^h)_q \qquad s = 1, \dots, n$$

of E^{nh} .

We have

$$(A^h u^h)_s = (\pi_h^* A \ \pi_h u^h \mid e_s^h) =$$

$$= (A \ \pi_h u^h, \ \pi_h e_s^h)$$

and since

$$\pi_h e_s^h = \varphi_s^h (x) \ ,$$

we find

$$(A^h u^h)_s = (Au_h(x), \ \varphi_s^h (x)) \ .$$

Therefore, $(A^h u^h)_s$ is expressed, in term of the components $(u_q^h)_q$ of u^h, by the same functions $A_s^h(u_1^h, \dots, u_n^h)$ introduced in Section 1 :

$$(44) \quad (A^h u^h)_s = A_s^h (u_1^h, \dots, u_n^h) = (A(\sum_q u_q^h \varphi_q^h), \varphi_s^h), \quad s = 1, \dots, n .$$

Thus, the discrete problem (41) is the same as the discrete problem of that section, namely

$$(45) \quad u^h \equiv (u_q^h) \in C^h \ : \ \sum_s A_s^h (u_1^h, \dots, u_n^h) \, . \, (v_s^h - u_s^h) \geqslant 0$$

$$\forall \ v^h \equiv (v_q^h) \in C^h \ .$$

We shall not discuss in our lectures which are the methods that can be used to solve numerically the discrete problem.

Let us only remark that if the map A^h is strongly monotone, then the iterative methods of Chapter 2 could be applied.

On the other hand, if A is the gradient of a convex functional and, therefore, the solution of (38) is also the vector that optimizes a convex function on a convex subset of E^n, then the numerical solution of (38) could be carried on by means of one of the several methods available in convex optimization, see for instance J. Céa [2] .

For a linear A, also pivoting methods as those typical of linear programming can be also tried. We shall see an example of that in the last section of the present Chapter.

Remark 10 The role played by the euclidean metric and the canonical basis $e_1 = (\delta_{q1})_q, \dots$ in associating the discrete problem (41) with the approximate problem (39) can be taken by any inner product $(. \,|\, .)$ and any orthonormal basis with respect to that inner product. This change will naturally affect the maps π_h and its transpose $\pi_h{}^*$, hence also the subset C^h of R^n and the map A^h of R^n into itself.

Since the discrete problem (45) will be modified in consequence, a suitable choice of the new metric and a new basis can facilitate the actual numerical solution of (45). In this regard, let us recall Remark 8 of Section 3 of Chapter 2 .

[The map π_h still is the linear map of R^n into X that brings the vector $\tilde{e}_s$ of the new basis $(\tilde{e}_s)_s$ chosen in R^n into the functions φ_s^h of the basis (40) of X_h ; hence $\pi_h v^h$, for any $v^h \in R^n$, is given again by (42), where $(v_q^h)_q$ are now the components of v^h , in the basis $(\tilde{e}_s)$. While the subset $C^h = \pi_h^{-1} K_h$ changes accordingly only to the change of π_h , the map $A^h = \pi_h^* A \pi_h$ will be also affected by the change of the metric, on which the transpose π_h^* depends. Let us also notice that the discrete problem is still given by (45) above, where the functions A^h are the same as before, provided the components $(v_q^h)_q$, $(u_q^h)_q$ of v^h and u^h are now taken with respect to the new basis in R^n .] ∎

<u>Remark 11</u> When A is the differential DF of a (convex) functional F on X , we know that the variational inequality (38) characterizes a solution u of the minimum problem :

(46) $u \in K$ minimizes F on K .

This suggests that we could directly solve this minimum problem, by taking an approximate problem

(47) $u_h \in K_h$ minimizes F on K_h

and then, once the basis $(\varphi_s^h)_s$ has been chosen in X_h, numerically solving the discrete euclidean problem

(48) $u^h \in C^h$ minimizes $\tilde{F}$ on C^h ,

Here $C^h = \pi_h^{-1} K_h$ and $\tilde{F} = F \cdot \pi_h$, that is

$$\tilde{F}(v_1^h, \ldots, v_n^h) = F\left(\sum_q v_q^h \varphi_q^h\right) , \quad v^h \equiv (v_q^h)_q \in E^n$$

It should be remarked, however, that the discrete variational inequality associated with the discrete minimum problem (48) above, that is

$$u^h \in C^h \; : \; (\nabla F(u^h) \mid v^h - u^h) \geq 0 \qquad \forall v^h \in C^h \; ,$$

is the same as the variational inequality (45), where

$$A_s^h(u_1^h, \ldots, u_n^h) = (DF\left(\sum u_q^h \varphi_q^h\right), \varphi_s^h) \; ,$$

that we may find by discretizing the variational inequality

$$u \in K \; : \; (DF(u), v - u) \geq 0 \qquad \forall v \in K \; ,$$

associated with the initial problem (46).

In fact, we have

$$\nabla \tilde{F} = \pi_h^* \, DF \; \pi_h$$

and the components of $\nabla\tilde{F}(u^h)$ in the canonical basis (e_s) of E^n are given by

$$\frac{\partial \tilde{F}}{\partial u_s^h}(u^h) = (\nabla\tilde{F}(u^h) \mid e_s) = (DF(\sum u_q^h \varphi_q^h), \varphi_s^h)$$

for $\pi_h e_s = \varphi_s^h$ and $\pi_h u^h = u_h = \sum u_q^h \varphi_q^h$. Thus, Riesz-Galerkin discretization and weak characterization of minimum problems are "commuting" operations

6. Finite-dimensional approximation, II : convergence of the approximate solutions .

We shall now give conditions on the map A and the approximants convex sets K_h in order that the approximate solutions $u_h(x)$ converge to the solution $u(x)$ of the initial problem.

As the stability theorem shows, the most natural convergence to be expected from the sequence $(u_h(x))$ is the weak convergence of $u_h(x)$ to $u(x)$ on the space X where A acts, together with the convergence to zero of the form $(Au_h - Au, u_h - u)$. As we said in Section 3, for a whole class of maps, by definition : those of type (S) , the convergence of $u_h(x)$ to $u(x)$ in the norm of the space X will then follows as a consequence.

Let us notice, however, that we shall not be able to give estimates of the error $\| u - u_h \|$ of the same type as those which are common in the analogue approximation of the equation $Au = f$. In this regard see Remark 15 below.

By taking the finite-dimensional existence theorem of Chapter 2 into account, we can state the stability theorem of Section 3 in the form of a "constructive" existence theorem, more suitable for our present aims.

THEOREM 6 Let A be a bounded, strictly monotone and hemicontinuous map of X into X^*.

Let K be a closed convex subset of X, $\neq \emptyset$, $(K_h)_h$ a sequence of finite-dimensional closed convex subset of X, such that

$$K = \lim K_h \qquad \text{in } X,$$

according to Definition 1 of Section 2.

Let us assume, furthermore, that the following coerciveness condition holds

There exists $v_0 \in \bigcap_h K_h$ such that

$$(\tilde{d}_1) \qquad (Av, v - v_0) \longrightarrow +\infty \qquad \text{as } \|v\| \longrightarrow +\infty.$$

Then, there exists for every h a unique solution u_h of the problem

$$u_h \in K_h \qquad (Au_h, v_h - u_h) \geqslant 0 \qquad \forall v_h \in K_h$$

and a unique solution u of the problem

$$u \in K : (Au, v - u) \geqslant 0 \quad \forall v \in K .$$

Moreover, u_h converges weakly to u in X and the form $(Au_h - Au, u_h - u)$ converges to zero.

COROLLARY If, in addition, A is of type (S), then u_h converges strongly to u in X.

Proof : Theorem 6 reduces to a special case of Theorem 2 once we have proved the existence of a bounded sequence of approximate solutions u_h. For given h, the existence of u_h can be proved by applying Theorem 3 to the map A and the set K_h. Let us only note that, if K_h is unbounded, then the coerciveness property (d_0) required in Theorem 3 is now an obvious consequence of our present assumption $(\tilde{d}_1)$ (see Remark 6 above). The boundedness of the sequence (u_h) is a further consequence of $(\tilde{d}_1)$: in fact, if v_0 is the vector that appears in $(\tilde{d}_1)$, $v_0 \in K_h$ hence

$$(Au_h, u_h - v_0) \leqslant 0 \quad ,$$

for every h, and this clearly implies, by $(\tilde{d}_1)$, that the sequence (u_h) is bounded in X. ∎

Remark 12 The assumption $(\tilde{d}_1)$ of Theorem 6 requires, in particular, that the approximants K_h have a non-empty intersection.

A more general condition avoiding this hypotesis can be found, for instance, in U. Mosco [4], namely Theorem 3. See also the following remark. ∎

Remark 13 Condition $(\tilde{d}_1)$ in Theorem 6 can be replaced by the following assumption on A : there exists a function

$$\gamma : [0, +\infty) \mapsto [0, +\infty] \quad ,$$

continuous and strictly increasing at 0^+, with $\gamma(0) = 0$ and $\gamma(r) \to +\infty$ as $r \to +\infty$, such that

$$\|v - w\| \, \gamma(\|v - w\|) \leq (Av - Aw, v - w), \qquad v, w \in X .$$

Let us also note, in particular, that a map A of this type satisfies the property (C) mentioned in the corollary of Theorem 6, see Remark 2 of Section 3 ∎

Remark 14 The approximate solutions considered in Theorems 6 are required to satisfy the initial inequality exactly. This corresponds, indeed, to the fact that the map A was left unchanged in the approximate problem. However, a simultaneous approximation of A could be also taken into account and in this regard we recall Remark 3 of Section 3 . ∎

As we said at the beginning, the problem we are concerned with in the present section is the convergence of the approximate solutions u_h to the initial solution u. In this respect, the meaning of Theorem 6 above is that, for inequalities involving a map A of the type considered in that theorem, the proof of the weak or strong convergence of u_h to u is reduced to the proof of the convergence of the approximate sets K_h to the convex set K of the initial problem, in the sense of Section 2 . The proof that $K = \lim K_h$ can be achieved, in some cases, by using one of the general convergence results given in Section 2 or otherwise by carrying it out directly in the specific situation at hand.

We shall now consider some examples.

(a) Ritz-Galerkin approximation. An example of the first circumstance mentioned above is the internal approximation of Ritz-Galerkin type of a convex set with a non-empty interior. In this case, given an initial convex set K in the space X, the approximants K_h are simply chosen to be the finite-dimensional sections.

$$K_h = K \cap X_h$$

of K, with respect to an increasing sequence of finite dimensional subspaces X_h of X such that X = closure of $\bigcup_h X_h$.

Then we know from (c) of Section 2, that we have convergence of K_h to K, hence, under the assumption of Theorem 6, convergence of u_h to u in the sense of that theorem, provided the interior of K is not empty

If $\overset{o}{K}$ is empty, then a condition of type mentioned in (a) of Section 2 could be cheked.

(b) Internal approximation of Sobolov spaces. A general scheme of internal approximation, which is typical, for instance, of the finite-element methods, can be described as follows.

K is a closed convex subset of X:

(V^h) is a sequence of finite-dimensional spaces,,

L^h is, for every h, a closed convex subset of V^h.

We assume that there exists for every h an injective map

$$p_h : V^h \longmapsto X$$

and a map

$$r^h : X \longmapsto V^h$$

such that (i) and (ii) below hold :

(i) $p_h L^h \subset K$ <u>for every</u> h

(ii) $\| v - p_h r^h v \| \to 0$ <u>for every</u> v <u>of</u> K

Under these assumption, if

$$K_h = p_h L^h$$

we obviously have

$$K = \lim K_h \qquad \text{in } X .$$

Differently then in example (a), this scheme requires that the convergence condition (ii) must be checked directly in the specific situation at hand. We shall see now a classical example of this kind of approximation, namely the internal approximation of the Sobolev space $H_0^1(\Omega)$. For this, as well as for the example of <u>external</u> approximation of $H_0^1(\Omega)$ we shall give in the subsection (c) below, we refer to J. Cea [1], J. P. Aubin [1-5], F. Di Guglielmo [1].

<u>Example 1. Internal approximation of</u> $H_0^1(\Omega)$.

Here $X = H_0^1(\Omega)$, where Ω is a bounded open subset of R^n, $H_0^1(\Omega)$ the Sobolev space of all real functions $v(x)$ on Ω,

$$x = (x_1, \ldots, x_n) \in R^n ,$$

such that $v \in L^2(\Omega)$ and v_{x_i}, distribution derivative of v, also belongs to $L^2(\Omega)$, for every $i = 1,\dots,n$.

$H^1_0(\Omega)$ is a reflexive Banch space with the norm

$$\|v\|_{H^1_0(\Omega)} = \left(\|v\|^2_{L^2(\Omega)} + \sum_{i=1}^{n} \|v_{x_i}\|^2_{L^2(\Omega)} \right)^{1/2}$$

Given a discretization parameter

$$h = (h_1,\dots,h_n) \quad , \qquad h_i > 0 \qquad \text{for every } i \; ,$$

we shall now define an injective map

$$p_h : V^h \longmapsto X_h \quad , \qquad X_h \text{ subspace of } X = H^1_0(\Omega) \; ,$$

where V^h is an n_h-dimensional vector space, n_h depending on h and on the given Ω.

Let us consider, indeed :

- $\alpha(\xi)$ = the characteristic function of the real interval $[-\frac{1}{2}, \frac{1}{2}]$

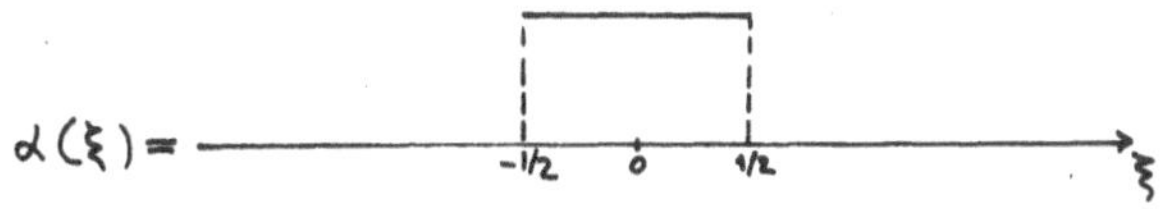

- $\alpha * \alpha\ (\xi)$ = the one-dimensional "tent" function

$\alpha * \alpha(\xi) =$

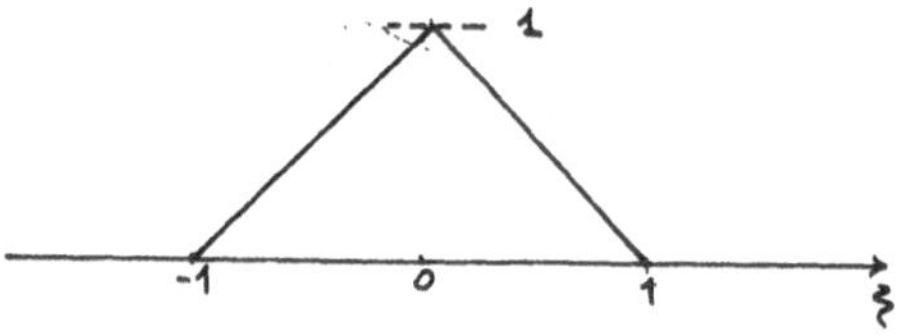

- for any multi-index $q = (q_1, \ldots, q_i, \ldots q_n)$, $q_i \in \mathbb{N}$ $\forall i$, let

$$\varphi_q^h(x) = \alpha * \alpha(\frac{x_1}{h_1} - q_1) \ldots \alpha * \alpha(\frac{x_i}{h_i} - q_i) \ldots$$

$$\ldots \alpha * \alpha(\frac{x_n}{h_n} - q_n) ,$$

i.e., the n-dimensional tent function whose support is the "n-cube"

$$\{x = (x_i) : (q_i - 1)h_i \leq x_i \leq (q_i + 1) h_i , \quad i = 1, \ldots, n\}$$

with center at the point

$$Q = (q_1 h_1, \ldots, q_i h_i, \ldots, q_n h_n)$$

and "edge lenght"

$$2h = (2h_1, \ldots, 2h_i, \ldots 2h_n) \qquad :$$

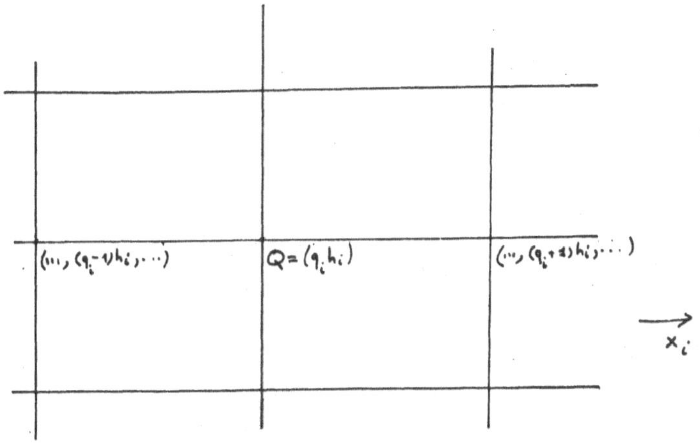
(..., (q_i-1)h_i, ...)
Q=(q_i h_i)
(..., (q_i+1)h_i, ...)
x_i

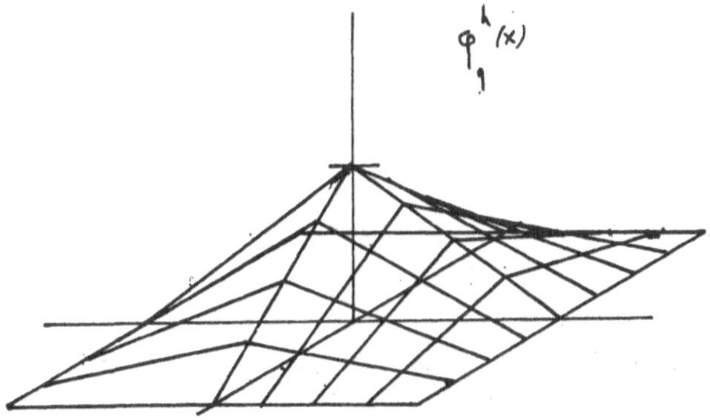
φ_q^h(x)

- Q^h = the set of all multi-indices $q = (q_1, \ldots, q_n)$, such that

$$\text{supp } \varphi_q^h \subset \Omega \quad ,$$

and

$$\Omega^h = \bigcup_{q \in Q^h} \text{supp } \varphi_q^h$$

Note that

$$\varphi_q^h(x) \in H_0^1(\Omega) \text{ for every } q \in Q^h \quad ;$$

- $V^h \equiv V^h(\Omega)$ = the space of all real vectors

$$v^h = (v_q^h)_{q \in Q^h} \quad ;$$

-the map

$$p_h : V^h = V^h(\Omega) \longmapsto X = H_0^1(\Omega)$$

given by

$$p_h v^h \equiv v_h(x) = \sum_{q \in Q^h} v_q^h \varphi_q^h(x) \quad , \quad v^h = (v_q^h)_{q \in Q^h} \quad ,$$

that is

$$v_h(x) = \sum_{q \in Q^h} v_q^h \left[\alpha * \alpha \left(\frac{x_1}{h_1} - q_1\right) \ldots \alpha * \alpha \left(\frac{x_n}{h_n} - q\right) \right]$$

- $X_h = p_h V^h(\Omega)$, the subspace of $X = H_0^1(\Omega)$

generated by the basis functions

$$\varphi_q^h(x) \quad , \qquad q \in Q^h$$

Since p_h is obviously injective, V^h and X_h are isomorphic and in many arguments they can be indeed identified ;

- the map

$$r^h \; : \; X = H_0^1(\Omega) \longmapsto V^h = V^h(\Omega)$$

which associates the function $v(x)$ with the vector $v^h = (v_q^h)_{q \in Q^h}$

whose q-component

$$v_q^h = \frac{1}{h_1 \cdots h_n} \int_{(q_1 - \frac{1}{2})h_1}^{(q_1 + \frac{1}{2})h_1} \cdots \int_{(q_n - \frac{1}{2})h_n}^{(q_n + \frac{1}{2})h_n} v(x)\, dx_1 \ldots dx_n$$

is the mean-value of $v(x)$ on the region

$$\left\{ x = (x_i) \; : \; (q_i - \tfrac{1}{2})h_i \leqslant x_i \leqslant (q_i + \tfrac{1}{2})h_i \; ; \qquad i = 1, \ldots, n \right\}$$

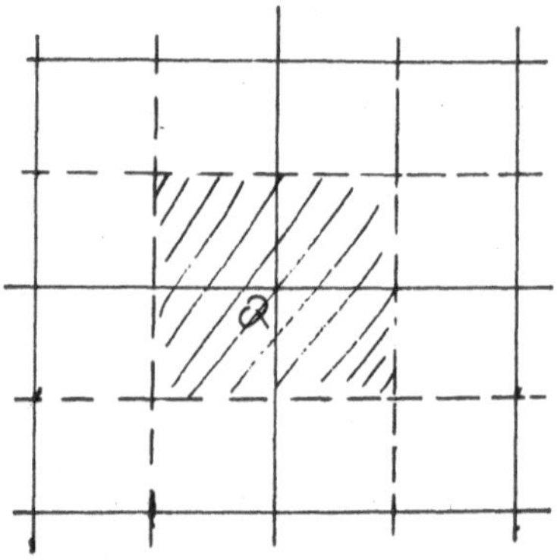

The main approximation results can then be summarized as follows :

<u>Lemma</u> 3 <u>For every</u> $v \in H_0^1(\Omega)$, $p_h \; r^h v \in H_0^1(\Omega)$ <u>and</u>

$$\| v - p_h \; r^h v \|_{H_0^1(\Omega)} \longrightarrow 0 \qquad \text{as} \quad |h| \to 0$$

<u>Corollary</u> 1 <u>If</u> $X = H_0^1(\Omega)$ <u>and</u> $X_h = p_h \, V^h(\Omega)$, <u>then</u> $X = \lim X_h$.

<u>Corollary</u> 2 <u>If</u>

$$K = \{ v \in H_0^1(\Omega) : \; v \geqslant 0 \;\; \text{a.e.} \;\; \text{in} \;\; \Omega \}$$

<u>and</u>

$$K_h = p_h \, L^h \; , \quad L^h = \{ v^h \equiv (v_q^h) \in V^h(\Omega^h) : \; v_q^h \geqslant 0 \quad \forall q \}$$

<u>then</u>

$$K = \lim K_h \quad \text{in} \quad H_0^1(\Omega)$$

For the estimate of the error $\| v - p_h r^h v \|$ we refer to the papers quoted above .

(c) <u>External approximation of Sobolev spaces</u> . The examples mentioned up to now have the common feature that the set K is approximated by sets K_h which are contained in K , and it is to stress this fact that we used the term internal approximation.

However, it is easy to adapt the scheme (b) to the more general situation in which the approximant set K_h is not required to be contained in K .

It suffices, indeed, to replace condition (i) with the condition

(ii) *If* $v_j \in p_{h_j} L^{h_j}$, *where* $(L^{h_j})_j$ *is a subsequence of* $(L^h)_h$, *and* v_j *converges weakly to* v *in* X , *then* $v \in K$

In fact, this condition together with condition (j) above is equivalent to the limit

$$K = \lim p_h L^h \quad \text{in} \quad X \quad .$$

Although improperly, we may call this kind of approximation an *external* approximation.

As it well known, the finite-difference methods for the approximation of partial differential operators can be put in the external framework of approximation we have just mentioned. As an example of that, let us briefly describe the external approximation of the Sobolev space $H_0^1(\Omega)$.

Example . *External approximation of* $H_0^1(\Omega)$

Let us consider :

- for each $i = 1,\dots,n$, the space

$$[H^1(\Omega)]_i = \{ v \in L^2(\Omega) : v_{x_i} \in L^2(\Omega) \}$$

with the norm

$$\| v \|_{[H^1(\Omega)]_i} = (\| v \|^2_{L^2(\Omega)} + \| v_{x_i} \|^2_{L^2(\Omega)})^{1/2} ;$$

- the product space

$$X = \prod_{i=1}^{n} [H^1(\Omega)]_i$$

of all vector functions

$$\underline{v}(x) = (v_1(x),\dots,v_n(x)) , \quad v_i(x) \in [H^1(\Omega)]_i , \quad i = 1,\dots,n,$$

with the product topology ;
-the closed subspace

$$X_0 = \{ \underline{v}(x) = (v_i(x))_i : v_i = v \in H^1_0(\Omega) , \quad i = 1,\dots,n \}$$

contained in the diagonal of X . We shall identify the space $H^1_0(\Omega)$ with X_0, that is, we shall identify the function $v(x) \in H^1_0(\Omega)$ with the vector function

$$\underline{v}(x) = (v(x),\dots,v(x)) \text{ in } X_0 ;$$

- the n-dimensional tent functions "smooth in the i-direction" :

$$\varphi_q^{h,i}(x) = \alpha(\frac{x_1}{h_1} - q_1)\ldots \quad \alpha * \alpha(\frac{x_i}{h_i} - q_i)\ldots\ldots \alpha(\frac{x_n}{h_n} - q_n),$$

$i = 1,\ldots,n,$ where $q = (q_1,\ldots,q_n)$ is a multi-index and $\alpha(\xi)$ is the one-dimensional tent function considered in the previous example :

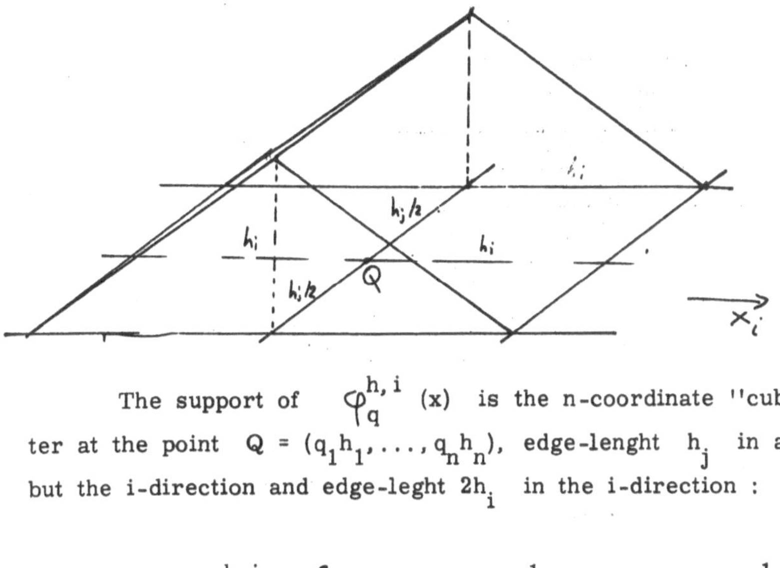

The support of $\varphi_q^{h,i}(x)$ is the n-coordinate "cube" with center at the point $Q = (q_1h_1,\ldots,q_nh_n)$, edge-lenght h_j in all directions but the i-direction and edge-leght $2h_i$ in the i-direction :

$$\text{supp } \varphi_q^{h,i}(x) = \Big\{x = (x_i)_i : (q - \tfrac{1}{2})h_r \leqslant x_r \leqslant (q_r + \tfrac{1}{2})h_r,$$

$$r = 1,\ldots,i-1,\ i+1,\ldots n, (q_i - 1)h_i \leqslant x_i \leqslant (q_i + 1)h_i \Big\}$$

- Q^h_{est} = the set of all multi-indices $q = (q_1\ldots q_n)$, such that the "cross region"

$$\bigcup_{i=1}^{n} \text{supp } \varphi_q^{h,i}(x) \quad ,$$

with center at the point $Q = (q_i\, h_i)$ and "arm-lenght" $2h_i$ in every direction, is contained in Ω ;

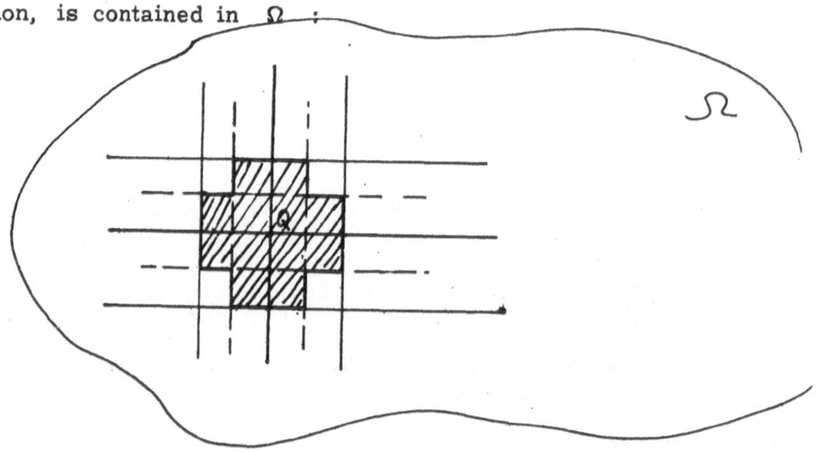

and

$$\Omega^h_{est} = \bigcup_{q \in Q^h_{est}} \Big(\bigcup_{i=1}^{n} \text{supp } \varphi_q^{h,i} \Big).$$

Clearly, for every $i = 1,\dots,n$

$$\varphi_q^{h,i}(x) \in [H^1(\Omega)]_i \qquad \text{for all} \quad q \in Q^h_{est} \quad ;$$

- $V^h_{est} = V^h_{est}(\Omega)$ = the space of all real vectors

$$v^h = (v_q^h)_{q \in Q^h_{est}} \qquad ;$$

- the map

$$\underline{p}_h \equiv (p_{h,i})_i \;:\; V^h_{est} = V^h_{est}(\Omega) \mapsto X = \prod_i \left[H^1(\Omega) \right]_i$$

$$v^h \mapsto \underline{p}_h v^h = \left(p_{h1} v^h , \; \ldots \; , p_{hn} v^h \right) ,$$

where, for every $i = 1, \ldots, n$, the map

$$p_{hi} \;:\; V^h_{est} \longmapsto \left[H^1(\Omega) \right]_i$$

is given by

$$p_{hi} v^h = v_{hi}(x) = \sum_{q \in Q^h_{est}} v^h_q \cdot \varphi^{hi}_q (x) , \quad v^h = (v^h_q)_{q \in \Omega^h_{est}} .$$

- $X_h = \underline{p}_h V^h_{est}(\Omega)$, the subspace of $X = \prod_i \left[H^1(\Omega) \right]_i$

generated by the basis vector-functions

$$\underline{\varphi}^h_q(x) = (\varphi^{h1}_q (x), \ldots , \varphi^{hn}_q (x)) , \qquad q \in Q^h_{est}$$

Clearly, $\underline{p}_h$ is an isomorphism of $V^h_{est}(\Omega)$ onto X_h, which associates the vector function of $X = \prod_i [H^1(\Omega)]_i$,

$$\underline{v}_h(x) = \underline{p}_h\, v^h = (v_{h1}(x), \dots, v_{hn}(x)) \quad ,$$

where $v_{hi}(x) = p_{hi}\, v^h$, $i = 1, \dots, n$, with the vector v^h of $V^h_{est}(\Omega)$.

Finally, we shall still denote by r^h the map with associates every function $v(x)$ $L^2(\Omega)$ with the vector

$$v^h = (v^h_q)_q \in Q^h_{est} \quad \text{of} \quad V^h_{est}(\Omega) \quad ,$$

whose q-component is given by the mean value of $v(x)$ on the region

$$\{ x = (x_i)_i : (q_i - \tfrac{1}{2})h_i \leq x_i \leq (q_i + \tfrac{1}{2})h_i \quad , \ i = 1, \dots, n \}$$

<u>Lemma</u> 4 <u>Let</u> $v \in H^1_0(\Omega)$. <u>For each</u> $i = 1, \dots, n$, $p_{hi}\, r^h v \in [H^1(\Omega)]_i$ <u>and</u>

$$\| v - p_{hi}\, r^h v \|_{[H^1(\Omega)]_i} \longrightarrow 0 \qquad \text{as} \qquad |h| \to 0 \ .$$

<u>Lemma</u> 5 <u>Let</u> $v^h = (v^h_q)_q \in Q^h_{est}$ <u>and suppose that for every</u> $i = 1, \dots, n$,

$$v_{hi}(x) = p_{hi}\, v^h$$

converges weakly to a function w_i in $[H^1(\Omega)]_i$ as $|h| \to 0$.

Then, we have

$$w_i(x) \equiv v(x) \quad \text{a.e.} \quad \text{for all} \quad i = 1, \dots, n$$

where $v(x)$ belongs to $H_0^1(\Omega)$. Moreover, $\underline{v}_h(x) = \underline{p}_h\, v^h$ converges weakly to $\underline{v}(x) = (v(x), \dots, v(x))$ in $X = \prod_i [H^1(\Omega)]_i$, as $|h| \to 0$.

Corollary 1 : Let $X_0 \simeq H_0^1(\Omega)$ and X_h the subspace of $X = \prod_i [H^1(\Omega)]_i$ spanned by the basis vector-functions $\underline{\varphi}_q^h(x)$, $q \in Q_q^h$. Then

$$X_0 = \lim X_h \quad \text{in} \quad X \quad \text{as} \quad |h| \to 0$$

Corollary 2 : Let $K = \{ v \in H_0^1(\Omega) : v \geq 0 \text{ a.e. in } \Omega \}$ be identified with a cone K in X and

$$\underline{K}_h = \underline{p}_h\, L^h \,, \quad L^h = \{ v^h = (v_q^h)_{q \in Q_{est}^h} : v_q^h \geq 0 \quad \forall q \}.$$

Then,

$$\underline{K} = \lim \underline{K}_h \quad \text{in} \quad X \quad \text{as} \quad |h| \to 0 .$$

(d) Projection methods. The approximation methods discussed

so far may be called of injective type : they are based on suitable injective mappings p_h of some finite-dimensional space V^h into X, carrying some convex subset L^h of V^h in a convex suset K_h of X that approximates the given K.

However, a conceptually different type of approximation is also possible, which is based instead on some projection mappings p_h of X onto a finite-dimensional subspace X_h of X : in this case, the map p_h carries the given set K onto an approximate K_h in X_h.

The most natural setting for these methods involves an Hilbert space X, an increasing sequence of finite-dimensional subspace X_h of X, with $\bigcup_h X_h$ dense in X, and, for each h, the orthogonal projection p_h of X onto X_h.

If $\mathcal{R}$ is a map of X into X and K is a bounded closed convex subset of X, then the problem

$$u \in K \; : \; (\mathcal{R}u \,/\, v - u) \geqslant 0 \qquad \forall v \in K$$

may be approximated by the sequence of problems

$$u_h \in K_h \; : \; (\mathcal{R}_h u_h \,|\, v_h - u_h) \geqslant 0 \qquad \forall v_h \in K_h$$

where

$$\mathcal{R}_h = p_h \mathcal{R} p_h$$

is a map of X_h into X_h and

$$K_h = p_h K$$

is a (bounded) closed convex subset of X_h (see (d') of Section 2) .

Let us remark that the approximate problem above is in the space X_h. However, its solution u_h is the same as that of the problem

$$u_h \in K_h \ : \ (\mathcal{A} u_h \mid v_h - u_h) \geqslant 0 \qquad \forall v_h \in K_h$$

in the space X . In fact, we have

$$(\mathcal{A}_h u \mid w) = (p_h \mathcal{A} p_h u \mid w) = (\mathcal{A} p_h u \mid p_h w)$$

hence

$$(\mathcal{A}_h u_h \mid w_h) = (\mathcal{A} u_h \mid w_h) \text{ for every } u_h, w_h \in X_h .$$

Therefore, the proof of the convergence of the approximate solutions u_h still can rely on Theorem 6, by taking now the convergence result stated in (e') of Section 2 into account. We refer to the author's paper [4] , e.g., Proposition 3.1, for more details on this point.

Projection methods for solving <u>equations</u> involving non-linear operators in Banch spaces have been extensively investigated by many authors, let us mention here F. E. Browder, [10] [12] , F. E. Browder and W. V. Petryshyn [1] and the review paper by R. I. Kachurovskii [3] , where further reference on the subject can be found. See also D. G. de Figuerido [1] .

<u>Remark</u> 15 . The usual estimates of the error $\|u - u_h\|$ in the Ritz-Galerkin approximation of an equation $Au = f$ involving a linear map A are based on the inequality

$$\| u - u_h \| \leq c. \text{ dist } (u, X_h) ,$$

X_h being the subspace of X, which the approximate solution u_h belongs to. This estimate, however, is in general false for variational inequalities, even if it refers to an internal approximation of convex cones.

This can be seen with the following simple example, due to G. Strang : $u \equiv (0,0)$ is the vector that minimizes the distance functional $F(v) = \frac{1}{2} \| v - z_0 \|^2$, $v = (v_1, v_2) \in E^2$ $z_0 = (-1, 0)$, on the half plane $v_1 \geq 0$ of the euclidean space E^2 , while $u_h = (0, -h)$ is for a given $h > 0$ the vector that minimizes $F(v)$ on the cone K_h described in the figure below

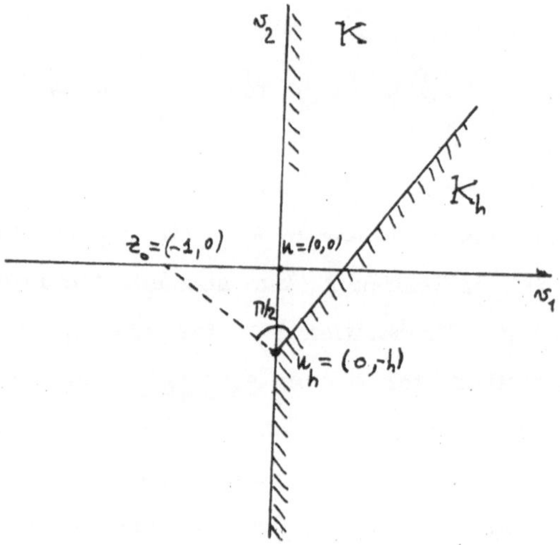

Then, $\| u - u_h \| = h$, whereas dist $(u, K_h) \simeq h^2$ for $h > 0$ small.

7 Dual variational inequalities and complementarity systems

It is well known that many minimum problems of the calculus of variations and optimization theory admit a "complementary" on "dual" formulation. On this matter, let us only refer here, for instance, to A. M. Arthurs [1], J. Stoer-Witzgal [1], J. Cea [2], Robinson [1], Moreau [2], U. Dieter [1][2].

For variational inequalities too, many "dual" characterizations of the solutions can be given, which are based, essentially, on separation or mini-max theorems. A discussion of some of these dual methods can be found in J. L. Lions, R. Glowinski, R. Tremolières, loc. cit.

As we shall see below, it is always possible, at least in principle, to associate a variational inequality in X^* - the "dual" inequality - with any given variational inequality in the space X - the "primal" inequality - in such a way that a vector u of X is the solution of the primal inequality if and only if the vector $u^* = -Au$ of X^* is a solution of the associated dual inequality. However, the explicit formulation of the dual inequality may often be in practice a difficult problem in itself.

We shall first consider variational inequalities on convex cones. The dual scheme we have in mind becomes then particularly simple and both the primal and dual inequalities can be characterized more symmetrically by means of a so-called (generalized) complementarity system.

Let indeed M be any map of X into X^*, H a convex cone with vertex at 0 in X, z a solution of the variational inequality

(48') $$z \in H : (Mz, w - z) \geq 0 \qquad \forall w \in H$$

Then, the pair z, $z^* = -Mz$ is a solution of the problem

$$(48'') \qquad z \in H \quad , \quad z^* \in H^* \quad , \quad (z^*, z) = 0$$

In fact, by putting $w = z + v$ into (48'), with v an arbitrary vector of H, we find

$$(Mz, v) \geq 0 \qquad \forall\ v \in H \quad ,$$

which is to say, $z^* \in H^*$. Moreover, by replacing now the vector w in (48') once with 0 and then with $2z$, we find

$$(Mz, -z) \geq 0 \qquad \text{and} \qquad (Mz, z) \geq 0$$

respectevely, therefore $(z^*, z) = 0$.

Conversely, if the pair z, $z^* = -Mz$ satsfies (48") above, then for every $w \in H$ we have

$$(Mz, w-z) = (-z^*, w-z) = (-z^*, w) \geq 0$$

since $z^* \in H^*$.

Therefore, we have proved the following

LEMMA 6 : Let M be any map of X into X^*, H a convex cone with vertex at 0 in X. A vector z is a solution of the veriational inequality

$$z \in H : (Mz, w - z) \geq 0 \qquad \forall\ w \in H$$

if and only if the pair z, $z^* = -Mz$ is a solution of the problem

$$z \in H \quad , \quad z^* \in H^* \quad , \quad (z^*, z) = 0$$

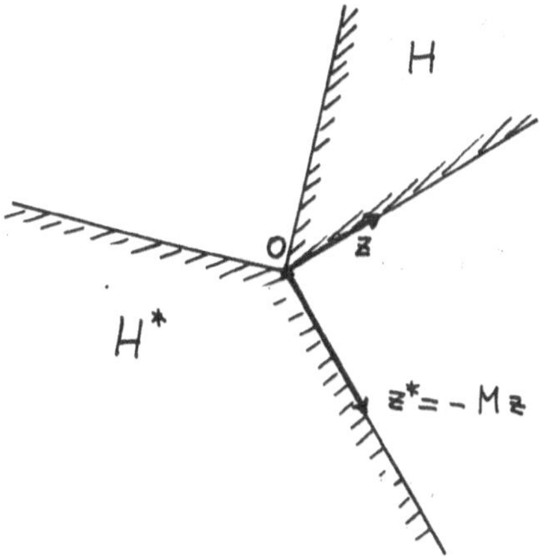

Let us suppose now that the map M is 1-1 of X onto X^* and let us define the map

$$M' : X^* \longmapsto X .$$

by

(49) $$M' w^* = - M^{-1}(-w^*) \qquad\qquad w^* \in X^* \quad .$$

Note that $M' = M^{-1}$ if M is linear. Let us also assume that H is a <u>closed</u> convex cone with vertex at 0. Then, if we denote by H^{**} the polar cone of H^* <u>in</u> X, i.e.

$$H^{**} = \{ z \in X : (z, z^*) \leq 0 \qquad \forall\, z^* \in H^* \}$$

a simple argument, based on the separation theorems for convex sets, shows that

$$H^{**} = H$$

Since the relation

$$z^* = - Mz$$

is clearly equivalent to

$$z = -M'z^*$$

by applying Lemma 6 once to the map M and the cone H and then to M' and H^*, we obtain the following

THEOREM 7 : Let M be a 1-1 map of X onto X^*, H a closed convex cone with vertex at 0 in X. If M' is the map of X^* on X given by (49) and H^* the polar cone of H in X^*, then the following three problems are equivalent

(i) $z \in H : (Mz, w - z) \geq 0 \quad \forall\, w \in H$

(ii) $z^* \in H^* : (M'z^*, w^* - z^*) \geq 0 \quad \forall\, w^* \in H^*$

(iii) $z \in H \;,\; z^* \in H^* \;,\; (z^*, z) = 0 \;,$

provided z and z^* are related by

$$z^* = - Mz \quad , \quad \text{i.e.,} \quad z = -M'z^* \quad .$$

Remark 16: If $M = DF$, F being a convex functional on X, then $M'z^* = -DF^*(-z^*)$, $z^* \in X^*$, where F^* is the conjugate functional of F (see Section 2). Then, the dual problems (i) and (ii) of Theorem 7 characterize the minimum problems

(i) z minimizes $F(w)$ on H

(ii) z^* minimizes $F^*(w^*)$ on $-H^*$

respectevely. Problems (i) and (ii) above are conjugate in the sense, for instance, of Fenchel's duality theorem, cfr. R. T. Rockafellar [2]. ■

Remark 17 : Let v_0 be a given vector of X, A a given 1-1 map of X onto X^*, A' defined as in (49). If we apply Theorem 7 to the map

$$Mz = A(z+v_0) \quad , \quad z \in X \quad ,$$

then we find that the following problems are equivalent

(i) $u \in v_0 + H : (Au, v - u) \geqslant 0 \quad \forall\ v \in v_0 + H$

(ii) $u^* \in H^* \quad : (A'u^* + v_0, v^* - u^*) \geqslant 0 \quad \forall\ v^* \in H^*$

(iii) $\quad u - v_o \in H \quad , \quad u^* \in H^* \quad , \quad (u^*, u - v_o) = 0$

provided

$$u^* = -Au \quad , \quad \text{i.e.} \quad , \quad u = -A'u^* \quad .$$

It suffices infact to make the change of variable $z = u - v_o$. ▨

When $X = \mathbb{R}^n \simeq X^*$ and $H \simeq - H^*$ is the non-negative ortant of $\mathbb{R}^n$, then problems such as (iii) above are known in the litarature as complementarity systems: linear c. s. if M is an affine map of $\mathbb{R}^n$ into itself; non linear c. s. in the general case. They arise in many problems of optimization and game theory, as well as in geometric or physical applications, and have been investigated by many authors: see R. Cottle-I. Dantzig [1] [2], R. Cottle [1], C.E. Lemke [1], S. Karamardian [1], where further reference on these systems and their applications can be found.

In these papers many algorithms for the numerical solution both of linear and non-linear complementarity systems have been given.

These algorithms, which are based mainly on suitable pivoting techniques, can thus be also used to solve discrete variational inequalities on convex cones. We shall see an example in the following Section 8.

In turn, the reduction of a complementarity system to a variational inequality, hence to a fixed-point problem, can be convenient in order to obtain more general existence results. Moreover, this reduction is also fruitful from an algorithmic point of view, for it makes it possible to use iterative methods of solution. For more details on the relation between variational inequalities and complementarity systems in finite-dimensional spaces we refer to Karamardian, loc. cit., I. Dolcetta [1], J. Moré [1].

Variational inequalities in connection with convex programming have been also investigated, from a computational point of view too, by O. G. Mancino-G. Stampacchia [1].

The duality for variational inequalities on convex cones considered above is a special case of a general dual scheme for variational inequalities of type

$$(50) \qquad u \in X : (Au, v - u) \geq F(u) - F(v) \quad , \quad \forall\, v \in X \ ,$$

where F is a l. s. c. convex functional on the normed space X, with values in $(-\infty, +\infty]$.

In this case, the dual variational inequality can be witten as

$$(51) \qquad u^* \in X^* : (A'u^*, v^* - u^*) \geq F^*(u^*) - F^*(v^*) \ , \qquad v^* \in X^*$$

where A' is defined as in (49) above and F^* is the Young-Fenchel conjugate of F, see Section 2.

It can be proved that a vector u is a solution of (50) if and only if the vector

$$(52) \qquad u^* = -Au \qquad\qquad (\text{i.e., } u = -A'u^*)$$

is a solution of (51). Moreover, both solutions u and u^* are characterized by the Young-Fenchel identity

$$(53) \qquad F(u) + F^*(u^*) = (u^*, u) \quad ,$$

where u and u^* are related as in (52) above.

The special case of theorem 7 is obtained by taking F to be the indicator function δ_H of the convex cone H in X (see Section 2), hence $F^* = (\delta_H)^* = \delta_{H^*}$ is the indicator function of the polar cone H^* of H in X^*.

[The dual prolems of Remark 17 are given, insteed, by $F = \delta_{v_o+H}$, hence $F^*(w^*) = \delta_{H^*}(w^*) + (w^*, v_o)$.] For more details we refer to U. Mosco[7].

Remark 18: When A is the differential of a convex functional G, than the dual problems (50) (51) characterise a pair of dual extrenum problems in the sense of Fenchel's duality theorem, see R. T. Rockafellar [2]. When X is a Hilbert space, $X^* \simeq X$, and A = identity map of X, then problems (50) (51) above and the equivalent system (53) are related to the so-called proximity mappings introduced by J. J. Moreau [2].

An application of a dual scheme of this type to prove the regularity of the solution has been given by H. Brezis [4].

Remark 19: The explicit formulation of the dual variational inequality (51) requires the knowledge of the inverse map A^{-1} and of the conjugate functional F^*. In particular, the calculation of F^* may be a difficult problem even for "simple" F. However, the dual scheme described in the present section can be modified in concrete situations, by making a sort of "change of variable" in the initial inequality before operating with the duality. In some cases this leads to a more feasible "partial" inversion of A and dualization of F. Dual extremum problems have been indeed investigated along these lines by R. Teman [1], by relying on a generalized form of Fenchel's duality theorem given by R. T. Rockafellar [7]. For a similar approach to dual variational inequalities see M. Matzeu [1].

In the following section we shall apply the dual scheme described above to the "obstacle problem" mentioned in Section 1 of Chapter 2 and we shall rely on it to give a method for the numerical approximation of the solution.

Let us also mention, in this respect, that a different approach to duality, based mainly on minimax techniques, has been also applied to the numerical solution of problems as those mentioned in Section 1 of Chapter 2, by J. Cea-R. Glowinski [1], [2], J. Cea-R. Glowinski-Nedelec [1], M. Nedelec [1], J. F. Bourgat [1]. See also J. Cea [2].

8. An example

Let us consider the obstacle problem of Section 1 of Chapter 3 (Example 3):

$$u \in H_0^1(\Omega) \quad , \quad u \geqslant \psi \quad \text{a.e. in } \Omega :$$

$$a(u, v-u) \geqslant 0 \quad \forall\, v \in H_0^1(\Omega)\ ,\ v \geqslant \psi \quad \text{a.e. in } \Omega\ ,$$

where $a(u,v)$ is the Dirichlet form

$$a(u,v) = \sum_{i=1}^{n} \int_{\Omega} u_{x_i} v_{x_i}\, dx \quad , \qquad \Omega \subseteq \mathbb{R}^n ,$$

and ψ is a fiven function in $H_0^1(\Omega)$.

We are in the situation described in Remark of the proceding section, with

$$X = H_0^1(\Omega)$$

$$A = -\Delta_2 : H^1_o(\Omega) \mapsto H^{-1}(\Omega)$$

$$H = \{ w \in H^1_o(\Omega) \cdot w \geq 0 \text{ a.e. in } \Omega \}$$

$$H^* = \{ \tau \in H^{-1}(\Omega) : \tau \text{ measure} \leq 0 \} .$$

Moreover,

$A' = A^{-1} = G : H^{-1}(\Omega) \mapsto H^1_o(\Omega)$ is the Green operator for the Dirichlet problem in Ω : for any measure τ in $H^{-1}(\Omega)$, the potential

$$v(x) = G\tau(x)$$

is given by

$$v(x) = \int_\Omega g(x, y) d\tau(y)$$

where $g(x, y)$ is the Green function for the Dirichlet problem in Ω.

The primal variational inequality now is

(50) $$u \in \psi + H : (-\Delta_2 u, v-u) \geq 0 \qquad \forall v \in \psi + H$$

while the dual ineouality now is

(51) $$\mu \in H^* : (G\mu + \psi, \tau - \mu) \geq 0 \qquad \forall \tau \in H^* .$$

Both these inequalities are equivalent to the complementarity sistem (see Remark 17 of the preceding section)

(52) $$z = u - \psi \in H \quad , \quad \mu \in H^* , \quad (\mu, z) = 0$$

where

(53) $\mu = \Delta_2 u$, i.e., $u = - G\mu$.

The approximate solution of problem (50), or of the equivalent direct minimum problem, has been studied by many authors, see R. Glowinski [2], M. Goursat [1], Sibony [2], Marzulli [1], G. Stampacchia [5], V. Comincioli-L. Guerra and G. Volpi [1] J. J. Moreau [7], J. F. Durand [1].

We shall summarize below the method followed in A. Fusciardi et al. [1]. The complementarity system (52) is approximated by a sequence of finite-dimensional complementarity systems, what gives a direct simultaneous approximation of the function u and the measure $\mu = \Delta u$, the solution of (50) and (51) respectevely, without any assumption of regularity.

This discretization is obtained by realizing an internal approximation of the cone of measures H^*, by means of unit masses supported by the (n-1)-dimensional meshes of a given coordinate subdivision of Ω.

We shall now describe this approximation, by taking, for sake of simplicity, n=2.

Let us consider a coordinate subdivision of $\mathbb{R}^2$ as that given in Ex. 1 of Section 6, $h=(h_1, h_2)$ being the discretization parameter and $Q = (q_1 h_1, q_2 h_2)$, $q = (q_1, q_2) \in Z^2$, the vertices of the subdivision.

For every $q=(q_1, q_2) \in Z^2$ we shall denote by $S^h_{q^1}$ and $S^h_{q^2}$ the <u>one-dimensional</u> (1=n-1) meshes of the subdivision which have Q as the left end point and the lowest end point, respectevely: that is,

$$S^h_{q^1} = \{ x = (x_1, q_2 h_2) : q_1 h_1 \leq x_2 \leq (q_1+1)h_1 \}$$

$$S^h_{q^2} = \{ x = (q_1 h_1, x_2) : q_2 h_2 \leq x_2 \leq (q_2+1)h_2 \}$$

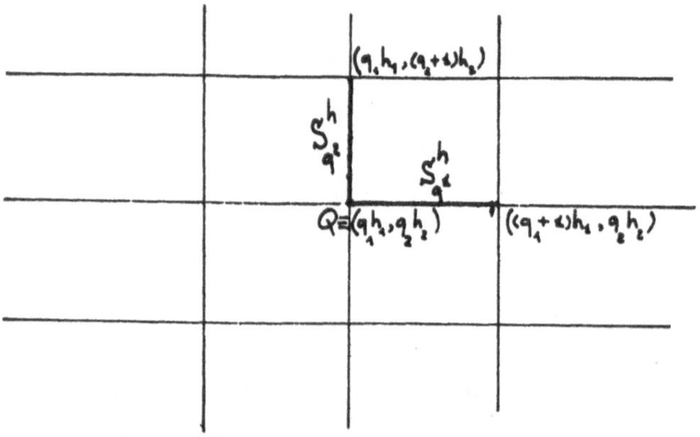

Let us now consider the functional $\sigma^h_{q^1}$ and $\sigma^h_{q^2}$ which associate the mean values on $S^h_{q^1}$ and $S^h_{q^2}$, respectevely:

$$\sigma^h_{q^1}(\varphi) = \frac{1}{|S^h_{q^1}|} \int_{S^h_{q^1}} \varphi \, dx_1 = \frac{1}{h_1} \int_{q_1h_1}^{(q_1+1)h_1} \varphi(x_1, q_2h_2)dx_1$$

$$\sigma^h_{q^2}(\varphi) = \frac{1}{|S^h_{q^2}|} \int_{S^h_{q^2}} \varphi \, dx_2 = \frac{1}{h_2} \int_{q_2h_2}^{(q_2+1)h_2} \varphi(q_1h_1, x_2)dx_2$$

with each function $\varphi \in C^\infty_o(\Omega)$.

It is easy to show that $\sigma^h_{q^1}$ and $\sigma^h_{q^2}$ satisy the estimates

$$|\sigma^h_{q^i}(\varphi)| \leq \left(\frac{\operatorname{diam}\Omega}{h_i}\right)^{1/2} \|\varphi\|_{H^1_0(\Omega)} \quad , \quad \forall \varphi \in C^\infty_o(\Omega) ,$$

for every q such that both $S^h_{q^1}$ and $S^h_{q^2}$ are contained in Ω. We shall denote by Q^h the set of all such q's.

[We have in fact

$$\left| \int_{x_1'}^{x_1''} \varphi(x_1, \bar{x}_2) dx_1 \right| \leq |x_1'' - x_1'|^{1/2} \left(\int_{x_1'}^{x_1''} |\varphi(x_1, \bar{x}_2)|^2 dx_1 \right)^{1/2} ;$$

for every $\varphi \in C_o^{\infty}(\Omega)$, we can find x_2^o such that

$$\varphi(x_1, \bar{x}_2) = \int_{x_2^o}^{\bar{x}_2} \frac{\partial \varphi}{\partial x_2}(x_1, x_2) \, dx_2$$

thus,

$$|\varphi(x_1, \bar{x}_2)|^2 \leq |x_2 - x_2^o| \int_{x_2^o}^{\bar{x}_2} \left| \frac{\partial \varphi}{\partial x_2}(x_1, x_2) \right|^2 dx_2$$

$$\leq (\operatorname{diam} \Omega) \int_{-\infty}^{+\infty} \left| \frac{\partial \varphi}{\partial x_2}(x_1, x_2) \right|^2 dx_2$$

hence

$$\int_{x_1'}^{x_1''} |\varphi(x_1, x_2)|^2 dx_1 \leq (\operatorname{diam} \Omega) \int_{\Omega} \left| \frac{\partial \varphi}{\partial x_2}(x_1, x_2) \right|^2 dx_1 \, dx_2$$

Therefore,

$$\frac{1}{|x_1'' - x_1'|} \left| \int_{x_1'}^{x_1''} \varphi(x_1, \bar{x}_2) dx_1 \right| \leq \left(\frac{\operatorname{diam} \Omega}{|x_1'' - x_1'|} \right)^{1/2} \|\varphi\|_{H_o^1} \quad .]$$

Then, $\sigma^h_{q^1}$ and $\sigma^h_{q^2}$, $q \in Q^h$, are both elements of the dual $H^{-1}(\Omega)$

of $H_0^1(\Omega)$ and since they are obviously non-negative (i.e. $\sigma_{qi}^h(\varphi) \geq 0$ provided $\varphi \geq 0$, $i=1,2$), we can conclude that σ_{q1}^h and σ_{q2}^h are non-negative measures belonging to $H^{-1}(\Omega)$.

We shall now denote by H_h the convex cone, with vertex at 0, generated by the non-positive measures $-\sigma_{qi}^h$, $q \in Q^h$, $i=1,2$:

$$H_h^* = \left\{ \tau \in H^{-1}(\Omega) : \tau = \sum_{\substack{q \in Q^h \\ i=1,2}} \tau_{qi}^h (-\sigma_{iq}^h), \quad \tau_{qi}^h \geq 0 \quad \forall q, i \right\}$$

Clearly,

$$H_h^* \subseteq H^* \qquad \text{for every } h,$$

where, let us recall it, H^* is the cone of all non-positive measures in $H^{-1}(\Omega)$.

The finite-dimensional cones H_h^* approximate the cone H^* in the sense of the following lemma:

LEMMA $H^* = \lim H_h^*$ in $H^{-1}(\Omega)$.

In view of the Corollary of Theorem 1 of Section 2, the convergence just stated is equivalent to the convergence of the polar cones: the polar cone of H^* in $H_0^1(\Omega)$ is the cone H of all non-negative functions of $H_0^1(\Omega)$ we started with, while the polar cone H_h of H_h^* is the cone of all functions $v \in H_0^1(\Omega)$ whose trace on each S_{qi}^h, $q \in Q^h$, $i=1,2$, has a non-negative mean value:

$$H = \left\{ v \in H_0^1(\Omega) : v \geq 0 \quad \text{a.e.} \quad \text{in } \Omega \right\},$$

$$H_h = \{ v \in H^1_0(\Omega) \quad : \quad (\sigma^h_{qi}, v) \geq 0 \ , \ \forall\, q \in Q^h, \ i=1,2 \}.$$

Now, it is not difficult to prove that

$$\lim H_h = \bigcap_h H_h = H \qquad \text{as } |h| \to 0 \ ,$$

we refer to A. Fusciardi et al., loc. cit.

We can thus apply the approximation sheme described in Sections 5 and 6 above.

The finite-dimensional problems that approximate problem (51) can be obtained by replacing H^* with H_h^* (we take $\psi_h = \psi$ for all h and we also leave the operator G unchanged):

$$(54) \qquad \mu_h \in H_h^* : (G\mu_h + \psi, \tau_h - \mu_h) \geq 0 \qquad \forall\, \tau_h \in H_h^*$$

which is equivalent to the complementarity system

$$(55) \qquad z_h \in H_h \ , \quad \mu_h \in H_h^* \ , \quad (\mu_h, z_h) = 0$$

where

$$z_h = u_h - \psi \quad , \qquad u_h = - G\mu_h \ .$$

[Note that whereas the cone H_h^* is finite-dimensional, its polar cone H_h is not such. However, z_h belongs to the finite-dimensional cone $-G(H_h) - \psi$ thus the complementarity system above is essentially a finite-dimensional one.]

We now write the approximate measure μ_h in terms of the basis

$\{-\sigma^h_{qi}\}$ that generates H^*_h:

$$\mu_h = \sum_{\substack{q \in Q^h \\ i=1,2}} \mu^h_{qi} (-\sigma^h_{qi}) \quad , \quad \mu^h_{qi} \geqslant 0 \ ,$$

If, similarly, we put

$$\tau_h = \sum_{\substack{q \in Q^h \\ i=1,2}} \tau^h_{qi} (-\sigma^h_{qi}) \quad , \quad \tau^h_{qi} \geqslant 0 \ ,$$

then the discrete variational inequality corresponding to (54), according to what we have seen in Section, is given by

$$(56) \quad \begin{cases} \mu^h \equiv (\mu^h_{qi})_{\substack{q \in Q^h \\ i=1,2}} \geqslant 0 \quad : \\ \displaystyle\sum_{\substack{q, r \in Q^h \\ i,j=1,2}} (G^h_{qi;rj} \mu^h_{qi} - \psi^h_{qi}) (\tau^h_{rj} - \mu^h_{rj}) \geqslant 0 \\ \forall \quad \tau^h \equiv (\tau^h_{rj})_{\substack{p \in Q^h \\ j=1,2}} \geqslant 0 \end{cases}$$

[by writing that a vector of $\mathbb{R}^N$ is non-negative, we mean that all its components are non-negative],

where for each $q \in Q^h$, $i=1,2$

$$\psi^h_{qi} = (\sigma^h_{qi}, \psi) = \text{mean value of } \psi \text{ on } S^h_{qi}$$

$$G^h_{qi;rj} = (\sigma^h_{qi}, G\, \sigma^h_{rj}) = \text{mean value on } S^h_{qi}$$

of the potential in Ω

(59) $$g^h_{rj}(x) = G\,\sigma^h_{rj}(x) = \frac{1}{|S^h_{rj}|}\int_{S^h_{rj}} g(x;y)dy_j$$

of the measure σ^h_{rj} carried by the mesh S^h_{rj}

[Note that

$$G^h_{qi;rj} = G^g_{rj;qi} \quad , \qquad q, r \in Q^h \ , \ i, j = 1, 2$$

because of the symmetry of the Green operator G.]

For example, if i=1, j=2, the matrix element $G^h_{q1,r2}$ is given by

$$G^h_{q_1,r_2} = \frac{1}{h_1h_2}\int_{q_1h_1}^{(q_1+1)h_1}\int_{q_2h_2}^{(q_2+1)h_2} g(x_1,q_2h_2;r_1h_1,y_2)dx_1dy_2$$

The discrete problem (56) is in turn equivalent to the discrete complementarity system

(58) $$\begin{cases} \mu^h_{qi} \geq 0 \quad , \quad z^h_{qi} \geq 0 \quad , \quad \mu^h_{qi}\, z^h_{qi} = 0 \quad ; \qquad q \in Q^h, \ i=1,2 \\ z^h_{qi} = \sum\limits_{\substack{r \in Q^h \\ j=1,2}} G^h_{qi;rj}\, \mu^h_{rj} - \psi^h_{qi} \end{cases}$$

which can be solved, for instance, by using pivoting thechniques (see F. Scarpini, A. Valdinoci [1])

Once this system has been solved, we can write the approximate measure

$$\mu_h = \sum_{\substack{q \in Q^h \\ i=1,2}} \mu_{qi}^h \, (-\sigma_{qi}^h)$$

and the approximate function

$$u_h(x) = \sum_{\substack{q \in Q^h \\ i=1,2}} \mu_{qi}^h \, g_{qi}^h(x) \quad .$$

By the convergence results of Section,6 we know that $u_h(x)$ converges strongly in $H_0^1(\Omega)$ to the solution $u(x)$ of problem (51) as $|h| \to 0$, while the measure μ_h converges strongly in the dual $H^{-1}(\Omega)$ to the solution μ of (52).

Let us also remark that the coefficients $z_{qi}^h + \psi_{qi}^h$, obtained by solving system (58), yield a direct approximation of the mean values of the solution $u(x)$. on each one-dimensional mesh of the subdivision used in the approximation. We refer to A. Fusciardi et al. for more details on this point. [Note that the mean-values are well defined for an arbitrary function $u \in H_0^1(\Omega)$, (n=2) whereas the point-values of $u(x)$ are not defined, unless some regularity of the solution $u(x)$, depending on the regularity of the obstacle ψ, is known]

Remark 20. The approximation of the one-dimensional obstacle problem in an interval (a, b) is particularly simple. Then, the basis measures σ_q^h can be taken to be the unit mass δ_q at a point $x_q = qh$ of a subdivision of $\mathbb{R}$. The approximate measures are finite-combinations of such Dirac measures

$$\mu_h = - \sum_q \mu_q^h \, \delta_q \quad , \qquad \mu_q^h \geq 0$$

and since the potential $g(x) = G\,\delta_q(x)$ in (a, b) is the "triangle" function

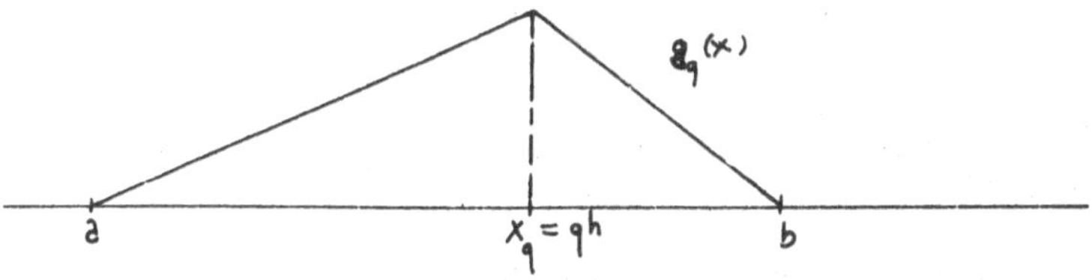

then the approximate functions

$$u_h(x) = \sum_q \mu_q^h \, g_q(x)$$

are piece-wise affine function in (a, b). The solution of the complementarity system in this case is particularly simple, see F. Scarpini, A. Valdinoci [1]. ∎

Remark 21. The approximation method described in this section requires the knowledge of the Green function g(x, y) for the Dirichlet problem in Ω It is still possible, however, to combine the dual approach discussed above with an approximation of the Laplace operator of finite-difference type (by replacing point-values with suitable mean values). This requires an external approximation of the measures in $H^{-1}(\Omega)$, in place of the internal one described above. We refer to U. Mosco-F. Scarpini [1]. ∎

REFERENCES

B. D. ANNIN, [1] Existence and uniqueness of the solution of the elastic-plastic torsion problem for a cylindrical bar of oval cross-section, Prikl. Mat. Meh. 29 (1965).

A. M. ARTHURS, [1] Complementary variational principles, Clarendon Press Oxford, 1970.

E. ASPLUND, [1] Positivity of duality mappings, Bull. Amer. Math. Soc. 73 (1967), 200-203.

J. P. AUBIN, [1] Approximation of variational inequalities, in "Functional Analysis and Optimization" (E. R. Caianiello Ed.), Acad. Press 1966, 7-14.

[2] Approximation des espaces de distributions et des opérateurs différentiels, Bull. Soc. Math. France, Mémoire 12 (1967), 1-139.

[3] Behavior of the error of the approximate solutions of boundary value problems for linear elliptic operators by Galerkin's and finite difference methods, Annali Scuola Norm. Sup. 21 (1967), 599-637.

[4] Evaluation des erreurs de troncature des approximations des espaces de Sobolev, J. Math. Anal. Appl. 21 (1968), 356-368.

[5] Approximation des problèmes aux limites non homogènes pour des opérateurs non linéaires, J. Math. Anal. Appl. 30 (1970), 510-521.

C. BAIOCCHI, [1] Su un problema di frontiera libera connesso a questioni di idraulica, to appear in Annali Mat. Pura Appl.

A. BEURLING and A. E. LIVINGSTONE, [1] A theorem on duality mappings in Banach spaces, Ark. Mat. 4 (1962), 405-411.

L. BOCCARDO, [1] Alcuni problemi al contorno con vincoli unilaterali dipendenti da un parametro, to appear.

J. F. BOURGAT, [1] Analyse numérique du problème de la torsion elastoplastique, Thèse, I. R. I. A. Paris, 1971.

H. BREZIS, [1] Une genéralisation des opérateurs monotones, Inéquations d'évolution abstraites, C. R. Acad. Sci. Paris, t. 264 (1967), 683-686 and 732-735.

[2] Sur certains problèmes non-linéaires, Séminaire Choquet n. 18 (1966-67), 1-18.

[3] Equations et inéquations non linéaires dans les espaces vectoriels en dualité, Ann. Inst. Fourier 18 (1968), 115-175.

[4] Problemes unilateraux, to appear.

H. BREZIS and M. SIBONY, [1] Méthodes d'approximation et d'iteration pour les opérateurs monotones, Arch. Rat. Mech. Anal. 28 (1969), 59-82.

[2] Equivalence de deux inéquations variationnelles et applications, Arch. Rat. Mech. Anal., to appear.

H. BREZIS and G. STAMPACCHIA, [1] Sur la régularité de la solution d'inéquations elliptiques, Bull. Soc. Math. France 96 (1968), 153-180.

F. E. BROWDER, [1] Nonlinear elliptic boundary value problems, Bull. Amer. Math. Soc. 69 (1963), 862-874.

[2] Nonlinear elliptic boundary value problems, II, Trans. Amer. Math. Soc. 117 (1965), 530-550.

[3] Continuity properties of monotone nonlinear operators in Banach spaces, Bull. Amer. Math. Soc. 70 (1964), 551-553.

[4] On a theorem of Beurling and Livingstone, Canad. J. Math. 17 (1965), 367-372.

[5] Nonlinear monotone operators and convex sets in Banach spaces, Bull. Amer. Math. Soc. 71 (1965), 780-785.

[6] Existence and uniqueness theorems for solutions of nonlinear boundary value problems, Proc. Amer. Math. Soc. Symp. Appl. Math. XVII (1965), 24-49.

[7] Problèmes non-linéaires, Les Presses de l'Univ. de Montréal, 1966, 1-48.

[8] Existence and approximation of solutions of nonlinear variational inequalities, Proc. Natl. Acad. Sci. U. S. 56 (1966), 1080-1086.

[9] On the unification of the calculus of variations and the theory of monotone nonlinear operators in Banach spaces, Proc. Natl. Acad. Sci. U. S. 56, 419-425.

[10] Non-linear accretive operators in Banach spaces, Bull. Amer. Math. Soc. 73 (1967), 470-476

[11] A new generalization of the Schauder fixed point theorem, Math. Annalen 174 (1967), 285-290.

[12] Approximation - solvability of nonlinear functional equations in normed linear spaces, Arch. Rat. Mech. Anal. 26 (1967), 33-42.

[13] Non-expensive nonlinear operators in a Banach space, Proc. Nat. Acad. Sci. USA, 54 (1965), 1041-1044

[14] Non-linear variational inequalites and maximal monotone mappings in Banach spaces, Math. Annalen 175 (1968), 89-113

[15] Nonlinear operators and nonlinear equations of evolution in Banach spaces, Proc. Amer. Math. Symp. Nonlinear Functional Analysis, Chicago, 1968.

F. E. BROWDER and W. V. PETRYSHYN, [1] The solution by iteration of nonlinear functional equations in Banach spaces, Bull. Amer. Math. Soc. 72 (1966), 571-575.

[2] Construction of fixed points of nonlinear mappings in Hilbert space, J. Math. Anal. Appl., 20 (1967), 197-228.

J. CEA, [1] Approximation variationnelle des problèmes aux limites, Ann. Inst. Fourier 14 (1964), 345-444.

[2] Optimisation, théorie et algorithmes, Dunod éd., Paris, 1971.

J. CEA and R. GLOWINSKI, [1] Minimisation des fonctionnelles non differentiables, to appear.

[2] Méthodes numériques pour l'écoulement laminaire d'un fluide rigide viscoplastique incompressible, to appear.

J. CEA, R. GLOWINSKI and J. L. NEDELEC, [1] Méthodes numériques pour la torsion élasto-plastique d'une barre cylindrique, to appear.

V. COMINCIOLI, L. GUERRA and G. VOLPI [1] Analisi numerica di un problema di frontiera libera connesso col moto di un fluido attraverso un mezzo poroso, Pubbl. n. 17 del Lab. Anal. Numer., Pavia, 1971.

R. W. COTTLE, [1] Nonlinear Programs with positively bounded Jacobians, SIAM J. Appl. Math. 14 (1966), 147-157.

R. W. COTTLE and G. B. DANTZIG, [1] Positive (semi-)definite programming, Symp. Math. Progr., H. W. Kuhn Ed., Princeton, 1970.

[2] Complementary pivot theory of mathematical programming, in Linear Algebra and its Applications, vol. I (1968), 103-125.

J. P. DIAS and M. SIBONY, [1] Méthodes d'approximation pour certains problèmes non linéaires non homogenes, to appear.

U. DIETER, [1] Optimierung soufgaben in topologischen Vektorräumen I: Dualitätstheorie, Z. Wahrscheinlichkeitheorie verw. Geb. 5 (1966), 89-117.

[2] Dual extremal problems in linear spaces with examples and applications in game theory and statistics, Proc. NATO Adv. Study Inst. on Theory and Appl. of Monotone Operators, Venezia 1968, Oderisi Ed., 1969, 1-9.

F. DI GUGLIELMO, [1] Construction d'approximations des espaces de Sobolev sur des reseaux en simplexes, Calcolo 6 (1969), 279-331.

I. DOLCETTA, [1] Sistemi di complementarità e disuguaglianze variazionali, Tesi, Università di Roma, 1972

J. F. DURAND, [1] Résolution numérique de problèmes aux limites sous-harmonique, Thèse à l'Univ. de Montpellier, 1968-69.

C. DUVAUT and J. L. LIONS, [1] Méchanique et inéquations, Dunod. ed., Paris, 1971

D. G. de FIGUERIDO, [1] Topics in non-linear functional analysis, Lecture Series N. 48, University of Maryland, 1967.

A. FUSCIARDI, U. MOSCO, F. SCARPINI and A. SCHIAFFINO, [1] A dual method for the numerical solution of some variational inequalities, to appear in J. Math. Anal. Appl. 40 (1972).

E. GIUSTI, [1] Superfici minime cartesiane con ostacoli discontinui, Arch. Rat. Mech. Anal. 40 (1971).

R. GLOWINSKI, [1] Méthodes numériques pour l'écoulement stationnaire d'un fluide rigide visco-plastique incompressible, to appear.

[2] La méthode de relaxation. Applications à la minimisation avec et sans contraintes de fonctionnelles convexes, Quaderni dei Rendiconti, Ist. Mat. Univ. Roma, 1971.

[3] Méthodes numeriques pour la torsion elasto-plastique d'une barrè cylindrique, formulation variationnelle, Colloq. Anal. Numer Supper Besses, 1970.

R. GLOWINSKI, J. L. LIONS and R. TREMOLIERES, [1] Méthodes numériques de résolution des problèmes d'inéquations variationnelles en mécanique, et en physique, Dunod Ed., Paris, to appear.

M. GOURSAT, [1] Analyse numérique de problèmes d'elasto-plasticité et de visco-plasticité, Thèse, I. R. I. A., Paris, 1971.

P. H. HARTMAN and G. STAMPACCHIA, [1] On some non linear elliptic differential functional equations, Acta Math. 115 (1966), 271-310.

A. IOFFE and V. TIKHOMIROV, [1] Duality of convex functions and extremun problems, Uspekhi Mat. Nauk. 23, 6 (1968), 51-116; Russian Math. Surveys 23, 6 (1968), 53-124.

J. L. JOLY, [1] Une famille de topologies et de convergences sur l'ensemble des fonctionnelles convexes, Thése à la Faculté des Sciences de Grenoble, 1970.

R. I. KACHUROVSKII, [1] On monotone operators and convex functionals, Uspehi Mat. Nauk. 15, 94 (1960), 213-215.

[2] Monotone non-linear operators in Banach spaces, Dokl. Akad. Nauk. SSSR 163 (1965), 559-562.

[3] Nonlinear monotone operators in Banach spaces, Uspehi Mat. Nauk. 23 (1968), 121-168 Russian Math. Surveys 23, 2 (1968), 117-165.

S. KANIEL, [1] Construction of a fixed-point for contractions in Banach space, Israel J. Math., 9 1971, 535-540.

S. KARAMARDIAN, [1] The nonlinear complementarity problem with application, JOTA 4 (1969), 87-98.

T. KATO, [1] Demicontinuity, hemicontinuity and monotonicity, Bull. Amer. Math. Soc. 70 (1964), 548-550; idem., Part. II, ibid. 73 (1967), 886-889.

H. LANCHON, [1] Solution du probléme de torsion élasto-plastique d'un barre cylindrique de section quelconque, C. R. Acad. Sc. Paris, 269 (1969), 791-794

H. L. LANCHON and C. DUVAUT, [1] Sur la solution du problème de la torsion élasto-plastique d'une barre cylindrique de section quelconque, C. R. Acad. Sci. Paris 264 (1967)

C. E. LEMKE, [1] Recent results on complementarity problems, Proc. Princeton Symp. on Math. Programming, H. W. Kuhn ed., Princeton Univ. Press, 1970, 349-384.

J. LERAY and J. LIONS, [1] Quelques résultats de Visik sur les problèmes elliptiques nonlineaires par les méthodes de Minty-Browder, Bull. Soc. Math. France 93 (1965), 97-107.

C. LESCARRET, [1] Cas d'addition des applications monotones maximales dans un espace de Hilbert, C. R. Acad. Sc. Paris 261 (1965), 1160-1163.

H. LEWY, [1] On a variational problem with inequalities on the boundary, J. Math. Mech. 17 (1968), 861-884.

[2] On a minimum problem for superharmonic functions, Int. Conf. on Functional Anal., Tokyo 1969.

H. LEWY and G. STAMPACCHIA, [1] On the regularity of a solution of a variational inequality, Comm. Pure Appl. Math. 22 (1969), 153-188.

[2] On the regularity of certain superharmonic functions, J. d'Analyse Math. 23 (1970), 227-236

[3] On existence and smoothness of solutions of some non-coercive variational inequalities, to appear.

J. L. LIONS, [1] Quelques méthodes de résolution des problémes aux limites non lineaires. Dunod et Gauthier-Villars Ed., Paris, 1969.

J. L. LIONS and G. STAMPACCHIA, [1] Variational inequalities, Comm. Pure Appl. Math. 20 (1967), 493-519.

W. LITTMAN, G. STAMPACCHIA and H. F. WEINBERGER, [1] Regular points for elliptic equations with discontinuous coefficients Ann. Scuola Normale Sup. Pisa, 17 (1963), 45-79

O. G. MANCINO, G. STAMPACCHIA, [1] Convex programming and variational inequalities, JOTA 9 (1972), 3-23.

P. MARZULLI, [1] Risoluzione alle differenze di equazioni alle derivate parziali di tipo ellittico oon condizioni su un contorno libero, Calcolo, Suppl. 1, 5 (1968), 1-22.

M. MATZEU, [1] Dualità nella teoria della capacità, Tesi, Ist. Mat. Univ. di Roma, 1972.

G. J. MINTY, [1] Monotone (nonlinear) operators in Hilbert space, Duke Math. J. 29 (1962), 341-346.

[2] On a "monotonicity" method for the solution of non linear equations in Banach spaces, Proc. Natl. Acad. Sci. US, 50 (1963), 1038-1041.

[3] On the monotonicity of the gradient of a convex function, Pacific J. Math. 14 (1964), 243-247.

[4] On the solvability of nonlinear functional equations of monotonic type, Pacif J. Math. 14 (1964), 243-247.

[5] On the generalization of a direct method of the calculus of variations, Bull. Amer. Math. Soc. 73 (1967), 315-321.

[6] On some aspects of the theory of monotone operators Proc. NATO Adv. Study Inst. on Theory and Appl. of Monotone Operators, Venezia 1968, Oderisi Ed., 1969, 67-82.

M. MIRANDA, [1] Frontiere minimali con ostacoli, to appear.

J. J. MORE, [1] The application of variational inequalities to complementarity problems end existence theorems, Tech. Rep. n. 71-110, Dept. Computer Science, Cornell Univ., 1972.

J. J. MOREAU, [1] Fonctionnelles convexes, Séminaire sur les équations aux dérivées partielles, Collége de France, Paris, 1966-1967, miltigraph, 1-108.

[2] Proximité et dualité dans un espace hilbertien, Bull. Soc. Math. France 93 (1965), 273-299.

U. Mosco

[3] One-sided constraints in hydrodinamics, in J. Abadie Ed., Nonlinear programming, Nort Holland Pub., Amsterdam(1967), 257-279.

[4] Principes extrémaux pour le probleme de la naissance de la cavitation, Journ. de Mécanique, 5 (1966), 439-470.

[5] La notion de sur-potentiel et les liaisons unilaterales en élastotatique, C. R. Acad. Sci. Paris, 267 (1968), 954-957.

[6] Sur les lois de frottement, de plasticité et de viscosité, C. R. Acad. Sci. Paris, 271 (1970), 608-611.

[7] Traitement numerique d'un probleme aux dérivées partielles de type unilateral, Publ. No.15, Départ. d'Informatique, Univ. de Montréal, 196..

U. MOSCO, [1] Approximation of the solutions of some variational inequalities, Ann. Scuola Normale Sup. Pisa, 21 (1967), 373-394.

[2] A remark on a theorem of F. E. Browder, J. Math. Anal. Appl. 20 (1967), 90-93.

[3] Convergence of solutions of variational inequalities, Proc. NATO, Adv. Study Inst. on Theory and Appl. of Monotone Operators, Venezia 1968, Oderisi Ed., 1969, 231-247.

[4] Convergence of convex sets and of solution of variational inequalities, Adv. in Math. 3, 4 (1969), 510-585.

[5] Perturbation of variational inequalities, Proc. Amer. Math. Soc. Symp. Pure Math. XVIII (1970), 182-194.

[6] On the continuity of the Young-Fenchel transform, J. Math. Anal. Appl. 35 (1971), 518-535.

[7] Dual variational inequalities to appear in J. Math. Anal. Appl. 39 (1972).

U. MOSCO, F. SCARPINI, [1] On the approximation of some complementarity systems in Sobolev spaces, to appear.

M. NEDELEC, [1] Un algorithme dual pour le problème de la torsion élasto-plastique d'une barre, Colloq. d'Analyse Numer., Supper Besses, 1969-70.

J. C. C. NITSCHE, [1] Variational problemswith inequalities as boundary conditions, Arch. Rat. Mech. Anal. 35 (1969), 83-113.

Z. OPIAL, [1] Non expansive and monotone mappings in Banach spaces Lecture Notes Divis. Appl. Math., Brown Univ., 1967.

W. V. PETRYSHYN, [1] Projection methods in non linear numerical functional analysis, J. Math. Mech. 17 (1967), 353-372.

P. D. ROBINSON, [1] Complementary Variational Principles, in Nonlinear Functional Analysis and applications, edited by L. B. Rall, Academic Press, 1971

R. T. ROCKAFELLAR, [1] Characterization of the subdifferentials of convex functions, Pacific J. Math. 17 (1966), 497-510.

[2] Extension of Fenchel's duality theorem for convex functions, Duke Math. J. 33 (1966), 81-90.

[3] On the virtual convexity of the domain and range of a non linear maximal monotone operator, Math. Annalen

[4] Local boundedness of nonlinear monotone operators Michigan Math. J.

[5] On the maximal monotonicity of subdifferential mappings, Michigan Math. J.

[6] Convex functions, monotone operators and variational inequalities, Proc. NATO Study Inst. on Theory and Appl. of Monotone Operators, Venezia 1968, Oderisi Ed., 1969, 231-247.

[7] Convex analysis, Princeton Univ. Press., Princeton, . 1970

F. SCARPINI and T. VALDINOCI, [1] Su alcuni sistemi di complementarità connessi a disequazioni variazionali di tipo ellittico, to appear in Calcolo.

A. SCHIAFFINO, [1] Su un problema di disequazioni variazionali per operatori differenziali ordinari, Boll. U. M. I. (1969), 25-35.

M. SIBONY, [1] Sur l'approximation d'équatios et inéquations aux dérivées partielles non linéaires de type monotone, to appear in J. Math. Anal. Appl.

[2] Méthodes itératives pour les équations et inequations aux dérivées partielles non linéaires de type monotone, Calcolo 7 (1970), 65-183.

G. STAMPACCHIA, [1] Formes bilinéaires coercitives sur les ensembles convexes, C. R. Acad. Sc. Paris 258 (1964), 4413-4416.

[2] On the regularity of solutions of variational inequalities Int. Conf. on Functional Anal., Tokyo, 1969.

[3] Variational inequalities, Proc. NATO Adv. Study Inst., Venezia 1968, Oderisi Ed., 1969, 101-191.

[4] Regularity of solutions of some variational inequalities, Proc. Amer. Math. Soc. Symp. Pure Math. XVIII (1970), 271-281.

[5] On a problem of numerical analysis connected with the theory of variational inequalities, I. E. I., CNR, Pisa, Nota interna B72/5, 1972.

J. STOER and C. WITZGALL, [1] Convexity and optimization in finite dimensions I, Springer Verlay, Berlin, 1970

R. TEMAM, [1] Solutions généralisées d'équations non linéaires non uniformément elliptiques, Publ. Math. d'Orsay, Univ. Paris XI, 1970-71.

T. W. TING, [1] Elastic-plastic torsion Problem, Arch. Rat. Mech. Anal. 25 (1967), 342-366.

M. M. VAINBERG, [1] Variational methods for the study of nonlinear operators, Holden Day; San Francisco Cal., 1964.

[2] Le problème de la minimisation des fonctionnelles non linéaires, in Problems in Non-linear Analysis, CIME Varenna 1970, Cremonese Ed., Roma, 1971.

M. M. VAINBERG and R. I. KACHUROVSKI, [1] On the variational theory of non-linear operators and equations, Dokl. Akad. Nank. SSSR 129 (1959), 1199-1202.

H. de VEIGA [1] Sulla holderianità delle soluzioni di alcune disequazioni variazionali con condizioni unilaterali al bordo, Annali Mat. Pura Appl. 83 (1969), 73-112.

[2] Régularité pour une clesse d'inequations non linéaires, C. R. Acad. Sci. Paris 271 (1970), 23-25.

R. A. WIJSMAN, [1] Convergence of sequences of convex sets, cones and functions, Bull. Amer. Math. Soc. 70 (1964), 186-188; idem, Part. II, Trans. Amer. Math. Soc. (1966), 32-45

U. Mosco

E. H. ZARANTONELLO, [1] Solving functional equations by contractive averaging, Tech. Rep. N. 160, U. S. Army Research Center Madison, Wisconsin, 1960.

CENTRO INTERNAZIONALE MATEMATICO ESTIVO

(C. I. M. E.)

I. SINGER

BEST APPROXIMATION IN NORMED LINEAR SPACES

Corso tenuto ad Erice dal 27 giugno al 7 luglio 1971

Best approximation in normed linear spaces

by

Ivan Singer

(University of Bucharest)

Contents

3.4. Semi-Čebyšev and Čebyšev sybspaces and quotient spaces.

3.5. Strongly unique elements of best approximation. Strongly Čebyšev subspaces. Interpolating subspaces.

3.6. Almost Čebyšev subspaces. k-semi-Čebyšev and k-Čebyšev subspaces. Pseudo-Čebyšev subspaces.

3.7. Very non-Cebysev subspaces.

4. Properties of metric projections.

4.1. Definition and some properties of metric projections.

4.2. Continuity of metric projections.

4.3. Weak continuity of metric projections.

4.4. Lipschitzian metric projections.

4.5. Differentiability of metric projections.

4.6. Linearity of metric projections.

4.7. Semi-continuity and continuity of set-valued metric projections.

4.8. Continuous selections and linear selections for setvalued metric projections.

5. Best approximation by elements of non-linear sets.

5.1. Best approximation by elements of convex sets.

5.2. Best approximation by elements of N-parameter sets.

5.3. Generalizations.

5.4. Best approximation by elements of arbitrary sets.

I. Singer

Introduction

Here we want to present briefly some results, problems and directions of research in the modern theory of best approximation, i.e. in which the methods of functional analysis are applied in a consequent manner. In this theory the functions to be approximated and the approximating functions are regarded as elements of certain normed linear (or, more generally, of certain metric) spaces of functions and best approximation amounts to finding "nearest points". The advantages and a brief history of this modern point of view have been described in the Introduction to the monograph [82] and we shall not repeat them here; the material which will be presented in the sequel will be convincing enough, we hope, to prove again that the theory of best approximation in normed linear spaces constitutes both a rigorous theoretical foundation for the existing classical and more recent results in various concrete spaces and a powerfull tool for obtaining new results, solving the new problems which appear.

Since June 1966, when the Romanian version of the monograph [82] has gone to print, the theory of best approximation in normed linear spaces has developed rapidly and the number of papers in this field is growing continuously. However, except the expository paper up to 1967, by A. L. Garkavi [31] , which appeared in 1969, and the bibliography [23] compiled by F. Deutsch and J. Lambert in 1970, we know of no other survey material on these new developments. One of the aims of our course is to fill this gap to a certain extent, by presenting much new material which appeared after the monograph [82] . In this respect the present course, though self-contained, may be regarded as an up to date complement to the monograph [82] ; however, the bibliography does not aim at begin complete, but wants merely to give useful orientation to the reader . Naturally, since another aim of the course is to introduce the non-specialits to this field , some overlapping with the material of the monograph [82] is unavoidable; however, even this part is presented here in a slightly improved way.

We shall give here only a few simple proofs, but for all results we shall give references. Even with this "economy" in proofs, some important topics had to be omitted.

The reader is assumed to know some elements of functional analysis and integration theory, but we shall recall, whenever necessary, the definition of the notions (especially those of the geometry of normed linear spaces) which will be used in the sequel.

We acknowledge with pleasure that we benefited from attending seminar lectures on metric projections by F. R. Deutsch (at Pennsylvania State University, in 1968) and G. Godini (at the Institute of Mathematics of the Academy, Bucharest, in 1970/71).

I. Singer

1 . Characterizations of elements of best approximation.

1.1. The first main theorem of characterization

Throughout the sequel, without any special mention, we shall denote by ρ the distance in a metric space E and, in particular, if E is a normed linear space, ρ will denote the distance in E induced by the norm, i. e.

$$\rho(x, y) = \|x - y\| \qquad (x, y \in E) . \tag{1.1.}$$

Definition 1.1. Let E be a metric space, G a set in E and $x \in E$. An element $g_0 \in G$ is called an element of best approximation of x (by the elements of the set G) if we have

$$\rho(x, g_0) = \inf_{g \in G} \rho(x, g) , \tag{1.2}$$

i. e., if g_0 is "nearest" to x among the elements of G ; we shall denote by $\mathcal{P}_G(x)$ the set of all such elements g_0, i. e.

$$\mathcal{P}_G(x) = \left\{ g_0 \in G \;\middle|\; \rho(x, g_0) = \inf_{g \in G} \rho(x, g) \right\} . \tag{1.3}$$

It is natural to consider first the problem of characterization of elements of best approximation, i. e. the probelm of finding necessary and sufficient conditions in order that $g_0 \in \mathcal{P}_G(x)$, since these results will be applied to solve the other problems on best approximation (e. g. those of existence and uniqueness of elements of best approximation, etc.).

Also, the characterization theorems in concrete spaces (see e.g. the "altemation theorem" 1.7 below) are convenient tools for verifying whether or not a given g_0 satisfies $g_0 \quad {}_G(x)$, since they are easier to use than (1.2).

Since we have obviously

$$\mathcal{P}_G(x) = \begin{cases} x & \text{for} \quad x \in G \\ \emptyset & \text{for} \quad x \in \overline{G} \setminus G, \end{cases} \tag{1.4}$$

it will be sufficient to characterize the elements of best approximation of the elements $x \in E \setminus \overline{G}$. In order to exclude the case when such elements do not exist, in the sequel we shall assume, without any special mention, that $\overline{G} \neq E$.

Unless otherwise stated, the field of scalars for all (general or concrete) normed linear spaces considered in the sequel can be either the field of complex numbers or the field of real numbers.

The first main theorem of characterization of elements of best approximation by elements of linear subspaces in normed linear spaces is the following (see [82], p. 18) :

Theorem 1.1. Let E be a normed linear space, G a linear subspace of E, $x \in E \setminus \overline{G}$ and $g_0 \in G$. We have $g_0 \in \mathcal{P}_G(x)$ if and only if there exists an $f \in E^*$ such that

(1.5) $$\|f\| = 1,$$

(1.6) $$f(g) = 0 \qquad (g \in G),$$

(1.7) $$f(x - g_0) = \|x - g_0\|.$$

We recall that E^* denotes the conjugate space of E, i.e. the space of all continuous linear functionals on E, endowed with the usual vector operations and with the norm

$$\|f\| = \sup_{\substack{x \in E \\ \|x\| \leqslant 1}} |f(x)|.$$

To prove theorem 1.1, assume that $g_0 \in \mathcal{P}_G(x)$. Then, since $x \in E \setminus \overline{G}$, we have $\rho(x, G) = \|x - g_0\| > 0$ and hence, by a corollary of the Hahn-Banach theorem (see e.g. [25], p. 64., lemma 12), there exists an $f_0 \in E^*$ such that

$$\|f_0\| = \frac{1}{\|x - g_0\|}, \quad f_0(g) = 0 \ (g \in G) \quad \text{and} \quad f_0(x) = 1.$$

Then the functional $f = \|x - g_0\| f_0 \in E^*$ satisfies (1.5)-(1.7). Conversely, if there is an $f \in E^*$ satisfying (1.5)-(1.7), then for any $g \in G$ we have

$$\|x - g_0\| = |f(x - g_0)| = |f(x - g| \leqslant \|f\| \|x - g\| = \|x - g\|$$

and hence $g_0 \in \mathcal{P}_G(x)$, which completes the proof.

It is easy to see that theorem 1.1 admits the following geometrical interpretation : We have $g_0 \in \mathcal{P}_G(x)$ if and only if there exists a closed hyperplane H in E (i.e., a closed linear subspace H such that $\dim E/H = 1$) containing G, which supports the cell

$$S(x, \|x - g_0\|) = \left\{ y \in E \,\middle|\, \|y - g_0\| \leqslant \|x - g_0\| \right\} \quad \text{(i.e.,}$$

$$\rho(H, S(x, \|x - g_0\|)) = 0 \quad \text{and} \quad H \cap \operatorname{Int} S(x, \|x - g_0\|) = \emptyset).$$

Any functional $f \in E^*$ satisfying (1.5) and (1.7) is called a "maximal functional" of the element $x - g_0$ (because we have

$$\|x - g_0\| = \sup_{\substack{h \in E^* \\ \|h\| = 1}} |h(x - g_0)|).$$

The usefulness of theorem 1.1 for applications in various concrete normed linear spaces is due to the fact that for these spaces the general form of maximal functionals of the elements of the space is well known and simple (see e.g. [73], [99]). Let us give now some examples of applications of theorem 1.1 in concrete spaces.

We shall use the word "compact" in the sense of N. Bourbaki, i.e.: bicompact Hausdorff. For compact space Q we shall denote by $C(Q)$, respectively $C_R(Q)$, the space of all complex or real, respectively of all real continuous functions on Q, endowed with the usual vector operations and with the norm

$$\|x\| = \max_{q \in Q} |x(q)|$$

Using the general form of maximal functionals of the elements of $C_R(Q)$, from theorem 1.1 we obtain (see [82], p. 33):

Theorem 1.2.. Let $E = C_R(Q)$ (Q compact), G a linear subspace of E, $x \in E \setminus \bar{G}$ and $g_0 \in G$. We have $g_0 \in \mathcal{P}_G(x)$ if and only if there exist two disjoint closed subsets $Y^+_{g_0}$, $Y^-_{g_0}$ of Q and a Radon measure μ on Q, such that

$$|\mu|(Q) = 1, \tag{1.8}$$

$$\int_Q g(q)d\mu(q) = 0 \qquad (g \in G), \tag{1.9}$$

$$\mu \geq 0 \text{ on } Y^+_{g_0}, \ \mu \leq 0 \text{ on } Y^-_{g_0} \text{ and } Y^+_{g_0} \cup Y^-_{g_0} \supset S(\mu), \tag{1.10}$$

$$x(q) - g_0(q) = \begin{cases} \|x - g_0\| & \text{for } q \in Y^+_{g_0} \\ -\|x - g_0\| & \text{for } q \in Y^-_{g_0} \end{cases} \tag{1.11}$$

where $S(\mu)$ denotes the carrier of the measure μ.

One can also give a characterization theorem in the spaces $E = C(Q)$ (see [82], p. 29). Theorem 1.2, which appeared in [79], has constituted the first theorem of characterization in $E = C_R(Q)$ (even in $E = C_R([a, b])$) of elements of best approximation by elements of linear subspaces G of arbitrary dimension.

For a positive measure space (T, ν) (we shall not specify the σ-field of subsets of T on which the measure ν is defined; this will cause no confusion) and for $1 \leq p \leq \infty$ (respectively $p = \infty$) we shall denote by $L^p(T, \nu)$ the space of all equivalence classes of functions with ν-integrable p-th power (respectively of ν-measurable and ν-essentially bounded functions on T), endowed with the usual vector operations and with the norm

$$\|x\| = \left(\int_T |x(t)|^p d\nu(t) \right)^{\frac{1}{p}}$$

(respectively, $\|x\| = \operatorname{ess\,sup}_{t \in T} |c(t)|$;
for simplicity, we use here the same notation for a function and for its equivalence class in L^p. Again the subscript R will mean, both here and for the spaces occurring in the sequel, that we restrict ourselves to real scalars. For a function x' on T we shall use the notation

$$(1.12) \qquad Z(x') = \{ t \in T \mid x'(t) = 0 \}$$

Using the general form of maximal functionals of the elements of $L^1(T, \nu)$, we obtain from theorem 1.1 the following theorem (see [82], p. 46), which was obtained initially by B. R. Kripke and J. T. Rivlin [51] with different (function-theoretic) methods and with the above functional analytic methods in [81] :

Theorem 1.3. Let $E = L^1(T, \nu)$ (where (T, ν) is a positive measure space), G a linear subspace of E, $x \in E \setminus \overline{G}$ and $g_0 \in G$. We have $g_0 \in \mathcal{P}_G(x)$ if and only if

(1.13) $$\left| \int_{T \setminus Z(x - g_0)} g(t) \operatorname{sign} [x(t) - g_0(t)] \, d\nu(t) \right| \leqslant \int_{Z(x - g_0)} |g(t)| \, d\nu(t) \qquad (g \in G) .$$

We recall that for a complex number $\alpha \neq 0$, by definition

$$\operatorname{sign} \alpha = e^{-i \arg \alpha} = \frac{\bar{\alpha}}{|\alpha|}$$

and that sign $0 = 0$.

For $E = L^p(T, \nu)$ with $i < p < \infty$ and for an abstract inner product space $\mathcal{H}$ we obtain from theorem 1.1 the following well known results (see [82], pp. 56-57) :

Theorem 1.4. a) Let $E = L^p(T, \nu)$ (with (T, ν) a positive measure space and $1 < p < \infty$), G a linear subspace of E, $x \in E \setminus \overline{G}$ and $g_0 \in G$. We have $g_0 \in \mathcal{P}_G(x)$ if and only if

(1.14) $$\int_T g(t) \, | x(t) - g_0(t) |^{p-1} \operatorname{sign} [x(t) - g_0(t)] \, d\nu(t) = 0 \qquad (g \in G) .$$

b) Let $E = \mathcal{H}$ be an inner product space, G a linear subspace of E, $x \in E \setminus \overline{G}$ and $g_0 \in G$. We have $g_0 \in \mathcal{P}_G(x)$ if and only if

(1.15) $$(g, x - g_0) = 0 \qquad (g \in G) ,$$

where (x, y) denotes the scalar product in $E = \mathcal{H}$.

Some applications of theorem i. 1 in other concrete spaces, e. g. in the spaces $C^1(Q, \nu)$ and $C_R^1(Q, \nu)$, are given in [82] .

We recall that for a compact space Q and a positive Radon measure ν on Q such that $S(\nu) = Q$, $C^1(Q, \nu)$ denotes the (dense) linear subspace of $L^1(Q, \nu)$ consisting of the equivalence classes of the continuous functions on Q, endowed with the usual vector operations and with the norm $\|x\| = \int_Q |x(q)| \, d\nu(q)$; the assumption that the carrier $S(\nu)$ is the whole space Q implies that each equivalence class of $C^1(Q, \nu)$ contains exactly one continuous function.

The foregoing results in concrete spaces illustrate the power of the methods of functional analysis in the theory of best approximation.

Indeed, although theorem 1. 1 in general normed linear spaces is obatined by a simple application of a corollary of the Hahn-Banach theorem, it gives in various concrete spaces theorems 1. 2-1. 4 and other results as particular cases, simply by using the general form of maximal functionals in these spaces. A direct proof of theorems 1. 2-1. 4 would require different methods in each of the concrete spaces involved, apparently having connection with each other; however, they all turn out to be particular cases of theorem 1. 1, and this unified method of obtaining them is simpler and clearer than the separate proofs in each concrete space. This unified method is carried out in a consequent manner for the whole theory of best approximation (e. g. for problems of uniqueness, etc.), in the monograph [82] . In the sequel we shall only indicate some examples of applications in concrete spaces (rather than mention all known applications) of some of the general results on best approximation in arbitrary normed linear spaces which we shall give.

A natural generalization of the problem of characterization of elements of best approximation is the problem of simultaneous characterization of a set of elements of best approximation : given E, $G \subset E$ and x as above and a subset M of G, what are the necessary and sufficient conditions in order that every element $g_0 \in M$ be an element of best approximation of x by the elements of G ? The answer is given (see [82] , p. 23) by

Theorem 1.5. Let E be a normed linear space, G a linear subspace of E, $x \in E \setminus \overline{G}$ and $M \subset G$. We have $M \subset \mathcal{P}_G(x)$ if and only there exists an $f \in E^*$ satisfying (1.5), (1.6) and

(1.16) $$f(x - g_0) = \| x - g_0 \| \qquad (g_0 \in M) .$$

In other words this says that one can find the functional $f \in E^*$ of theorem 1.1 to be common for all $g_0 \in M$. Theorem 1.5 is an immediate consequence of theorem 1.1 and the observation that $\|x - g_1\| = \|x - g_2\|$ for all pairs $g_1, g_2 \in M \subset \mathcal{P}_G(x)$; naturally, the converse is also true, since theorem 1.1 is even a particular case of theorem 1.5. We shall see in § 3 that theorem 1.5 has applications in the study of the uniqueness of elements of best approximation.

Now we shall consider the important particular case when $\dim G = n < \infty$, i.e. when $G = [x_1, \ldots, x_n]$ = the (closed) linear subspace of E spanned by n linearly independent elements $x_1, \ldots, x_n$.

Naturally, the preceding results are also applicable in this particular case; however, by using effectively the assumption $\dim G = n < \infty$, we can obtain additional information. We recall that an element x of a closed convex set A in a topological linear L is called an extremal point of A if the relations $y, z \in A$, $0 < \lambda < 1$, $x = \lambda y + (1 - \lambda) z$ imply $y = z = x$. Using the classical theorem of Minkowski that the elements of the unit cell $S_{E^*} = \left\{ f \in E^* \,\middle|\, \|f\| \leqslant 1 \right\}$ in a finite-dimensional space E^* can be expressed as finite convex combinations of extremal points of S_{E^*} and that if E_0 is an arbitrary linear subspace of a normed linear space E, then every extremal point of the unit cell $S_{E_0^*}$ can be extended to an extremal point of S_{E^*} (see e.g. [82], p. 168), from theorem 1.1 we obtain (see [82], p. 170):

Theorem 1.6. Let E be a normed linear space, $G = [x_1, \ldots, x_n]$ an n-dimensional linear subspace of E, $x \in E \setminus G$ and $g_0 \in G$. We have $g_0 \in \mathcal{P}_G(x)$ if and only if there exist h extremal points $f_1, \ldots, f_h$ of the unit cell S_{E^*}, where $1 \leq h \leq n+1$ if the scalars are real and $1 \leq h \leq 2n+1$ if the scalar are complex and h numbers $\lambda_1, \ldots, \lambda_h > 0$ with

$$\sum_{j=1}^{h} \lambda_j = 1,$$

such that

$$(1.17) \qquad \sum_{j=1}^{h} \lambda_j f_j(g) = 0 \qquad (g \quad G),$$

$$(1.18) \qquad \sum_{j=1}^{h} \lambda_j f_j(x - g_0) = \| x - g_0 \| .$$

In other words, the additional information to theorem 1.1 which we obtain for $\dim G = n < \infty$ is that for such subspaces G one can take the functional f of theorem 1.1 to be a convex combination

$$f = \sum_{j=1}^{h} \lambda_j f_j$$

of $h \leqslant n + 1$ (respectively $h \leqslant 2n + 1$) extremal points of the unit cell $S_{E^*} = \left\{ f \in E^* \middle| \|f\| \leqslant 1 \right\}$. Theorem 1.6 is also convenient for applications in the usual concrete spaces, because for these spaces the general form of the extremal points of S_{E^*} is well known and simple. For example (see e.g. [25], p. 441, lemma 6) for $E = C_R(Q)$ (Q compact), a functional $f \in E^*$ is an extremal point of S_{E^*} if and only if there exist a $q \in Q$ and a scalar α with $|\alpha| = 1$ such that

$$(1.19) \qquad f(y) = \alpha\, y(q) \qquad (y \in E = C(Q)).$$

We recall that a system of n functions $x_1, \ldots, x_n \in C(Q)$ (Q compact) is called a Čebyšev system (on Q) if every non-zero linear combination

$$\sum_{i=1}^{n} \alpha_i x_i$$

has at most n-1 zeros on Q. For $Q = [a, b]$ and for real scalars we have the following classical "alternation theorem" of P. L. Čebyšev-S. N. Bernstein (see e.g. [82], p. 184):

<u>Theorem 1.7. Let</u> $G = [x_1, \ldots, x_n]$ <u>be an n-dimensional linear subspace of</u> $e = C_R([a, b])$ <u>such that</u> $x_1, \ldots, x_n$ <u>form a Čebyšev system and let</u> $x \in E \setminus G$ <u>and</u> $g_0 \in G$. <u>We have</u> $g_0 \in \mathcal{P}_G(x)$ <u>if and only if there exist</u> n + 1 <u>points</u> $q_1 < q_2 < \ldots < q_{n+1}$ <u>of</u> $[a, b]$, <u>at which the difference</u> $x(q) - g_0(q)$ <u>takes the value</u> $\|x - g_0\|$ <u>with alternating signs</u> <u>(i.e., with opposite signs at consecutive points</u> q_j, q_{j+1} $(j = 1, \ldots, n)$).

This classical theorem follows as a particular case both from theorem 1.2 for $\dim G = n < \infty$ anf from theorem 1.6 for $E = C_R(Q)$, using the general form (1.19) of the extremal points of S_{E^*} (naturally, since the scalars are real, we have now $\alpha = \pm 1$ in (1.19) .

1.2. The second main theorem of characterization .

The second main theorem of characterization of elements of best approximation by elements of linear subspaces in normed linear spaces is the following (see [82] , p. 62) :

Theorem ..8. Let E be a normed linear space, G a linear subspace of E, $x \in E \setminus \overline{G}$ and $g_0 \in G$. We have $g_0 \in \mathcal{P}_G(x)$ if and only if for every $g \in G$ there exists an extremal point f^g of the unit cell $S_{E^*} = \{f \in E^* \, \|f\| \leq 1\}$ such that

(1.20) $$\operatorname{Re} \; f^g(g - g_0) \leq 0$$

(1.21) $$f^g(x - g_0) = \|x - g_0\| .$$

For a geometrical int erpretation of theorem 1.8 see [82] , p. 75.

From theorem 1.8, using the general form (1.19) of the extremal points of S_{E^*} for the space $E = C(Q)$, one obtains immediately the following classical theorem of characterization of elements of best approximation, due A. N. Kolmogorov (see [82] , p. 69):

Theorem 1.9. Let $E = C(Q)$ (Q compact), G a linear subspace of E, $x \in E \setminus \overline{G}$ and $g_0 \in G$. We have $g_0 \in \mathcal{P}_G(x)$ if and only if for every $g \in G$ there exists a $q = q^g \in Q$ such that

(1.22) $$\operatorname{Re} \left[x(q) - g_0(q)\right] g(q) \geq 0 ,$$

(1.23) $$|x(q) - g_0(q)| = \|x - g_0\| .$$

For applications of theorem 1.8 in other concrete spaces, see [82].

Returning to arbitrary normed linear spaces, it is easy to see that the sufficiency part of theorem 1.8 remains valid for an arbitrary set G in E. Indeed, if the condition is satisfied, then for every $g \in G$ we have

$$\|x - g_0\| = \operatorname{Re} f^g (x - g_0) \leqslant \operatorname{Re} f^g (x - g) \leqslant |f^g (x - g)| \leqslant$$

$$\leqslant \|f^g\| \; \|x - g\| = \|x - g\|,$$

whence $g_0 \in \mathcal{P}_G(x)$, which proves the assertion. The problem of characterizing those sets $G \subset E$ for which the condition in theorem 1.8 is also necessary, is important in non-linear approximation (see §5).

1.3. Differential characterizations.

Since best approximation amounts, by definition, to the minimization of the convex functional $\chi = \chi_{G,x}$ on the linear subspace G of a normed linear space E defined by

$$(1.24) \qquad \chi(g) = \|x - g\| \qquad (g \in G),$$

it arises naturally the problem of obtaining characterizations of the elements of best approximation $g_0 \in \mathcal{P}_G(x)$ with the aid of differential calculus. The main difficulty is that in general the norm in E is not necessarily Gâteaux differentiable at each non-zero point of E. Nevertheless, it is known (see e.g. [25], p. 445, lemma 1) that the limits

$$(1.25) \qquad \tau(x, y) = \lim_{t \to 0^+} \frac{\|x + ty\| - \|x\|}{t} \qquad (x, y \in E)$$

always exist and one can use them to give the following characterizations of elements of best approximation (see [ ] , pp. 88-90) :

Theorem 1.10. Let E be a normed linear space, G a linear subspace of E, $x \in E \setminus \bar{G}$ and $g_0 \in G$. We have $g_0 \in \mathcal{P}_G(x)$ if and only if

$$(1.26) \qquad \tau(x - g_0, g) \geq 0 \qquad (g \in G) .$$

If the norm in E is Gâteaux differentiable at $x - g_0$, this condition is equivalent to the following :

$$(1.27) \qquad \tau(x - g_0, g) = 0 \qquad (g \in G)$$

For some remarks on the uses of the theory of convex functions (subdifferentials, etc.) for characterizations of elements of best approximation by elements of more general sets, see §5 .

1.4. Other characterizations .

We have the following characterization theorem, proved by Y. Ikebe [41] for $E = C(Q)$ and in [85] for arbitrary normed linear spaces :

Theorem 1.11. Let E be a normed linear space, G a linear subspace of E, $x \in E \setminus \bar{G}$ and $g_0 \in G$. We have $g_0 \in \mathcal{P}_G(x)$ if and only if 0 belongs to the $\sigma(E^*, E)$- closure of the convex hull of the set

$$(1.28) \qquad A = \left\{ \overline{f(x - g_0)} f \big|_G \in G^* \;\middle|\; f \in \mathcal{E}(S_{E^*}),\ f(x - g_0) = \|x - g_0\| \right\},$$

where $\mathcal{E}(S_{E^*})$ denotes the set of all extremal points of the unit cell S_{E^*} and $f|_G$ denotes the restriction of f to the subspace G.

A related result in $E = C_R(Q)$ has been obtained by E. W. Cheney and A. A. Goldstein (see [82], p. 44). In the particular case when $\dim G < \infty$, from theorem 1.11 one obtains

Corollary 1.1. Let E be a normed linear space, $G = [x_1, \ldots, x_n]$ an n-diemnsional linear subspace of E, $x \in E \setminus G$ and $g_0 \in G$. We have $g_0 \in \mathcal{P}_G(x)$ if and only if 0 belongs to the convex hull of the following set in the n-dimensional euclidean space :

$$D = \left\{\{\overline{f(x-g_0)}f(x_1), \ldots, \overline{f(x-g_0)}f(x_n)\}\,\middle|\, f \in \mathcal{E}(S_{E^*}),\ |f(x-g_0)| = \right. \tag{1.29}$$

$$\left. = \|x - g_0\|\right\}$$

In the particular case when $E = C(Q)$, taking into account (1.19) one obtains from corollary 1.1 a result of E. W. Cheney [19], p. 73); the necessity part of this latter result was observed, essentially, by T. J. Rivlin and H. S. Shapiro (see [82], p. 181).

For a characterization of elements of best approximation in terms of fixed points of a set-valued mapping, see § 5, theorem 5.5. For some other characterizations of elements of best approximation, see also [82].

2 . Existence of elements of best approximation

2.1. Characterizations of proximinal linear subspaces

The basic notion in connection with the existence of elements of best approximation is the following :

Definition 2.1. A set G in a metric space E is said to be proximinal if every element $x \in E$ has at least one element of best approximation in G, i.e. if

$$\mathcal{P}_G(x) \neq \emptyset \qquad (x \in E) . \tag{2.1}$$

The term "proximinal" set (a combination of "proximity" and "minimal") was proposed by R. Killgrove and used first by R. R. Phelps ([68] , p. 790). Some authors use for such sets the term distance set, or existence set, or (E)-set.

Since by 1 , formula (1.4), $\mathcal{P}_G(x) \neq \emptyset$ for all $x \in G$, condition (2.1) is equivalent to

$$\mathcal{P}_G(x) \neq \emptyset \qquad (x \in E \setminus G) . \tag{2.2}$$

Let us also observe that every proximinal set is necessarily closed , since otherwise no $x \in \overline{G} \setminus G$ would have an element of best approximation in G .

Now we shall consider the problem of characterization of proximinal linear subspaces G of a normed linear space E , i. e. the problem of giving necessary and sufficient conditions in order that G be proximinal . The basic observation is the following : A linear subspace G of a normed linear space E is proximinal if and only if G is proximinal in every linear subspace $E_0 \subset E$ such that G is a closed hyperplane in E_0 (i.e. a closed linear subspace of E_0 such that $\dim E_0/G = 1$). In fact, although this observation is obvious, it is useful since it reduces the problem of characterization of proximinal linear subspaces to that of the characterization of proximinal closed hyperplanes.

This latter problem is solved by the observation that a closed hyperplane G in E_0 is proximinal if and only if there exists an element $z \in E_0 \setminus \{0\}$ such that $0 \in \mathcal{P}_G(z)$ (Indeed, if $x \in E_0 \setminus G$ and $g_0 \in \mathcal{P}_G(x)$, then $0 \in \mathcal{P}_G(x - g_0)$ and conversely, if $0 \in \mathcal{P}_G(z)$ and $x \in E_0 \setminus G$, then, since G is a closed hyperplane in E_0, we have $x - \alpha z \in G$ for a suitable scalar $\alpha \neq 0$ whence, since also $0 \in \mathcal{P}_G(\alpha z)$, we infer $x - \alpha z \in \mathcal{P}_G(\alpha z + x - \alpha z) = \mathcal{P}_G(x)$: note that this proof is somewhat shorter than that given in [82], pp. 93-94). Using these observations, and the characterization of best approximations given in § 1, theorem 1.1, we obtain the following main theorem of characterization of proximinal linear subspace :

Theorem 2.1. A linear subspace G of a normed linear space E is proximinal if and only if G is closed and for every linear subspace $E_0 \subset E$ such that G is a closed hyperplane in E_0 and every functional $\varphi \in E_0^*$ with $\varphi|_G = 0$, there exists an element $z \in E_0$ such that

$$\varphi(z) = \|\varphi\| , \qquad \|z\| = 1 . \tag{2.3}$$

Any element z satisfying (2.3) is called a maximal element of φ.

In the usual concrete normed linear spaces the general form of functionals which admit maximal elements and the general form of maximal elements of such a functional are well known and simple (see e.g. [73], [99]) and therefore theorem 2.1 is suitable for applications in concrete spaces.

Let us also mention another characterization of proximinal linear subspaces, which appears in Cheney-Wulbert [20] :

Proposition 2.1. A linear subspaces G of a normed linear space E is proximinal if and only if

(2.4) $$E = G + \pi_G^{-1}(0) ,$$

where we use the notations

(2.5) $$\pi_G^{-1}(0) = \{x \in E \mid 0 \in \mathcal{P}_G(x)\},$$

(2.6) $$A + B = \{a + b \mid a \in A,\ b \in B\} .$$

Indeed, if G is proximinal and $x \in E$, $g_0 \in \mathcal{P}_G(x)$, then $x = g_0 + (x - g_0) \in G + \pi_G^{-1}(0)$. Conversely, if we have (2.4) and $x \in E$, $x = g_0 + y$, where $g_0 \in G$, $y \in \pi_G^{-1}(0)$, then $0 \in \mathcal{P}_G(y) = \mathcal{P}_G(x - g_0)$, whence $g_0 \in \mathcal{P}_G(x)$.

Some other characterizations of proximinal linear subspaces of normed linear spaces are given in [82], pp. 94-95. From the theorems in concrete spaces, let us mention the following characterization of proximinal linear subspaces G of finite codimension of $E = C_R(Q)$ (we recall that, by definition, codim G = dim E / G), due to A. L. Garkavi (see [82], p. 302) :

Theorem 2.2. A closed linear subspace G of codimension n of $E = C_R(Q)$ (Q compact) is proximinal if and only if the following three conditions are satisfied :

α) For every $\mu \in G^{\perp} \setminus \{0\}$ the carrier $S(\mu)$ admits a Hahn-decomposition into two closed sets $S(\mu)^{+}$ and $S(\mu)^{-} = S(\mu) \setminus S(\mu)^{+}$.

β) For every pair of measures $\mu_1, \mu_2 \in G^{\perp} \setminus \{0\}$ the set $S(\mu_1) \setminus S(\mu_2)$ is closed.

γ) For every pair of measures $\mu_1, \mu_2 \in G^{\perp} \setminus \{0\}$ the measure μ_1 is absolutely continuous with respect to μ_2 on the set $S(\mu_2)$.

In the particular case when $n = 1$ (i.e., when G is a closed hyperplane), conditions β) and γ) are automatically satisfied (since $\dim G^{\perp} = 1$) and hence condition α) is necessary and sufficient in order that G be proximinal; one can also show directly that this condition is equivalent to that of theorem 2.1, i.e. to the existence of a maximal element for each $\mu \in G^{\perp} \setminus \{0\}$.

Problem 2.1. Let $E = L^1_R(T, \nu)$, where (T, ν) is a positive measure space and let G be a closed linear subspace of codimension n of E. What conditions are necessary and sufficient in order that G be proximinal ?

Some necessary conditions, which are also sufficient when $n = 1$, are known (see [82], p. 325, theorem 2.10).

The notation $\pi_G^{-1}(0)$ for the sets (2.5) is motivated by § 4, definition 4.1 a). Since in the subsequent sections the set $\pi_G^{-1}(0)$ will be a useful tool in the study of problems of best approximation, let us mention here some properties of these sets (see [82], p. 143 and [66], [38]):

Proposition 2.2. Let G be a linear subspace of a normed linear space E. Then

a) The set $\pi_G^{-1}(0)$ is closed.

b) The set $\pi_G^{-1}(0)$ is "star-shaped", i.e. $x \in \pi_G^{-1}(0)$ implies $\alpha x \in \pi_G^{-1}(0)$ for all scalars α (in particular, $0 \in \pi_G^{-1}(0)$).

c) We have

$$\mathcal{P}_G(x) = (x + \pi_G^{-1}(0)) \cap G \qquad (x \in E) . \tag{2.7}$$

d) if $\dim E = \infty$, $\dim G < \infty$, then $\pi_G^{-1}(0)$ is homeomorphic to E.

e) If G is a semi-Čebyšev subspace of E (see § 3, definition 3.1), then $\pi_G^{-1}(0)$ is nowhere dense in E.

2.2 Some classes of proximinal linear subspaces.

Whenever a new class (= family) of subspaces is introduced, it is natural to ask whether it is non-void; in particular, one can ask whether proximinal linear subspaces exist in every normed linear space. The answer is affirmative, since from theorem 2.1 it follows that whenever $f \in E^*$ has a maximal element, $G = \{x \in E \mid f(x) = 0\}$ is a proximinal hyperplane. Now we shall give some other important classes of proximinal linear subspaces.

Theorem 2.3. Let E be a normed linear space and let G be a linear subspace of E such that the unit cell $S_G = \{g \in G \mid \|g\| \leq 1\}$ is sequentially compact for the weak topology $\sigma(E, E^*)$. Then G is proximinal.

This theorem (due to V. Klee [45]) can be deduced from theorem 2.1 but it admits also a simple direct proof ; a similar remark is also valid for theorem 2.4 below. For the proofs see [82] Ch. I, § 2. An immediate consequence of theorem 2.3 is the following :

Corollary 2.1. Let E be a normed linear space and let G be a linear subspace of E with the property that G is a reflexive Banach space. Then G is proximinal. In particular, every, finite-dimensional linear subspace G of a normed linear space E is proximinal.

Thoerem 2.4. Let E^* be the conjugate space of a normed linear space E. Then

a) Every linear subspace Γ of E^* having the unit cell $S_\Gamma = \{f \in \Gamma \mid \|f\| \leqslant 1\}$ compact for $\sigma(E^*, E)$ is proximinal. In particular, every $\sigma(E^*, E)$-closed linear subspace Γ of E^* is proximinal.

b) The same holds when S_Γ is sequentially compact for $\sigma(E^*, E)$.

Note that the first statement in a) is indeed more general than the second, since we did not assume E complete. Also, it can be shown by examples that between a) and b) there is no relation of implication.

Recently J. Blatter [6] has shown that the usual concrete Banach spaces E are proximinal in their second conjugate space E^{**} (identifying E with its canonical image in E^{**}) and has asked whether all Banach spaces have this property. W. Pollul has shown [69] that the answer is negative, for example the space c_0 endowed with the equivalent norm

$$(2.8) \qquad \|x\| = \sup_n |\xi_n| + \sum_{i=1}^{\infty} \frac{1}{2^i} |\xi_i| \quad (x = \{\xi_n\} \in c_0)$$

is a Banch space which is not proximinal in its second conjugate space.

2.3. Normed linear spaces in which all closed linear subspaces are proximinal.

Along with every new class of linear subspaces of normed linear spaces which we introduce, it is natural to consider also the complementary class, i.e. the family of all linear subspaces which do not belong to that class; we shall denominate the linear subspaces of this family by the prefix "non-"followed by the name of the original class, e.g. the linear subspaces which are not proximinal will be called "non-proximinal".

Naturally, in every infinite dimensional normed linear space there exist non-proximinal linear subspaces, for example, the non-closed linear subspaces. However, now we shall show that the problem of existence of non-proximinal closed linear subspaces may have a negative answer, i.e. there exist normed linear spaces in which all closed linear subspaces are proximinal, and we shall give some characterizations of such spaces.

Theorem 2.5. For a normed linear space E the following statements are equivalent :

1°. All closed linear subspaces of E are proximinal.

2°. The restriction of each $f \in E^*$ to every closed linear subspace of E has a maximal element.

If E is a Banach space, these statements are equivalent to the following :

3°. All closed linear subspaces of E, of a certain fixed finite codimension m, where $1 \leq m \leq \dim E - 1$, are proximinal.

4°. Every $f \in E^*$ has a maximal element.

5°. E is reflexive.

The equivalence $1^o \Leftrightarrow 2^o$ follows from theorem 2.1, and the other equivalence are a consequence of the profound theorem of R. C. James [43] (for which only difficult proofs are known to-day) that $4^o \Rightarrow 5^o$.

Note that a normed linear space E satisfying 4° need not be a Banach space (and thus it need not satisfy 5°), as shown by the following (unpublished) example of R. C. James: Let B be the space of all sequences of real numbers

$$(2.9) \qquad x = \{ \xi_1^1 ; \xi_1^2 , \xi_2^2 ; \xi_1^3 , \xi_2^3 , \xi_3^3 ; \dots$$

such that

$$(2.10) \qquad \|x\| = \Big\{ |\xi_1^1|^2 + \big[\max(|\xi_1^2|, |\xi_2^2|)\big]^2 + \big[\max(|\xi_1^3|, |\xi_2^3|, |\xi_3^3|)\big]^2 + \dots \Big\}^{\frac{1}{2}} < \infty ,$$

endowed with the usual operations and with the norm (2.10), and let E be the linear span of all members x of B for which

$$(2.11) \qquad |\xi_1^n| = |\xi_2^n| = \dots = |\xi_n^n| \quad \text{for } n = 1, 2, \dots$$

Then B is a reflexive Banach space and E is a dense linear subspace of B with $E \neq B$ (hence E is not a Banach space), having the property 4° of theorem 2.5 above.

2.4. Normed linear spaces which are proximinal in every superspaces.

We have seen in corollary 2.1 above that if G is a reflexive Banach space, then G is proximinal in every superspace E (i.e., in every normed linear space E containing G as a subspace). It is natural to raise the problem of characterization of all normed linear spaces G with this property. It is obvious that a normed linear space G which is proximinal in every superspace must be complete, i.e., a Banach space, since every non-complete normed linear space G is non-proximinal in its completion. Recently W. Pollul [69] has proved that each nonreflexive Banach space G can be embedded isometrically as a non-proximinal closed hyperplane in another Banach space E.

Thus we have

Theorem 2.6. A normed linear space G is proximinal in every superspace E if and only if G is a reflexive Banach space.

The proof of Pollul [69] has used James' theorem (the implication $4^{\circ} \Rightarrow 5^{\circ}$ of theorem 2.5 above), by observing that if $\varphi \in G^{*}$ has no maximal element, then $\{G, 0\}$ is a non-proximinal hyperplane in $E = G \times R$ endowed with the norm

$$\| \{x, \lambda\} \| = \max(\|x\| , | \varphi(x) + \lambda |)$$

and G is isometric to $\{G, 0\}$. A more elementary proof of theorem 2.6, which does not make use of James' theorem, was given recently in [86].

2.5. Transitivity of proximinality.

Let us denote the statement "G is a proximinal linear subspace of the normed linear space E" by : $G \overset{(p)}{\subset} E$. We shall consider now the problem, to what extent is this relation transitive.

Theorem 2.7. Let E be a Banach space. The following statements are equivalent :

1°. For all Banach spaces F, G

$$E \overset{(p)}{\subset} F, \quad F \overset{(p)}{\subset} G \quad \text{implies} \quad E \overset{(p)}{\subset} G. \tag{2.12}$$

2°. Same as 1°, with $\dim F/E = \dim G/F = 1$.

3°. For all Banach spaces F, G

$$F \overset{(p)}{\subset} E, \quad E \overset{(p)}{\subset} G \quad \text{implies} \quad F \overset{(p)}{\subset} G. \tag{2.13}$$

4°. Same as 2°, with $\dim E/F = \dim G/E = 1$.

5°. E is reflexive.

The implications $5^{\circ} \Longrightarrow 1^{\circ} \Longrightarrow 2^{\circ}$ and $5^{\circ} \Longrightarrow 3^{\circ} \Longrightarrow 4^{\circ}$ are immediate consequences of corollary 2.1 and the fact that every closed linear subspace of a reflexive space is reflexive. The implications $2^{o} \Longrightarrow 5^{o}$ and $4^{o} \Longrightarrow 5^{o}$ have been proved recently by W. Pollul [69], using again Jame's theorem .

One can also consider a third type of transitivity property, with E "on the last place" , but the following theorem of W . Pollul [69] shows that some non-reflexive Banach spaces also have this property.

Theorem 2.8. Let $E = c_0$. Then for all F, G

(2.14) $G \overset{(p)}{\subset} F, \quad F \overset{(p)}{\subset} E, \quad \dim F/G < \infty, \quad \dim E/F < \infty$ imply

$$G \overset{(p)}{\subset} E .$$

Naturally, by corollary 2.1, every reflexive Banach space E also has property (2.14). On the other hand, there exist non-reflexive Banach spaces E which do not have property (2.14), even with

$$\dim F/G = \dim E/F = 1,$$

e.g. $E = C_R(Q)$ (Q compact) and $E = L^1_R(T, \nu)$ ((T, ν) a positive measure space) whenever $\dim E = \infty$ (W. Pollul [69]) .

Problem 2.2. (W. Pollul [69]). Wich non-reflexive Banach spaces E have property (2.14) with $\dim F/G = \dim E/F = 1$?

Problem 2.3. (W. Pollul [69]). Does $E = c_0$ have the property

(2.15) $$G \overset{(p)}{\subset} F, \quad F \overset{(p)}{\subset} E \quad \text{imply} \quad G \overset{(p)}{\subset} E \ ?$$

or, alternatively: is every Banach space E with property (2.15) reflexive ?

2.6. Proximinality and quotient spaces

The following theorem of Cheney and Wulbert [20] shows how

proximinality is transmitted to and from quotient spaces:

Theorem 2.8. If G_1 is a proximinal linear subspace of a normed linear space E and G_2 an arbitrary closed linear subspace of G_1, then G_1/G_2 is proximinal in E/G_2.

Conversely, if G_1 is a closed linear subspace of E and G_2 a closed linear subspace of G_1 such that G_1/G_2 is proximinal in E/G_2 (in particular, if G_1/G_2 is reflexive) and that G_2 is proximinal in E, then G_1 is proximinal in E.

2.7. Very non-proximinal linear subspaces.

Definition 2.2. A set G in a metric space E is said to be very non-proximinal if G is closed and no element $x \in E \setminus G$ has an element of best approximation in G, i.e. if $\overline{G} = G$ and

$$\mathcal{P}_G(x) = \emptyset \qquad (x \in E \setminus G). \tag{2.16}$$

Obviously, every very non-proximinal set is non-proximinal. We have the following simple characterization of very non-proximinal subspaces of normed linear spaces:

Proposition 2.3. A linear subspace G of a normed linear space E is very non-proximinal if and only if there is no element $z \in E \setminus \{0\}$ such that

$$0 \in \mathcal{P}_G(z) \tag{2.17}$$

Indeed, if a $z \in E \setminus \{0\}$ satisfies (2.17), then $z \in E \setminus G$ and $\mathcal{P}_G(z) \neq \emptyset$. Conversely, if $x \in E \setminus G$, $\mathcal{P}_G(x) \neq \emptyset$, $g_0 \in \mathcal{P}_G(x)$, then $x - g_0 \in E \setminus G$ and $0 \in \mathcal{P}_G(x - g_0)$.

From this proposition and from the observation made before theorem 2.1 it follows that a closed hyperplane G in E is very non-proximinal if and only if it is non-proximinal. Consequently, by theorem 2.5, a Banach space E contains very non-proximinal linear subspaces if and only if E is non-reflexive; moreover, in this case, for every $f \in E^*$ which has no maximal element, the closed hyperplane $G = \{x \in E \mid f(x) = 0\}$ is very non-proximinal.

3 Uniqueness of elements of best approximation.

3.1 Characterizations of semi-Čebyšev and Čebyšev subspaces.

The basic notions in connection with the uniqueness of elements of best approximation are given in

Definition 3.1. A set G in a metric space E is said to be

a) a semi-Čebyšev set if every element $x \in E$ has at most one element of best approximation in G, i.e. if

$$(3.1) \qquad x \in E, \quad g_1, g_2 \in \mathcal{P}_G(x) \quad \text{imply} \quad g_1 = g_2 \quad ;$$

b) a Čebyšev set, if it simultaneously a proximinal and semi-Čebyšev set i.e. if every element $x \in E$ has exactly one element of best approximation in G.

Obviously, in definition 3.1 the condition $x \in E$ can be replaced by : $x \in E \setminus G$.

The term "semi-Čebyšev" set has been proposed in the monograph [82] . The term "Čebyšev" set has been used previously by several authors, e.g. by N. Efimov and S. B. Stečkin [28] . Some outhors use for such sets the term Haar set.

Let us consider now the problem of characterization of semi-Čebyšev and Čebyšev (linear) subspaces G of a normed linear space E.

From § 1, theorem 1.5, it follows

Theorem 3.1. A linear subspace G of a normed linear space E is a semi-Čebyšev subspace if and only if there do not exist $f \in E^*$, $x \in E$ and $g_0 \in G \setminus \{0\}$ such that

$$(3.2) \qquad \|f\| = 1 \ ,$$

$$(3.3) \qquad f(g) = 0 \qquad (g \in G),$$

$$(3.4) \qquad f(x) = \| x \| = \| x - g_0 \| .$$

Although this characterization of semi-Čebyšev subspaces is not intrinsic (i.e., it involves also elements of $E \setminus G$), it is convenient for applications, since one can deduce from it intrinsic characterizations of semi-Čebyšev subspaces in the usual concrete normed linear spaces, as we shall see below.

Let us mention some characterizations of Čebyšev subspaces related to §2, proposition 2.1, which can be found in [20], [77] and [38] respectively :

Proposition 3.1. For a closed linear subspace G of a normed linear space E the following statements are equivalent :

1°. G is a Čebyšev subspace .

2°. We have

$$(3.5) \qquad E = G \oplus \pi_G^{-1}(0) ,$$

where $\pi_G^{-1}(0)$ is the set defined by §2, formula (2.6) and where $\oplus$ means that the sum decomposition of each element $x \in E$ is unique.

3°. G is proximinal and

$$(3.6) \qquad (\pi_G^{-1}(0) - \pi_G^{-1}(0)) \cap G = \{0\} .$$

4°. G is proximinal and the restriction $\omega_G \big|_{\pi_G^{-1}(0)}$ of the canonical mapping $\omega_G : E \longrightarrow E/_G$ to the set $\pi_G^{-1}(0)$ is one-to-one .

Indeed, the equivalence $1^\circ \Longleftrightarrow 2^\circ$ is essentially proved by the same arguments as those used to prove § 2 , proposition 2.1. Now, if 3° does not hold, then either G in non-proximinal, contradicting (3.5) (by § 2, proposition 2.1) or (3.6) does not hold, say

$$y_1, y_2 \in \pi_G^{-1}(0), \quad y_1 - y_2 \in G \setminus \{0\}$$

and then $y_1 = 0 + y_1 = (y_1 - y_2) + y_2$, contradicting (3.5); thus, $2^\circ \Rightarrow 3^\circ$.

Furthermore, if $y_1, y_2 \in \pi_G^{-1}(0)$, $y_1 \neq y_2$, $\omega_G(y_1) = \omega_G(y_2)$, then $y_1 - y_2 \in (\pi_G^{-1}(0) - \pi_G^{-1}(0)) \cap G \setminus \{0\}$, contradicting (3.6) ; thus, $3^\circ \Longrightarrow 4^\circ$. Finally, if $x \in E$, $g_1, g_2 \in \mathcal{P}_G(x)$, $g_1 \neq g_2$, then $x - g_1$, $x - g_2 \in \pi_G^{-1}(0)$, $x - g_1 \neq x - g_2$, $\omega_G(x - g_1) = \omega_G(x - g_2)$ (because $x - g_1 - (x - g_2) = g_2 - g_1 \in G$), contradicting 4° ; thus $4^\circ \Longrightarrow 1^\circ$, which completes the proof. Naturally, one obtains corresponding characterizations of semi-Čebyšev subspaces G by requiring instead of 2° that each element $x \in E$ have at most one sum decomposition and by omitting in 3° and 4° the condition of proximinality of G .

In connection with 4° above let us observe that for any closed linear subspace $G \subset E$ we have, by definition, $\| \omega_G(x) \| = (x, G) (x \in E)$ and hence, by § 2 , formula (2.5),

$$\pi_G^{-1}(0) = \{ x \in E \mid \| \omega_G(x) \| = \| x \| \} ; \tag{3.7}$$

in other words, $\pi_G^{-1}(0)$ is that subset of E , on which the restriction of ω_G is norm-preserving.

We recall that a linear subspace G of a normed linear space E is said to have property (U) if every functional $\varphi \in G^*$ has a unique extension with the same norm to the whole space E . From proposition 3.1 we obtain the following relations of duality between Čebyšev subspaces and subspaces with property (U) :

Corollary 3.1. Let E be a normed linear space. Then

a) A $\sigma(E^*, E)$-closed linear subspace Γ of E^* is a Čebyšev subspace if and only if $\Gamma_\perp = \{x \in E \mid f(x) = 0 \ (f \in \Gamma)\} \subset E$ has property (U) .

b) If G is a proximinal linear subspace of E and $G^\perp = \{f \in E^* \mid f(x) = 0 \ (x \in G)\} \subset E^*$ has property (U), then G is a Čebyšes subspace of E .

Indeed, a) follows from (3.7) (for G, E replaced by Γ and E^* respectively) and the observation that $\omega_\Gamma : E^* \longrightarrow E^*/\Gamma$ carries each $f \in E^*$ into the set of all extensions of $f|_{\Gamma_\perp} \in (\Gamma_\perp)^*$ to the whole space E (we have $\Gamma = (\Gamma_\perp)^\perp$, since Γ is $\sigma(E^*, E)$-closed). Furthermore, if $G^\perp = (G^{\perp\perp})_\perp \subset E^*$ has property (U), then by part a) $G^{\perp\perp}$ is a Čebyšev subspace, in E^{**}, whence, since $G \subset G^{\perp\perp}$, b) follows. However, one can show that $E = c_0$ has finite-dimensional Čebyšev subspaces and that for no such subsapce G has $G^\perp$ the property (U) ,(see [82] , p. 108), and thus the converse of b) is not true.

Let us return now to applications of theorem 3.1. In $E = C_R(Q)$ from theorem 3.1 one obtains the following theorem of [78] which has been the first intrinsic characterization of semi-Čebyšev subspaces G of an arbitrary (finite or infinite) dimension of $E = C_R(Q)$ (see also [82] , p. 117) :

Theorem 3.2. A linear subspace G of $E = C_R(Q)$ is a semi-Čebyšev subspace if and only if there do not exist a Radon measure μ on Q , two disjoint closed sets $Y^+, Y^- \subset Q$ and an element

$g_0 \in G \setminus \{0\}$ such that

(3.8) $$|\mu|(Q) = 1 ,$$

(3.9) $$\int_Q g(q) d\mu(q) = 0 \qquad (g \in G)$$

(3.10) $\mu \geqslant 0$ on Y^+, $\mu \leqslant 0$ on Y^- and $Y^+ \cup Y^- \supset S(\mu)$,

(3.11) $$g_0(q) = 0 \qquad (q \in S(\mu)),$$

where $S(\mu)$ denotes the carrier of the measure μ.

Recently E. W. Cheney and D. E. Wulbert [20] have proved that a linear subspace G of $E = C_R(Q)$ is a semi-Čebyšev subspace if and only if 0 is the only element of G which vanishes on an "α_G--set" i.e., on a subset of Q of the form

$$\{q \in Q \mid |x(q)| = \|x\|\}$$

for some $x \in \pi_G^{-1}(0)$. Naturally, one can also deduce this easily from theorem 3.1 or directly from theorem 3.2.

Some other characterizations of semi-Čebyšev and Čebyšev subspaces of arbitrary dimension, in general normed linear spaces and in some other concrete normed linear spaces, are given in [82], Ch. I, § 3 and in [20].

Let us consider now the particular case when $\dim G = n < \infty$. By § 2, corollary 2.1, every such G is proximinal. Naturally, the preceding results on uniqueness of elements of best approximation are also valid in this particular case, but exploiting the assumption of finite dimensionality, we can obatin additional information. Thus, from theorem 3.1 one obtains (see [82], pp. 210-211) :

Theorem 3.3. An n-dimensional linear subspace G of a normed linear space E is a Čebyšev subspace if and only if there do not exist h extremal points $f_1, \ldots, f_h$ of S_{E^*}, where $1 \leq h \leq n$ if the scalars are real and $1 \leq h \leq 2n-1$ if the scalars are complex, h numbers $\lambda_1, \ldots, \lambda_h > 0$ with

$$\sum_{j=1}^{h} \lambda_j = 1 ,$$

and $x \in E$, $g_0 \in G \setminus \{0\}$, such that we have

$$\sum_{j=1}^{h} \lambda_j f_j (x_h) = 0 \qquad (k = 1, \ldots, n) \tag{3.12}$$

$$f_j(x) = \|x\| = \|x - g_0\| \qquad (j = 1, \ldots, h) \tag{3.13}$$

The main difficulty in the proof of this result consists in establishing the bounds $h \leq n$ and $h \leq 2n-1$ respectively (note that in § 1, theorem 1.6 we had only the bounds $h \leq n+1$ and $h \leq 2n+1$ respectively).

In particular, for $E = C(Q)$ (Q compact), we have the following classical theorem, given for real scalars by A. Haar [36] and for complex scalars by A. N. Kolmogorov [49] .

Theorem 3.4. An n- dimensional linear subspace $G = [x_1, \ldots, x_n]$ of $E = C(Q)$ (Q compact) is a Čebyšev subspace if and only if $x_1, \ldots, x_n$ form a "Čebyšev system" (i.e., every

$$\sum_{i=1}^{n} \alpha_i x_i \not\equiv 0$$

has at most n-1 zeros on Q).

This theorem follows as a simple particular case both from theorem 3.2 for $\dim G = n < \infty$ (respectively, a complex version of it) and from theorem 3.3 for $E = C(Q)$. These two proofs of theorem 3.4, compared with the other function-theoretic proofs of theorem 3.4 given previously by several authors, show clearly the advantages of applying the methods of functional analysis in the theory of best approximation.

We mention also the following characterization of finite-dimensional Čebyšev subspaces of $C_R(Q)$, due (for $Q = [a, b]$) to Y. Ikebe [40] :

<u>Proposition 3.2. A finite-dimensional subspace</u> G <u>of</u> $E = C_R(Q)$ <u>(Q compact) is a Čebyšev subspace if and only if</u>

$$\|g_0\| < 2\,\|x\| \qquad (x \in E \setminus \{0\},\ g_0 \in \mathcal{P}_G(x)). \tag{3.14}$$

Let us consider now closed linear subspaces G of finite codimension . From theorem 3.1 one obtains the following result, due to A. L. Garkavi (see [82] , p. 296) :

<u>Theorem 3.5. A closed linear subspace</u> G <u>of codimension</u> n <u>of a normed linear space</u> E , <u>say</u> $G = \{x \in E \mid f_1(x) = \dots = f_n(x) = 0\}$ <u>(with</u> $f_1, \dots, f_n \in E^*$ <u>linearly independent), is a semi-Čebyšev subspace if and only if for every</u> $f_0 \in G^\perp \setminus \{0\}$ <u>the set</u> $\mathfrak{M}_{f_0}$ <u>of all maximinal elements of</u> f_0 <u>is of dimension</u> $r = r(f_0) \leq n-1$ <u>and contains</u> $r+1$ <u>elements</u> $x_0, x_1, \dots, x_r$ <u>such that</u>

(3.15) $$\operatorname{rank}(f_k(x_i))_{\substack{k=1,\dots,n\\ i=0,1,\dots,r}} = r+1 \quad ;$$

moreover, in this case for any $r+1$ linearity independent elements $x_0, x_1, \dots, x_r \in \mathcal{M}_{f_0}$ we have (3.15).

Applying this result in the space $E = C(Q)$ (Q compact) and combinig it with § 2, theorem 2.2, one obtains the following result, due to A. L. Garkavi (see [82], pp. 315-320):

Theorem 3.6. A closed linear subspace G of codimension n of $E = C_R(Q)$ (Q compact), say $G = \{x \in C(Q) \mid \mu_1(x) = \dots = \mu_n(x) = 0\}$ (with $\mu_1, \dots, \mu_n \in E^* \equiv \mathcal{M}^1_R(Q)$ linearly independent), is a Čebyšev subspace if and only if the following four conditions are satisfied:

a) For every $\mu \in G^\perp \setminus \{0\}$ the space Q admits a Hahn decomposition with respect to μ into two closed sets Q^+ and Q^-.

b) Any two measures $\mu_1, \mu_2 \in G^\perp \setminus \{0\}$ are equivalent (i.e., each is absolutely continuous with respect to the other) on the set Q' of all limit points of Q.

c) For every $\mu \in G^\perp \setminus \{0\}$ the set $Q \setminus S(\mu)$ consists of at most $n-1$ points.

d) For any r isolated points $q_1, \dots, q_r$ of Q, where $1 \leqslant r \leqslant n-1$, we have

(3.16) $$\operatorname{rank}(\mu_k(\{q_i\}))_{\substack{k=1,\dots,n\\ i=1,\dots,r}} = r .$$

If Q contains at least n isolated points and G is proximinal, then consition c) is necessary and sufficient in order that G be a Čebyšev subspace. If Q has no isolated point and G is proximinal, then G is a Čebyšev subspace if and only if $S(\mu) = Q$ for all $\mu \in G^{\perp} \setminus \{0\}$.

Some other characterizations of semi-Čebyšev and Čebyšev subspaces of finite codimension in general normed linear spaces and in various concrete spaces are given in [82], Ch. III and in [20], [66], [32] - [34]. In particular, characterizations of semi-Čebyšev subspaces of finite codimension in $L^1_R(T, \nu)$ are given in [82], p. 326 and in [66]; theorem 5. At the time when [82] appeared, no characterization of Čebyšev subspaces of finite codimension in $L^1_R(T, \nu)$ was known (only some necessary and some sufficient conditions were known, see [82], p. 328-330); this problem has been solved recently by A. L. Garkavi [34].

3.2. Existence of semi-Čebyšev and Čebyšev subspaces.

It is obvious that every normed linear space E contains semi--Čebyšev linear subspaces; for example, every linear subspace G dense in E, with $G \neq E$, satisfies $\mathcal{P}_G(x) = \emptyset$ $(x \in E \setminus G)$ and hence is a semi-Čebyšev subspaces.

The problem of existence of semi-Čebyšev closed linear subspaces of Banach spaces in less trivial, but still has an affirmative answer, namely, every Banach space has at least one semi-Čebyšev closed hyperplane. Indeed, by § 2, every non-reflexive Banach space E contains very-non-proximinal closed hyperplanes and any such closed hyperplane is clearly semi-Čebyšev. On the other hand, by a result of J. Lindenstrauss [59], in every reflexive Banach space E the unit cell S_E has an exposed point, i.e., a point $x \in \mathrm{Fr}\, S_E$ with the property that there exists a support hyperplane H of S_E such that $H \cap S_E = \{x\}$, and hence E obviously has a Čebyšev hyperplane (namely, H-x).

For Čebyšev subspaces of Banach spaces the situation is different, as shown by the following example of A. L. Garkavi (see [82], p. 114):

If I is a set of cardinality $> c$, then the (non-separable) Banach space $E = E(I)$ of all bounded families $x = \{\xi_i\}_{i \in I}$ of scalars which have at most a countable number of non-zero coordinates ξ_i, endowed with the usual vector operations and with the norm

$$\|x\| = \sup_{i \in I} |\xi_i|$$

has no Čebyšev subspace. However, the following problem, raised essentially by A. L. Garkavi, in 1964 (see [82], p. 116 and [83]) is open :

Problem 3.1. Does there exist a separable Banach space E which has no Čebyšev subspace ?

We have seen above that the assumption of separability of E in problem 3.1 is essential. The assumption of completeness of E is also essential, since V. Klee and myself have given (see [83]) the following example of a separable non-complete normed linear space E which has no Čebyšev subspace : the dense linear subspace E of c_0 consisting of all almost-zero sequences (i.e. sequences with all coordinates $= 0$, except a finite number of them).

It is well known that the space $E = c_0$ has no Čebyšev subspace of infinite dimension, but it has Čebyšev subspaces of any finite dimension. It is also known that $E = L^1_R([0, 1])$ has no Čebyšev subspace of finite dimension or of finite codimension, but still it has Čebyšev subspaces. Therefore one might perhaps find a solution to problem 3.1 by combining in some way the spaces c_0 and $L^1_R([0, 1])$

Furthermore, since in every separable conjugate Banach space $E = B^*$ the unit cell has at least one exposed point and hence a Čebyšev subspace, one may expect a positive solution of problem 3.1 only among those separable Banach spaces which are not isometric to any conjugate Banach space (the spaces c_0 and $L^1_R([0, 1])$ do have this property; moreover, they are even not isomorphic to any conjugate Banach space) .

Now we shall consider the problem of existence of Čebyšev subspaces G with various restrictions on dim G or on codim G .

For some time the following problem, raised by V. Klee [46] in 1957, has been open : Does every m-dimensional Banach space E, where $3 \leq m < \infty$, possess a one-dimensional Čebyšev subspace G ? Or, geometrically : does every E with $3 \leq \dim E = m < \infty$ possess a line G through the origin such that there exists no segment on $\mathrm{Fr}\, S_E = \{x \in E \mid \|x\| = 1\}$, parallel to G ? Affirmative answers have been given by various authors for m = 3, 4 and for arbytrary m when S_E is sufficiently smooth and the problem has continued to preoccupy many mathematicians. Recently an affirmative solution for arbitrary m has been given by G. Ewald, D. G. Larman and C. A. Rogers [29] .

Let us consider now the following problem : What are (i.e., characterize topologically) the compact spaces Q for which $E = C(Q)$ has n-dimensional Čebyšev subspaces ? This problem presents interest also from the following general point of view : The classical theorem of Banch-Stone (according to which two compact spaces Q_1 and Q_2 are homeomorphic if and only if $C(Q_1)$ and $C(Q)_2$ are linearly isometric) shows that, theoretically, the metric-linear properties of the spaces $C(Q)$ are completely determined by the topological properties of the compact spaces Q and conversely; but, the effective, explicit study of this interpendence still presents many open problems and any non-trivial answer to the above question may be regarded also as a contribution to this study. Let us first observe that by the Haar-Kolmogorov theorem (theorem 3.4 above) the problem is equivalent to the following :

What are the compact spaces Q which admit a real (or complex) Čebyšev system $x_1, \ldots, x_n$? (naturally, we assume that Q consists of at least $n+1$ points) .

For $n = 1$ the answer is obvious, since every compact space admits a Čebyšev system consisting of one function x_1 for example the function $x_1 \equiv 1$. For $n \geqslant 2$ the answer is given by

Theorem 3.7. A compact space Q admits a real Čebyšev system $x_1, \ldots, x_n$ (or, what is equivalent: $C_R(Q)$ has an n-dimensional Čebyšev subspace) with $n \geqslant 2$ if and only if Q is homeomorphic to a subset of the unit circumference

$$\left\{ \{\xi_1, \xi_2\} \mid \xi_1^2 + \xi_2^2 = 1 \right\} .$$

Moreover, if Q is homeomorphic to the whole unit circumference, then every real Čebyšev system on Q consists of an odd number of elements.

This result has been conjectured by S. Mazur and proved by J. C. Mairhuber under the assumption that Q is a subset of a finite-dimensional euclidean space; for general compact spaces it has been proved by K. Siekluki and P. C. Curtis Jr. (see [82] , pp. 218-222, where a proof due to I. J. Schoenberg and C. T. Yang is presented; for the last statement of theorem 3.7 see e.g. [61] , p. 26).

For the case of complex scalars the problem is still open (only partial results are known, see [82], p. 222) .

From theorems 3.7 and 3.4 one can deduce the following result, due to R. R. Phelps (see [82], p. 222) :

Corollary 3.2. The space $E = L_R^{\infty}(T, \nu)$, where (T, ν) is a σ-finite positive measure space such that $\dim E = \infty$, has no Čebyšev subspace of finite dimension ≥ 2 (however, it does have Čebyšev subspaces of dimension 1, even in the case of complex scalars) .

It is natural to consider the similar problem for subspaces of finite codimension, i.e. the problem of characterizing those compact spaces Q for which C(Q) contains a semi-Čebyšev or a Čebyšev subspace of finite codimension $n \geq 1$. The problem is solved for metric compact spaces Q and real scalars, by the following results, due to A. L. Garkavi (see [82], pp. 314 and 325, footnote, and [33]) :

Theorem 3.8. a) For every infinite compact metric space Q and every integer n with $1 \leq n < \infty$, the space $C_R(Q)$ contains semi-Čebyšev subspaces G of codimension n and Čebyšev hyperplanes.

b) For a compact metric spaces Q and any integer n with $2 \leq n < \infty$, the space $C_R(Q)$ has a Čebyšev subspaces G of codimension n if and only if

$$Q = \overline{Q \setminus Q'} \tag{3.17}$$

(i. e., Q coincides with the closure of the set of its isolated points).

For non-metrizable compact spaces Q only partial results are known, namely, conditions which are necessary (e.g. that Q have at most a countable number of disjoint open subsets and that Q contain no open connected infinite subset) or which are sufficient in order that $C_R(Q)$ have Čebyšev subspaces of finite codimension $n \geq 1$ or $n \geq 2$

(see [82], Ch. III).

One can also consider the analogous problem for the spaces

$$L^1_R(T, \nu) \quad (\text{or } L^1(T, \nu))$$

on positive measure spaces (T, ν). For real scalars we have the following results corresponding to theorems 3.4 and 3.8 b), which are due, in the case when (T, ν) is σ-finite, to A. L. Garkavi (see [82], p. 233 and p. 331):

Theorem 3.9. *Let* (T, ν) *be a positive measure space such that* $L^1_R(T, \nu)^* \equiv L^\infty_R(T, \nu)$ *and let* $n \geqslant 1$. *The following statements are equivalent*:

1°. $L^1_R(T, \nu)$ *has an n-dimensional Čebyšev subspace.*

2°. $L^1_R(T, \nu)$ *has a Čebyšev subspace of codimension* n.

3°. (T, ν) *has at least* n *atoms*.

We recall that an atom of (T, ν) is a measurable set $A \subset T$ with $\nu(A) > 0$, such that if B is any measurable subset of A, then either $\nu(B) = 0$ or $\nu(A \setminus B) = 0$. From theorem 3.9 it follows, in particular, that *if* (T, ν) *has no atoms* (*e.g. if* $T = [0, 1]$ *and* ν *is the Lebesgue measure*) *then* $L^1_R(T, \nu)$ *has no Čebyšev subspaces of finite dimension or codimension.* For Čebyšev subspaces of finite dimension this latter result is known to hold also for complex scalars (see [82], p. 230-232), but no extension of theorem 3.9 to complex $L^1(T, \nu)$ spaces is known.

E. W. Cheney and D. E. Wulbert have proved ([20], theorem 34) that $E = C^1_R(Q, \nu)$ (Q *compact,* $S(\nu) = Q$, *contains a Čebyšev subspace of codimension* n *if and only if* Q *has at least* n *iso-*

lated points.

We conclude with the following problem of A. L. Garkavi (see 83 , p. 2 and 31 , p. 96) :

Problem 3.2. Does the space $C_R(\ 0,\ 1\ )$ possess a Cebysev subspace G of infinite dimension and infinite codimension?

Recently D. E. Wulbert 95 has shown that there exist compact spaces Q such that the analogue of problem 3.2 for $C_R(Q)$ has an affirmative answer . It is also known (see 82 , p. 332) that $L^1_R(\ 0, 1\ )$ has Cebysev subspaces of infinite dimension and infinite codimension.

3.3. Normed linear spaces in which all linear (respectively, all closed linear) subspaces are semi-Čebyšev (respectively, Čebyšev) subspaces .

We recall that a normed linear space E is said to be strictly convex (or rotund) if the relations

$$(3.18)\qquad x,\ y \in E \setminus \{0\},\ \ \|x + y\| = \|x\| + \|y\|$$

imply the existence of a $c > 0$ such that

$$(3.19)\qquad y = cx\ .$$

It is well known and easy to show that this property is equivalent to each of the following properties : a) Fr $S_E = \mathcal{E}(S_E)$; b) Fr S_E contains no segment ;
c) each $f \in E^*$ has at most one maximal element.

Using theorem 3.1, one obtains (see [82] , p. 110) :

Theorem 3.10. For a normed linear space E the following statements are equivalent:

1°. All linear subspaces of E are semi-Čebyšev subspaces.

2°. All linear subspaces of E of a certain fixed finite dimension n, where $1 \leqslant n \leqslant \dim E - 1$, are semi-Čebyšev (or, what is equivalent, Čebyšev) subspaces.

3°. All closed linear subspaces of E of a certain fixed finite codimension m, where $1 \leqslant m \leqslant \dim E - 1$, are semi-Čebyšev subspaces.

4°. E is strictly convex.

Combining this with § 2, theorem 2.5, we obtain (see [82], p. 111):

Theorem 3.11. For a Banach space E the following statements are equivalent:

1°. All closed linear subspaces of E are Čebyšev subspaces.

2°. All closed linear subspaces of E of a certain fixed finite codimension m, where $1 \leqslant m \leqslant \dim E - 1$, are Čebyšev subspaces.

3°. E is reflexive and strictly convex.

In particular, since the spaces $E = L^p(T, \nu)$ $(1 < p < \infty)$, where (T, ν) is a positive measure space, and the Hilbert (= complete inner product) spaces $E = \mathcal{H}$ satisfy 3°, it follows that all closed linear subspaces of these spaces are Čebyšev subspaces.

3.4. Semi-Čebyšev and Čebyšev subspaces and quotient spaces.

The following results, corresponding to § 2, theorem 2.8, have been proved by E. W. Cheney and D. E. Wulbert [20]:

Theorem 3.12. a) If G_1 is a semi-Čebyšev linear subspace of a normed linear space E and G_2 a closed linear subspace of G_1, which is proximinal in E, then G_1/G_2 is a semi-Čebyšev subspace of E/G_2.

b) If G_1 is a linear subspace of E and $G_2 \subset G_1$ a closed linear subspace of E such that G_1 / G_2 is semi-Čebyšev in E/G_2 and that G_2 is semi-Čebyšev in E, then G_1 is a semi-Čebyšev subspace of E.

c) If G_1 is a closed linear subspace of E and G_2 is a closed linear subspace of G_1, such that G_1/G_2 is Čebyšev in E/G_2 and that G_2 is Čebyšev in E, then G_1 is a Čebyšev subspace of E.

3.5. Strongly unique elements of best approximation. Strongly Čebyšev subspaces. Interpolating subspaces.

Definition 3.2. Let G be a set in a metric space E. An element $g_0 \in G$ is said to be a strongly unique element of best approximation of an element $x \in E$ if there exists a constant $r = r(x, G)$ with $0 < r \leqslant 1$, such that

$$\rho(x, g) \geqslant \rho(x, g_0) + r\,\rho(g_0, g) \qquad (g \in G). \tag{3.20}$$

In this case $\mathcal{P}_G(x) = \{g_0\}$, i.e. g_0 is the unique element of best approximation of x, since by $r > 0$ for every $g \in G \setminus \{g_0\}$ we have $\rho(x, g) > \rho(x, g_0)$. The following characterization of such elements is due, essentially (namely, for $g = 0$ and $\|x\| = 1$), to D. E. Wulbert [97]:

Proposition 3.3. Let G be a linear subspace of a real normed linear space E. An element $g_0 \in G$ is a strongly unique element of best approximation of an element $x \in E \setminus \overline{G}$ if and only if there exists a constant $r = r(x, G)$ with $0 < r \leqslant 1$ such that

(3.21) $$\sup_{f \in \mathcal{M}_{x - g_0}} f(g) \geqslant r \, \| g \| \quad (g \in G) ,$$

where

(3.22) $$\mathcal{M}_{x - g_0} = \left\{ f \in E^* \,\middle|\, \| f \| = 1, \quad f(x - g_0) = \| x - g_0 \| \right\} .$$

Definition 3.3. A Čebyšev set G in a metric space E is said to be a strongly Čebyšev set if every $x \in E$ has a strongly unique element of best approximation in G .

D. J. Newman and H. S. Shapiro (see [19], p. 80) and, respectively, D. E. Wulbert [97] have proved

Theorem 3.13. In the spaces $C_R(Q)$ (Q compact) and $L^1_R(T, \nu)$ ((T, ν) a positive measure space) every finite-dimensional Čebyšev subspace is a strongly Čebyšev subspace.

On the other hand, D. E. Wulbert [97] has observed that in a smooth normed linear space E no Čebyšev subspace is strongly Čebyšev .

We recall that E is said to be smooth if for every $x \in E$ there exists only one $f = f_x \in E^*$ such that $\| f \| = 1$ $f(x) = \| x \|$.

Definition 3.4. An n-dimensional linear subspace G of normed linear space E is called an interpolating subspace if for any n linearly independent extremal points $f_1, \dots, f_n$ of S_E and any n numbers $c_1, \dots, c_n$ there exists exactly one $g \in G$ such that

(3.23) $$f_j(g) = c_j \qquad (j = 1, \dots, n) .$$

In arbitrary normed linear spaces such subspaces have been first considered in [76], where it was proved that they are Čebyšev subspaces (indeed, this is a consequence of theorem 3.3; see [82], pp. 213--214) and that the converse need not hold even if $\dim E < \infty$. Recently D.A. Ault, F. R. Deutsch, P. D. Morris and J. E. Olson [3] have studied best approximation by elements of interpolating subspaces, proving, among other results, the following :

Theorem 3.14. Every interpolating subspace G of a normed linear space E is a strongly Čebyšev subspace.

From the Haar-Kolmogorov theorem (theorem 3.4 above) it follows that a finite-dimensional subspace G of $E = C(Q)$ is a Čebyšev subspace if and only if it is an interpolating subspace ; this, together with theorem 3.14, implies again the first part of theorem 3.13.

On the other hand, from theorems 3.4 and the observation made after theorem 3.13 it follows that a smooth normed linear space E (in particular, the $L^p(T, \nu)$ spaces, for $1 < p < \infty$) contains no interpolating subspace. For the L^1_R - spaces, D. A. Ault, F. R. Deutsch, P. D. Morris and J. E. Olson [3] have proved the following result, which should be compared with theorem 3.9 :

Theorem 3.15. For a σ-finite positive measure space (T, ν) the space $E = L^1_R(T, \nu)$ contains an interpolating subspace of dimension $n > 1$ if and only if T is the union of at least n atoms (or, equivalently, $L^1_R(T, \nu)$ is linearly isometric either to l^1_R or to some $(l^1_m)_R$. Also, $L^1_R(T, \nu)$ contains a one-dimensional interpolating subspace if and only if T contains an atom.

For complex scalars and $n > 1$ the situation is quite opposite, namely, the complex spaces l^1 and l^1_m contain no proper interpolating subspace of any finite dimension $n > 1$ (J. H. Biggs, F. R. Deutsch, R. E. Huff, P. D. Morris and J. E. Olson [4]); on the other hand, it is clear that the unit vector $\{1, 0, 0, \dots\}$ in the complex spaces

l^1 or l^1_m spans a one-dimensional interpolating subspace.

3.6. Almost Čebyšev subspaces. k-semi-Čebyšev and k-Čebyšev subspaces. Pseudo-Čebyšev subspaces.

We shall consider now some generalizations of semi-Čebyšev and Čebyšev subspaces.

Definition 3.5. A set G in a metric space E is called an almost Čebyšev set if the set of all $x \in E$ for which $\mathcal{P}_G(x)$ does not consist of a single element forms a set at most of the first category in E.

Almost Čebyšev linear subspaces of normed linear spaces have been introduced by A. L. Garkavi (see [82], p. 116), since they have the advantage that in every separable Banach space E there exist almost Čebyšev subspaces of any finite dimension. However, the Banach space $E(I)$ of section 3.2 has no almost Čebyšev subspace of infinite dimension. For results on finite-dimensional almost Čebyšev subspaces of $C_R(Q)$ (Q compact) see [82], p. 224-225.

Definition 3.6. A linear subspace G of a normed linear space E is called a k-semi-Čebyšev subspace, respectively a k-Čebyšev subspace (where k is an integer with $0 \leq k < \infty$), if

$$-1 \leq \dim \mathcal{P}_G(x) \leq k \qquad (x \in E), \tag{3.24}$$

respectively if

$$0 \leq \dim \mathcal{P}_G(x) \leq k \qquad (x \in E) . \tag{3.25}$$

We recall that $\mathcal{P}_G(x)$ is a convex set, since it is the intersection of the two convex sets G and $S(x, \rho(x, G))$ and that for a non-void convex set A in a linear space E the dimension dim A is defined as the dimension of the linear subspaces of E spanned by $A-y$, where y is an arbitrary element of A; if $A = \emptyset$, then, by definition, $\dim A = -1$. Thus, the 0-semi-Čebyšev and 0-Čebyšev subspaces are nothing else than the usual semi-Čebyšev and, respectively, Čebyšev subspaces.

Most of the preceding results (e.g. theorems 3.1, 3.2, 3.3, 3.4 and 3.10) admit extensions to k-semi-Čebyšev and k-Čebyšev sunspaces (see [82]). However, only partial extensions of theorem 3.7 are known, i.e., it is not known which are the compact spaces Q such that $C_R(Q)$ has k-Čebyšev subspaces of finite dimension n, where $n \geq 2$ and $0 \leq k \leq n-1$ (or, equivalently, such that Q admits systems $x_1, \ldots, x_n \in C_R(Q)$ " of Čebyšev rank $\leq k$ ", i.e. with the property that there do not exist $n-k$ distinct points $q_1, \ldots, q_{n-k} \in Q$ and $k+1$ linearly independent elements $g_0, g_1, \ldots, g_k \in G = [x_1, \ldots, x_n]$ such that $g_i(q_j) = 0$ for $j = 1, \ldots, n-k$ and $i = 0, 1, \ldots, k$).

Definition 3.7. A linear subspace G of a normed linear space E is called a pseudo-Čebyšev subspace if

$$(3.26) \qquad 0 \leq \dim \mathcal{P}_G(x) < \infty \qquad (x \in E).$$

In particular, every finite-dimensional linear subspace and every k-Čebyšev subspace ($0 \leq k < \infty$) is pseudo-Čebyšev. P. D. Morris [66] has constructed examples of pseudo-Čebyšev subspaces of finite codimension of $E = l_R^\infty$ which are not Čebyšev subspaces.

The following characterization of pseudo-Čebyšev subspaces of $E = C_R(Q)$, due to P. D. Morris [66], should be compared with theorems 3.2 and 3.4 (although the latter is only for codim $G < \infty$) :

Theorem 3.16. A proximinal linear subspace G of $E = C_R(Q)$ is a pseudo-Čebyšev subspace if and only if for every $\mu \in G^{\perp} \setminus \{0\}$ the set $Q \setminus S(\mu)$ is finite.

3.7. Very non-Čebyšev subspaces.

One can introduce the following notion, corresponding to §2, definition 2.2 :

Definition 3.8. A set G in a metric space E is said to be a very non-Čebyšev set if G is closed and if for no element $x \in E \setminus G$ does the set $\mathcal{P}_G(x)$ consist of a single element.

One can show (see e.g. [82], pp. 114-116) that all closed linear subspaces G of cardinality $> \mathfrak{c}$ of the space $E(I)$ of section 3.2 and all infinite-dimensional closed linear subspaces of $E = c_0$ are very non-Čebyšev subspaces.

4. Properties of metric projections.

4.1. Definition and some properties of metric projections

Definition 4.1. a) If G is a set in a metric space E, we shall denote by π_G the multi-valued mapping $D(\pi_G) \rightarrow G$ defined by

$$\pi_G(x) \in \mathcal{P}_G(x) \qquad (x \in D(\pi_G)) \; ; \tag{4.1}$$

this mapping should be distinguished from the "set-valued metric projection" $\mathcal{P}_G$ defined in section 4.7 (the reader may compare π_G with the function $y(t) = \sqrt{t}$ on $[0, \infty)$).

b) In the particular case when $D(\pi_G) = E$ and π_G is one-valued (i.e. when G is a Čebyšev set), π_G is called the metric projection of E onto G.

Some properties of the mappings π_G (and hence, in particular, of metric projections) onto linear subspaces of normed linear spaces are collected in

Theorem 4.1. Let E be a normed linear space and G a linear subspace of E. Then

a) $G \subset D(\pi_G)$ and π_G is one-valued on G, namely, $\pi_G(g) = g$ for all $g \in G$. Hence, if $x \in D(\pi_G)$, then $\pi_G(x) \in D(\pi_G)$ and we have

$$\pi_G^2(x) = \pi_G(x) \qquad (x \in D(\pi_G)), \tag{4.2}$$

i.e. the mapping π_G is idempotent

b) We have

$$\Big| \|x - \pi_G(x)\| - \|y - {}_G(y)\| \Big| \leq \|x-y\| \qquad (x, y \in D(\pi_G)), \tag{4.3}$$

(4.4) $\| x - \pi_G(x) \| \leq \| x \|$ $\quad (x \in D(\pi_G))$,

(4.5) $\| \pi_G(x) \| \leq 2 \| x \|$ $\quad (x \in D(\pi_G))$.

c) π_G is continuous at every point $g \in G$ (i.e, $x_n \to g \in G$ implies that for any $\pi_G(x_n) \in \mathcal{P}_G(x_n)$ we have $\pi_G(x_n) \to \pi_G(g) = g$)

d) If G_1 is a linear subspace of G, we have

(4.6) $\| x - \pi_G(x) \| \leq \| x - \pi_{G_1}(x) \|$ $\quad (x \in D(\pi_G) \cap D(\pi_{G_1}))$

If, in addition, the mapping π_G is one-valued on $D(\pi_G)$ (i.e. if G is a semi-Čebyšev subspace), then

e) If $x \in D(\pi_G)$ and $g \in G$, then $x+g \in D(\pi_G)$ and we have

(4.7) $\pi_G(x+g) = \pi_G(x) + \pi_G(g) = \pi_G(x)+g$ $\quad (x \in D(\pi_G), g \in G)$,

i.e. π_G is quasi-additive.

f) If $x \in D(\pi_G)$ and α is an arbitrary scalar, then $\alpha x \in D(\pi_G)$ and we have

(4.8) $\pi_G(\alpha x) = \alpha \pi_G(x)$ $\quad (x \in D(\pi_G), \alpha = \text{scalar})$,

i.e. π_G is homogeneous.

g) If G is closed and $x_n \in D(\pi_G)$, $\lim_{n\to\infty} x_n = x$, $\lim_{n\to\infty} \pi_G(x_n) = g$, then $x \in D(\pi_G)$ and $\pi_G(x) = g$, i.e. π_G is closed.

The proofs are straightforward ; see [82], pp. 140-142 and 390. Part. a) shows that in the particular case when $D(\pi_G) = E$ and π_G is one-valued, the metric projection π_G is indeed a (non-linear) closed projection of E onto G. Some authors use for the metric projection π_G

the term normal projection, or best approximation operator, or nearest point map, or Čebyšev map.

4.2. Continuity of metric projections.

The main characterization of Čebyšev subspaces G with continuous metric projection π_G is the following result, due to R. B. Holmes ([38], theorem 6; in the particular case when E/G is reflexive, this result also follows from [84], theorem 3):

Theorem 4.2. For a Čebyšev subspace G of normed linear space E the metric projection π_G is continuous if and only if the restriction $\omega = \omega_G \big|_{\pi_G^{-1}(0)}$ of the canonical mapping $\omega_G : E \to E/G$ to the set $\pi_G^{-1}(0)$ is a homeomorphism of $\pi_G^{-1}(0)$ onto E/G.

Proof. Since G is a Čebyšev subspace, by § 3, proposition 3.1 ω is one-to-one. Furthermore, ω is always continuous and a mapping onto E/G, since for any $x+G \in E/G$ we have $x-\pi_G(x) \in \pi_G^{-1}(0)$ and $\omega_G(x-\pi_G(x)) = x+G$. Thus, the condition that $\omega = \omega_G \big|_{\pi_G^{-1}(0)}$ be a homeomorphism onto E/G is equivalent to the continuity of ω^{-1}. Assume now that ω^{-1} is continuous and let $x_n, x \in E$, $\lim_{n\to\infty} \|x_n - x\| = 0$. Then by the preceding remark, $x_n - \pi_G(x_n) = \omega^{-1}(x_n+G)$, $x-\pi_G(x) = \omega^{-1}(x+G)$, whence

$$\|\pi_G(x_n) - \pi_G(x)\| \leq \|\pi_G(x_n) - \pi_G(x) - (x_n - x)\| + \|x_n - x\| =$$

$$= \|\omega^{-1}(x+G) - \omega^{-1}(x_n+G)\| + \|x_n - x\| \to 0 \quad \text{as} \quad n \to \infty,$$

and thus π_G is continuous.

Conversely, assume now that π_G is continuous and let x_n+G, $x+G \in E/G$, $\lim (x_n+G) = x+G$ and $\varepsilon > 0$. Since π_G is continuous at the point $\omega^{-1}(x+G) \in \pi_G^{-1}(0)$, there exists a $\delta > 0$ such that

$$\| z - \omega^{-1}(x+G) \| < \delta \text{ implies } \|\pi_G(z)\| = \|\pi_G(z) - \pi_G(\omega^{-1}(x+G)\| < \tfrac{1}{2}\varepsilon.$$

Consider the open cell

$$V = \operatorname{Int} S(\omega^{-1}(x+G), \min(\delta, \tfrac{1}{2}\varepsilon)) = \left\{ z \in E \;\middle|\; \|z - \omega^{-1}(x+G)\| < \min(\delta, \tfrac{1}{2}\varepsilon) \right\}.$$

Since the canonical mapping ω_G is open; the set $\omega_G(V)$ is open and obviously $x+G \in \omega_G(V)$. Hence, $x_n+G \in \omega_G(V)$ for $n > N = N(\varepsilon)$ and thus there exist elements $z_n \in V$ such that $z_n+G = \omega_G(z_n) = x_n+G (n>N)$.

Therefore $x_n - z_n \in G$ and hence, by the quasi-additivity of π_G, we have $\pi_G(x_n) = \pi_G(x_n - z_n + z_n) = x_n - z_n + \pi_G(z_n)$. Consequently, since $z_n \in V$ $(n>N)$, we obtain

$$\| \omega^{-1}(x_n+G) - \omega^{-1}(x+G) \| = \| x_n - \pi_G(x_n) - \omega^{-1}(x+G) \| \leq$$

$$\leq \| x_n - \pi_G(x_n) - z_n \| + \| z_n - \omega^{-1}(x+G) \| =$$

$$= \| \pi_G(z_n) \| + \| z_n - \omega^{-1}(x+G) \| < \frac{\varepsilon}{2} + \frac{\varepsilon}{2} = \varepsilon \qquad (n>N),$$

and thus π_G is continuous, which completes the proof of theorem 4.2.

Note that ω^{-1} is nothing else than the mapping $E/G \to \pi_G^{-1}(0)$ induced by $I - \pi_G$, where I denotes the identical mapping of E onto E.

Since for every $x \in \pi_G^{-1}(0)$ we have $\omega^{-1}(x+G) = x - \pi_G(x) = x$, theorem 4.2 can be also rephrased as follows; <u>π_G is continuous if and only if the relations</u> $x_n, x \in \pi_G^{-1}(0)$, $\lim_{n\to\infty} \rho(x_n - x, G) = 0$

imply $\lim_{n\to\infty} \| x_n - x \| = 0$.

Corollary 4. 1. Let E be a normed linear space. Then

a) A $\sigma(E^*, E)$ - closed Čebyšev subspace Γ of E admits a continuous metric projection π_Γ if and only if the (uniquely determined) extension map $\varphi \in (\Gamma_\perp)^* \to f \in E^*$ with $f|_{\Gamma_\perp} = \varphi$, $\|f\| = \|\varphi\|$, is continuous.

b) If G is a Čebyšev subspace of E, such that $G^\perp \subset E^*$ has property (U) and that the extension map $\varphi \in (G^\perp)^*$ $\Phi \in E^{**}$ with $\Phi|_{G^\perp} = \varphi$, $\|\Phi\| = \|\varphi\|$, is continuous, then π_G is continuous.

Indeed, this follows from theorem 4. 2 by the arguments of § 3, proof of corollary 3. 1. A direct proof of corollary 4. 1 a) has been given by J. Lindenstrauss ([57], § 7), but the proof sketched here is simpler.

Some other, more elementary, characterizations of the continuity of π_G, due to E. W. Cheney-D. E. Wulbert [20] and R. B. Holmes [38] are collected in

Proposition 4. 1. For a Čebyšev subspace G of a normed linear space E the following statements are equivalent :

1° The metric projection π_G is continuous.

2° π_G is continuous at each point of $\pi_G^{-1}(0)$.

3° The direct sum decomposition $E = G \oplus \pi_G^{-1}(0)$ is topological (i. e. , $\lim_{n\to\infty} x_n = x$ if and only if $\lim_{n\to\infty} \pi_G(x_n) = \pi_G(x)$ and $\lim_{n\to\infty} (x_n - \pi_G(x_n)) = x - \pi_G(x)$).

4° $\pi_G|_{A_1(G)}$ is continuous, where

$$A_1(G) = \{x \in E \mid \|x - \pi_G(x)\| = 1\} = \{x \in E \mid \rho(x, G) = 1\} \tag{4.9}$$

5° <u>The functional</u>

$$\Phi_G(x) = \| \pi_G(x) \| \qquad (x \in E) \tag{4.10}$$

<u>is continuous.</u>

6° <u>The mapping</u> $\psi_G : E \setminus G \longrightarrow \pi_G^{-1}(0) \cap \mathrm{Fr}\, S_E$ <u>defined by</u>

$$\psi_G(x) = \frac{1}{\|x - \pi_G(x)\|}(x - \pi_G(x)) \qquad (x \in E \setminus G) \tag{4.11}$$

<u>is continuous.</u>

<u>Proof.</u> The implication $1° \rightarrow 2°$ is obvious. Conversely, if $1°$ does not hold, say $x_n \longrightarrow x$, $\pi_G(x_n) \not\rightarrow \pi_G(x)$, then $x_n - \pi_G(x) \rightarrow$ $\rightarrow x - \pi_G(x) \in \pi_G^{-1}(0)$, but $\pi_G(x_n - \pi_G(x)) = \pi_G(x_n) - \pi_G(x) \not\rightarrow 0$, contradicting $2°$ Thus, $1° \Rightarrow 2°$

The equivalence $1° \Leftrightarrow 3°$ and the implication $1° \Longrightarrow 5°$ are obvious. Conversely, if we have $5°$ and $x_n \longrightarrow x \in \pi_G^{-1}(0)$, then

$$\| \pi_G(x_n) - \pi_G(x) \| = \| \pi_G(x_n) \| \longrightarrow \| \pi_G(x) \| = 0,$$ which proves that $5° \Longrightarrow 2°$

The implication $1° \Rightarrow 6°$ is also obvious. Furthermore, if we have $6°$ and if $x_n, x \in A_1(G)$, $x_n \longrightarrow x$, then

$$\| \pi_G(x_n) - \pi_G(x) \| \leq \| x_n - \pi_G(x_n) - (x - \pi_G(x)) \| + \| x_n - x \| =$$

$$= \| \psi_G(x_n) - \psi_G(x) \| + \| x_n - x \| \longrightarrow 0 \quad \text{as } n \rightarrow \infty,$$

which proves that $6° \Longrightarrow 4°$

Assume finally that we have $4°$ and let $x_n \rightarrow x$. Since by theorem 4.1 c) π_G is always continuous at the points $x \in G$, we may assume that $x \notin G$, hence $x_n \notin G$ for $n > N$. Then $\frac{x_n}{\|x_n - \pi_G(x_n)\|}, \frac{x}{\|x - \pi_G(x)\|} \in A_1(G)$

for $n > N$ and by theorem 4. 1 b), formula (4. 3), $\|x_n - \pi_G(x_n)\| \to$ $\to \|x - \pi_G(x)\|$. Therefore $\frac{x_n}{\|x_n - \pi_G(x_n)\|} \to \frac{x}{\|x - \pi_G(x)\|}$ whence, by 4°,

$$\pi_G(x_n) = \|x_n - \pi_G(x_n)\| \pi_G \left(\frac{x_n}{\|x_n - \pi_G(x_n)\|}\right) \to \|x - \pi_G(x)\| \pi_G\left(\frac{x}{\|x - \pi_G(x)\|}\right) =$$

$$= \pi_G(x) ;$$

thus, $4^\circ \Rightarrow 1^\circ$, which completes the proof of proposition 4. 1.

No theorem is knoun in concrete spaces about characterization of Čebyšev subspaces G of arbitrary dimension with a continuous metric projection.

Let us consider now, in arbitrary normed linear spaces, the problem of characterization of Čebyšev subspaces G with continuous metric projection π_G, when we have restrictions on dim G or codim G, or restrictions on the quotient space E/G.

Theorem 4. 3. For every finite-dimensional Čebyšev subspace G of a normed linear space E, the metric projection π_G is continuous.

The proof is straightforward, using a compactness argument (see [82], pp. 251 and 386-390).

For Čebyšev subspaces G of codimension 1, we have even a stronger property of π_G (see [82], pp. 142-145):

Theorem 4. 4. For every Čebyšev hyperplane G in a normed linear space E, the metric projection π_G is linear, and hence (by theorem 4. 1 c)) continuous.

We have the following characterizations of Čebyšev subspaces of finite codimension with continuous metric projections, due to Cheney-Wulbert [20] and Holmes [38] respectively :

Theorem 4.5. For a Čebyšev subspace G of finite codimension of a normed linear space E, the following statements are equivalent:

1°. π_G is continuous.

2°. $\pi_G^{-1}(0)$ is boundedly compact, that is, interesects every cell $S(x,r) \subset E$ in a compact set.

3°. $\pi_G^{-1}(0) \cap \mathrm{Fr}\, S_E$ is compact.

Proof. Since $\dim E/G < \infty$, $S_{E/G}$ and $\mathrm{Fr}\, S_{E/G}$ are compact. Furthermore, by § 3, formula (3.7), $\pi_G^{-1}(0) \cap \mathrm{Fr}\, S_E = \{x \in E \,|\, \|\omega_G(x)\| = 1\}$, whence

(4.12) $$\omega_G(\pi_G^{-1}(0) \cap \mathrm{Fr}\, S_E) = \mathrm{Fr}\, S_{E/G},$$

(4.13) $$\omega_G(\pi_G^{-1}(0) \cap S_E) = S_{E/G},$$

where ω_G is the canonical map $E \to E/G$. Now, if π_G is continuous, then, by theorem 4.2, $\omega = \omega_G \big|_{\pi_G^{-1}(0)}$ is a homeomorphism and hence, by (4.13), we have 2°. Finally, since the implication 2° ⟹ 3° is obvious, let us assume that we have 3°. Then, by (4.12) and by the remarks made at the beginning of the proof of theorem 4.2, $\omega_G \big|_{\pi_G^{-1}(0) \cap \mathrm{Fr}\, S_E}$ is a one-to-one continuous mapping of the compact set $\pi_G^{-1}(0) \cap \mathrm{Fr}\, S_E$ onto the compact set $\mathrm{Fr}\, S_{E/G}$ and hence a homeomorphism. Therefore, since both $\pi_G^{-1}(0)$ and E/G are star-shaped (by § 2, proposition 2.2) and since ω_G is homogeneous (i.e., $\omega_G(\alpha x) = \alpha \omega_G(x)$), it follows easily that $\omega_G \big|_{\pi_G^{-1}(0)}$ is a homeomorphism of $\pi_G^{-1}(0)$ onto E/G and hence, by theorem 4.2, π_G is continuous, which completes the proof of theorem 4.5.

Naturally, one can also prove theorem 4.5 directly, i.e. without using theorem 4.2. Moreover, with a direct argument one can prove that the relations $3^\circ \Leftrightarrow 2^\circ \Rightarrow 1^\circ$ remain valid for arbitrary Čebyšev subspaces G [20].

For reflexive strictly convex Banach spaces E we have also another useful characterization of Čebyšev subspaces G of finite codimension with continuous π_G, due to R. B. Holmes [38], in terms of the " spherical image map " $v: E^* \to E$ defined as follows : for $f \in E^*$, $v(f)$= the (unique) element $x \in E$ such that $f(x) = \|f\| \cdot \|x\|$, $\|x\| = \|f\|$.

Theorem 4.6. For a Čebyšev (or, equivalently, closed linear) subspace G of finite codimension of a reflexive strictly convex Banach space E the metric projection π_G is continuous if and only if the restriction $v|_{G^\perp}$ of the spherical image map $v: E^* \to E$ to $G^\perp$ is continuous.

In connection with theorem 4.6, note that we have always $v(G^\perp) = \pi_G^{-1}(0)$ and, if E is smooth, then also $v^{-1}(\pi_G^{-1}(0)) = G^\perp$. Furthermore, $v|_{G^\perp}$ is one-to-one if and only if E/G is smooth. Using that v is a " duality map" and hence a "maximal monotone operator" in the sense of F. Browder (see e.g. [16]), R. B. Holmes [38] has proved that the sufficiency part of theorem 4.6 remains valid if instead of codim $G < \infty$ we assume only that the norm of E/G is Fréchet differentiable at every non-zero point.

Let us mention the following results on characterization of Čebyšev subspaces of finite codimension with continuous metric projection in some concrete spaces, due to P. D. Morris [66] :

Theorem 4.7. a) For a Čebyšev subspace G of finite codimension of $E = C_R(Q)$ (Q compact infinite), π_G is continuous (if and) only if

G is a closed hyperplane.

b) For Čebyšev subspace G of finite codimension of $E = L^1_R(T, \nu)$, where (T, ν) is a σ- finite positive measure space, π_G is continuous if and only if the set

$$\bigcup_{\beta \in G^{\perp}\setminus\{0\}} \text{at} \ \{t \in T \mid |\beta(t)| = \|\beta\|\}$$ (where at A denotes the set of all atoms of A) is finite.

A characterization of such subspaces in $E = C^1_R(Q, \nu)$ has been given by E. W. Cheney and D. E. Wulbert ([20] , theorem 35).

Various extensions of theorem 4. 7 a) have been given by A. Lazar , D. E. Wulbert and P. D. Morris [56] . They imply, in particular, the following sharpening of the theorem 4. 7 b) for the space $E=c_R$ ([56], corollary 3. 10): For a Čebyšev subspace G of $E = c_R$, π_G is continuous (if and) only if either G is finite-dimensional or G is a hyperplane.

By theorem 4. 7 a), for every Čebyšev subpsace G of finite codimension $n \geq 2$ of $E = C_R(Q)$ (Q compact infinite) π_G is discontinuous, and theorem 4. 7 b) shows how to construct Čebyšev subspaces G of $L^1_R(T, \nu)$-spaces (with (T, ν) σ-finite) with discontinuous π_G. The first example of a Čebyšev subspace G of a normed linear space E (namely, a subspace G of codimension 2 of $C([0,1])^*$) with discontinuous metric projection π_G has been given in 1964, by J. Lindenstrauss [57], and then examples in other spaces by E. W. Cheney-D. E. Wulbert [20] (in ℓ^1_R), R. B. Holmes-B. R. Kripke [39] and others.

Let us give now some classes of Banach spaces E in which for every Čebyšev subspace G the metric projection π_G is continuous and of spaces E such that all closed linear subspace G are Čebyšev subspaces with continuous π_G

Using corollary 4. 1 a) , J. Lindenstrauss [57] has proved that if

E^* is locally uniformly convex (i. e., if the relations f_n, $f \in \mathrm{Fr}\, S_{E^*}$, $\lim_{n\to\infty} \| f_n + f \| = 2$ imply $\lim_{n\to\infty} \| f_n - f \| = 0$), then every $\sigma(E^*, E)$-closed linear subspace Γ of E^* is a Čebyšev subspace with continuous π_Γ. From this result it follows.

Proposition 4.2. In a uniformly convex Banach space E all closed linear subspaces G are Čebyšev subspaces with continuous π_G.

We recall that a Banach space E is called uniformly convex if for every $\varepsilon > 0$ there is a $\delta(\varepsilon) > 0$ such that the relations $\|x\| = \|y\| = 1$, $\|x-y\| \geq \varepsilon$ imply $\|x+y\| \leq 2(1-\delta(\varepsilon))$. It is well known (see e. g. [21]) that every uniformly convex space is reflexive and strictly convex.

On the other hand, we have proved in [80], corollary 4, that if E is a reflexive Banach space with property (H) (i. e. such that the relations $x_n \longrightarrow x$ weakly and $\lim_{n\to\infty} \| x_n \| = \| x \|$ imply $\lim_{n\to\infty} \| x_n - x \| = 0$), then the metric projection π_G onto any Čebyšev subspace G of E is continuous; for spaces E which are, in addition, strictly convex, this was obtained by KyFan-I. Glicksberg [30]. This again implies proposition 4.2, since it is well known that every uniformly convex space has property (H) (see e. g. [21]).

We mention that for subspaces G of finite codimension (and even for G such that the norm of E/G is Fréchet differentiable at every non-zero point) proposition 4.2 also follows from results of M. I. Kadec, who has proved ([44], lemma 2) that for a uniformly convex space E the spherical image map $v : E^* \to E$ is continuous and ([44], lemma 3) for codim $G < \infty$ the set $\pi_G^{-1}(0) \cap \mathrm{Fr}\, S_E$ is compact (and thus we can apply theorem 4.6 or 4.5, combined with § 3, theorem 3.11).

There also exist some results giving conditions in order that for all Čebyšev subspaces G of finite codimension the metric projection π_G

be continuous. For example, R. B. Holmes [38] has observed that from theorem 4. 6 it follows

Proposition 4. 3. In a reflexive strictly convex Banach space E the metric projection onto every Čebyšev (or, equivalently, closed linear) subspace G of finite codimension is continuous if and only if the restriction of the spherical image map $v : E^* \to E$ to every finite-dimensional linear subspace of E^* is continuous (or, in other words, if and only if v is continuous on E^* endowed with " finite topology").

Similarly, P. D. Morris [66] has used theorem 4. 7 b) to deduce

Proposition 4. 4. If (T, ν) is a positive measure space such that at T is finite, then for every Čebyšev subspace G of finite codimension of $E = L^1_R (T, \nu)$, the metric projection π_G is continuous.

Naturally, when applying the results on continuity of π_G (for example, theorem 4. 7 or proposition 4. 4), one should take into account the results of §3 on characterization and existence of Čebyšev subspaces (in particular, of finite codimension, for example, theorems 3.6, 3. 8 b), 3. 9).

We conclude this section with the following theorem of Cheney and Wulbert [20] on continuity of metric projections in quotient spaces, corresponding to §2, theorem 2. 8 and §3, theorem 3. 12.

Theorem 4. 8. Let G_1 be a linear subspace of a normed linear space E and G_2 a subspace of G_1 such that G_2 is a Čebyšev subspace of E with continuous π_{G_2} . Then G_1 is a Čebyšev subspace of E with continuous π_{G_1} if and only if G_1/ G_2 is a Čebyšev subspace of E/G_2 with continuous π_{G_1/G_2}

The converse of theorem 4. 8 is not valid : one can give [20] an example of $G_2 \subset G_1 \subset E$, G_2 Čebyšev in E with codim $G_2 = 2$, G_1 Čebyšev in E with continuous π_{G_1} and G_1/G_2 Čebyšev in E/G_2 with continuous π_{G_1/G_2}, but π_{G_2} discontinuous.

4.3. Weak continuity of metric projections.

a.) Weak sequential continuity of π_G. The following analogues of some results of section 4.2 are due to R. B. Holmes [38] :

Theorem 4.9. a) If G is a Čebyšev subspace of a normed linear space E, such that $\omega_G|_{\pi_G^{-1}(0)}$ is a weak sequential homeomorphism of $\pi_G^{-1}(0)$ onto E/G (where $\omega_G : E \longrightarrow E/G$ is the canonical mapping), then π_G is weakly sequentially continuous.

b) If G is a Čebyšev subspace of finite codimension, and if π_G is continuous, then π_G is weakly sequentially continuous.

c) If G is any closed linear subspace of finite codimension of a normed linear space E, then $\pi_G^{-1}(0)$ is weakly sequentially closed.

Theorem 4.10. For a Čebyšev subspace G of a reflexive Banach space E the following statements are equivalent :

1°. π_G is weakly sequentially continuous.

2°. $\pi_G^{-1}(0)$ is weakly sequentially closed.

3°. $\omega_G|_{\pi_G^{-1}(0)}$ is a weak sequential homeomorphism.

Corollary 4.2. For every Čebyšev subspace G of finite codimension of a reflexive Banach E, the matric projection π_G is weakly sequentially continuous.

Indeed, this follows immediately from theorem 4.10, implication $2° \Longrightarrow 1°$ and theorem 4.9 c).

Theorem 4. 11. *If E is a reflexive strictly convex smooth Banach space such that the mapping* $x \to \gamma_x$ *of* $E \setminus \{0\}$ *into* Fr S_{E^*} *defined by* $\gamma_x(y) = \lim_{t \to 0^+} \frac{\|x+ty\| - \|x\|}{t}$ *for all* $y \in E$, *is weakly sequentially continuous, then for every Čebyšev (or, equivalently, closed linear) subspace G of E the metric projection* π_G *is weakly sequentially continuous.*

In particular, it is known that the spaces $E = \ell^p$, where $1<p<\infty$, satisfy the conditions of theorem 4. 11.

B) *Weak continuity* of π_G. V. Klee [47] has given conditions on E which guarantee the weak continuity of π_G if dim $G < \infty$, for example the following ([47], proposition 2. 5) : *If for any pair of elements* $x, y \in E$ *the " equidistant set "*

$$P(x,y) = \{z \in E \mid \|x-z\| = \|y-z\|\} \tag{4 14}$$

is weakly closed then for every finite-dimensional Čebyšev subspace G *of* E, π_G *is weakly continuous.* For the case when dim $G = 1$ C. A. Kottman and Bor-Luh Lin [50] have given the following sharpening in this result : *If G is a Čebyšev subspace with* dim $G = 1$ *and if for some* $g \in G$ *the set* $P(g,-g)$ *is weakly closed, then* π_G *is weakly continuous.* Let us also mention the following partial analogue of theorem 4. 10, due to Kottman and Lin [50]: *If G is a finite dimensional Čebyšev subspace of a Banach space* E, *such that* ${}_G^{-1}(0)$ *is weakly closed, then* ${}_G$ *is weakly continuous.*

A similar result for the bw-topology was also given in [50] theorem 3.

An example of a one-dimensional Čebyšev subspace G of $E = c_0$ such that π_G is not weakly or bw-continuous, was given by Kottman and Lin [50].

4.4. Lipschitzian metric projections.

A) Pointwise Lipschitzian metric projections. The following result is due, essentially, to G. Freud, and E. W. Cheney (see [19], p. 82):

Proposition 4.5. For every strongly Čebyšev subspace (hence, in particular, for every interpolating subspace) G of a normed linear space E the metric projection π_G is pointwise Lipschitzian, i.e. for each $x \in E$ there exists a constant $\lambda = \lambda(G,x)$ such that

$$\| \pi_G(x) - \pi_G(y) \| \leq \lambda \| x - y \| \qquad (y \in E). \tag{4.15}$$

Proof. If $r = r(G,x)$ is as in §3, formula (3.20), then, putting there $g_0 = \pi_G(x)$, $g = \pi_G(y)$, we obtain

$$r \| \pi_G(x) - \pi_G(y) \| \leq \| x - \pi_G(y) \| - \| x - \pi_G(x) \| \leq$$

$$\leq \| x - y \| + \| y - \pi_G(y) \| - \| x - \pi_G(x) \| \leq$$

$$\leq \| x - y \| + \| y - \pi_G(x) \| - \| x - \pi_G(x) \| \leq$$

$$\leq \| x - y \| + \| y - x \| + \| x - \pi_G(x) \| - \| x - \pi_G(x) \| = 2 \| x - y \|,$$

and thus we may take $\lambda = \frac{2}{r}$, which completes the proof.

The converse of proposition 4.5 is not valid, since e.g. in $E = \ell^2$ we have (4.15) even with $\lambda = 1$, independent of G and x (since π_G is the orthogonal projection onto G), but $E = \ell^2$ has no strongly Čebyšev subspace G (by the remark made after §3, theorem 3.13, about smooth spaces).

Combining proposition 4.5 with §3, theorem 3.13, it follows that in the spaces $E = C_R(Q)$ and $E = L^1_R(T, \nu)$ for every finite-dimensional

Čebyšev subspace G the metric projection π_G is pointwise Lipschitzian.

On the other hand, R. B. Holmes and B. R. Kripke [39] have given an example of a Čebyšev line G in a 3-dimensional uniformly convex space E, such that π_G is not pointwise Lipschitzian.

B) Lipschitzian metric projections. The main characterization of Čebyšev subspaces G with Lipschitzian metric projection π_G is the following analogue of theorem 4. 2, due to R. B. Holmes [38].

Theorem 4. 12. For a Čebyšev subspace G of a normed linear space E the metric projection π_G is Lipschitzian if and only if $\omega = \omega_G |_{\pi_G^{-1}(0)}$ is a Lipschitzian homeomorphism of $\pi_G^{-1}(0)$ onto E/G.

The proof is similar to that of theorem 4. 2 (the condition amounts to ω^{-1} being Lipschitzian). One can also give a corollary similar to corollary 4. 1.

Some other, more elementary, characterizations of Lipschitzian π_G, due to R. B. Holmes and B. R. Kripke [39], are collected in

Proposition 4. 6. For a Čebyšev subspace G of a normed linear space E, the following statements are equivalent :

1°. π_G is Lipschitzian.

2°. π_G is uniformly continuous on E.

3°. $\pi_G | A_1(\pi_G^{-1}(0))$ is bounded, where

$$A_1(\pi_G^{-1}(0)) = \{x \in E \mid \rho(x, \pi_G^{-1}(0)) = 1\}. \tag{4.16}$$

4°. π_G is "uniformly locally pointwise Lipschitzian ", i. e. there exist two constants $\lambda = \lambda(G) > 0$ and $\delta = \delta(G) > 0$ such that the relations $x \in \pi_G^{-1}(0) \cap \mathrm{Fr}\, S_E$, $\|x - y\| \leq \delta$ imply $\|\pi_G(x) - \pi_G(y)\| \leq$

$\leq \lambda \|x - y\|$.

R. B. Holmes and B. R. Kripke [39] have also observed that from a result of Lindenstrauss [58] one obtains, as a particular case, the following important necessary condition for π_G to be Lipschitzian :

Theorem 4. 13. If the metric projection π_G onto a Čebyšev subspace G of a reflexive Banach space is Lipschitzian, then G is complemented in E.

However, the condition in theorem 4. 13 is not sufficient. Indeed, R. B. Holmes and B. R. Kripke [39] have given even an example of a one-dimensional Čebyšev subspace G of ℓ^p, where $2 < p < \infty$, such that π_G is not Lipschitzian.

Combining theorem 4. 13 with a recent characterization of Banach spaces isomorphic to Hilbert spaces, due to J. Lindenstrauss and L. Tzafriri [60], we obtain

Corollary 4. 2. If for all closed linear subspaces G of a strictly convex reflexive Banach space E, π_G is Lipschitzian (or, in particular, uniformly Lipschitzian, i. e., with a constant λ independent of G), then E is isomorphic to a Hilbert space.

R. B. Holmes and B. R. Kripke [39] have proved that in the finite-dimensional $L^p(T, \nu)$ spaces, where $2 < p < \infty$, π_G is Lipschitzian for all Čebyšev (or equivalently, all linear) subspaces G, but not uniformly Lipschitzian, if dim $L^p(T, \nu) \geq 3$. If E is a strictly convex space of dimension 2 , then, by theorem 4. 4, π_G is uniformaly Lipschitzian on E. R. B. Holmes and B. R. Kripke [39] have constructed examples of non-Hilbert spaces E of any finite dimension such that π_G is uniformly Lipschitzian on E.

In this connection we also have (see [82] , pp. 247-249 and 350):

Theorem 4. 14. If E is normed linear space of (finite or infinite) dimension $\geq$ 3 with the property that for every linear (not necessarily Čebyšev) subspace G of a certain fixed finite dimension n, or codimension n, where $1 \leqslant n \leqslant \dim E-1$, the (generally multi-valued) mapping π_G satisfies $\|\pi_G(x)\| \leq \|x\|$ for all $x \in D(\pi_G)$ (hence, in particular, if π_G is" contractive " on $D(\pi_G)$ i. e. Lipschitzian with constant 1), then E is linearly isometric to a Hilbert space.

4. 5. Differentiability of metric projections.

The notions and results of this section are due to R. B. Holmes and B. R. Kripke [39].

Definition 4. 2. If G is a Čebyšev subspace of a normed linear space E and $x, y \in E$ and if the limit

$$(4.17) \qquad \pi'_G(x,y) = \lim_{t \to 0} \frac{\pi_G(x+ty)-\pi_G(x)}{t}$$

exists, then $\pi'_G(x,y)$ is called the Gâteaux derivative of π_G at x in the direction y.

The following observations are immediate :

a) $\pi'_G(x, cy) = c\,\pi'_G(x,y)$ if either side exists.

b) $\pi'_G(g,y) = \pi_G(y)$ for all $g \in G$, $y \in E$.

c) If $x \in E \setminus G$ and either $\pi'_G(x,y)$ or $\pi'_G(\psi_G(x), y)$ exist, where ψ_G is as in proposition 4. 1, then both exist and are equal.

Theorem 4. 15. If for a Čebyšev subspace G of a normed linear space E $\pi'_G(x,y)$ exists for all $x \in \pi_G^{-1}(0) \cap \mathrm{Fr}\, S_E$, $y \in \mathrm{Fr}\, S_E$ and if
$$\sup_{\substack{x \in \pi_G^{-1}(0) \cap \mathrm{Fr}\, S_E \\ y \in \mathrm{Fr}\, S_E}} \|\pi'_G(x,y)\| < \infty,$$
then π_G is Lipschitzian.

If $x \in E \setminus \{0\}$ and if for any y, $z \in E$ the function $N(s,t) = \|x+sy+tz\|$

is twice continuously differentiable in a neighbourhood of (0,0), then one can define a functional on $E \times E$ by

(4 18) $$\langle y, z\rangle_x = \frac{\partial^2 N}{\partial s\, \partial t}(0,0)$$

and one can show that $\langle y, z\rangle_x$ is a continuous symmetric bilinear form on $E \times E$, satisfying $\langle y, z\rangle_{cx} = \frac{1}{|c|}\langle y, z\rangle_x$ whenever $c \neq 0$.

This form is used in

Theorem 4.16. *Let $G = [x_1, \dots, x_n]$ be an n-dimensional Čebyšev subspace of a normed linear space E and suppose that for some $x \in \pi_G^{-1}(0) \cap \mathrm{Fr}\, S_E$, $y \in \mathrm{Fr}\, S_E$, the function $f(s_1, \dots, s_n, t) = \|x + ty - \sum_{k=1}^{n} s_k x_k\|$ is twice continuously differentiable in a neighbourhood of the origin in R^{n+1}. Let π_x be the $n \times n$ matrix $(\pi_x)_{i,j} = \langle x_i, x_j\rangle_x$ and let $q_x(y)$ be the n-vector whose i-th component is $\langle x_i, y\rangle_x$. Then, if π_x is invertible, $\pi'_G(x, y)$ exists and is given by*

(4 19) $$\pi'_G(x, y) = \sum_{k=1}^{n} (\pi_x^{-1} q_x(y))_k x_k$$

A straightforward computation shows that if $E = L_R^p(T, \nu)$, $2 < p < \infty$, $x \in \mathrm{Fr}\, S_E$, $y, z \in E$ and $\pi_{[y]}(x) = 0$, then $\langle y, z\rangle_x = (p-1)\int_T y(t) z(t)\, |x(t)|^{p-2}\, d\nu(t)$ and that the functional $x \to \langle y, z\rangle_x$ is continuous on E for any fixed $y, z \in E$. Moreover, using this observation and theorem 4.16, one obtains

Theorem 4.17. *For every finite-dimensional linear subspace G of $E = L_R^p(T, \nu)$, where $2 < p < \infty$, $\pi'_G(x, y)$ exists for all $x \in E \setminus G$, $y \in E$.*

The assumption here that $\dim G < \infty$, is essential, since one can give an example of an infinite-dimensional closed linear (hence Čebyšev) subspace G of ℓ_R^p, where $1 < p < \infty$, $p \neq 2$, such that π_G is non-differentiable and non-pointwise Lipschitzian.

Definition 4.3. If G a Čebyšev subspace of a normed linear space E, π_G is said to be Fréchet C^1 on the open set $E \setminus G$ if there exists a continuous mapping u: $E \setminus G \rightarrow L(E, G)$ (where L(E, G) denotes the space of all continuous linear mapping of E into G, with the uniform norm), such that $\pi'_G(x, y)$ exists and

$$\pi'_G(x, y) = (u(x))(y) \qquad (x \in E \setminus G, y \in E). \tag{4.20}$$

Using theorem 4.15 one can show that if dim $E < \infty$ and if G is a Čebyšev subspace of E such that π_G is Fréchet C^1 on $E \setminus G$, then π_G is Lipschitzian.

Some more results on the existence of $\pi'_G(x, y)$ whenever $\|x-x_o\| < \delta$ (for some $x_o \in E \setminus G$) and on π_G satisfying a Lipschitz condition for all x in a neighbourhood of x_o and all $y \in E$, are given in [39].

4.6. Linearity of metric projections.

By theorem 4.1. f), for any semi-Čebyšev subspace G the linearity of π_G on $D(\pi_G)$ is equivalent to its additivity on $D(\pi_G)$.

The main characterization of Čebyšev subspaces G with linear metric projection π_G is the following analogue of theorem 4.2, due to R. B. Holmes [38].

Theorem 4.18. For a Čebyšev subspace G of a normed linear space E the metric projection π_G is linear if and only if $\omega = \omega_G|_{\pi_G^{-1}(0)}$ is an isometric (i.e., distance-preserving) mapping of $\pi_G^{-1}(0)$ onto E/G.

Note that, as was observed in the proof of theorem 4.2 and in §3, formula (3.7), for any Čebyšev subspace G, $\omega = \omega_G|_{\pi_G^{-1}(0)}$ is a one-to-one continuous norm-preserving mapping of $\pi_G^{-1}(0)$ onto E/G. One can also give a corollary of theorem 4.18, similar to corollary 4.1.

Some other characterizations of the linearity of π_G, given in [82], p. 144 and in [39], theorem 3, are collected in

Proposition 4.7. For a semi-Čebyšev subspace G of a normed linear space E the following statements are equivalent :

1°. π_G is one-valued and linear on $D(\pi_G)$.

2°. $\pi_G^{-1}(0)$ is a closed linear subspace of E.

3°. $\pi_G^{-1}(0)$ is convex.

If, in addition, G is proximinal (and hence a Čebyšev subspace), these statements are equivalent to the following :

4°. $\pi_G^{-1}(0)$ contains a linear subspace F of E such that $E = G + F$.

5°. There exists a constant K_G such that

$$\| \pi_G(x+y) \| \leq K_G(\|\pi_G(x)\| + \|\pi_G(y)\|) \quad (x, y \in E) \tag{4.21}$$

6°. π_G is continuously Gateaux differentiable.

In theorem 4.4 it was established that for every Čebyšev hyperplane G in a normed linear space E, π_G is linear. There we also observed that whenever π_G onto a Čebyšev subspace G is linear, it is also continuous and hence a bounded linear projection ; thus a necessary condition in order that π_G be linear is that G be complemented in E. Moreover, in this case, by theorem 4.1 b) we have $1 \leq \|\pi_G\| \leq 2$ and $\|I - \pi_G\| = 1$.

Some simple characterizations of the situation when $\|\pi_G\| = 1$, due to R. B. Holmes and B. R. Kripke [39], are collected in

Proposition 4.8. For a Čebyšev subspace G of a normed linear space E, such that π_G is linear, the following statements are equivalent :

1°. $\|\pi_G\| = 1$.

2°. $(I-\pi_G)(x) \in \mathcal{P}_{\pi_G^{-1}(0)}(x)$ for all $x \in E$ (recall that $\pi_G^{-1}(0)$ is now a closed linear subspace, by proposition 4. 7).

If, in addition, $\pi_G^{-1}(0)$ is a Čebyšev subspace, these statements are equivalent to the following :

3°. $\pi_{\pi_G^{-1}(0)} = I - \pi_G$

4°. $\pi^{-1}_{\pi_G^{-1}(0)}(0) = G.$

The following sufficient condition for the linearity of π_G, related to theorem 4. 6, was observed by R. B. Holmes [38]: If for a Čebyšev (or, equivalently, closed linear) subspace G of a reflexive strictly convex Banach space E the restriction $v|_{G^\perp}$ of the spherical image map $v : E^* \to E$ to $G^\perp$ is linear, then π_G is linear. However, the converse is not valid.

There arises naturally the problem of characterizing in the usual concrete normed linear spaces E the Čebyšev subspaces G for which π_G is linear. In this direction we have the following result of P. D. Morris ([66], theorem 9), related to theorem 4. 7 b):

Theorem 4. 19. For a Čebyšev subspace G of finite codimension of $E = L^1_R(T, \nu)$, where (T, ν) is a positive measure space, π_G is linear if and only if there exists a $\beta_0 \in G^\perp$ such that

$$(4.22)\quad \operatorname{at}\{t \in T \mid |\beta_0(t)| = \|\beta_0\|\} = \bigcup_{\beta \in G^\perp \setminus \{0\}} \operatorname{at}\{t \in T \mid |\beta(t)| = \|\beta\|\}$$

An example of a Čebyšev subspace G of codimension 2 of $E = \ell^1_R$, with π_G continuous but not linear, has been given by P. D. Morris [66].

The following result (see [82] pp. 249, and 351 and [84]), somewhat related to corollary 4. 2 and theorem 4.14, gives, in particular, a characterization of the spaces E in which all π_G are linear :

Theorem 4.20. If E is a normed linear space of (finite or infinite) dimension ≥ 3, with the property that for every closed linear subspace G of a given fixed finite dimension n, or codimension n, where $1 \leq n \leq \dim E-2$, respectively where $2 \leq n \leq \dim E-1$ the mapping π_G is one-valued on $D(\pi_G) = E$ and linear, then E is linearly isometric to an inner product space.

For codim $G \geq 3$ one can prove more, namely (see [82], p. 352 and [84]):

Theorem 4.21. If E is a normed linear space of (finite or infinite) dimension ≥ 4, such that for every closed linear subspace G of a certain fixed finite codimension n, where $3 \leq n \leq \dim E - 1$, the mapping π_G is one-valued on $D(\pi_G)$ and satisfies

$$\pi_G(x+y) = \pi_G(x) + \pi_G(y) \qquad (x, y, \quad x+y \in D(\pi_G)), \tag{4.23}$$

then E is linearly isometric to an inner product space.

However [84], it is not known what happens if the same conditions are satisfied for every closed linear subspace G of codimension 2.

Finally, let us mention that in some cases for a certain (increasing or decreasing) sequence $\{G_n\}$ of Čebyšev subspaces of a space E, each π_{G_n} is linear, but not for all Čebyšev subspaces G of E is π_G linear. For example, in $E = \ell^1$, the increasing sequence $G_n = [e_1, \dots, e_n]$ (where $e_k = \{\underbrace{0, \dots, 0}_{k-1}, 1, 0, \dots\}$) and the decreasing sequence $G_n = \{x = \{\xi_k\} \in E \mid \xi_1 = \dots = \xi_n = 0\}$ $(n = 1, 2, \dots)$ have this property; in this example we also have $\dim G_n = \operatorname{codim} G_n = n$ for all $n = 1, 2, \dots$ In the general case, this property is related to the existence of a "Schauder basis" in E.

4. 7. Semi-continuity and continuity of set-valued metric projections

Definition 4. 4. For a proximinal set G in a metric space E, the mapping $\mathcal{P}_G : x \to \mathcal{P}_G(x)$ of E into 2^G (= the collection of all closed non-void subsets of G) is called the set-valued metric projection of E onto G.

Semi-continuity and continuity properties of set-valued metric projections in normed linear spaces and, more generally, in metric spaces, have been first investigated by K. Tatarkiewicz [87] and in [80].

We recall that a mapping $U : E \longrightarrow 2^G$ is called upper semicontinuous, respectively lower semi-continuous, if the set $\{x \in E \mid U(x) \subset M\}$ is open for each open subset M of G, respectively closed for each closed subset M of G, or, equivalently if the set $\{x \in E \mid U(x) \cap N \neq \emptyset\}$ is closed for each closed subset N of G, respectively open for each open subset N of G. Furthermore, $U : E \longrightarrow 2^G$ is called continuous in the Hausdorff metric, if $x_n \longrightarrow x$ implies

$$\max \left\{ \sup_{g_o \in U(x)} \rho(g_o, U(x_n)), \sup_{g_n \in U(x_n)} \rho(g_n, U(x)) \right\} \to 0 \tag{4.24}$$

We also recall the following well known (and easily proved) facts: a) Continuity of $\mathcal{P}_G$ in the Hausdorff metric implies the lower semi-continuity and, if $\mathcal{P}_G(x)$ is compact for every $x \in E$, then also the upper semi-continuity of $\mathcal{P}_G$. b) If G is boundedly compact, then $\mathcal{P}_G$ is always upper semi-continuous and Hausdorff metric continuity of $\mathcal{P}_G$ is equivalent to its lower semi-continuity. c) If G is Čebyšev (i. e., for every $x \in E$, $\mathcal{P}_G(x)$ is the one-point set $\{\pi_G(x)\}$, the upper semi-continuity, the lower semi-continuity and the Hausdorff metric continuity of $\mathcal{P}_G$ are all equivalent to the continuity of π_G.

There arises naturally the problem of the characterization of those proximinal linear subspaces G of a normed linear space E, for which $\mathcal{P}_G$ is upper or lower semi-continuous or Hausdorff metric continuous. From b) above it follows , in particular, that for every finite-dimensional linear subspace G of a normed linear space E, $\mathcal{P}_G$ is upper semi-continuous. For subspaces of finite codimension, we have the following generalization of theorem 4. 5, due P. D. Morris [66]:

Theorem 4. 22. For proximinal linear subspace G of finite codimension of a normed linear space E the following statements are equivalent :

1° . $\mathcal{P}_G$ is upper semi-continuous and $\mathcal{P}_G(x)$ is compact for every $x \in E$.

2° . $\pi_G^{-1}(0)$ is boundedly compact.

On the other hand, A. L. Brown ([18] , proposition 1. 1) has observed.

Proposition 4. 9. For a proximinal linear subspace G of a normed linear space E, $\mathcal{P}_G$ is lower semi-continuous if and only if the relations $x_n \longrightarrow x$, $g_0 \in \mathcal{P}_G(x)$ imply $\rho(g_0, \mathcal{P}_G(x_n)) \longrightarrow 0$.

For the spaces $E = C_R(Q)$ we have the following results, due to P. D. Morris [66] and respectively to J. Blatter, P. D. Morris and D. E. Wulbert [7] (recently, the proof of the latter has been simplified by A. L. Brown [18]) :

Theorem 4. 23. a) If G is a pseudo-Čebyšev subspace of finite codimension $n \geq 2$ of $E = C_R(Q)$ (Q compact), then $\mathcal{P}_G$ is not upper semi-continuous.

b) For a pseudo-Čebyšev subspace G of $E = C_R(Q)$, in order that $\mathcal{P}_G$ be lower semi-continuous it is necessary, and if $\mathcal{P}_G$ is upper semi-continuous , also sufficient , that for every $x \in \pi_G^{-1}(0)$ the set.

$$(4\ 25)\quad Z(\mathcal{P}_G(x)) = \{ q \in Q \mid g_o(q) = 0 \text{ for all } g_o \in \mathcal{P}_G(x)\}$$

be open.

Part a) is a generalization of theorem 4. 7 a). Blatter, Morris and Wulbert [7] have proved the following corollary of part b) :

Corollary 4. 3. For a compact space Q the following continuous are equivalent :

1°. Q is connected.

2°. Every pseudo-Čebyšev subspace G of $E = C_R(Q)$, such that $\mathcal{P}_G$ is lower continuous, is a Čebyšev subspace.

3°. Every one-dimensional subspace G of $E = C_R(Q)$, such $\mathcal{P}_G$ is lower semi-continuous , is a Čebyšev subspace.

For some related results see also B. Brosowski, K. -H. Hoffmann, E. Schäfer and H. Weber [14].

For the spaces $E = L^1_R(T, \nu)$, A Lazar, D. E. Wulbert and P. D. Morris [56] have proved

Theorem 4. 24. For an n-dimensional linear subspace G of $E=L^1_R(T,\nu)$ where (T, ν) is a σ-finite positive measure space, $\mathcal{P}_G$ is lower semi-continuous if and only if there do not exist $\beta \in G^{\perp} \setminus \{0\}$ and $g \in G$ with the following three properties :

α) The set $S(\beta) = \{ t \in T \mid |\beta(t)| < \|\beta\| \}$ is purely atomic (i. e.., $\nu(S(\beta) \setminus \bigcup_{t \in \text{at } S(\beta)} \{t\}) = 0$) and contains at most n-1 atoms.

β) $S(\beta) \subset Z(g) = \{ t \in T \mid g(t) = 0 \}$.

γ) $T \setminus Z(g)$ is not the union of a finite family of atoms.

Corollary 4.4. If G is a finite-dimensional non-Čebyšev linear subspace of $E=L^1_R(T, \nu)$ (where (T, ν) is a σ-finite positive measure space) such that no $g \in G \setminus \{0\}$ has $T \setminus Z(g)$ purely atomic, consisting of a finite number of atoms (in particular, if (T, ν) has no atoms and hence any finite-dimensional G is non-Čebyšev by §3, theorem 3.9), then $\mathcal{P}_G$ is not lower semi-continuous.

Let us consider now the normed linear spaces in which for every proximinal subspace G set-valued metric projection $\mathcal{P}_G$ has one of the above semi-continuity or continuity properties. From [80], theorem 1 and implication $5^\circ \Longrightarrow 3^\circ$ of theorem 3, it results the following extension of a remark made after proposition 4.2 above :

Theorem 4.25. For all proximinal linear subspaces G of a reflexive Banach space with property (H), $\mathcal{P}_G$ is upper semi-continuous.

On the other hand, A. L. Brown [17] has proved the equivalence $1^\circ \Longrightarrow 3^\circ$ (and introduced "property (P)") and Blatter, Morris and Wulbert [7] observed tho other equivalence of

Theorem 4.26. For a normed linear space E the following statements are equivalent :

1°. For every finite-dimensional linear subspace G of E, $\mathcal{P}_G$ is lower semi-continuous.

2°. For every one-dimensional linear subspace G of E, $\mathcal{P}_G$ is lower semi-continuous.

3°. E has the following property :

(P) For every pair of elements $x, z \in E$ such that $\|x+z\| \leq \|x\|$, there exist constants $b = b(x, z) > 0$, $c = c(x, z) > 0$ such that

$$\|y + cz\| \leq \|y\| \qquad (y \in S(x, b)). \tag{4 26}$$

It is natural to ask, which normed linear spaces E have property (P). A. L. Brown [17] has proved a) and b), and Blatter [5] and Blatter-Morris -Wulbert [7] have proved the other statements of

Proposition 4. 10. a) Every strictly convex normed linear space E has property (P).

b) Every finite-dimensional normed linear space E, in which the unit cell S_E is a polyhedron (i. e., the intersection of a finite number of half-spaces or, what is equivalent, the convex hull of a finite number of points) has property (P).

c) $C_R(Q$ compact) has property (P) if and only if Q is finite.

d) c_o has property (P)

e) If (T, ν) is a σ-finite positive measure space such that T is not the union of a finite number of atoms, then $L^1_R(T, \nu)$ does not have property (P)

4. 8. Continuous selections and linear selections for set-valued metric projections.

We recall that if G and E are metric spaces, a continuous mapping $u : E \longrightarrow G$ is said to be a continuous selection for a set-valued mapping $U : E \longrightarrow 2^G$ if $u(x) \in U(x)$ for all $x \in E$. If G is a linear subspace of a normed linear space E, one can define a linear selection for U in a similar way. By theorem 4. 1 b) or c), if G is proximinal, every linear selection for $\mathcal{P}_G$ is continuous.

In order to characterize the proximinal linear subspaces G of a normed linear space E, for which $\mathcal{P}_G$ admits a continuous selection or a linear selection, we define a set-valued mapping

$$V_G : E/G \longrightarrow 2^{\pi_G^{-1}(0)} \quad \text{by}$$

$$(4.27) \quad V_G(x+G) = \left\{ x - \pi_G(x) \;\middle|\; \pi_G(x) \in \mathcal{P}_G(x) \right\} \qquad (x+G \in E/G).$$

It is easy to see that $V_G(x+G) \in 2^{\pi_G^{-1}(0)}$, i. e.., is non-void and closed. Indeed, $V_G(x+G) \neq \emptyset$ since G is proximinal. Furthermore, if $x_n - \pi_G(x_n) \longrightarrow x \in E$, then since $\pi_G^{-1}(0)$ is closed, we have $x \in \pi_G^{-1}(0)$, whence $x = x - \pi_G(x) \in V_G(x+G)$, which proves that $V_G(x+G)$ is closed. Observe that if G is a Čebyšev subspace of E, then $V_G(x+G)$ is the one-point set $\{\omega^{-1}(x)\} = \{x - \pi_G(x)\}$, where $\omega = \omega_G|\pi_G^{-1}(0)$ (see section 4. 2). Also, V_G is nothing else than the set-valued mapping $E/G \rightarrow 2^{\pi_G^{-1}(0)}$ induced by $I - \mathcal{P}_G$, where I is the identical mapping of E onto itself.

We have the following generalization of theorem 4. 2 (which, in the particular case when E/G is reflexive, was essentially proved in [84], theorem 3):

<u>Theorem 4 27</u> <u>For proximinal linear subspace</u> G <u>of a normed linear space</u> E, $\mathcal{P}_G : E \rightarrow 2^G$ <u>admits a continuous selection if and only if the mapping</u> $V_G : E/G \rightarrow 2^{\pi_G^{-1}(0)}$ <u>defined by (4 27) admits a continuous selection</u>

Indeed, this can be proved either similarly to the above proof of theorem 4. 2, or using, in the necessity part, a theorem of Bartle and Graves (see [63]) according to which the mapping $w_G : E/G \rightarrow 2^E$ defined by

$$(4.28) \quad w_G(x+G) = \{x+g \mid g \in G\} = x + G$$

always admits a continuous selection w_G and then putting

$$(4.29) \quad v_G(x+G) = w_G(x+G) - \pi_G^{(0)} w_G(x+G) \qquad (x+G \in E/G),$$

where $\pi_G^{(0)}$ is a continuous selection for $\mathcal{P}_G$.

From theorem 4. 27 we obtain the following generalization of corollary 4. 1, the first part of which is due to J. Lindenstrauss [57] and the second part to [84] :

Corollary 4. 5. Let E be a normed linear space. Then a) For a $\sigma(E^*, E)$-closed (hence proximinal) linear subspace Γ of E^*, $\mathcal{P}_\Gamma$ admits a continuous selection $\pi_\Gamma^{(0)}$ if and only if the (set-valued) extension map $\varphi \in (\Gamma_\perp)^* \to \{f \in E^* | f|_{\Gamma_\perp} = \varphi, \|f\| = \|\varphi\|\}$ admits a continuous selection.

b) If G is a proximinal linear subspace of E , such that the extension map $\varphi \in (G^\perp) \to \{\Phi \in E^{**} | \Phi|_{G^\perp} = \varphi, \|\Phi\| = \|\varphi\|\}$ admits a continuous selection, then $\mathcal{P}_G$ admits continuous selection.

Let us observe that one can also obtain relations between the semi-continuity properties of the set-valued mapping $\mathcal{P}_G : E \to 2^G$ and $V_G : E/G \to 2^{\pi_G^{-1}(0)}$ (as well as corollaries of the above type).

The results on lower semi-continuity are particularly useful because of the following theorem on continuous selections, due to E. A. Michael ([63]; theorem 3. 2") : If E , G are Banach spaces, every lower semi-continuous $U : E \to 2^G$ such that U(x) is convex for each $x \in E$, admits a continuous selection. Hence, in particular, if G is a proximinal linear subspace of a Banach space E, such that $\mathcal{P}_G$ is lower semi-continuous then $\mathcal{P}_G$ admits a continuous selection; the converse is not true, even if dim G = 1. From this observation and from the results of the preceding section there follow sufficient conditions on a given G in order that $\mathcal{P}_G$ admit a continuous selection and sufficient conditions on E oi order that for all subspaces G of E, $\mathcal{P}_G$ admit a continuous selection. Conversely, in some cases the results on non-lower semi-continuity of $\mathcal{P}_G$ can be sharpened to non-existence of continuous selections for $\mathcal{P}_G$. For example , A. Lazar , D. E. Wulbert and P. D. Morris have proved the following partial sharpening of corollary 4. 4 ([56] , theorem 1. 4) :

Theorem 4.28. If G is any finite-dimensional subspace of $E = L^1_R(T, \nu)$, where (T, ν) is a positive measure space having no atoms, then $\mathcal{P}_G$ admits no continuous selection.

We have the following characterizations of the one-dimensional linear subspace G of $E = \ell^1_R$ and $E = C_R(Q)$ for which $\mathcal{P}_G$ admits a continuous selection, due to A. Lazar ([55] , lemma 5.2) and respectively A. Lazar, D. E. Wulbert and P. D. Morris ([56] , proposition 2.6):

Proposition 4.11. a) For the one-dimensional linear subspace G of $E = \ell^1_R$ spanned by an element $g = \{\gamma_n\}$, $\mathcal{P}_G$ admits a continuous selection if and only if there do not exist two disjoint subsets N_1, N_2 of $N = \{1, 2, 3, \dots\}$ such that

α) $N_1 \cup N_2 = \{i \in N \mid \gamma_i \neq 0\}$.

β) $\sum_{i \in N_1} |\gamma_i| = \sum_{i \in N_2} |\gamma_i|$

b) For the one-dimensional linear subspace $G = [g]$ of $E = C_R(Q)$ (Q compact), spanned by an element g of norm 1, $\mathcal{P}_G$ admits a continuous selection if and only if

α) card Fr $Z(g) \leq 1$, where $Z(g) = \{q \in Q \mid g(q) = 0\}$.

β) $q \in \mathrm{Fr}\, Z(g)$ implies that there exists a neighborhood of q on which g is either non-positive or non-negative.

Let us also mention the following recent results of A. L. Brown ([18] , theorems 2.8 and 3.10) :

Theorem 4.29. a) If G is a " Z-subspace" of $E = C_R(Q)$ (Q compact), i.e. a closed linear subspace such that Int $Z(g) = \emptyset$ for all $g \in G \setminus \{0\}$, then either there is no continuous selection for $\mathcal{P}_G$ or there is a unique one.

b) There exists a 5-dimensional Z-subspace G of E = $C_R([-1,+1])$ which contains the constants, is non-Čebyšev and such that $\mathcal{P}_G$ admits a unique continuous selection.

The latter result (which disproves a claim of A. Lazar, D.E. Wulbert and P.D. Morris : [56], theorem 2.1) shows that in the particular case when dim G < ∞ (and hence $\mathcal{P}_G$ is upper semi-continuous), the implication $1^\circ \Longrightarrow 2^\circ$ of corollary 4.3 cannot be sharpened so as to assume only existence of a continuous selection for $\mathcal{P}_G$ instead of the lower semi-continuity of $\mathcal{P}_G$.

Concerning linear selections for $\mathcal{P}_G$ we have (see [82] ,p. 142):

Theorem 4.30. For every proximinal hyperplane G in a normed linear space E, $\mathcal{P}_G$ admits a linear selection.

By theorem 4.1, if for a proximinal linear subspace G of a normed linear space E, $\mathcal{P}_G$ admits a linear selection $\pi_G^{(0)}$, then $\pi_G^{(0)}$ is a continuous linear projection of E onto G and hence G is complemented. Obviously, the converse is not valid.

Finally, let us mention that one can give a characterization of proximinal linear subspaces G for which $\mathcal{P}_G$ admits a linear selection, generalizing theorem 4.18 in a similar way as we generalized theorem 4.2 by theorem 4.27 and one can than prove also a corollary corresponding to corollary 4.5. Also, one can define weakly continuous, Lipschitzian and differentiable selections for $\mathcal{P}_G$ and obtain for then similar extensions of the preceding results.

5. Best approximation by elements of non-linear sets

5. 1. Best approximation by elements of convex sets.

By a non-linear set in a normed linear space E we mean any set $G \subset E$ which is not a " linear manifold ", i. e. which is not of the form $x+G_0$, where $x \in E$ and where G_0 is a linear subspace of E. Since best approximation by elements of linear manifolds can be reduced, by a simple translation, to best approximation by elements of linear subspaces, we shall not consider here this problem, but refer the reader to [82], pp. 135-140 and 242-246. We want to present here, briefly, some directions of research on best approximation by elements of non-linear sets. Note that the existing results in this field do not yet constitute a unified theory (as is the theory of best approximation by elements of linear subspaces) and the construction of such a theory in general normed linear spaces is only at its beginnings.

The first natural step when passing from best approximation in normed linear spaces E by elements of linear sets $G \subset E$ to non-linear sets is to take as G a convex set in E. The following extension of §1, theorem 1. 1, to this case has been given, for real scalars, by G. Š. Rubinstein [75] and Ch. Roumieu ([74], proposition 5) and for complex scalars in [82], pp. 360-361 and [22], [37] (independently) :

Theorem 5. 1. Let G be a convex set in a normed linear space E, and let $x \in E \setminus \overline{G}$, $g_0 \in G$. We have $g_0 \in \mathcal{P}_G(x)$ if and only if there exists an $f \in E^*$ with the following properties :

(5, 1) $$\|f\| = 1,$$

(5, 2) $$\operatorname{Re} f(g_0-g) \geq 0 \qquad (g \in G),$$

(5, 3) $$\operatorname{Re} f(x-g_0) = \|x-g_0\|.$$

This theorem admits the following geometric interpretation, observed by V. N. Burov (see [82], p. 362): $g_0 \in \mathcal{P}_G(x)$ <u>if and only if there exists a real hyperplane</u> H <u>which separates</u> G from $S(x, \|x-g_0\|)$. Clearly, such a hyperplane H must pass through g_0 and support the cell $S(x, \|x-g_0\|)$. The particular case of theorem 5.1, when G is a convex cone, was also considered by G. Š. Rubinstein (see [75], pp. 362-363); another characterization theorem for best approximation by elements of convex cones has been given by G. Godini [35].

For finite-dimensional convex sets F. R. Deutsch and P. H. Maserick [22] and, independently, S. Ia. Havinson [37] have proved the following extension of §1, theorem 1.6:

<u>Theorem 5.2. Let G be an n-dimensional convex set in a normed linear space E and let</u> $x \in E \setminus \bar{G}$, $g_0 \in G$. <u>We have</u> $g_0 \in \mathcal{P}_G(x)$ <u>if and only if there exist h extremal points</u> $f_1, \ldots, f_h$ <u>of</u> S_{E^*}, <u>where</u> $1 \leq h \leq n+1$ <u>if the scalars are real and</u> $1 \leq h \leq 2n+1$ <u>if the scalars are complex, and h numbers</u> $\lambda_1, \ldots, \lambda_h > 0$ <u>with</u> $\sum_{j=1}^{h} \lambda_j = 1$, <u>such that</u>

$$(5.4) \qquad \sum_{j=1}^{h} \lambda_j \operatorname{Re} f_j (g_0 - g) \geq 0 \qquad (g \in G),$$

$$(5.5) \qquad \sum_{j=1}^{h} \lambda_j \operatorname{Re} f_j (x-g_0) = \|x-g_0\|.$$

Actually, this follows from theorem 5.1 in the same way as §1, theorem 1.6 follows from §1, theorem 1.1.

The second main characterization theorem of §1 (theorem 1.8) remains valid in the case when G is a convex set in E; this was observed by A. L. Garkavi (see [82], p. 360) and, independently, by G. Choquet (umpublished).

Some other characterizations of elements of best approximation by elements of convex sets G have been obtained by F. J. Laurent [52] (see also [54]), who has used the " convex cones of displacement " (introduced by A. J. Dubovitskii and A. A. Miliutin [24]) and by J. J. Moreau [65], who has used the tools of the theory of convex functions (e. g. subdifferentials, indicatrices, etc. , see [64] and [42]).

It was observed in [82], p. 360, that several results on existence and uniqueness of elements of best approximation (for example, § 2, theorem 2. 3, with S_G replaced by " all bounded subsets of G"; § 2, theorem 2. 6 with the following addition : 6° All closed convex subsets of E are proximinal ; § 3 , theorem 3. 11 with the following addition ; 4°. All closed convex subsets of E are Čebyšev sets) and on properties of the mapping π_G not involving the linearity of G, remain valid for the case when G is a convex set in E. However, a systematic extension of the results on best approximation by elements of linear subspaces G to convex sets G has not yet been accomplished. We mention that some results on existence and uniqueness of elements of best approximation have been extended to convex cones by G. Godini [35]. Some difficulties which arise at the best approximation by elements of certain finite-dimensional convex cones in $E = C_R([0, 1])$ have been pointed out by J. R. Rice (see [82] , p. 363). Some results on the upper semi-continuity of $\mathcal{P}_G$ for closed convex sets G in reflexive Banach spaces have been given by E. V. Ošman [67].

5. 2. Best approximation by elements of N-parameter sets.

One of the most important classical problems of best approximation by elements of non-linear sets is that of best rational approximation in $E = C_R([a,b])$, raised by P. L. Čebyšev (see N. I. Ahiezer [1], Ch. II), i. e. the problem of best approximation in $E = C_R([a, b])$ by elements of the set

$$G=R_{n,m} = \left\{ g\in E \;\middle|\; g(t) = z(t) \frac{\sum_{i=1}^{n} \alpha_i t^{i-1}}{\sum_{i=1}^{m} \beta_i t^{i-1}}, \right. \tag{5.6}$$
$$\left. \sum_{i=1}^{m} \beta_i t^{i-1} \neq 0 \ (t\in [a,b]) \right\},$$

where n, m are given positive integers and z is a given function in $E = C_R([a,b])$, such that $z(t) \geq 0$ $(t\in[a,b])$; obviously, in the particular case when m=1 and $z(t) \equiv 1$, this problem reduces to that of best approximation by elements of the n-dimensional linear subspace $G_n = [1, t, \dots t^{n-1}]$ of $E = C_R([a,b])$. A slightly more general problem is that of best approximation in $E=C_R([a,b])$ by elements of the set

$$G=R_{G_1,G_2} = \left\{ g\in E \;\middle|\; g(t) = z(t) \frac{g_1(t)}{g_2(t)}, g_1\in G_1, g_2\in G_2, \right. \tag{5.7}$$
$$\left. g_2(t) \neq 0 \quad (t\in[a,b]) \right\},$$

where G_1, G_2 are given finite-dimensional linear subspaces of $E=C_R([a,b])$ and where z is as above. One can also replace the condition $g_2(t) \neq 0$ $(t\in[a,b])$ by weaker ones. A further generalization of the problem consists in replacing the interval $[a,b]$ by a compact space Q and the set R_{G_1, G_2} by an "N-parameter set", i. e. by a set of the form

$$G=\left\{ g\in E = C_R(Q) \;\middle|\; g = g(\alpha_1,\dots,\alpha_N, q) \ (\{\alpha_1,\dots,\alpha_N\} \in P, q\in Q)\right\} \tag{5.8}$$

where P is a subset of a real N-dimensional Banach space $(N<\infty)$, say B_N. The aim is to find classes of sets (5.8) such that the known results of the theory of rational approximation or of convex approximation (in particular, of linear approximation), e. g. the alternation theorem, the characterization theorem of Kolmogorov (see §1, theorem 1. 9) , uniqueness theorems, etc. , remain valid for the sets G of these classes. For this purpose, there have been introduced, by various authors, " interpolating" N-parameter sets $G\subset E=C_R(Q)$, (i. e. having the property described in §3, definition 3. 4, with n=N ; they are also called " unisolvent " N-parameter sets), "locally unisolvent " and " assymptotically convex" N-parameter sets $G\subset E=C_R(Q)$, and, in an attempt to include also other important non-linear sets G of approximating functions (e. g. of functions of the type $g(t)=\alpha_1\alpha_2^t+\alpha_3$), " varisolvent" N-parameter sets. The problem of necessary and sufficient conditions on (5. 8) in order that a certain known theorem on rational approximation remain valid for (5. 8) has been also studied; for example, local unisolvence is necessary and sufficient in order that the alternation theorem remain valid. The literature of these approximation problems in $E=C_R([a,b])$ and in $E=C_R(Q)$ (Q compact) is very vast ; the reader may consult the monographs of N. I. Ahiezer [1] , E. W. Cheney [19], G. Meinardus [62] , J. R. Rice [70] and B. Brosowski [9] , and the papers in the bibliographies of these monographs.

From the above it is clear that it would be important to develop a theory of best approximation by N-parameter sets G in a general normed space E, i. e. by sets.

$$(5.9)\qquad G=\left\{g=g(\alpha_1,\ldots,\alpha_N)\in E \,\middle|\, \{\alpha_1,\ldots,\alpha_N\}\in P\right\},$$

where P is a subset of a real N-dimensional Banach space B_N ; naturally , such a theory would include as particular cases, the above theories. This problem was first raised in [82], p. 137, and it is difficult even when dim $E < \infty$. Two different approaches to this problem have been proposed by J. R. Rice [71], [72], [70] (see also [82], pp. 371-374) and D. E. Wulbert [96], [97], [98], respectively. Both authors agree in pointing out the importance of the particular case when (5. 9) is a manifold and obtain more results for this case ; unfortunately, this does not include completely the rational approximation, since the set $G=R_{n,m}$ defined by (5. 6) need not be a manifold in $E=C_R([a,b])$, even when $n=m=2$. While the approach of J. R. Rice points out the importance of the concept of " curvature", and insists more on the case when dim $E < \infty$, that of D. E. Wulbert emphasizes the utility, of "boundedly connectedness". Among other results, D. E. Wulbert [98] has obtained a characterization of those n-dimensional C^1-submanifolds G of $E=C_R(Q)$ (Q compact) which are Čebyšev sets and satisfy a certain additional condition. We shall not enter here into more details. Let us only mention that often differentiability is used to linearize the problem and to draw from the known linear results for the non-linear case, by observing that " local " best approximation (i. e. minimizing $\|x-g\|$ on a neighbourhood of g_0) is equivalent to best approximation by elements of the "tangent" linear manifold to G passing through g_0; naturally, this contains best approximation by elements of linear subspaces G as a particular case, since it is known (see [82], p. 90) that for linear subspaces G any element of local best approximation is already in $\mathcal{P}_G(x)$, i. e. is a " global" best approximation.

5. 3. Generalizations.

The problem of best approximation by elements of N-parameter sets admits further generalizations. We shall mention here two directions of such generalizations. Both of them have been considered first in $E = C_R(Q)$ and then in general normed linear spaces.

A) The set P of parameters $\{\alpha_1, \ldots, \alpha_N\}$ in (5.8) can be replaced by a subset P of an infinite-dimensional normed linear space P; usually P is assumed to be open in F. For this case, assuming also Fréchet differentiability with respect to the parameter, G.Meinardus and D. Schwedt have given in $E = C_R(Q)$ a necessary condition for an element of best approximation (see [62], p. 140, theorem 89, or [10], p. 28, theorem 5), which extends the necessity part of §1, theorem 1.9. In general this condition is not a sufficient one and the problem of characterization of the sets $G \subset E = C_R(Q)$ for which this condition is also sufficient, raised by B. Brosowski [10], has been solved recently by B. Brosowski and R. Wegmann [15], P. J. Laurent [53] has given in an arbitrary normed linear space E, under similar assumptions, necessary conditions for an element of best approximation which extend the necessity parts of §1, theorems 1.1 and 1.6.

B) The set P of parameters can be completely omitted and one can consider the problem of finding classes of sets $G \subset E$ such that the results of the theory of N-parameter approximation or of convex approximation (in particular, of linear approximation) remain valid for the sets G of these classes. The notions of interpolating (=unisolvent), locally unisolvent, assymptotically convex and varisolvent sets G, mentioned in section 5.2, do not solve the problem in $E = C_R(Q)$, since they assume that G is an N-parameter set. B. Brosowski [9] has introduced the notion of a "regular" set G in $E = C_R(Q)$, which is independent of the notion of N-parameter set, and has proved that <u>for a set $G \subset C_R(Q)$ Kolmogorov's criterion</u> (§1, theorem 1.9) <u>gives a necessary and sufficient condition in order to have</u> $g_0 \in \mathcal{P}_G(x)$ <u>if and only if</u> G <u>is a regular set</u> (we recall that, as shown in §1, for any set G Kolmogorov's criterion gives a <u>sufficient</u> condition in order that $g_0 \in \mathcal{P}_G(x)$.

A set $G \subset C_R(Q)$ is called [9] regular, if for every pair of elements $g, g_0 \in G$, every $\lambda > 0$ and every closed subset $A \subset Q$ such that $g(q) - g_0(q) \neq 0$ $(q \in A)$, there exists an element $g_\lambda \in G$ such that.

(5.10) $\quad \operatorname{sign}(g(q)-g_0(q)) = \operatorname{sign}(g_\lambda(q)-g_0(q)) \qquad (q \in A),$

(5.11) $\quad \|g - g_0\| < \lambda$

From the above result on Kolmogorov's criterion it follows, in particular, that every convex set G, every set $G=R_{G_1, G_2}$ of the form (5.7), every varisolvent N-parameter set G and every assymptotically convex set G in $E=C_R(Q)$ is regular (naturally, this can be deduced also directly from the definitions ; see [9] . Some other problems of best approximation in $E=C_R(Q)$ by elements of regular sets G (for example, uniqueness), have been also studied by B. Brosowski [9] .

The similar problem (of finding suitable classes of sets G and studying best approximation by elements of the sets G belonging to these classes) in arbitrary normed linear spaces has been also attacked successfully by B. Brosowski in a series of papers. Let us give here an example of the results obtained in this direction. We recall that a set G in a normed linear space E is called [48] an α-sun, respectively a β-sun, if for every $x \in E$ and every $g_0 \in \mathcal{P}_G(x)$ we have

(5.12) $\quad g_0 \in \mathcal{P}_G(\lambda x + (1-\lambda) g_0) \qquad (1 \leq \lambda < \infty),$

respectively if for every $x \in E$ there exists a $g_0 \in \mathcal{P}_G(x)$ such that we have (5.12).

Theorem 5.3. For a set G in a normed linear space E the following statements are equivalent :

1°. $g_0 \in \mathcal{P}_G(x)$ implies

$$(5\ 13)\qquad \min_{f\in \mathcal{E}(S_{E^*})\cap \mathfrak{M}_{x-g_o}} \operatorname{Re} f(g-g_o) \leq 0 \qquad (g \in G),$$

where $\mathcal{E}(S_{E^*})$ denotes the set of all extremal points of S_{E^*} and where

$$\mathfrak{M}_{x-g_o} = \left\{ f\in E^* \middle| \|f\| = 1,\quad f(x-g_o) = \|x-g_o\| \right\} .$$

2°. G is an α-sun.

3°. For every pair of elements $g, g_o \in G$, every $\lambda > 0, x \in E\setminus G$ and every $\sigma(E^*, E)$- closed set $A\subset \mathcal{E}(S_{E^*})$ containing $\mathfrak{M}_{x-g_o}$ and such that $\operatorname{Re} f(g-g_o) > 0$ for all $f\in A$, there exists an element $g_\lambda \in G$ such that

$$(5.14)\quad \operatorname{Re} f(g_\lambda-g_o) > \operatorname{Re} f(x-g_o) - \|x-g_o\| \qquad (f\in A),$$

$$(5.15)\qquad \|g_\lambda - g_o\| < \lambda .$$

The equivalence $1^\circ \Longleftrightarrow 2^\circ$ has been proved by B. Brosowski [11] and the equivalence $1^\circ \Longleftrightarrow 3^\circ$ by B. Brosowski and R. Wegmann [15]. Naturally, a set $G\subset E=C_R(Q)$ has the above properties if and only if it is regular.

Theorem 5. 4. For a set G in a normed linear space E the following statements are equivalent :

1°. $\mathcal{P}_G(x)=\emptyset$ implies the existence of an element $g_o \in \mathcal{P}_G(x)$ satisfying (5. 13).

2°. G is a β-sun.

If G is proximinal and if $\mathcal{P}_G(x)$ is compact for every $x\in E$, these statements are equivalent to the following :

3°. For every $x \in E$ and $r > 0$ the set-valued mapping $\Lambda_{x,r}$: $E \to 2^E$ defined by

$$(5.16)\quad \Lambda_{x,r}(y) = x + \frac{r}{r+\rho(x,G)}\,(y - \mathcal{P}_G(y)) \qquad (y \in E)$$

has a fixed point (i. e., a point $y_0 \in E$ such that $y_0 \in \Lambda_{x,r}(y_0)$).

The equivalence 1°$\Leftrightarrow$2° has been proved by B. Brosowski [11] and the equivalence 1°$\Leftrightarrow$3° by B. Brosowski, K.-H. Hoffmann, E. Schäfer and H. Weber [13]. Some other characterizations of the above classes of sets in terms of fixed points of set-valued mapping have been given by B. Brosowski [12]. A localized Kolmogorov type criterion, in terms of cones of displacement, has been also given by B. Brosowski [11] and the sets $G \subset E$ for which this criterion is necessary and sufficient in order that $g_0 \in \mathcal{P}_G(x)$ have been characterized by B. Brosowski and R. Wegmann [15].

For further results on best approximation by elements of the above classes of sets G in arbitrary and in some concrete normed linear spaces E we refer the reader to the papers of B. Brosowski.

5.4. Best approximation by elements of arbitrary sets.

A) A characterization of elements of best approximation.

The following characterization of elements of best approximation in terms of fixed points of a set-valued mapping, was given by O. Brandt [8] and B. Brosowski [12]:

Theorem 5.5. Let E be a normed linear space, G an arbitrary set in E, $x \in E \setminus \overline{G}$ and $g_0 \in G$. We have $g_0 \in \mathcal{P}_G(x)$ if and only if g_0 is a fixed point of the mapping $\mathcal{B}_x$. $G \to 2^G$ defined by

$$(5.17)\quad \mathcal{B}_x(g) = G \cap S\left(x, \frac{\rho(x,G) + \|x-g\|}{2}\right) \qquad (g \in G),$$

(i. e., $g_o \in \mathcal{B}_x(g_o)$). Moreover, in this case we have

(5.18) $$\mathcal{B}_x(g_o) \subset \mathcal{P}_G(x) .$$

B) Some problems on existence of elements of best approximation by elements of closed sets.

One of the problems studied recently is that of finding the Banach spaces E with the property that for every closed set $G \subset E$ the set

(5.19) $$D(\pi_G) = D(\mathcal{P}_G) = \{x \in E \mid P_G(x) \neq \emptyset\}$$

is dense in E. S. B. Stečkin and M. Edelstein [26] have proved that every uniformly convex Banach space E has this property. This result has been slightly extended by D. E. Wulbert [96], who has proved that every Banach space E "with property (2R)" also has the above property. We recall (see e.g. [21]) that a Banach space E is said to have property (2R) if every sequence $\{x_n\} \subset E$ such that $\lim_{n,m\to\infty} \|x_n + x_m\| = 2$ is a Cauchy sequence (and hence convergent); clearly, every space with property (2R) is uniformly convex, but the converse is not true. Since every uniformly convex space (and hence every space with property (2R) is strictly convex, it is natural to ask whether there exist non-strictly convex spaces E with the above property (i. e., such that for every closed set $G \subset E$ the set $D(\pi_G)$ is dense in E. D. E. Wulbert [96] has given an affirmative answer, by proving that every uniformly smooth Banach space E with property (H) (see §4, section 4.2) also has the above property. We recall (see e. g. [21]) that E is called uniformly smooth if for every $\eta > 0$ there exists an $\varepsilon = \varepsilon(\eta)$ such that the relation $\|x-y\| \leq \varepsilon$ implies $\|x\| + \|y\| \leq (1+\eta)\|x+y\|$; D. E. Wulbert [96] has shown that there exist uniformly smooth spaces with property (H) which are not strictly convex.

By the remark made at the end of §2 (on very non-proximinal subspaces) a Banach space E with the above property must be reflexive. D. E. Wulbert 96 has raised the problem whether the converse is true, i. e. :

Problem 5.1 Does there exists a reflexive Banach space E containing a closed set G such that $D(\pi_G)$ is not dense in E ?

Some other problems related to existence of elements of best approximation are concerned with very non-proximinal sets (see § 2, definition 2.2). M. Edelstein [27] has proved that in a separable conjugate space E^* no closed bounded set Γ is very non-proximinal. He has also shown [27] that in the separable space $E = c_0$ (which is not is isomorphic to any conjugate Banach space) there do exist bounded very non-proximinal sets.

V. Klee (see [82], p. 371) has considered the problem of characterization of the classes N_i $(i = 0, 1, 2, 3, 4)$ of all normed linear spaces E which contain a very non-proximinal set G having respectively the following properties : (0) no additional property ; (1) G is convex; (2) G is bounded and convex; (3) $E \setminus G$ is convex; (4) $E \setminus G$ is bounded and convex. V. Klee has made the following remarks : a) N_1 is the class of all non-reflexive spaces ; b) $N_3 \supset N_1$; c) no Banach space is in N_2, but $N_2 \neq \emptyset$; d) $N_4 \neq \emptyset$; e) it is possible that N_4 (whence also N_3, N_0) coincides with the class of all normed linear spaces.

C) Some problems on uniqueness of elements of best approximation.

For an arbitrary set G in a normed linear space E, S. B. Stečkin (see [82], p 375) has studied the set

$$U_G = \{x \in E \mid g_1, g_2 \in \mathcal{P}_G(x) \Rightarrow g_1 = g_2\}, \tag{5.20}$$

i. e. the set of all elements $x \in E$ which have at most one element of best approximation in G, and has obtained, among other results, the following "constructive characterization" of strictly convex spaces :

Theorem 5.6. A Banach space E has the property that for every set $G \subset E$ the set U_G is dense in E if and only if E is strictly convex.

Furthermore, S. E. Stečkin has proved that if E is a strictly convex Banach space, then for every boundedly compact set $G \subset E$ the set U_G is of the second category in E. However , it is not known whether this also holds for every set $G \subset E$;it is also unknown whether from the fact that for every compact $G \subset E$ the set U_G is either dense in E or of the second category in E it follows that E is strictly convex. Finally, we mention the following results of Stečkin : If E is a locally uniformly convex Banach space, then for every $G \subset E$ the set U_G is of the second category and if E is a uniformly convex Banach space, then for every closed set G in E the set $D(\pi_G) \cap U_G$ is of the second category. However, it is not known whether the second result remains valid also for locally uniformly convex spaces.

We conclude this section with a famous classical problem, namely, the problem of convexity of Čebyšev sets. We have seen in section 5.1 that a Banach space E has the property that every closed convex set $G \subset E$ is a Čebyšev set if and only if E is reflexive and strictly convex. It is natural to ask what are the Banach spaces E in which the converse property holds, i.e. in which every Čebyšev set $G \subset E$ is convex. This problem has been solved only for 3-dimensional spaces E(see [82], p. 364), namely, E has this property if and only if every exposed point of S_E (see §3, section 3.2) admits a unique maximal functional of norm 1. For Banach spaces E of finite dimension $m \geq 4$ it is only known that the smoothness of E is a sufficient but not necessary condition for the convexity of all Čebyšev sets $G \subset E$. For infinite-dimensional Banach spaces E the problem is considerably more difficult, even the answer to the following problem being unknown :

Problem 5. 2 In a Hilbert space $\mathcal{H}$, is every Čebyšev set necessarily convex ?

V. Klee has conjectured that the answer is negative and has proved (see [82], p. 370) that in every infinite-dimensional Hilbert space $\mathcal{H}$ there exist non-convex closed semi-Čebyšev sets. On the other hand, much work has been done towards a positive answer. L. P. Vlasov has observed (see [82], p. 366) that in a smooth normed linear space E every Čebyšev set G which is an α-sun (see section 5. 3) is convex (the converse is immediate) and thus the problem reduces to prove that every Čebyšev set is an α-sun. With an ingenious application of Schauder's fixed point theorem, L. P. Vlasov has proved (see [82], p. 365) that in an arbitrary Banach space E every boundedly compact Čebyšev set G is an α-sun and hence, if E is smooth, G is convex.

The assumption of boundedly compactness of G in this result was weakened by N. V. Efimov and S. B. Stečkin and others (see [82], pp. 368-369), under additional restrictions on the space E (e. g. uniform convexity). An important step in this direction was the idea of V. Klee of imposing continuity conditions on the metric projection π_G onto G rather than imposing conditions directly on the Čebyšev set G; in this way, for all classes of Čebyšev sets G for which π_G has the required continuity properties, it follows that the sets G in those classes are convex. L. P. Vlasov [93] has proved.

Theorem 5. 7. If E is a Banach space such that the conjugate space E^* is strictly convex (in particular, if E is a smooth Banach space), then every Čebyšev set $G \subset E$ with continuous metric projection π_G is convex.

For Hilbert spaces E. Asplund [2] has shown that it is sufficient here to assume that π_G is continuous from the norm topology to the weak topology. Also, E. Asplund [2] has proved.

Theorem 5.8. If G is a Čebyšev set in a Hilbert space $\mathcal{H}$ such that every closed half-space intersects G in a proximinal set, then G is convex.

These two theorems contain as particular cases the previously known results, since e. g. every boundedly compact Čebyšev set G satisfies the above hypotheses. Let us note that the arguments of E. Asplund [2] lean heavily on the tools of the theory of convex functions; some other uses of the theory of convex functions to problems of best approximation have been mentioned in section 5.1. For the continuity of metric projections onto Čebyšev sets (see also D. E. Wulbert [94].).

We mention that the above problems can be generalized in several ways, e. g. some of the above results remain valid if we replace the assumption that G is a Čebyšev set by the weaker assumption that G is a proximinal set such that for every $x \in E$ the set $\mathcal{P}_G(x)$ is convex. For these problems and for other related results we refer the reader to [82], pp. 364-371, [83] and to the recent papers of L. P. Vlasov [88]-[93]. Finally, for some results and problems on best approximation in metric (not necessarily normed linear) spaces we refer to [82], pp. 377-391.

I. Singer

References

[1] N. I. Ahiezer, Lectures on the theory of approximation. Second edition, Moscow (1965) [Russian].

[2] E. Asplund, Čebyšev sets in Hilbert space. Trans. Amer. Math. Soc. 144 (1969), 236-240.

[3] D. A. Ault, F. R. Deutsch, PD. Morris and J. E. Olson, Interpolating subspaces in approximation theory. J. Approx. Theory 3 (1970), 164-182.

[4] J. H. Biggs, F. R. Deutsch, R. E. Huff, P. D. Morris and J. E. Olson, Interpolating subspaces in ℓ_1 -spaces (to appear).

[5] J. Blatter, Zur Stetigkeit von mengenwertigen metrischen Projektionen. Schriften des Rheinisch-Westf. Inst. für Instrum. Math. Univ. Bonn, Ser. A, No. 16(1967), 17-38.

[6] J. Blatter, Approximation und Selection. Habilitationsschrift, Bonn (1969).

[7] J. Blatter, P. D. Morris and D. E. Wulbert, Continuity of the set-valued metric projection. Math. Ann. 178 (1968), 12-24.

[8] O. Brandt, Geometrische Approximations theorie in normierten Vektorräumen. Schriften des Rheinisch-Westf. Inst. für Instrum. Math. Univ. Bonn. ser. A, No. 18 (1968), 1-36.

[9] B. Brosowski, Nicht-lineare Tschebyscheff-Approximation. Bibliogr. Inst. Hochschulskripten Bd. 808/808a, Mannheim (1968).

[10] B. Brosowski, Einige Bemerkungen zum verallgemeinerten Kolmogoroffschen Kriterium. Funktionalanalytische Methoden der numerischen Mathematik. ISNM 12, Birkhäuser Verlag (1969), 25.-34.

[11] B. Brosowski, Nichtlineare Approximation in normierten Vektorräumen. Abstract spaces and approximation. ISNM 10, Birkhäuser-Verlag (1969), 140-159.

[12] B. Brosowski, Fixpunktsätze in der Approximations theorie. Mathematica II(34) (1969), 195-220.

[13] B. Brosowski, K. -H. Hoffmann, E. Schäfer und H. Weber, Stetigkeitssätze für Metrische Projektionen. Iterationsverfahren. Numerische Mathematik. Approximations-theorie. ISNM 15, Birkhäuser-Verlag (1970), 11-17.

[14] B. Brosowski, K.-H. Hoffmann, E. Schäfer und H. Weber, Metrische Projektionen auf lineare Teilräume von $C_o[Q, H]$. Iterationsverfahren. Numerische Mathematik. Approximationstheorie. ISNM 15, Birkäuser-Verlag (1970), 19-27.

[15] B. Brosowski und R. Wegmann, Charakterisierung bester Approximationen in normierten Vektorräumen. J. Approx. Theory 3 (1970), 369-397.

[16] F. Browder, Multivalued monotone nonlinear mappings and duality mappings in Banach spaces. Trans. Amer. Math. Soc. 118(1965), 338-351.

[17] A. L. Brown, Best n-dimensional approximation to sets of fuctions. Proc. London Math. Soc. 14(1964), 577-594.

[18] A. L. Brown, On continuous selections for metric projections in spaces of continuous functions (to appear).

[19] E. W. Cheney, Introduction to approximation theory. Mc Graw Hill, New York (1966).

[20] E. W. Cheney and D. E. Wulbert, Existence and unicity of best approximations. Math. Scand. 24(1969), 113-140.

[21] M. M. Day, Normed linear spaces. Springer-Verlag, Berlin-Göttingen-Heidelberg (1962).

[22] F. R. Deutsch and P. H. Maserick. Applications of the Hann-Banach theorem in approximation theory. SIAM Rev. 9 (1967), 516-530.

[23] F. R. Deutsch and J. Lambert, A bibliography on metric projections (mimeographed).

[24] A. I. Dubovitskii and A. A. Miliutin, Extremum problems in the presence of restrictions. Ž. Vycisl. Mat. i Mat. Fiz. 5(1965), 395-453 [Russian].

[25] N. Dunford and J. Schwartz, Linear operators. Part. I :General theory. Interscience Publ., New York (1958).

[26] M. Edelstein, Nearest points of sets in uniformly convex Banach spaces. J. London Math. Soc. 43(1968), 375-377.

[27] M. Edelstein, A note on nearest points. Quarterly J. Math. 21 (1970), 403-406.

[28] N. V. Efimov and S. B. Stečkin, Some properties of Čebyšev sets. Doklady Akad. Nauk SSSR 18 (1958), 17-19 [Russian].

[29] G. Ewald, D. G. Larman and C. A. Rogers, The directions of the line segments and of the n-dimensional balls on the boundary of a convex body in Euclidean space (to appear).

[30] Ky Fan and I. Glicksberg, Some geometric properties of the spheres in a normed linear space. Duke Math. J. 52(1958), 553-568.

[31] A. L. Garkavi, The theory of best approximation in normed linear spaces. Mathematical analysis 1967, Moscow (1969), 75-132 [Russian].

[32] A. L. Garkavi, The problem of Helly and best approximation in the space of continuous functions. Izevestija Akad. Nauk SSSR 31 (1967), 641-656 [Russian].

[33] A. L. Garkavi, Compact admitting Čebyšev systems of measures. Matem. Sbornik 74 (116) (1967), 209-217 [Russian].

[34] A. L. Garkavi, Characterization of Čebyšev subspaces of finite codimension L_1. Matem. Zametki 7(1970), 155-163 (1970) [Russian].

[35] G. Godini, Best approximation in normed linear spaces by elements of convex cones. Studii si cercet. mat. 21(1969), 931-936 (1969) [Romanian].

[36] A. Haar, Die Minkowskische Geometrie und die Annäherung an stetige Funktionen. Math. Ann. 78(1918), 294-311.

[37] S. Ia. Havinson, On approximation by elements of convex sets. Doklady Akad. Nauk SSSR 172(1967), 294-297 [Russian].

[38] R. B. Holmes, On the continuity of best approximation operators. Symp. on infinite dimensional topology, Princeton University Press (to appear).

[39] R. B. Holmes and B. R. Kripke, Smoothness of approximation. Michigan Math. J. 15(1968), 225-248.

[40] Y. Ikebe, A characterization of Haar subspaces in C [a, b]. Proc. Japan Acad. 44(1968), 219-220.

[41] Y. Ikebe, A characterization of best Tchebycheff approximations in function spaces. Proc. Japan Acad. 44(1968) 485-488.

[42] A. D. Ioffe and V. M. Tihomirov, Duality of convex functions and extremal problems. Uspehi Mat. Nauk 23, 6(144) (1968), 51-116 [Russian].

[43] R. C. James, Characterizations of reflexivity. Studia Math. 23(1964), 205-216.

[44] M. I. Kadec, Topological equivalence of uniformly convex spaces. Uspehi Mat. Nauk. 10, 4(66) (1955), 137-141 [Russian].

[45] V. Klee, The support property of a convex set in a linear normed space. Duke Math. J. 15 (1948), 767-772.

[46] V. Klee, Research problem no. 5, Bull. Amer. Math. Soc. 63 (1957), 419.

[47] V. Klee, Convexity of Chebyshev sets, Math. Ann. 142 (1961), 292-304.

[48] V. Klee, Remarks on nearest points in normed linear spaces. Proc. Coll. on convexity (Copenhagen, 1965), Univ. of Copenhagen (1966), 168-176.

[49] A. N. Kolmogorov, A remark on the polynomials of P. L. Čebyšev deviating the least from a given function. Uspehi Mat. Nauk 3, 1(23) (1948), 216-221 [Russian].

[50] C. A. Kottman and Bor-Luh Lin, On the weak continuity of metric projections. Michigan Math. J. 17(1970), 401-404.

[51] B. R. Kripke and T. J. Rivlin, Approximation in the metric of $L^1(X,\mu)$ Trans. Amer. Math. Soc. 115 (1965), 101-122.

[52] P. J. Laurent, Théorèms de caractérisation en approximation convexe. Mathematica 10 (33) (1968), 95-111.

[53] P. J. Laurent, Conditions nécessaires pour une meilleure appromation non linéaire dans un espace normé. Compte rendus Acad. Sci. (Paris) Sér. A-B 269(1969), A 245-A 248.

[54] P. J. Laurent and Pham-Dinh-Tuan, Global approximation of a compact set by elements of a convex set in a normed space. Numer. Math. 15 (1970), 137-150.

[55] A. J. Lazar, Spaces of affine continuous functions on simplexes. Trans. amer. Math. Soc. 134(1968), 503-525.

[56] A. J. Lazar, D. E. Wulbert and P. D. Morris, Continuous selections for metric projections. J. Functional Anal. 3(1969), 193-216.

[57] J. Lindenstrauss, Extension of compact operators. Memoirs. . Amer. Math. Soc. 48 (1964).

[58] J. Lindenstrauss, On nonlinear projections in Banach spaces. Michigan Math. J. 11(1964), 263-287.

[59] J. Lindenstrauss, On nonseparable reflexive Banach spaces. Bull. Amer. Math. Soc. 72(1966), 967-970.

[60] J. Lindenstrauss and L. Tzafriri, On the complemented subspace problem (to appear).

[61] G. G. Lorentz, Approximation of functions. Holt, Rinehart and Winston, New York (1966).

[62] G. Meinardus, Approximation of functions, theory and numerical methods. Springer-Verlag, Berlin-Heidelberg New Jork (1967).

[63] E. A. Michael, Continuous selections. I. Ann. Math. 63(1956), 361-382.

[64] J. J. Moreau, Fonctionnelles convexes. Séminaire sur les équaions aux dérivées partielles. Collège de France, Paris (1966-67).

[65] J. J. Moreau, Distance à un convexe d'un espace normé et caractérisation des points proximaux. Séminaire d'analyse unilatérale. Fac. Sci. Montpellier, 2(1969), exposé no 6.

[66] P. D. Morris, Metric projections onto subspaces of finite codimension. Duke Math. J. 35 (1968), 799-808.

[67] E. V. Ošman, Continuity of metric projections and some geometric properties of the unit sphere in a Banach space, Doklady Akad. Nauk SSSR (1969), 34-36. [Russian].

[68] R. R. Phelps, Convex sets and nearest points. Proc. Amer. Math. Soc. 8(1957), 790-797.

[69] W. Pollul, Reflexivität und Existenz-Teilräume in der linearen Approximationstheorie. Schriften der Gas. für Math. und Datenverarbeitung, Bonn (to appear)

[70] J. R. Rice, The approximation of functions. Vol. I; Linear theory. Vol. II: Non-linear and multivariate theory. Addison-Wesley, Reading, Mass. -London-Don Mills, Ont. (1964 and 1969).

[71] J. R. Rice, Nonlinear approximation. Approximation of functions (Ed. by H. L. Garabedian). Elsevier. Amsterdam-London-New York (1965), 111-133.

[72] J. R. Rice, Non-linear approximation. II. Curvature in Minkowski geometry and local uniqueness. Trans. Amer. Math. Soc. 128(1967), 437-459.

[73] W. W. Rogosinski, Continuous linear functionals on subspaces of $\mathcal{L}^p$ and $\mathcal{C}$. Proc. London Math. Soc. 6, 22(1956), 175-190.

[74] Ch. Roumieu, Sur quelques problèmes d'approximation. Sémin. math. fac. sci. Montpellier (1966).

[75] G. Š. Rubinstein, On an extremal problem in a linear normed space. Sibirak. Mat. Ž. 6(1965), 711-714 [Russian]

[76] I. Singer, Properties of the surface of the unit cell and applications to the solution of the problem of uniqueness of the polynomial of best approximation in arbitrary Banach spaces. Studii si cercet. mat. 7(1956), 95-145 [Romanian]

[77] I. Singer, Caractérisation des éléments de meilleure approximation dans un espace de Banach quelconque. Acta Sci. Math. 17(1956), 181-189

[78] I. Singer, On best approximation of continuous functions. Math. Ann. 140 (1960), 165-168.

[79] I. Singer, On best approximation of continuous functions. II. Rev. math. pures et Appl. 6(1961), 507-511.

[80] I. Singer, Some remarks on approximative compactness. Rev. roum. math. pures et appl. 9(1964), 167-177.

[81] I. Singer, On the extension of continuous linear functionals and best approximation in normed linear spaces. Math. Ann. 159(1965), 344-355.

[82] I. Singer, Best approximation in normed linear spaces by elements of linear subspaces. Publ. House Acad.

Soc. Rep. Romania, Bucharest (1967) [Romanian]. English translation : Publ. House Acad. Soc. Rep. Romania, Bucharest and Springer-Verlag, Berlin-Heidelberg-New York (1970).

[83] I. Singer, Some open problems on best approximation in normed linear spaces. Séminaire Choquet, 6e année. Université de Paris (1966/67), exposé no. 12.

[84] I. Singer, On metric projections onto linear subspaces of normed linear spaces. Proc. Confer. on "Projections and related topics" held in Clemson. Aug. 1967 Preliminary Edition (January 1968).

[85] I. Singer, Remark on a paper of Y. Ikebe. Proc. Amer. Math. Soc. 21 (1969), 24-26.

[86] I. Singer, On normed linear spaces which are proximinal in every superspace. J. Approx. Theory (to appear).

[87] K. Tatarkiewicz, Une théorie généralisée de la meilleure approximation. Ann. Univ. Mariae Curie-Sklodowska 6 (1952), 31-46.

[88] L. P. Vlasov, On Čebyšev sets. Doklady Akad. Nauk SSSR 173 (1967), 491-494 [Russian].

[89] L. P. Vlasov, Approximatively convex sets in uniformly smooth spaces. Mat. Žametki 1(1967), 443-449 [Russian].

[90] L. P. Vlasov, On Čebyšev and approximatively convex sets. Mat. Zametki 2(1967), 191-200 [Russian].

[91] L. P. Vlasov, Čebyšev sets and some generalizations of them Mat. Zametki 3(1968), 59-69 [Russian].

[92] L. P. Vlasov, Approximative properties of sets in Banach spaces. Mat. Zametki 7 (1970), 593-604 [Russian].

[93] L. P. Vlasov, Almost convex and Čebyšev sets. Mat. Zametki 8 (1970) 545-550 [Russian].

[94] D. E. Wulbert, Continuity of metric projections. Trans. Amer. Math. Soc. 134 (1968), 335-343.

[95] D. E. Wulbert, Convergence of operators and Korovkin's theorem. J. Approx. Theory 1(1968), 8-18.

[96] D. E. Wulbert, Differential theory for non-linear approximation. (preprint).

[97] D. E. Wulbert, Uniqueness and differential characterization of approximations from manifolds of functions. Amer. J. Math. (to appear).

[98] D. E. Wulbert, Nonlinear approximation with tangential characterization (to appear).

[99] S. I. Zuhovitskii, On minimal extensions of linear functionals in the space of continuous functions. Izvestija Akad. Nauk SSSR 21(1957), 409-422 [Russian] .

CENTRO INTERNAZIONALE MATEMATICO ESTIVO

(C. I. M. E.)

G. STRANG

AND

G. FIX

A FOURIER ANALYSIS OF THE FINITE ELEMENT VARIATIONAL METHOD

Corso tenuto ad Erice dal 27 giugno al 7 luglio 1971

These lectures were prepared for a CIME Advanced Summer Institute held in 1971.
The first author has been supported by the Office of Naval Research and the National Science Foundation (GP - 13778), and the second by AEC Contract 7158 - 2 .

§0. Apologia

This paper has been taken from a preliminary draft of our book "An Analysis of the Finite Element Method", to be published by Prentice-Hall about the end of 1972. In this first draft we developed the theory of finite elements on a regular mesh, with Fourier analysis as the principal tool, and we were able to discuss the connections with finite difference equations and to include a part of the theory of splines. This framework we now call the "abstract finite element method".

In our book, the emphasis will be shifted to the "nodal finite element method" as developed by structural engineers, in which irregular elements are more the rule than the exception. In this case splines are much less convenient, and Fourier analysis is impossible. However the basic theorems remain valid - we refer to a forthcoming paper of the first author in Numerische Mathematik on "Approximations in the Finite Element Method". Furthermore it becomes possible to examine the errors due to the presence of curved boundaries and inhomogeneous boundary data.

We hope that the book will give a reasonably complete and realistic treatment of the essential finite element theory for linear problems, and that the reader will accept the present paper as an interim report.

§1. Introduction

The development of approximate methods for the solution of elliptic problems has apparently entered a third stage. The first stage employed approximations of the form

$$u(x) \approx \sum_{1}^{N} v_j \varphi_j(x) ,$$

with trial functions φ_j selected in advance, and coefficients v_j determined by a variational principle, collocation, or least squares. The associated mathematical questions were the completeness and uniform linear independence of the sequence φ_j in the space of admissible functions. However, since N could not in practice be taken very large, the analysis was clearly secondary to a clever choice of the φ_j. The computer reversed this position completely, and introduced a second stage: a million unknowns became feasible, provided they were determined by equations of an extremely simple and systematic form. Finite difference equations have precisely this property, and the construction of fast iterative methods (and more recently also direct methods) for their solution has dominated the numerical analysis literature on elliptic problems since 1950.

The third stage, which we hope to analyze, seems to be due at least in part to the persistence of those who never left stage one. (There may be readers with similar intentions about stage two.) It lies precisely at the intersection of the first two stages:

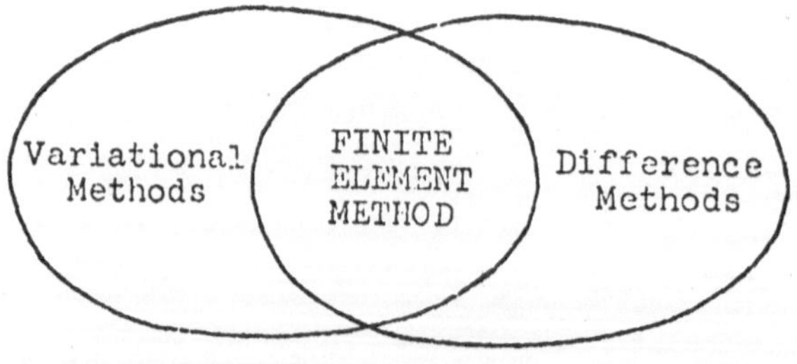

The crucial idea in this synthesis is to choose the basis φ_j in such a way that <u>the variational equations for the</u> v_j <u>turn out to be difference equations</u>. The customary description of the resulting method is in variational terms, and we shall introduce it in this way below. To think of it at the same time as a difference scheme will require a certain tolerance on the reader's part, since at first sight the finite element method seems to depart at several points from conventional difference equations. In fact it is just these points which represent for us the main contributions of the method to finite difference theory; they are innovations which <u>could</u> have been devised independently, but never were.

Our goal will be to decide when the finite element method is convergent and numerically stable, and to estimate the error. Although we do not discuss at this point the irregular meshes and general boundary conditions which are met in applications, we have tried to retain the mathematical essentials of the method; these we study in some generality. Of course the ultimate question is whether the finite element method is more effective than its competitors, namely those techniques which lie outside the above intersection. The evidence suggests that although difference schemes can be constructed which require fewer operations for a given order of accuracy, nevertheless the variational approach has an important coherence which derives from the fact that, once the basis φ_j is chosen, the rest is largely automatic. This coherence seems to be reflected in a more regular behavior both of the error and of the user, who has otherwise to make a separate choice of finite difference replacement for each term in the differential equation and boundary conditions. The evidence in this comparison is still very limited, however, and we shall try to remain neutral.

The name we have adopted was originally chosen by engineers [1], who decompose a continuous structure, for numerical purposes, into a set of "finite elements". The history of the underlying mathematics is less clear. Both Courant [2] and Pólya [3] commented on the merits, in certain variational problems, of seeking approximate solutions which are linear within each (triangular) element; accuracy is improved by increasing the number of elements rather than the complexity of the approximating functions. With this trial space, the Laplace operator acting on u induces its familiar 5-point difference analogue, acting on the coefficient vector v. Such trial functions therefore make the Ritz method especially simple to execute, and it seems very likely that this idea was proposed even earlier.

The development of the method has led naturally from piecewise linear functions to splines and other piecewise polynomials of fixed degree p; each increase in p adds both to the accuracy and to the complexity of the method. As usual, the extra accuracy is initially worth the price; but just as Newton's method is more popular than its higher-order analogues, questions of convenience soon become paramount. In applications to second-order equations, cubic approximants $(p = 3)$ are apparently close to the turning point. The essential features of the method are the subdivision of the region into finite elements, and the choice of a so-called local basis for the space of approximating functions - that is, a basis composed of functions which vanish over all but a few elements.

We analyze in this paper the case of subdivision by a regular mesh, of width $h \to 0$, with a local basis constructed in the following systematic way. We start with a fixed set of trial functions $\varphi_1(x), \ldots, \varphi_N(x)$; N will be the number of basis functions for each meshpoint. To ensure a local basis, we insist that these functions φ_i vanish for large $|x|$, and rescale the independent variable x by the mesh width h. The eventual Rayleigh-Ritz equations will take the form of difference equations if the basis functions $\varphi^h_{i,j}$ associated with each meshpoint $(j_1 h, \ldots, j_n h)$ are simply translates of these rescaled functions $\varphi_i(x/h)$. The basis is thus composed, for each h, of the functions

$$(1.1) \qquad \varphi^h_{i,j}(x) = h^{-n/2}\, \varphi_i\left(\frac{x_1}{h} - j_1, \ldots, \frac{x_n}{h} - j_n\right).$$

Of course there have to be modifications at boundaries.

In the piecewise linear case there is a single parameter for each meshpoint, namely the value of the function at that point; thus $N = 1$. The graph of φ_1 is a pyramid with vertex at the origin and with base formed from the neighboring triangular elements.

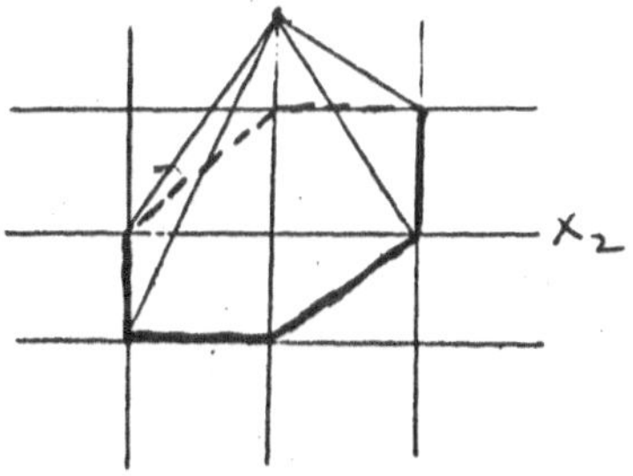

It should be clear that this φ_1 and its translates span the space of all continuous functions in the plane which are linear within each element. In some examples our description (1.1) of the basis will seem less transparent than a description of the space itself, in terms of the functions admitted within each element and the compatibility conditions across element boundaries. Nevertheless, both for practical computations and for the general theory, the definition of the basis is crucial; it is from combinations of these functions $\varphi_{i,j}^h$ that the Rayleigh-Ritz-Galerkin principle will select an approximation u^h

The fundamental questions for a numerical analyst are those of convergence and stability:

i) <u>how accurate is the approximate solution</u> u^h , <u>and</u>

ii) <u>how well conditioned are the equations from which</u> u^h <u>is determined numerically</u>?

The answers can only come from the connections between the given differential problem, the trial functions $\varphi_1,\dots,\varphi_N$, and the norms in which accuracy and condition number are measured. We want to study these questions for elliptic operators of arbitrary order $2m$, for quite general φ_i , and for the norms associated with the spaces $\mathcal{H}^s$ and W_∞^s . These norms measure the error $e^h = u^h - u$ and its derivatives

$$D^\alpha e^h = \left(\frac{\partial}{\partial x_1}\right)^{\alpha_1} \dots \left(\frac{\partial}{\partial x_n}\right)^{\alpha_n} e^h$$

of order $|\alpha| = \Sigma\alpha_i \leq s$ in the mean-square and pointwise senses respectively:

(1.2)
$$\|e\|^2_{\mathcal{H}^s} = \sum_{|\alpha| \leq s} \|D^\alpha e\|^2_{L^2}$$

$$\|e\|_{W^s_\infty} \equiv \sum_{|\alpha| \leq s} \|D^\alpha e\|_{L^\infty}$$

In Sobolev's notation the space $\mathcal{H}^s$ is written W^s_2.

Our arguments make constant use of the Fourier transform, which operates at full strength only on problems which are either periodic, or defined on the whole of Euclidean space R^n. We are convinced that (as in the theory of elliptic differential operators) the investigation of these special problems is fundamental to the understanding of more general boundary conditions. Thus we regard the present work as a necessary first step in analyzing the wide variety of problems, with irregular meshes and boundaries, which are actually being solved. Fortunately, most of the second step in the analysis is already complete; J.-P. Aubin, following the work of Céa and others in the French school, has successfully analyzed the solution of boundary problems by means of splines. In his terms, our contribution is to determine all trial functions to which his theory can extend. (We mention, in addition to his forthcoming manuscript, the reference [4].) The third step is the study of more general meshes, particularly those formed by an arbitrary triangulation of the region. This is a major point in our forthcoming book.

For problems on the whole space R^n, the index j in (1.1) runs over the set Z^n of all multi-integers $(j_1,\dots,j_n)$. We adopt the definition

$$\hat{\varphi}(\xi) = \int_{R^n} \varphi(x) e^{-ix\xi}\, dx$$

for the Fourier transform, where $\xi = (\xi_1,\dots,\xi_n)$ and $x\xi$ denotes $x_1\xi_1 + \dots + x_n\xi_n$. As a first application of the Fourier transform, Parseval's formula can be used to replace (1.2) by the equivalent and more convenient norm

$$\|e\|^2_{\mathcal{H}^s} = \int |\hat{e}(\xi)|^2 (1+|\xi|^2)^s\, d\xi . \tag{1.3}$$

We first describe the application of the Ritz-Galerkin method abstractly, to a linear elliptic problem on R^n. Using the conventional inner product $(f,g) = \int f(x)\overline{g}(x)\,dx$, the problem begins with a bilinear form

$$a(u,w) = \sum_{|\alpha|,|\beta|\le m} (q_{\alpha\beta}(x) D^\alpha u, D^\beta w) .$$

If the coefficients $q_{\alpha\beta}$ are bounded, the form a is defined whenever u and w lie in the space $\mathcal{H}^m$, that is, whenever all partial derivatives of order not exceeding m lie in $L_2(R^n)$. The form is called $\mathcal{H}^m$-elliptic provided that

$$\operatorname{Re} a(u,u) \ge \rho \|u\|^2_{\mathcal{H}^m} , \quad \rho > 0 . \tag{1.4}$$

The most familiar example is the form associated with the Laplace equation,

$$a(u,w) = \int \frac{\partial u}{\partial x_1}\frac{\partial \bar{w}}{\partial x_1} + \ldots + \frac{\partial u}{\partial x_n}\frac{\partial \bar{w}}{\partial x_n}\, dx .$$

As it stands this is not $\mathcal{H}^1$-elliptic, corresponding to the fact that Laplace's equation has non-zero solutions, e.g. $u =$ constant. To satisfy (1.4), and thereby eliminate this non-uniqueness of the solution, we need to add some positive multiple of the zero-order term (u,w) .

An elliptic form induces the following variational problem: given f , find u in $\mathcal{H}^m$ so that

$$a(u,w) = (f,w) \quad \text{for all} \quad w \quad \text{in} \quad \mathcal{H}^m . \tag{1.5}$$

This problem has one and only one solution u , provided the inhomogeneous data f is such that the right side makes sense; since w ranges over $\mathcal{H}^m$, this places f in the adjoint space $\mathcal{H}^{-m}$, so that

$$\int |\hat{f}(\xi)|^2 \, (1+|\xi|^2)^{-m}\, d\xi < \infty .$$

The elliptic problem (1.5) can equally well be put into the more familiar operational form

$$L\,u = f . \tag{1.6}$$

For this we integrate the left side of (1.5) by parts, shifting derivatives from w onto $q\, D^{\alpha} u$. The result is $(Lu, w) = (f, w)$, which is equivalent to (1.6); L is the map from $\mathcal{H}^m$ to $\mathcal{H}^{-m}$ given by

$$L = \sum_{|\alpha|, |\beta| \leq m} (-1)^{|\beta|} D^{\beta} q_{\alpha\beta} D^{\alpha} .$$

Many applications lead also to problems of <u>minimization</u>. In such cases the form a is self-adjoint, and the minimization of $a(u,u) - (f,u) - (u,f)$ leads exactly to the same variational problem (1.5). In fact the operational equation $Lu = f$ is nothing but the <u>Euler equation</u> from the calculus of variations.

The Ritz-Galerkin technique is now simple and very familiar; the space $\mathcal{H}^m$ is replaced in the variational statement (1.5) by a sequence of closed subspaces S^h. Thus the approximating problem, written variationally, is to find u^h in S^h so that

$$(1.7) \qquad a(u^h, w^h) = (f, w^h) \quad \text{for all } w^h \text{ in } S^h .$$

Ellipticity implies the existence and uniqueness of u^h; we are concerned with its computation. Therefore we put the approximate problem also into operational form. Choosing a basis φ_{μ}^h for S^h, we expand

$$(1.8) \qquad u^h = \sum v_{\mu}^h \varphi_{\mu}^h$$

and compute the vector v^h of unknown coefficients. Substituting into (1.7),

$$a\left(\sum v_\mu^h \varphi_\mu^h , \sum w_\nu^h \varphi_\nu^h\right) = \left(f, \sum w_\nu^h \varphi_\nu^h\right) .$$

Since this holds for all coefficients w_ν^h,

$$a\left(\sum v_\mu^h \varphi_\mu^h , \varphi_\nu^h\right) = \left(f , \varphi_\nu^h\right) .$$

Thus the vector v^h satisfies the discrete operational equation

$$A^h v^h = f^h , \tag{1.9}$$

where the entries in the coefficient matrix and the inhomogeneous vector are given by

$$(A^h)_{\nu\mu} = a(\varphi_\mu^h , \varphi_\nu^h) , \quad (f^h)_\nu = (f , \varphi_\nu^h) . \tag{1.10}$$

Since all these entries have to be calculated, either analytically or by numerical quadrature, one wants as simple a basis as possible. The use of the classical special functions, in other words a return to stage one, is by no means obsolete; both Urabe and Clenshaw have made successful application of Chebyshev polynomials. We are interested, however, in the basis functions $\varphi_{i,j}^h$ defined earlier; for problems on R^n the index j runs over Z^n, whereas if $f, u, \ldots$ all have period 1, we require that h^{-1} be an integer and that $0 \leq j_\nu < h^{-1}$ for each component of j. In this periodic case S^h has finite dimension $N h^{-n}$

The index i assumes the values $1,\ldots,N$, so it is natural to take the basis functions in groups

$$(1.11) \qquad \phi_j^h = (\varphi_{1,j}^h, \ldots, \varphi_{N,j}^h)$$

of N at a time. This partitions the matrix A^h into blocks of order N , a typical block being

$$(1.12) \qquad (A^h)_{jk} = \sum \int dx\, q_{\alpha\beta}(D^\alpha \phi_j^h)\,(D^\beta \phi_k^h)^* .$$

Thus the finite element relation $A^h v^h = f^h$ is a coupled system of N discrete equations. Normally this system is analogous to a continuous one, in which the original differential equation $Lu = f$ is coupled to some of its differentiated forms $D(Lu) = Df$ In the Hermite case the discrete system can be formally recombined to yield a single spline-like equation, just as the Hermite basis functions, with small support, can be combined with their translates to yield the spline basis.

We want now to summarize our results. A more extended summary has already appeared [5] in Studies in Applied Mathematics, a reincarnation of M.I.T.'s Journal of Mathematics and Physics. We hope that our discussion there, in terms of the model problem $-\Delta u + u = f$ and its 5- and 9- point difference analogues, will be a useful supplement to the present paper. For the moment we set aside extensions to eigenvalue problems and parabolic equations, and describe our conclusions only for the problems of convergence and stability stated above.

ii) The problem of stability is the simpler. Here the fundamental question is whether or not the independence of the basis elements $\varphi_{i,j}^{h}$ is uniform as $h \to 0$; in L_2, with our normalization (1.1) of the basis elements, this means

$$\|\sum v_{i,j} \varphi_{i,j}^{h}\|_{L_2}^{2} \geq \sigma \sum |v_{i,j}|^2 , \sigma > 0 . \tag{1.13}$$

For this uniform independence we find the following necessary and sufficient condition: <u>there exists no non-trivial combination $\Psi = \sum c_i \varphi_i$ and real ξ_0 such that the Fourier transform of</u> Ψ <u>satisfies</u>

$$\hat{\Psi}(\xi_0 + 2\pi j) = 0 \quad \text{for all} \quad j \in Z^n . \tag{1.14}$$

The reader will notice that everything depends on the φ_i ; only gross features of the differential problem, its order and ellipticity, are relevant to the condition number. This is an attraction, at least to the analyst, which should not disappear in more general boundary problems. For difference equations which are <u>not</u> derived variationally, Schaeffer's powerful work [6] has shown how deep the stability problem actually is, even in comparison with the corresponding question in the general theory of elliptic boundary problems. In Thomée's terminology [7] , (1.14) is necessary and sufficient for the difference equations (1.9) to be elliptic.

i) The rate of convergence of u^h depends on the "density" of the spaces S^h ; therefore we begin with a discussion of approximation theory. Our main result for the case $N=1$ takes the following form: smooth functions can be approximated from S^h with error $O(h^{p+1-s})$ in the $\mathcal{H}^s$ norm, $s \leq p$, if and only if all polynomials in $x_1,\ldots,x_n$ of degree $\leq p$ can be written as linear combinations of φ and its translates. Fourier analysis leads to an equivalent condition on the transform $\hat{\varphi}$; it must have zeros of order $p+1$ at all the points $\xi = 2\pi j$, $j \neq (0,\ldots,0)$. Here we have assumed the conditions most commonly met in practice, that φ is in $\mathcal{H}^p$ and stability holds; more precise results are proved in §2.

With $N>1$, the φ_i and linear combinations of their translates may have differing degrees of smoothness. In fact there are important cases in which this is bound to happen. Suppose for example that S^h is comprised of piecewise cubic functions $(n=1)$ which have continuous first derivatives at each joint. Since this Hermite space contains the spline subspace, whose elements have continuous second derivatives as well, the spline basis functions must be combinations of the Hermite basis; the former is in $\mathcal{H}^3$ and the latter only in $\mathcal{H}^2$. Thus, in the general case, we assume only that the φ_i are in $\mathcal{H}^q$, $q \leq p$, and again we prove that approximation of order h^{p+1-s} is possible in $\mathcal{H}^s$ (for $s \leq q$) if and only if all polynomials of degree $\leq p$ can be produced from the φ_i and their translates.

These approximation results are of course already known for many specific choices of S^h. We mention in particular the early estimates for splines by Birkhoff, de Boor, and others, the approximations constructed by Aubin in [8], and the detailed treatment of multi-dimensional spline and Hermite functions in Birkhoff, Schultz, and Varga [9] and in recent papers by Schultz (cf. [10]). The order $p+1-s$ of approximation, starting with a single arbitrary φ, has also been established elsewhere; in fact this problem was so striking that it was attacked independently by di Guglielmo [11], Babuška [12], and the two of us. (We regret that the order just given is chronological, although Δt is small.) Our own contributions to the theory of approximation on $\mathcal{H}^s$ with $N=1$ seem to be these: we have found the exact lower bound of possible constants in the error estimate $c\, h^{p+1-s} \|u\|_{\mathcal{H}^{p+1}}$, and we have proved the converse result, that such an estimate is <u>only</u> possible when polynomials can be manufactured from φ. Obviously this gives a special place to piecewise polynomial approximating functions; they not only lead to simple computations, which has until now been their real justification, but they also provide the most efficient basis from which to produce polynomials. (Of course one may use polynomials themselves in the Ritz method, but this produces major difficulties. If monomials x^j are used as the basis, then the matrix A^h is both hopelessly ill-conditioned and full of non-zero entries. The alternative, a basis of orthogonal polynomials, is awkward to compute numerically.) We note that our converse result was anticipated by Goël [13], who considered only a very limited class of approximating functions, and that di Guglielmo has discovered [11] a basis which makes A^h as sparse as possible. Splines are optimal in one dimension, as is the pyramid function drawn above when $n=2$, $p=1$. In general he "triangulates" R^n and uses splines whose continuity properties are imposed between the triangular elements, or simplexes; this leads to

$$(1.15) \qquad \hat{\varphi} = \left(\frac{\sin \xi_1/2}{\xi_1/2}\right)^p \cdots \left(\frac{\sin \xi_n/2}{\xi_n/2}\right)^p \frac{\sin \Sigma \xi_j/2}{\Sigma \xi_j/2} .$$

In the splines over rectangular elements introdued by de Boor and Birkhoff, the last factor disappears and the remaining exponents have to be increased to $p+1$. S^h becomes essentially a tensor product of the one-dimensional case. In either case, the zeroes at the points $\xi = 2\pi j \neq 0$ are evident.

For approximation in the pointwise norms W_∞^s, the proofs are different but the results are the same. Here we construct an explicit approximation $\tilde{u}^h$, a quasi-interpolate, by combining interpolation with a generalized Taylor expansion:

$$(1.16) \qquad \tilde{u}^h(x) = \sum_j \sum_{|\alpha| \le p} D^\alpha u(jh)\, h^{|\alpha|}\, \psi_\alpha(x/h - j) .$$

The ψ_α are those combinations of the basic functions $\varphi_1, \ldots, \varphi_N$ which are used in the construction of polynomials up to degree p; if $N = 1$ they are suitable multiples of φ_1. The estimate

$$\|\tilde{u}^h - u\|_{W_\infty^s} = O(h^{p+1-s})$$

then follows for smooth u by comparing Taylor series. Again this bound is already known in special cases.

We can now state our principal result for problem (i): <u>if f is sufficiently smooth, the error in either the $\mathcal{H}^s$ or the W_∞^s norm satisfies</u>

$$(1.17) \qquad \|u^h - u\|_s = O(h^r) \ , \ r = \min(p+1-s \ , \ 2(p+1-m)).$$

Apparently this exponent r is new, although in $\mathcal{H}^s$, at least for $s \leq m$, it can be deduced by variational arguments from the order of approximation given above. It is remarkable that the error in the Ritz method often <u>is not of the same order</u> as the error in the best approximation; the second expression in r may well be smaller than the first. For the special value $s = m$ the two orders automatically coincide, since ellipticity implies that for any choice of the Ritz spaces S^h,

$$(1.18) \qquad \|u^h - u\|_{\mathcal{H}^m} \leq \text{constant} \inf_{s \in S^h} \|u - s\|_{\mathcal{H}^m} .$$

We always assume $p \geq m$, which together with stability is <u>necessary and sufficient for the finite element method to succeed</u>; this applies to eigenvalue problems as well. We note that for splines of the minimal degree $p = m$, the error is $O(h)$ in $\mathcal{H}^m$ and regardless of m is only $O(h^2)$ in $\mathcal{H}^0 = L^2(R^n)$.

Our proof of the error estimate (1.17) is severely classical; we regard (1.8-1.9) as inducing a finite difference scheme and estimate the local truncation error. This allows us to treat the non-reflexive spaces W_∞^s, which in variational arguments are awkward at best; of course Sobolev's imbedding theorems may be used to deduce pointwise estimates, but the cost in powers of h is unacceptably high. Furthermore, the local error is well-defined even in the absence of global properties like ellipticity; our technique should extend for example to parabolic and hyperbolic operators. It may also be used to justify "Richardson extrapolation", which is a useful tool for increasing the accuracy. We note that the order of this local error is normally trivial to compute, by a comparison of Taylor expansions; but for a system of N difference equations, this is no longer so.

We cannot resist pointing to a rather paradoxical consequence of our estimates. It is concerned with the approximation of <u>derivatives</u> of u, in which Ritz methods are generally thought to be superior; they provide solutions which can be differentiated analytically, whereas more conventional difference schemes yield only mesh functions. However, Thomée has shown [14] that the order of accuracy is <u>not reduced</u> when these mesh functions are differenced, while our present estimates (which are best possible) imply that the accuracy is generally decreased by each differentiation and finally disappears. The paradox enters into the finite element method, which is both a Ritz system and a difference scheme: apparently the derivatives of u at x_0 should not be computed by differentiating u^h, but rather by regarding x_0 as the center of an h-mesh and applying an accurate difference operator! Unfortunately for the theory, round-off error generally reverses this recommendation.

One further problem, of obvious interest but questionable importance, is this: among all difference schemes consistent with a given differential equation, which can be produced by the finite element method? Certainly not all, since self-adjoint nonnegative-definite problems $(L = L^* \geq 0)$ lead only to self-adjoint nonnegative-definite finite element equations $(A^h = A^{h*} \geq 0)$. Furthermore, the inhomogeneous data f enters the Ritz equations, not through its values at meshpoints, but through the integrals $\int f\varphi_j^h$. This smoothing bears on a remark of Zlamal [15], which we discussed with Widlund and Morton, to the effect that variational error bounds require fewer derivatives of f than some of the standard finite difference estimates. We note that Herbold, Schultz, and Varga [16] have observed how the usual numerical quadratures of these integrals lead to familiar difference schemes.

If we consider only the leading term $-u_{xx}$, for second-order equations in one dimension, a complete answer is possible: all consistent self-adjoint nonnegative-definite difference matrices A^h can be produced. The proof depends on translating these properties of A^h, whose action is typically given by

$$(A^h v)_j = \sum A^h_{jk} v_k = \sum_{-M}^{M} \alpha_{k-j} v_k ,$$

into properties of its symbol

The correspondence is a familiar one:

Consistency <=> $a(\theta) = \theta^2 + O(\theta^3)$, so that $a(0) = a'(0) = 0$

Self-adjointness <=> a is real

Non-negativity <=> $a(\theta) \geq 0$ for all θ.

The Féjer-Riesz theorem allows us to factor such a nonnegative trigonometric polynomial into

$$(1.19) \qquad a(\theta) = \left| \sum_0^M \beta_\ell \, e^{i\ell\theta} \right|^2 .$$

Now we let φ be the piecewise linear function which vanishes for $x \leq 0$ and has slope β_j in $j < x < j+1$. Consistency gives $\Sigma \beta_j = 0$, so that φ returns to the value 0 at $x = M+1$. For larger x we keep it zero, so that φ satisfies the essential condition for the finite element method: it vanishes outside a compact set. Comparing coefficients of $e^{i\ell\theta}$ in (1.19), the Ritz-Galerkin coefficients are indeed the α_ℓ:

$$A_{jk} = \int \varphi_x(x-k)\, \bar{\varphi}_x(x-j)\, dx = \sum \beta_{\ell-k} \bar{\beta}_{\ell-j} = \alpha_{k-j} :$$

Unhappily, the result we have just proved is misleading; the proper analogue of the differential equation $Lu = f$ is <u>not</u> simply the explicit system $A^h v^h = f^h$. Instead it is the combination of this system with the expansion (1.18), expressing u^h in terms of the given basis for S^h; in our context this expansion is

(1.20) $$u^h(x) = \sum_{i,j} v^h_{i,j}\, \varphi_i\, (x/h - j)\ .$$

Suppose we consider only the values $u^h(kh)$ assumed at the mesh-points $x = kh$; these values are the components of a vector which we denote by $\underline{u}^h$. On this mesh the previous equation becomes

(1.21) $$\underline{u}^h = B^h\, v^h\ ,$$

where the entries in the matrix B^h are the values $\varphi_i(k-j)$.

Altogether, then, $\underline{u}^h$ is computed from the inhomogeneous data f by solving

(1.22) $$A^h (B^h)^{-1}\, \underline{u}^h = f^h\ .$$

This is the <u>implicit finite difference equation</u> derived by the finite element method; it is the truncation error associated with <u>this</u> equation which yields an estimate of $u^h - u$. There is a similar equation corresponding to every choice of h-mesh in the plane; our matrix B^h refers only to the mesh through the origin. Thus the finite element method is finally exposed as a family of difference equations, each an implicit system of N equations but otherwise more or less conventional, and all descended from the common ancestors $\varphi_1, \ldots, \varphi_N$.

§2. Approximation

In this section we shall isolate those conditions on the finite element spaces S^h which determine the accuracy of approximation. Although the proofs are rather technical, the conditions themselves are remarkably simple. Starting with a single function φ , that is with $N = 1$, we consider approximation in $\mathcal{H}^s$ by expansions of the form

$$\sum_{j\in Z^n} w_j^h \varphi_j^h(x) \ , \ \varphi_j^h(x) = h^{-n/2} \varphi(\frac{x}{h} - j) \ .$$

In order that each φ_j^h shall be non-zero within only a finite number of elements, we assume that φ vanishes for large $|x|$ This property of compact support ensures that A^h is a band matrix, since all products $\varphi_j^h \varphi_k^h$ are identically zero whenever $|j-k|$ is sufficiently large. We let $\mathcal{H}_c^q$ denote the space of functions with compact support whose derivatives of order $\leq q$ lie in L^2 ; in many applications the derivatives of order less than q will be continuous, and the q th derivatives piecewise continuous.

We adopt the standard notation for multi-integers, writing $j,k,\ldots$ for elements of Z^n and $\alpha,\beta,\ldots$ for elements of Z_+^n . Thus each component j_ν is an integer, each α_ν is a non-negative integer, and

$$|\alpha| = \alpha_1 + \ldots + \alpha_n \ , \ x^\alpha = x_1^{\alpha_1} \ldots x_n^{\alpha_n}$$

$$\alpha! = \alpha_1! \ldots \alpha_n! \quad , \quad \binom{\alpha}{\beta} = \frac{\alpha!}{\beta!\,(\alpha-\beta)!}$$

$$\alpha \geq \beta \text{ if and only if each } \alpha_\nu \geq \beta_\nu .$$

The range of summation is understood to be Z^n, and the range of integration is R^n, when nothing is said to the contrary.

THEOREM I. Suppose φ is in $\mathcal{H}_c^p$. Then the following conditions are equivalent:

(i) $\hat{\varphi}(0) \neq 0$, but φ has zeros of order at least $p+1$ at the other points of $2\pi Z^n$:

$$D^\alpha \hat{\varphi}\,(2\pi j) = 0 \quad \text{if} \quad 0 \neq j \in Z^n, \; |\alpha| \leq p \tag{2.1}$$

(ii) for $|\alpha| \leq p$, $\sum_{j \in Z^n} j^\alpha \varphi(t-j)$ is a polynomial in $t_1, \ldots, t_n$ with leading term Ct^α, $C \neq 0$

(iii) for each u in $\mathcal{H}^{p+1}$ there are weights w_j^h such that as $h \to 0$,

$$\|u - \sum w_j^h \varphi_j^h\|_{\mathcal{H}^s} \leq c_s h^{p+1-s} \|u\|_{\mathcal{H}^{p+1}}, \quad 0 \leq s \leq p, \tag{2.2}$$

$$\sum |w_j^h|^2 \leq K \|u\|^2_{\mathcal{H}^0} . \tag{2.3}$$

The constants c_s and K are independent of u.

Proof.

(i) <=> (ii). This equivalence, like much of the analysis later in this paper, depends on the Poisson formula: the values of a function Ψ on the lattice Z^n are connected with those of its Fourier transform $\hat{\Psi}$ on the lattice $2\pi Z^n$ by

$$\sum_{j\in Z^n} \Psi(j) = \sum_{j\in Z^n} \hat{\Psi}(2\pi j) .$$

If Ψ has compact support, the first sum involves only a finite number of terms, and the second is absolutely convergent. Applied to the function $\Psi(x) = x^\alpha \varphi(t-x)$, this yields

$$(2.4) \qquad \sum j^\alpha \varphi(t-j) = \sum i^{|\alpha|} D^\alpha(e^{-i\xi t} \hat{\varphi}(-\xi))|_{2\pi j}$$

Suppose first that (i) holds. Then for any $|\alpha| \leq p$, the terms on the right side with $j \neq 0$ all vanish. Therefore we have only to compute the contribution from $j = 0$:

$$i^{|\alpha|} D^\alpha(e^{-i\xi t} \hat{\varphi}(-\xi))|_0 = \sum_{\beta\leq\alpha} \binom{\alpha}{\beta} \frac{t^\beta D^{\alpha-\beta}\hat{\varphi}(0)}{i^{|\alpha|-|\beta|}}$$

The leading term is clearly Ct^α , with $C = \hat{\varphi}(0) \neq 0$ as required.

Now we suppose that (ii) holds. Taking $\alpha = 0$, this means from (2.4) that

$$\sum e^{-2\pi ijt} \hat{\varphi}(-2\pi j)$$

is a non-zero constant function. Therefore $\hat{\varphi}(0) \neq 0$, $\hat{\varphi}(2\pi j) = 0$ for $j \neq 0$.

Next we consider $\alpha = (1,0,\ldots,0)$; the right side of (2.4) is

$$i \sum \frac{\partial}{\partial \xi_1}(e^{-i\xi t}\,\hat{\varphi}(-\xi))|_{2\pi j} = i \sum e^{-2\pi i j t}\,\frac{\partial \hat{\varphi}}{\partial \xi_1}\,(2\pi j) + t_1\,\hat{\varphi}(0)\ .$$

According to (ii) this is a polynomial, and therefore this time we have $\partial\hat{\varphi}/\partial\xi_1\,(2\pi j) = 0$ for $j \neq 0$. Proceeding in the same way, in order of increasing α, we establish the remaining conditions in (2.1).

(i) => (iii). Our first step is to convert the inequalities (2.2) and (2.3), by Parseval's formula, into inequalities for Fourier transforms. We note first that

$$\hat{\varphi}_j^h(\xi) = \int h^{-n/2}\,\varphi(\tfrac{x}{h} - j)\,e^{-i\xi x}\,dx \tag{2.5}$$

$$= h^{n/2} \int \varphi(y)\,e^{-i\xi h(j+y)}\,dy = h^{n/2}\,e^{-i\xi jh}\,\hat{\varphi}(h\xi)\ .$$

Therefore the transform of $\Sigma\, w_j^h\, \varphi_j^h$ is

$$\left[h^{n/2} \sum w_j^h\, e^{-i\xi jh}\right] \hat{\varphi}(h\xi)\ .$$

We denote the function in brackets by $W^h(\xi)$, and remark that it has period $2\pi/h$ in each variable ξ_ν. Thus if we let C/h denote the cube $-\pi/h < \xi_\nu \leq \pi/h$, the stability condition (2.3) on the weights becomes

$$\int_{C/h} |W^h(\xi)|^2\, d\xi \leq K \cdot \int_{R^n} |\hat{u}\,(\xi)|^2\, d\xi\ . \tag{2.6}$$

Using the definition (1.3) of the $\mathcal{H}^s$ norm, the required estimate (2.2) is converted into

$$(2.7) \qquad \int_{R^n} |\hat{u}(\xi) - W^h(\xi)\hat{\varphi}(h\xi)|^2 \, (1+|\xi|^2)^s d\xi \leq c_s^2 h^{2(p+1-s)} \|u\|^2_{\mathcal{H}^{p+1}} .$$

To construct W^h satisfying these conditions, we first choose coefficients q_α , $|\alpha| \leq p$, so that

$$(2.8) \qquad \sum q_\alpha \, \eta^\alpha \, \hat{\varphi}(\eta) = 1 + O(|\eta|^{p+1}) .$$

Recalling that $\hat{\varphi}(0) \neq 0$ by (i) , we begin with

$$q_0 \, \hat{\varphi}(0) = 1 .$$

Then each remaining q_β is chosen to make the βth derivative of the left side of (2.8) vanish at $\eta = 0$. In fact, since only the αth derivative of η^α is non-zero at $\eta = 0$, this rule determines the q_β in order of increasing β by

$$\sum_{\alpha \leq \beta} \frac{q_\alpha \, D^{\beta-\alpha} \hat{\varphi}(0)}{(\beta-\alpha)!} = 0 .$$

Now we choose the weights w_j^h so that

$$(2.9) \qquad W^h(\xi) = \hat{u}(\xi) \sum_{|\alpha| \leq p} q_\alpha (h\xi)^\alpha \quad \text{in the period cube } C/h .$$

The stability condition (2.3) is satisfied, since $|W^h| \leq c|\hat{u}|$ in C/h . (We shall denote all constants in the following estimates by c .) Therefore we have to show that the integrals

$$I_1 = \int_{C/h} |\hat{u}(\xi) - W^h(\xi)\hat{\varphi}(h\xi)|^2 (1 + |\xi|^2)^s \, d\xi$$

$$I_2 = \int_{R^n - C/h} |\hat{u}(\xi)|^2 (1+|\xi|^2)^s \, d\xi$$

$$I_3 = \int_{R^n - C/h} |W^h(\xi)\hat{\varphi}(h\xi)|^2 (1+|\xi|^2)^s \, d\xi$$

are all bounded by $c\, h^{2(p+1-s)} \|u\|^2_{\mathscr{H}^{p+1}}$ We begin with

$$I_1 = \int_{C/h} |\hat{u}(\xi)|^2 |1 - \sum q_\alpha (h\xi)^\alpha \hat{\varphi}(h\xi)|^2 (1+|\xi|^2)^s d\xi$$

$$\leq \int_{C/h} |\hat{u}(\xi)|^2 |h\xi|^{2(p+1)} (1+|\xi|^2)^s d\xi \quad \text{by } (2.7)$$

$$\leq c \int_{C/h} |\hat{u}(\xi)|^2 |h\xi|^{2(p+1-s)} (1+|\xi|^2)^s d\xi \quad \text{since } |h\xi_\nu| \leq \pi$$

$$\leq c\, h^{2(p+1-s)} \int |\hat{u}(\xi)|^2 (1+|\xi|^2)^{p+1} d\xi \, .$$

For ξ outside C/h, we notice that $|h\xi| \geq 1$; therefore

$$I_2 \leq \int_{R^n - C/h} |u(\xi)|^2 \, |h\xi|^{2(p+1-s)} (1+|\xi|^2)^s d\xi$$

$$\leq c\, h^{2(p+1-s)} \, \|u\|^2_{\mathscr{H}^{p+1}} \, .$$

In I_3 we use the periodicity of W^h to change variables:

$$I_3 = \int_{C/h} \sum_{j \neq 0} |W^h(\xi)\hat{\varphi}(h\xi + 2\pi j)|^2 (1 + |\xi + \frac{2\pi j}{h}|^2)^s d\xi$$

$$(2.10) \qquad \leq c \int_{C/h} |\hat{u}(\xi)|^2 h^{-2s} \sum_{j\neq 0} |\hat{\varphi}(h\xi + 2\pi j)|^2 |h\xi + 2\pi j|^{2s} d\xi .$$

To show that the sum in this integral is $O(|h\xi|^{2p+2})$, we expand $\hat{\varphi}$ in a Taylor series around $2\pi j$; since by condition (i) it has a $p+1$-fold zero there, there is only the usual remainder term

$$\hat{\varphi}(h\xi + 2\pi j) = \sum_{|\alpha|=p+1} \frac{(h\xi)^\alpha D^\alpha \hat{\varphi}(\theta_j)}{\alpha!} .$$

The evaluation point $\theta_j = \theta_j(h\xi)$ lies on the line between $2\pi j$ and $2\pi j + h\xi$; since ξ is in C/h,

$$|h\xi + 2\pi j| \leq c|\theta_j| .$$

Therefore the sum in (2.10) is bounded by $c|h\xi|^{2p+2} S(\xi)$, where

$$S(\xi) = \sum_{j\neq 0} \sum_{|\alpha|=p+1} |D^\alpha \hat{\varphi}(\theta_j)|^2 |\theta_j|^{2s} .$$

To bound S uniformly in ξ we apply first the Paley-Wiener theorem: $\varphi \in \mathcal{H}_0^p$ implies that $\hat{\varphi}$ and its derivatives are entire functions of exponential type, and that for any $s \leq p$ they satisfy

$$S' = \int_{R^n} |D^\alpha \hat{\varphi}(\xi)|^2 |\xi|^{2s} d\xi < \infty .$$

The theory of entire functions allows us to estimate the sums S in terms of the integral S', as long as the evaluation points are sufficiently well spaced; this is a straightforward extension of Theorem 6.7.15 in Boas [17] . In our case θ_j always lies in the cube centered at $2\pi j$, with sides of length 2π , and we do get $S(\xi) \leq c$. Substituting back into (2.10) ,

$$I_3 \leq c \int |\hat{u}(\xi)|^2 h^{-2s} |h\xi|^{2p+2} d\xi$$

$$\leq c\, h^{2(p+1-s)} \|u\|^2_{\mathcal{H}^{p+1}} .$$

(iii) => (i) . At this step we use, not the full strength of (iii) , but only the fact that approximation in $\mathcal{H}^p$ is possible for the particular function U defined by

$$\hat{U}(\xi) = \begin{cases} 1 & \text{in the ball } B = \{|\xi| \leq 1\} \\ 0 & \text{outside } B \end{cases}$$

Since $U \in \mathcal{H}^{p+1}$, (iii) provides W^h such that

$$\int_{C/h} |W^h(\xi)|^2 d\xi \leq \text{constant} \tag{2.11}$$

$$\int |\hat{U}(\xi) - W^h(\xi)\hat{\varphi}(h\xi)|^2 (1+|\xi|^2)^p d\xi \to 0 \quad \text{as} \quad h \to 0 . \tag{2.12}$$

If we consider these two integrals only over B , where $\hat{\varphi}(\xi h) = \hat{\varphi}(0) + O(h)$, it follows that

$$\int_B |1 - W^h(\xi)\hat{\varphi}(0)|^2 d\xi \to 0 \quad \text{as} \quad h \to 0 . \tag{2.13}$$

Appealing again to the crucial constraint (2.11), we must have $\hat{\varphi}(0) \neq 0$.

Next we consider (2.12) over $R^n - C/h$, where $\hat{U} \equiv 0$:

$$\int_{R^n - C/h} |W^h(\xi)\hat{\varphi}(h\xi)|^2 (1+|\xi|^2)^p \, d\xi$$

$$= \sum_{j\neq 0} \int_{C/h} |W^h(\xi)\hat{\varphi}(h\xi + 2\pi j)|^2 \left(1 + \left|\frac{h\xi + 2\pi j}{h}\right|^2\right)^p d\xi \quad \text{(periodicity of } W\text{)}$$

$$\geq \int_B |W^h(\xi)\hat{\varphi}(h\xi + 2\pi j)|^2 \, h^{-2p} \, d\xi \to 0$$

for every $j \neq 0$. Therefore

$$\int_B |W^h(\xi)|^2 \left| \sum_{|\alpha|\leq p} \frac{D^\alpha\hat{\varphi}(2\pi j)}{\alpha!} (h\xi)^\alpha \right|^2 h^{-2p} d\xi \to 0 \; .$$

Comparing each power of h in turn, and recalling (2.11), we conclude that

$$D^\alpha\hat{\varphi}(2\pi j) = 0 \quad \text{for} \quad |\alpha| \leq p \, , \; j \neq 0 \; .$$

Remark 1. In one dimension, $n = 1$, the simplest function which has the zeros required by condition (i) is the one which generates the local basis for splines of degree p :

$$\hat{\sigma}_p(\xi) = \left(\frac{\sin \xi/2}{\xi/2}\right)^{p+1} \; .$$

Furthermore, any $\hat{\varphi}$ which satisfies (i) is a multiple of $\hat{\sigma}_p$,

$$\hat{\varphi} = \hat{E}\,\hat{\sigma}_p ,$$

where E is a suitable entire function. Thus every such φ, which <u>may or may not be a piecewise polynomial</u>, can be found from convolution with the "B-spline" σ_p. If $n > 1$ (or $N > 1$) no such common divisor exists, but condition (i) should make it possible to identify all useful bases. We have learned that this condition was known much earlier to Schoenberg; it appears in his fundamental paper [18] as the condition for a smoothing formula to map onto itself the space of polynomials of degree p The Poisson formula, which is the crucial connection between transforms on Z^n and R^n, also figures in the valuable recent work of Bramble and Hilbert [19] .

Remark 2. Theorem I differs a little from the result stated in the introduction, where we assumed stability but gave weaker forms for the conditions in the theorem. In the third condition, for example, the weaker statement imposed no restriction (2.3) on the weights; given the stability condition (1.14), however, this restriction is automatic. In the second condition, we assumed no particular form for the expansion of polynomials - only that they could somehow be represented as combinations of φ and its translates, e.g.

$$1 \equiv \sum \rho_j \,\varphi(t-j) . \tag{2.14}$$

We therefore want to show that stability forces these weights ρ_j to be equal. The role of stability is to make any representation (2.14) unique; the only representation of the zero function has all weights zero. Now we use translation invariance, shifting both sides of (2.14) through the unit vector e_ν in the t_ν direction:

$$1 = \sum \rho_j \varphi(t-j-e_\nu) = \sum \rho_{j-e_\nu} \varphi(t-j) .$$

By uniqueness $\rho_{j-e_\nu} = \rho_j$ for all j and ν, and the weights are equal

For an expansion

$$t_1 = \sum \sigma_j \varphi(t-j)$$

the same argument gives $\sigma_{j-e_k} = \sigma_j$ for $\nu = 2,\ldots,n$; after shifting through e_1 we add 1 to find

$$t_1 = \sum \sigma_j \varphi(t-j-e_1) + 1 = \sum (\sigma_{j-e_1} + \rho)\varphi(t-j) .$$

Uniqueness now gives $\sigma_j = \sigma_0 + j_1\rho$, so that

$$t_1 = \sum (\sigma_0 + j_1\rho)\varphi(t-j) = \frac{\sigma_0}{\rho} + \rho \sum j_1 \varphi(t-j) .$$

This means that the last sum is a polynomial of the form required by (ii) ; the induction is obvious. Thus the statement in the introduction may be regarded as a corollary to Theorem I .

Remark 3. The condition $\hat{\varphi}(0) \neq 0$ is in general not necessary for approximation to be possible in $\mathcal{H}^s$, if the restriction (2.3) on the weights is removed. Suppose for example that $n = 1$, and $\hat{\varphi}$ has a zero of order μ at the origin. Then if (2.1) holds with p replaced by $p + \mu$, one can construct weights w_j^h of order $h^{-\mu}$ such that (2.2) holds. A converse of this result is also possible. We emphasize that in such a situation the transform $\hat{\varphi}$ vanishes at <u>all</u> $2\pi j$, and (1.14) is satisfied at $\xi_0 = 0$. Therefore the associated Ritz-Galerkin system will be numerically unstable, and such a choice of φ is to be avoided.

Remark 4. Theorem I can be made much more precise in a number of directions. (These refinements may be of little interest to the sensible reader, who wants to get on with the plot; he can safely disregard Theorem I'.) First, we can show that the exponent $p+1-s$ in the error estimate (2.2) is best possible for any $u \neq 0$, and find the infimum of constants c_s for which this estimate holds. Second, we note that the smoothness of φ and the order of the derivatives of $\hat{\varphi}$ which vanish at the points $2\pi j$ were specified in Theorem I by the same integer p. This relation is the most efficient in practice, and consequently the most common, but there is no <u>a priori</u> reason why the two indices must agree. In the following we allow φ to be less smooth, say φ in $\mathcal{H}_c^q$ where $q \leq p$. There is no use in permitting extra smoothness, $q > p$, since the estimates are not improved. Third, we strengthen the converse part of the theorem by deducing this smoothness of φ rather than assuming it.

A further generalization, which we shall forego, is to give estimates also in fractional and negative norms. In fact the reader can verify that such results follow directly from our proofs; the norm (1.3), involving the Fourier transform, applies equally well to all real s. We also omit any discussion of L^p estimates for $p \neq 2, \infty$.

The more precise version of Theorem I is

THEOREM I'. For any integers $p \geq q \geq 0$, the following conditions are equivalent:

(i) φ lies in $\mathcal{H}_c^q$, $\hat{\varphi}(0) \neq 0$, and

$$D^\alpha \hat{\varphi}(2\pi j) = 0 \quad \text{if} \quad 0 \neq j \in Z^n, \ |\alpha| \leq p \tag{2.1}$$

(ii) φ lies in $\mathcal{H}_c^q$, and for $|\alpha| \leq p$, the function $\Sigma\, j^\alpha\, \varphi(t-j)$ is a polynomial in $t_1, \dots, t_n$ with leading term Ct^α, $C \neq 0$.

(iii) φ is a distribution with compact support, and for each u in $\mathcal{H}^{p+1}$ there are weights w_j^h such that as $h \to 0$,

$$\|u - \sum w_j^h \varphi_j^h\|_{\mathcal{H}^s} \leq c_s\, h^{p+1-s}\, \|u\|_{\mathcal{H}^{p+1}}, \quad s \leq q, \tag{2.15}$$

$$\sum |w_j^h|^2 \leq K\, \|u\|^2_{\mathcal{H}^0}. \tag{2.16}$$

The exponent $p+1-s$ is best possible for every s and every $u \neq 0$, if p is the largest integer for which (i) holds. In one dimension, the greatest lower bound of possible constants c_s is

$$(2.17)\qquad C_s = \left(|\hat{\varphi}(0)|\,(p+1)!\right)^{-1}\left(\sum_{j\neq 0} |2\pi j|^{2s}\, \left|\frac{\partial^{p+1}\hat{\varphi}}{\partial \xi^{p+1}}(2\pi j)\right|^2\right)^{1/2}$$

With $n > 1$ the last factor becomes

$$(2.18)\qquad \sup_{\omega} \left(\sum_{j\neq 0} |2\pi j|^{2s} |\partial_{\omega}^{p+1}\, \hat{\varphi}(2\pi j)|^2\right)^{1/2} ,$$

where ∂_{ω}^{p+1} is the derivative of order $p+1$ in the direction ω . (This constant C_s is computed in [20] for several φ .)

We must emphasize that we have not established that the estimate (2.2) holds, with $c_s = C_s + \epsilon$, for each h ; we know only that for every u in $\mathcal{H}^{p+1}$, it will hold in some range $h < h_0(\epsilon, u)$. Our constants C_s are in fact nothing but the constants involved in approximating polynomials of degree $p+1$. We know from condition (ii) in Theorem I that the polynomials of degree p can be reproduced exactly, and the point is that for small h any smooth function will resemble a polynomial locally. Therefore it should be no surprise that the error is asymptotically attributable to the terms of degree $p+1$ in a local Taylor expansion of u . In fact we make the following conjecture: for every smooth function u of one variable, the ratio of error in best approximation in $\mathcal{H}^s$ to $h^{p+1-s}\|D^{p+1}u\|_{L^2}$ approaches this same constant C_s.

This means that the constants C_s are a very useful criterion for the comparison of two different finite element functions φ (given that they have the same value of p); the constants are relevant not just to some extreme choice of u, but rather to all choices. With $n > 1$, the situation seems to be essentially the same; there <u>will</u> be a tiny subset of u for which the asymptotic constant exceeds C_s, namely those with $\hat{u}$ supported away from the optimal direction ω. (We hope to show elsewhere that this conjecture extends also to approximation in the maximum norm.)

Now we consider the approximation problem when the space S^h is generated by several functions $\varphi_1,\ldots,\varphi_N$. In this case there are N unknowns $v^h_{i,j}$, $i = 1,\ldots,N$, to be computed at each meshpoint from the finite element equations $A^h v^h = f^h$. The merit of such an extension is to make high accuracy p possible with relatively simple functions φ_i - their support can be small, so that frequently the required inner products are easier to compute and boundary conditions simpler to match, and they can have additional interpolating properties. In the one-dimensional Hermite case, for example, each φ_i is supported on the interval $[-1,1]$, and the $i-1$ st is the only one of the first $N-1$ derivatives to be non-zero at the origin. This means that the quantities $v_{i,j}^h$ which satisfy the finite element equation have physical significance in themselves, as the "displacement", "slope", "stress", etc. of the approximate solution at the meshpoint $x = jh$. This has been found very attractive by users.

THEOREM II. Suppose $\varphi_1, \ldots, \varphi_N$ are in $\mathcal{H}_c^q$. Then the following conditions are equivalent:

(i) there are linear combinations Ψ_α of the φ_i which satisfy

$$\hat{\Psi}_0(0) = 1, \quad \hat{\Psi}_0(2\pi j) = 0 \quad \text{for} \quad j \neq 0 \tag{2.19a}$$

$$\sum_{\beta \leq \alpha} \frac{D^\beta \hat{\Psi}_{\alpha-\beta}(2\pi j)}{\beta!\, i^{|\beta|}} = 0 \quad \text{for all} \quad j \in Z^n, \; 1 \leq |\alpha| \leq p \tag{2.19b}$$

(ii) there are linear combinations Ψ_α of the φ_i which satisfy

$$\frac{t^\alpha}{\alpha!} = \sum_{\beta \leq \alpha} \sum_j \frac{j^\beta \Psi_{\alpha-\beta}(t-j)}{\beta!}, \quad |\alpha| \leq p. \tag{2.20}$$

(iii) for each u in $\mathcal{H}^{p+1}$ there are weights $w_{i,j}^h$ such that for $s = 0, 1, \ldots q$,

(2.21) $$\|u-\sum_{i,j} w_{i,j}^h \varphi_{i,j}^h\|_{\mathcal{H}^s} \leq c_s h^{p+1-s} \|u\|_{\mathcal{H}^{p+1}}$$

(2.22) $$\sum_{i,j} |w_{i,j}^h|^2 \leq K\|u\|_{\mathcal{H}^0}^2 .$$

Remark 5. Babuska has asked us whether the following condition on the φ_i is equivalent to those in Theorem II:

(iv) there is a finite linear combination Ω of $\varphi_1,\dots,\varphi_N$ and their translates $\varphi_{i,j}$ which satisfies the conditions imposed in the case $N = 1$:

(2.23) $$D^{\alpha}\hat{\Omega}(2\pi j) = 0 \quad \text{for} \quad |\alpha| \leq p , \; j \neq 0 .$$

We shall prove that (i) => (iv) => (iii) , so that Babuska's insight allows the reduction of many N-dimensional problems to the simpler case $N = 1$. We note that the translates admitted in (iv) are the functions

$$\varphi_{i,j}(x) = \varphi_i(x-j) , \; 1 \leq i \leq N , \; j \in Z^n .$$

It is obvious that (iv) => (iii): if the function Ω satisfies the conditions of Theorem I, then combinations

$\Sigma\, w_j^h\, \Omega_j^h$ can be used to approximate u as required in (2.21-22).

To prove that (i) => (iii), we let t_α denote the unique polynomial of degree p jointly in $e^{i\xi_1},\dots,e^{i\xi_n}$ such that for $|\gamma| \le p$,

$$D^\gamma\, t_\alpha(0) = \begin{cases} \alpha!\; i^{|\alpha|} & \text{if} \quad \alpha = \gamma \\ 0 & \text{if} \quad \alpha \ne \gamma \end{cases}$$

Then we define the required Ω by giving its Fourier transform:

$$\hat{\Omega} = \sum_{|\alpha| \le p} t_\alpha\, \hat{\Psi}_\alpha \ . \tag{2.24}$$

Since $t_\alpha \hat{\Psi}_\alpha$ is the transform of a finite combination of translates of Ψ_α, this Ω is indeed a finite linear combination of the original φ_i and their translates.

By the periodicity of the t_α,

$$\hat{\Omega}(2\pi j) = \sum t_\alpha(0)\hat{\Psi}_\alpha(2\pi j) = \hat{\Psi}_0(2\pi j) \ ,$$

which by (2.19a) equals one if $j = 0$ and zero otherwise. To verify (2.23), we use the Leibniz rule for the derivative of a product:

$$D^\gamma\, \hat{\Omega}(2\pi j) = \sum_{|\alpha| \le p} D^\gamma (t_\alpha\, \hat{\Psi}_\alpha)\big|_{2\pi j} \tag{2.25}$$

$$= \sum_{\alpha \le \gamma} \binom{\gamma}{\alpha}\, \alpha!\; i^{|\alpha|}\, D^{\gamma-\alpha}\, \hat{\Psi}_\alpha\, (2\pi j)$$

$$= \gamma!\, i^{|\gamma|} \sum_{\beta \leq \gamma} \frac{D^{\beta}\, \hat{\Psi}_{\gamma-\beta}(2\pi j)}{\beta!\, i^{|\beta|}} ,$$

replacing α by $\gamma - \beta$. This vanishes, by the fundamental property (2.19b) of the $\hat{\Psi}_{\alpha}$, and condition (iv) is verified.

We note that (2.23) holds even for $j = 0$, when $1 \leq |\alpha| \leq p$. Thus the q_{α} which were constructed in the proof of Theorem I reduce to $\delta_{0\alpha}$, when we are approximating u by combinations of the functions Ω_j^h.

We turn now from the mean-square approximation of a function u and its derivatives to the problem of pointwise approximation. The latter is crucial numerically, even though the original differential problem (e.g. that of minimizing a quadratic functional) leads more naturally to the former. It is a fundamental property of the finite element method that the two go together. By this we mean not only that the order $p + 1 - s$ of the best approximation is the same in the two norms, as the next theorem shows, but also that the order r of the actual approximation by the Ritz-Galerkin function u^h

is the same in both norms. This exponent r is the smaller of $p+1-s$ and $2(p+1-m)$.

The following estimate includes known results for spline approximation, in the special case of equally spaced knots. Splines are typical of many important choices of the φ_i, in that their derivatives of some order q have jump discontiunities; we note that this leaves them safely in W_∞^q. (In the spline case, q is the degree of the polynomial in each interval, and coincides with the accuracy exponent p.)

THEOREM III. Suppose $\varphi_1,\ldots,\varphi_N$ satisfy the conditions of the previous theorem, and have bounded derivatives of order q. Then if

$$(2.26) \qquad \tilde{u}^h(x) = \sum u(jh)\ \Omega\left(\frac{x}{h} - j\right),$$

it follows that

$$(2.27) \qquad \|u - \tilde{u}^h\|_{W_\infty^s} \leq c_s\ h^{p+1-s}\ \|u\|_{W_\infty^{p+1}} .$$

Proof. We write $x = kh + th$, where k is the integral part of x/h and t lies in the unit cube $0 \leq t_\nu < 1$. Expanding in powers of h,

We know from the start that u^h cannot be closer to u than the optimal approximation from the space S^h. Therefore the error, measured in H^s or W^s_∞, will be at best of the order h^{p+1-s} determined in Theorems I and II. The question is whether the approximation produced by the finite element is actually of this optimal order. There is no a priori reason why this should always be so; in fact, if we fix φ in H^p and increase the order 2m of the equation, there is every reason to think otherwise. It seems unlikely that the error in H^0 would stay of order p + 1 until m exceeds p, at which point φ would no longer lead to admissible trial functions and the method would collapse. Therefore we anticipate that the order of accuracy will depend on m as well as p and s.

The correct order can in fact be determined over the range $0 \leq s \leq m$ by an elegant variational argument which we learned from Martin Schultz. A special but still typical case of this argument has been published by Nitsche [21], and Aubin has shown us an alternative route to the same result.

We begin with the fundamental result (cf. Varga [22]) that in the case s = m, the error $u^h - u$ is indeed of the optimal order h^{p+1-m} allowed by the approximation theorems. Repeating the standard argument, we deduce from the variational equations (1.5) and (1.7) that

$$a(u^h - u, w^h) = (f, w^h) - (f, w^h) = 0 \qquad \text{for all } w^h \text{ in } S^h \tag{2.31}$$

Therefore by ellipticity

$$\begin{aligned}\rho\|u^h - u\|^2_{H^m} &\leq \operatorname{Re} a(u^h - u, u^h - u)\\ &= \operatorname{Re} a(u^h - u, w^h - u)\\ &\leq K\|u^h - u\|_{H^m}\|w^h - u\|_{H^m}\end{aligned}$$

for all w^h. Cancelling the first factor on the right, and choosing w^h as the optimal approximation to u,

$$\|u^h - u\|_{H^m} \leqslant \frac{Kc_m}{\rho} h^{p+1-m} \|u\|_{H^{p+1}} \tag{2.32}$$

Here we need u in H^{p+1} in order to apply Theorems I and II, and therefore we assume that f lies in H^{p+1-2m}. If f is less smooth, the estimates of this section are not difficult to revise. It is to the estimate (2.32) that we may apply our calculation in Theorem I' of the minimal constant C_m in approximation.

Now we give Schultz's argument for $s < m$. It begins with the adjoint problem $L^*v = g$, which is equivalent to the variational equation

$$a(y, v) = (y, g) \text{ for all } y \text{ in } H^m \tag{2.33}$$

Again the ellipticity of a guarantees a unique solution v for g in H^{-m}, and furthermore that

$$\|v\|_{H^{2m-s}} \leqslant C\|g\|_{H^{-s}} \tag{2.34}$$

Taking $y = u^h - u$, and recalling (5.1),

$$(u^h - u, g) = a(u^h - u, v) = a(u^h - u, v - w^h) \tag{2.35}$$

for all w^h in S^h. Therefore

$$|(u^h - u, g)| \leqslant K\|u^h - u\|_{H^m} \|v - w^h\|_{H^m} . \tag{2.36}$$

To estimate the last term, we choose w^h as the best H^m approximation to v, and appeal to Theorems I and II:

$$\|v - w^h\|_{H^m} \leqslant \begin{cases} c_m h^{p+1-m} \|v\|_{H^{p+1}} & \text{if } p+1 \leqslant 2m-s \\ c_m h^{m-s} \|v\|_{H^{2m-s}} & \text{if } p+1 > 2m-s \end{cases} \tag{2.37}$$

(In the second case we reduced p to $2m - s' - 1$ before applying the approximation theorems; if their hypotheses hold for a given p they certainly hold for a smaller one.) Substituting (2.32) and (2.37) into (2.36) and using (2.34), we have

$$|(u^h - u, g)| \leq K' h^r \|u\|_{H^{p+1}} \|g\|_{H^{-s}},$$

$$r = \min(p + 1 - s, 2(p + 1 - m)) \quad . \tag{2.38}$$

Now as g runs over the unit ball in H^{-s}, the supremum of the left side is exactly the norm of $u^h - u$ in the dual space H^s. Therefore the final error estimate is

$$\|u^h - u\|_{H^s} \leq K' h^r \|u\|_{H^{p+1}} \tag{2.39}$$

We notice that with $s = m$ the first expression in r is the smaller, and $r = p + 1 - m$ in agreement with (2.32).

[1] O. C. Zienkiewicz, "The finite element method in structural and continuum mechanics," London: McGraw-Hill (1967).

[2] R. Courant, "Variational methods for the solution of problems of equilibrium and vibrations," Bull. Amer. Math. Soc. 49, 1-23(1943).

[3] G. Polya, "Sur une interpretation de la methode des differences finies qui peut fournir des bornes superieures ou inferieures," Comptes Rendus 235, 995-997 (1952).

[4] J. P. Aubin, "Behavior of the error of the approximate solutions of boundary value problems for linear elliptic operators by Galerkin's and finite difference methods," Rem. Sem. Mat. Padova.

[5] G. Fix and G. Strang, "Fourier analysis of the finite element method in Ritz-Galerkin theory," Studies in Appl. Math. 48, 265-273 (1969).

[6] D. Schaeffer, "Approximation of elliptic boundary value problem by difference equations; I. Factorization of the symbol," J. Functional Analysis 1970.

[7] V. Thomee, "Elliptic difference operators and Dirichlet's problem," Contr. Diff. Eqns. 3, 301-324 (1964).

[8] J. P. Aubin, "Approximation des espaces de distributions et des operateurs differentials," Bull. Soc. Math. France, Memoire 12 (1967).

[9] G. Birkhoff, M. H. Schultz and R. S. Varga, "Hermite interpolation in one and more variables with applications to partial differential equations," Numer. Math., 11, 232-256(1968).

[10] M. H. Schultz, "Rayleigh-Ritz-Galerkin methods for multidimensional problems," SIAM Numer. Anal. 6, 523-538 (1969).

[11] F. DiGuglielmo, "Methode des elements finis: une famille d'approximations des espaces de Sobolev par les translates de p-fonctions," Manuscript, 1970.

[12] I. Babuska, "Approximation by Hill functions," to appear.

[13] J. J. Goel, "Construction of basic functions for numerical utilization of Ritz's method," Numer. Math. 12, 435-447 (1968).

[14] V. Thomee, "On the convergence of difference quotients in elliptic problems," Univ. of Maryland, Note BN-537 (1968).

[15] M. Zlamal, "On the finite element method," Numer. Math. 12, 394-409 (1968).

[16] R. J. Herbold, M. H. Schultz, and R. S. Varga, "Quadrature schemes for the numerical solution of boundary value problems by variational techniques," Aequationes Mathematicae 3, 96-119 (1969).

[17] R. Boas, "Entire functions," New York, Academic Press (1954).

[18] I. J. Schoenberg, "Contributions to the problem of approximation of equidistant data by analytic functions, Parts A and B.," Quart. Appl. Math. 4, 45-99, 112-141 (1946).

[19] J. H. Bramble, and S. R. Hilbert, "Bounds for a class of linear functionals with applications to Hermite interpolation." Numer. Mathematik.

[20] G. Strang, "The finite element method and approximation theory, Numerical Solution of Partial Differential Equations II (SYNSPADE), Academic Press, 1971.

[21] J. Nitsche, "Ein Kriterium fur die quasi-optimilitat des Ritzschen verfahrens," Numer. Math. 11, 346-348 (1968).

[22] R. S. Varga, "Hermite interpolation-type Ritz methods for two-point boundary value problems," in: "Numerical solution of partial differential equations," J. H. Bramble, Ed., New York: Academic Press, 365-373 (1965).

CENTRO INTERNAZIONALE MATEMATICO ESTIVO
(C. I. M. E.)

M. ZERNER

CARACTERISTIQUES D'APPROXIMATION DES COMPACTS DANS LES ESPACES FONCTIONNELS ET PROBLEMES AUX LIMITES ELLIPTIQUES

Corso tenuto a Erice dal 27 giugno al 7 luglio 1971

CARACTERISTIQUES D'APPROXIMATION DES COMPACTS DANS LES ESPACES FONCTIONNELS ET PROBLEMES AUX LIMITES ELLIPTIQUES

par M. Zerner
(Université de Nice)

1. Considérons le problème aux limites :

$$(1)\qquad \begin{cases} Au = 0 \ \text{ sur } \Omega \\ B_j u = \varphi_j \quad \text{sur } \partial\Omega \qquad j = 1, \ldots, m \end{cases}$$

où Ω est un ouvert borné suffisamment régulier de $\mathbb{R}^n$. A est un opérateur elliptique à coefficients $\mathscr{C}^\infty$ d'ordre 2m, les B_j des opérateurs d'ordre , les φ_j des fonctions données dans des boules des $W^p_{k-m_j-1/2}$.

Nous faisons expressément l'hypothèse que ce problème est bien posé et que si $u = G(\varphi_1, \ldots \varphi_m)$ est la solution, alors G est un isomorphisme de $V_k = W^p_{k-m_1-1/p}(\partial\Omega)$ sur $W^p_k(\Omega) \cap A^{-1}(0)$ et celà pour tout k assez grand.

Nous voulons des indications sur la façon de discrétiser ce problème de façon à garantir une précision de ε au sens de $W^p_k(\Omega)$ $(1<k)$ pour toute donnée $\varphi = (\varphi_1, \ldots\ldots, \varphi_m)$ avec $\|\varphi\|_k \leq$ où

$$\|\varphi\|_k = \max_{1\leq j\leq m} \|\varphi_j\|_{W^p_{k-m_j-1/p}}$$

Nul n'ignore que si l'on veut arriver à ce résultat sans gaspillage inconsidéré de volume de mémoire et de temps machine, il faut construire un réseau de discrétisation plus serré près de la frontière qu'à l'intérieur.

Cependant si l'on veut aller plus loin, il faut donner une description axiomatique de ce qu'on entend par là.

Nous supposerons désormais k assez grand pour que tous les $W^p_{k-m_j-1/p}$ soient des espaces de fonctions continues.

On cherche alors des sous-ensembles finis $\mathcal{R}$ de $\bar{\Omega}$ et $\mathcal{S}$ de $\partial\Omega$, une application linéaire L de $(\mathbb{R}^{\mathcal{S}})^m$ dans $\mathbb{R}^{\mathcal{R}}$, une application J de $\mathbb{R}^{\mathcal{R}}$ dans $W^p_1(\Omega)$ telles que

(i) pour tout $v \in \mathbb{R}^{\mathcal{R}}$

$$(Jv)_{|\mathcal{R}} = v$$

(ii) pour toute $\varphi \in V_k$ vérifiant $\|\varphi\|_k \leq 1$,

$$\| J \circ L \circ \rho_{\mathcal{S}} \varphi - G\varphi \|_{W^p_1(\Omega)} \leq \varepsilon$$

où $\rho_{\mathcal{S}}$ désigne la restriction à $\mathcal{S}$

Remarque 1 : Cette version du formalisme suppose que la discrétisation ne met en jeu que les valeurs de la fonction elle-même sur les points du réseau. Or certaines méthodes d'éléments finis par exemple mettent en jeu des dérivées. Il n'y a là rien d'essentiel et tous les résultats généraux peuvent être conservés moyennant une légère complication du formalisme. On peut aussi mettre en jeu des moyennes sur des éléments du réseau, etc....

Rappel de définitions : Soit K un sous-ensemble d'un espace normé E.

On appelle n-ème épaisseur de K (dans E) et on note d_n (K) le nombre:

$$d_n(K) = \inf_{L \in G_n} \sup_{x \in K} \inf_{y \in L} \| x - y \|$$

où G_n est l'ensemble des sous-espaces de dimension n de E. En d'autres termes, on prend la distance des éléments de K à L et on la maximise sur K. On doit enfin faire varier le sous-espace L de dimension n de façon à minimiser le résultat obtenu.

Si F est un autre espace vectoriel normé et $F \hookrightarrow E$, on note d_n (F, E) la n-ème épaisseur de la boule unité de F dans E. Notons N_o le nombre des éléments de $\mathcal{S}$. L'application $J_o L_o \rho \mathcal{S}$ étant de rang N_o et G étant un isomorphisme de V_1 sur son image, on a :

<u>Proposition 1</u> : Pour que les conditions (i) et (ii) ci-dessus aient lieu, il est nécessaire que

(2) $$d_{N_o}(V_1, V_k) \leq C \varepsilon$$

où le nombre C ne dépend pas de ε.

Moyennant les évaluations connues de n-ème épaisseur, (Kolmogorov [4], Birman et Solomjak [2], El Kolli [3]), on en déduit

(3) $$N_o \geq C \varepsilon^{-(n-1)/(k-1)}$$

(ici et par la suite, C désignera une " constante" au sens des analystes, c'est-à-dire qui ne dépend pas de tous les autres nombres qui interviennent, mais elle pourra dépendre du numéro de la formule où elle intervient).

Dans des cas simple, on peut vérifier l'inégalité inverse de (3).

Le problème est donc maintenant de regarder $\mathcal{R}$. Dans la pratique $\mathcal{S}$ a autant de points qu'il y a de points de $\mathcal{R}$ " proches " de $\partial\Omega$. Supposons qu'on ait un réseau $\mathcal{R}$ régulier de pas h, au moins au voisinage de la frontière, et que $\mathcal{S}$ ait autant de points que $\mathcal{R}$ a de points dont la distance à $\partial\Omega$ est inférieure à Ah (A une certaine constante). On a alors :

(4) $$N_o \approx h^{-(n-1)}$$

($a \approx b$ signifie $a=0(b)$ et $b=0(a)$; il s'agit ici de relations pour $h \to 0$).

Nous appellerons N le nombre d'éléments de $\mathcal{R}$. Si le réseau est régulier $N \approx h^{-n} \approx N_o^{\frac{n}{n-1}}$ ce qui montrerait, s'il en était besoin, qu'il faut construire un réseau irrégulier.

On voudrait bien en effet que $N = 0\ (N_o)$.

Nous pouvons en dire un tout petit peu plus long en introduisant un "pas local".

Dans la partie 2 nous examinerons des réseaux $\mathcal{R}$ pour lesquels ce concept est simple. On construit une suite finie d'ouverts $\Omega=\Omega_o \supset \Omega_1 \supset \ldots \supset \Omega_\alpha$. Sur $\Omega_i - \Omega_{i+1}$, la trace de $\mathcal{R}$ est un réseau régulier de pas h_i, h_i est le pas local en $x \in \Omega_i - \Omega_{i+1}$. On prendra naturellement $h_i < h_{i+1}$. Si alors

(5) $$h_\alpha \approx h^\beta$$

il faudra, si on veut que

(6) $$N = 0\ (N_o)$$

avoir

$$\beta = \frac{n-1}{n} \tag{7}$$

De plus, si on pose :

$$a = k - 1$$

il faudra utiliser une méthode dont l'ordre de précision soit au moins :

$$b = \frac{an}{n-1}$$

2. Gardant les notations ci-dessus, nous allons donner une construction des Ω_i et des h_i que nous appellerons "construction de Bahvalov simplifiée" (en abrégé C. B. S). Nous dirons un mot des modifications à faire pour passer aux constructions de Bahvalov [1]. Ces modifications n'enlèvent pas leur validité aux calculs d'ordres de grandeur ci-dessous.

On se donne deux paramètres : $\delta \in]0,1[$ et $\tau > 1$. On pose :

$$s(i) = \min\{t \text{ entier}, \ \delta t \geq i\}.$$

Pour tout multi-entier j, $J(j,s)$ désignera le pavé :

$$\{x \,;\, |x_\mu - 2^{s-1} j_\mu| \leq h\, 2^{s-1}, \quad \mu = 1, \ldots\ldots, n\}.$$

Nous dirons que $J(j,s) \in \mathfrak{J}_s$ si le pavé de même centre τ fois plus grand

$$\{x \,;\, |x_\mu - 2^{s-1} j_\mu| \leq \tau\, 2^{s-1}, \quad \mu = 1, \ldots, n\}$$

est contenu dans $\overline{\Omega}$. Nous poserons alors

$$\overline{\Omega}_i = \bigcup_{J \in \mathfrak{J}_s} J$$

et Ω_i sera l'intérieur de son adhérence.

Enfin :

$$h_i = 2^i h.$$

On a alors, Λ désignant le côté du plus grand cube contenu dans Ω et pourvu que h soit assez petit pour que $2\tau h < \Lambda$:

(8) $$\frac{\Lambda}{2} \leq \tau h 2^{s(\alpha)} \leq \Lambda$$

d'où l'on déduit que (5) est vérifiée avec :

(9) $$\beta = 1 - \sigma$$

de sorte que (7) devient :

(10) $$\sigma \geq \frac{1}{n} .$$

Proposition 2 : Dans une C. B. S. vérifiant $\sigma > 1/n$, on a $N \approx h^{-(n-1)}$ (et par suite (6) est vérifié).

Lemme : Notons :

$$\Omega^{(r)} = \{x;\ x \in \Omega, d(x, \partial\Omega) \leq r\}.$$

Il existe C tel que :

$$\text{mes } (\Omega^{(r)}) \leq C\, r.$$

Démonstration : Comme $\Omega^{(r)} \subset \Omega$ qui est borné, il suffit de démontrer que

$$\lim_{r \to 0} \sup \frac{\text{mes } (\Omega^{(r)})}{r} < \infty .$$

Par compacité de la frontière, on se ramène au domaine d'une carte locale et à démontrer que mes $(U^r) \leq C'\, r$ où

$$U^r = \{ x;\ x = (x', t),\quad x' \in V,\quad 0 < t \leq \varphi(x', r)\}$$

où C' ne dépend pas de r, non plus que l'ensemble V borné dans $\mathbf{R}^{n-1}$ et

$$\varphi(x', r) \leq C'' r . \qquad (*)$$

c q f d.

<u>Démonstration de la proposition 2</u> - Si un point de Ω n'est pas dans Ω_i, sa distance à $\partial\Omega$ est au plus $\sqrt{n}\,(1+\tau)\,2^{s(i)-1}h$. On utilise alors le lemme pour voir que

$$\text{mes}\,(\Omega_{i-1} - \Omega_i) \leq C_1\, 2^{s(i)} h$$

Il y a donc au plus

$$C_2\, 2^{s(i) - ni} h^{-(n-1)}$$

points de $\mathcal{R}$ dans $\Omega_{i-1} - \Omega_i$, ce qui, d'après la définition de s(i) est majoré par

$$C_3\, 2^{-(n - 1/\sigma) i} h^{-(n-1)}$$

d'où enfin :

$$N \leq C_3 \sum_i 2^{-(n-1/\sigma)i} h^{-(n-1)}$$

or la série est convergente, <u>cqfd</u>

cqfd.

Bahvalov [1] donne deux constructions correspondant à $\tau = 2$ et 3 respectivement. Il est amené à ajouter des points au réseau près de la frontière des Ω_i. Il donne aussi les équations discrétisées correspondantes pour le problème :

$$(11) \qquad \begin{cases} \Delta u = 0 \\ u\big|_{\partial\Omega} = \varphi \end{cases}$$

et une méthode d'interpolation, l'ensemble assurant que la solution approchée diffère de la solution exacte de moins de ε en norme du sup.. Il donne ensuite une méthode de résolution itérative du sy-

(*) Le lemme s'applique à un ouvert borné lipschitzien et par conséquent la proposition 2 aussi.

stème discrétisé (celle de Jacobi semble - t - il) qui permet d'atteindre la précision ε en $0(h^{2\sigma-2}|\log h|)$ itérations dans un cas, en $0(|\log h|^2)$ itérations dans l'autre cas avec n = 2 variables indépendantes. Dans ce deuxième cas, la situation est donc très satisfaisante, mais je dois dire que la partie " résolution des systèmes discrétisés " m'a paru particulièrement obscure.

3. Nous revenons au problème général mais en prenant $p = +\infty$. Soit $x \in \Omega$, posons $R = d(x, \partial\Omega)$ et soit $r \in]0, R[$. Nous voudrions évaluer M, le nombre d'éléments de $\mathcal{R}$ appartenant à la boule U_r de centre x et rayon r. Il faut penser pour comprende la suite que r est petit devant le diamètre de Ω mais grand devant $h \approx N_o^{-1/(n-1)}$

Définition. Soit K un sous - ensemble borné d'un espace normé E, on appelle n $\underline{\text{ème}}$ épaisseur au sens de Gelfand de K le nombre:

$$e_n(K) = e_n = \inf_{L^{(n)}} \sup_{x \in K \cap L(n)} \|x\|$$

où $L^{(n)}$ parcourt l'ensemble des sous -espaces vectoriels fermés de codimension n.

Nous rappelons les deux propriétés faciles suivantes :

a) Soit D un sous-ensemble dense de E'.

Dans la définition de la n $\underline{\text{éme}}$ épaisseur au sens de Gelfand, il suffit de faire parcourir à $L^{(n)}$ l'ensemble des sous-espaces définis par des systèmes d'equations:

$$< x, \xi_j > = 0 \qquad j = 1, \ldots, n$$

où $\xi_j \in D$, $j = 1, \ldots, n$.

b) Soit L un sous-espace de dimension n+1 dans E et K la boule de rayon ρ dans L. On a :

$$\begin{cases} e_k(K) = \rho & k \leq n \\ e_k(K) = 0 & k > n. \end{cases}$$

Soit K l'ensemble parcouru par les solutions de (1) lorsque φ parcourt la boule unité de V_k. Nous voulons qu'une fonction $u \in K$ qui s'annule aux M points de $\mathcal{R} \cap U_r$ soit majorée par ε dans $W_1^\infty(U_r)$, d'où l'idée de déterminer M par la relation

(12) $$1_M(K) \leq \varepsilon \qquad \text{pour la norme } W_1^\infty(U_r)$$

Pour montrer que cette idée n'est pas entièrement utopique, nous allons nous limiter au cas n = 2, k > 1, 1 = 0, Ω simplement connexe et démontrer deux inégalités très simples.

Lemme : Sous les hypothèses ci-dessus il existe A tel que u se décompose en

$$u = u_1 + u_2$$

u_1 holomorphe, u_2 antiholomorphe, et

$$\|u_j\|_{L^\infty} \leq A \qquad j = 1,2$$

Si u est un polynome de degré k, u_1 et u_2 aussi, si les dérivées d'ordre $\leq \nu$ de u s'annulent en x, celles de u_1 et u_2 aussi.

Nous admettrons ce lemme qui résume des résultats classiques de la théorie des fonctions.

Proposition 3 : Il existe C telle que

$$\forall k \quad e_k(K) \leq C(r/R)^{k/2}$$

où $R = d(x, \complement\Omega)$.

<u>Démonstration</u> : Nous prenons pour $L^{(k)}$ l'ensemble des fonctions qui s'annulent avec leurs dérivées d'ordre $\leq k_1$. Compte-tenu des relations

$$\frac{\partial^{j_1+j_2+2j_3} u}{\partial x^{j_1} \partial y^{j_2+2j_3}} = (-1)^{j_3} \frac{\partial^{j_1+j_2+2j_3}}{\partial x^{j_1+2j_3} \partial y^{j_2}}$$

c'est un sous-espace de codimension au plus $2k_1 + 1$ d'où $k_1 \leq \frac{k-1}{2}$.

On a alors , d'après le fait que $u_1/(x_1+ix_2)^{k_1}$ et $u_2/(x_1-ix_2)^{k_1}$ sont encore respectivement holomorphe et antiholomorphe, et en utilisant le principe du maximum,

$$|u_j(y)| \leq C_1 (r/R)^{k_1}$$

on en déduit :

$$e_k(K) \leq C_2 \, (r/R)^{k/2} .$$

<u>Proposition 4</u> : Soit R' tel que $\overline{\Omega}$ soit contenu dans le disque ouvert de centre x et de rayon R'. Il existe C tel que, pour tout k :

$$C \, e_k(K) \geq (r/R')^{k_2}$$

<u>Démonstration</u> : On utilise la propriété b). On prend comme sous-espace de dimension k_1 l'ensemble des polynômes harmoniques de degré au plus k_1. On remarque qu'il existe $B > 0$ tel que u harmonique et majorée par B pour $\|y - x\| < R'$ implique $u \in K$. On divise $u_1(y_1, y_2)$ par $(y_1-x_1+iy_2-ix_2)^{k_1+1}$, $u_2(y_1, y_2)$ par $(y_1-x_1-iy_2+ix_2)^{1+1}$. Enfin on fait une inversion de centre x et on raisonne comme dans la démonstration de la proposition 3.

<u>Remarque en guise de conclusion</u> : Les densités prévues par les propositions 2 et 3 sont beaucoup plus faibles que celles des réseaux construits en 2. Pour les utiliser il faudrait trouver des formules de

M. Zerner

discrétisation utilisant toute la régularité de la solution , donc en ordre variable. Outre qu'on n'en est pas là ., on a vu que les réseaux du 2 suffisent à assurer que l'ordre de grandeur du nombre total de points est celui du nombre de points frontières.

BIBLIOGRAPHIE

[1] N. S. Bahvalov : 0 čislenom rešenij zadači Dirihle dlja uraňenja Laplasa, Vestnik Moskovsk, Un. 5 (1959).

[2] Birman et Solomonjak: Approximation polynômiale par morceaux des fonctions de classe $W^{\alpha,p}$, Mat. Sbornik 73 (115): 3, (1967), 331-355.

[3] A. El Kolli : n-ème épaisseur dans les espaces de Sobolev. Note aux C. R. Acad. Sc. Paris t. 272 (1971), 537-539.

[4] A. N. Kolmogorov : Math. Ann. (2) 37 (1936); 107-111.